# Radiofrequency Radiation Standards

Biological Effects, Dosimetry, Epidemiology, and Public Health Policy

# NATO ASI Series

## Advanced Science Institutes Series

*A series presenting the results of activities sponsored by the NATO Science Committee, which aims at the dissemination of advanced scientific and technological knowledge, with a view to strengthening links between scientific communities.*

The series is published by an international board of publishers in conjunction with the NATO Scientific Affairs Division

| | | |
|---|---|---|
| **A** | **Life Sciences** | Plenum Publishing Corporation |
| **B** | **Physics** | New York and London |
| | | |
| **C** | **Mathematical and Physical Sciences** | Kluwer Academic Pulishers |
| **D** | **Behavioral and Social Sciences** | Dordrecht, Boston, and London |
| **E** | **Applied Sciences** | |
| | | |
| **F** | **Computer and Systems Sciences** | Springer-Verlag |
| **G** | **Ecological Sciences** | Berlin, Heidelberg, New York, London, |
| **H** | **Cell Biology** | Paris, Tokyo, Hong Kong, and Barcelona |
| **I** | **Global Environmental Change** | |

### *Recent Volumes in this Series*

*Series A: Life Sciences*

# Radiofrequency Radiation Standards

Biological Effects, Dosimetry, Epidemiology, and Public Health Policy

Edited by

## B. Jon Klauenberg

Armstrong Laboratory
Radiofrequency Radiation Division
Brooks Air Force Base, Texas

## Martino Grandolfo

Istituto Superiore di Sanità
Rome, Italy

and

## David N. Erwin

Armstrong Laboratory
Radiofrequency Radiation Division
Brooks Air Force Base, Texas

Plenum Press
New York and London
Published in cooperation with NATO Scientific Affairs Division

Proceedings of a NATO Advanced Research Workshop on
Developing a New Standardization Agreement (STANAG) for Radiofrequency Radiation,
held May 17–21, 1993,
at Pratica di Mare, Italian Air Force Base, Pomezia (Rome), Italy

**NATO-PCO-DATA BASE**

The electronic index to the NATO ASI Series provides full bibliographical references (with keywords and/or abstracts) to more than 30,000 contributions from international scientists published in all sections of the NATO ASI Series. Access to the NATO-PCO-DATA BASE is possible in two ways:

—via online FILE 128 (NATO-PCO-DATA BASE) hosted by ESRIN, Via Galileo Galilei, I-00044 Frascati, Italy

—via CD-ROM "NATO Science and Technology Disk" with user-friendly retrieval software in English, French, and German (©WTV GmbH and DATAWARE Technologies, Inc. 1989). The CD-ROM also contains the AGARD Aerospace Database.

The CD-ROM can be ordered through any member of the Board of Publishers or through NATO-PCO, Overijse, Belgium.

Library of Congress Cataloging-in-Publication Data

On file

ISBN 0-306-44919-6

© 1995 Plenum Press, New York
A Division of Plenum Publishing Corporation
233 Spring Street, New York, N. Y. 10013

10 9 8 7 6 5 4 3 2 1

Printed in the United States of America

# PREFACE

The North Atlantic Treaty Organization (NATO) has sponsored research and personnel safety standards development for exposure to Radiofrequency Radiation (RFR) for over twenty years. The Aerospace Medical Panel of the Advisory Group For Aerospace Research and Development (AGARD) sponsored Lecture Series No. 78 *Radiation Hazards*,[1] in 1975, in the Netherlands, Germany, and Norway, on the subject of Radiation Hazards to provide a review and critical analysis of the available information and concepts. In the same year, Research Study Group 2 on Protection of Personnel Against Non-Ionizing Electromagnetic Radiation (Panel VIII of AC/243 Defence Research Group, NATO) proposed a revision to Standardization Agreement (STANAG) 2345. The intent of the proposal was to revise the STANAG to incorporate frequency-dependent-RFR safety guidelines. These changes are documented in the NATO STANAG 2345 (MED), *Control and Recording of Personnel Exposure to Radiofrequency Radiation,*[2] promulgated in 1979.

Research Study Group 2 (RSG2) of NATO Defense Research Group Panel VIII (AC/243) was organized, in 1981, to study and contribute technical information concerning the protection of military personnel from the effects of radiofrequency electromagnetic radiation. A workshop at the Royal Air Force Institute of Aviation Medicine, Royal Aircraft Establishment, Farnborough, U.K. was held to develop and/or compile sufficient knowledge on the long-term effects of pulsed RFR to maintain safe procedures and to minimize unnecessary operational constraints. That workshop brought together eighteen scientists from six NATO countries and resulted in Aeromedical Review 3-81; *A Workshop On The Protection Of Personnel Against Radiofrequency Electromagnetic Radiation.*[3] Also in 1981, a NATO Advanced Studies Institute (ASI) on *Advances in Biological Effects and Dosimetry of Low Energy Electromagnetic Fields* was held in Erice, Sicily, Italy (Also the fourth course of the International School of Radiation Damage and Protection of the Ettore Majorana Center for Scientific Culture). This meeting resulted in an ASI publication; *Biological Effects and Dosimetry of Non-ionizing Radiation: Radiofrequency and Microwave Energies.*[4]

In 1984, the Research Study Group Panel VIII NATO AC/243, held another workshop in Wachtberg-Werthoven, Federal Republic of Germany with over 40 scientists from five NATO countries attending. The consensus of the participants was that the workshop provided a significant and effective update on the state-of-knowledge regarding the biological effects of RFR and current developments in the setting of new RFR safety standards. The proceedings of the second workshop were published as USAFSAM-TP-85-14: *Proceedings Of A Workshop On Radiofrequency Radiation Bioeffects.*[5] That same year, a NATO Advanced Research Workshop (ARW) examined the *Interaction Between Electromagnetic Fields and Cells,*[6] in Ettore Majorana Center for Scientific Culture, Erice, Sicily, Italy. The 1984 workshop concluded that "It is important that today's RFR bioeffects research results, emerging from many countries, continue to be disseminated as efficiently as possible for consideration and use by the NATO military organization." The theme of the most recent AGARD lecture series, presented in Italy, Portugal, and France in 1985 was *The Impact Of Proposed Radiofrequency Standards On Military Operations.*[7]

A decade has elapsed since the last NATO sponsored RFR workshop and much research has been conducted; investigations into some entirely new areas of biological effects have developed and new guidelines and standards have been issued by several member states and by a few consensus organizations (IEEE/ANSI, IRPA, NRPB). Rapidly expanding technologies require that systems hardware and operation be integrated and compatible within NATO.

The organizing committee (see list below) and the directors developed a list of participants that met several criteria. The participants included experts with research and/or RFR standards development experience, membership on international RFR standards setting bodies, and if possible, prior experience with NATO workshops. Members of the military research and standards setting activities were solicited.

On 16-21 May 1993, a NATO Advanced Research Workshop on Developing a New Standardization Agreement (STANAG) for Radiofrequency Radiation was held at the Divisione Aerea Studi Ricerche e Sperimentazioni (DASRS) at the Aeroporto Pratica di Mare Italian Air Force Base, Pomezia (Rome) Italy. An international group of 47 specialists working in the field of biological effects of electromagnetic fields and standards development attended this Workshop to make presentations and participate in discussions on developing standards for human exposures. The presentations covered a wide range of subjects reflecting the wide range of interests of the participants. It was an exciting and stimulating five days as the state of knowledge in this field of research was examined in great detail. The program was divided into three major sections: Review of Standards, Scientific Basis for New Standards, and Public Health Policy Concerns.

Although the major theme was the development of a STANAG for NATO military forces with regard to human exposure to RFR, it was apparent to the participants that there was the opportunity to begin to think about the development of a general international standard. The deliberations and discussions that emerged during the ARW naturally form a basis for further deliberations for such a standard.

In keeping with NATO objectives for ARWs, this working meeting was designed to assess the state-of-the-art in a given scientific area (RFR) and to formulate recommendations for future. Several recommendations were issued by the ARW working groups and are attached as an Appendix. This ARW continues the NATO program goal to enhance security through scientific dialogue and to encourage peaceful exploitation of scientific skills and discoveries. We believe that the NATO Science Program objective of enhancing scientific and technological capabilities of Alliance countries was fulfilled. We hope that this proceedings will stimulate interest and support development of a unified NATO standard for control and evaluation of personnel exposure to radiofrequency fields.

Directors and Editors

B. Jon Klauenberg, Martino Grandolfo, David N. Erwin

Organizing Committee

G. Mariutti, P. Vecchia, G. Pecci, C. Gabriel, J. Leal, B. Servantie

**REFERENCES**

1. AGARD. Radiation hazards, Lecture Series AGARD-LS-78 (1975).
2. NATO STANAG 2345 (MED). Control and recording of personnel exposure to radiofrequency radiation, MAS (ARMY) 2345 (79) 060, (Edition 1), (16 Feb. 1979).
3. "Proceedings of a Workshop on the Protection of Personnel against Radiofrequency Electromagnetic Radiation," Research Study Group 2, Panel VIII, Defence Research Group, NATO, John C. Mitchell,

(ed.) at the Royal Air Force Establishment, Farnborough, U.K., 6-8 April, 1981, USAFSAM-TR-81-28, (Sept. 1981)

4.   M. Grandolfo, S.M. Michaelson and A. Rindi, (eds.) "Biological Effects and Dosimetry of Nonionizing Radiation: Radiofrequency and Microwave Energies," Plenum Press, New York and London (1983).

5.   "Proceedings of a Workshop on Radiofrequency Radiation Bioeffects," Defense Research Group, Panel VIII, NATO AC/243, J.C. Mitchell, (ed.), Research Establishment for Applied Science, D-5307 Wachtberg-Werthoven, Federal Republic of Germany, 11-13 Sept 1984, USAFSAM-TP-85-14, (April 1985).

6.   A. Chiabrera, C. Nicolini and H.P. Schwan (eds.) "Interactions between Electromagnetic Fields and Cells." Plenum Press, New York and London (1985).

7.   AGARD. The impact of proposed radiofrequency radiation standards on military operations, Lecture Series AGARD-LS-138 (1985).

# ACKNOWLEDGMENTS

This NATO Advanced Research Workshop would not have been possible without the personal commitment and attention of numerous organizations and individuals. We are indebted to the following sponsoring organizations: NATO Scientific Affairs Division; Armstrong Laboratory, United States Air Force; European Office of Aerospace Research and Development, United States Air Force; Divisione Aerea Studi Ricerche e Sperimentazioni (DASRS), Italian Air Force; National Institute of Nuclear Physics, Italy; and the Istituto Superiore di Sanità, Rome, Italy.

The organizers and participants of this ARW are deeply indebted to numerous individuals for their outstanding contributions that made the ARW such a success. Brigadier General Gianfranco Pecci and his staff at the DASRS provided support that exceeded all expectations. The beautiful facilities at the Pratica di Mare Air Force Base and the gracious hospitality of the Italian Air Force personnel made the conference extremely pleasant and the long working days enjoyable. The conference center is truly outstanding. Similarly, the contributions of the Istituto Superiore di Sanità, and the local arrangements committee (Drs Gianni Mariutti and Paolo Vecchia) and technical assistants (Marco Sabatini and Mimmo Monteleone) are greatly acknowledged. The thorough editing of Ms. Janet Trueblood of Systems Research Laboratory is gratefully acknowledged. Three individuals should be singled out for particular thanks for their untiring work both in preparation for and during the ARW: Ms Alma Paoluzi of the Istituto Superiore di Sanità and Staff Sergeant Patrick J. Karshis and Mrs. Minnie Marconi of the Armstrong Laboratory. Lastly, the careful and scholarly efforts of each of the participants is recognized and evidenced in these Proceedings.

# CONTENTS

## SESSION C:
## MILITARY OPERATIONS AND RFR STANDARDS
## CHAIR: L. COURT

## SESSION D:
## EVALUATION OF THE EPIDEMIOLOGIC DATABASE
## CHAIR: P. VECCHIA

## SESSION E:
## EVALUATION OF THE BIOEFFECTS DATABASE I:
## CELLULAR
## CHAIR: A. CHIABRERA

## SESSION F:
## EVALUATION OF THE BIOEFFECTS DATABASE II:
## SYSTEMS PHYSIOLOGY
## CHAIR: B. VEYRET

**Session A:   Standards and Guidelines: Present and Proposed**

**Chair: D. Erwin**


The Standardization Agreement (STANAG) on the Protection
of NATO Personnel against Radiofrequency Radiation
*Martino Grandolfo*


International Commission on Non-Ionizing Radiation Protection
Progress towards Radiofrequency Field Standards
*Michael H. Repacholi*


European Communities Progress towards Electromagnetic
Fields Exposure Standards in the Workplace
*Gianni F. Mariutti*


ANSI/IEEE Exposure Standards for Radiofrequency Fields
*James C. Lin*


Radiofrequency Radiation Safety Guidelines in the Federal
Republic of Germany
*Klaus W. Hofmann*

# THE STANDARDIZATION AGREEMENT (STANAG)
# ON THE PROTECTION OF NATO PERSONNEL
# AGAINST RADIOFREQUENCY RADIATION

Martino Grandolfo

National Institute of Health
Department of Physics
Rome, Italy

## INTRODUCTION

A standard is a general term incorporating both regulations and guidelines and can be defined as a set of specifications or rules to promote the safety of an individual or group of people. Absolute assurances are rarely if ever attainable and specifying permissible exposure limits for different hazards depends on the degree of risk that is scientifically and socially acceptable.[1] Among the many factors that go into the development of an exposure standard, the selection of a good scientific biological effects data base plays, quite obviously, the most important role.

After World War II, powerful and reliable magnetrons operating at 2.45 GHz allowed extensive medical efforts to begin at the Mayo Clinic (undertaken by Krusen and his colleagues) as well as at a number of other institutions. The Office of Naval Research had started a small but effective program, and during the late fifties and early sixties the Tri-service meetings took place.

Daily[2] conducted the first studies on US Navy personnel who were exposed over a period of time in the operation and testing of relatively low-power radar. No evidence of radar-induced pathology was found. Lidman and Cohn[3] examined the blood of 124 men who had been exposed to microwave (MW) radiation for periods from 2 to 36 months. They found no evidence of stimulation or depression of erythropoiesis or leukocytopoiesis. A decade later, Barron, Love and Baraff[4] and Barron and Baraff[5] reported on a large group of radar workers who, along with a control group, were put under a four-year surveillance program. During this period, they underwent repeated physical, laboratory, and eye examinations. The examinations failed to detect any significant changes in the physical inventories of the subjects.

Bioengineering efforts began at the University of Pennsylvania (Schwan) and somewhat later at the University of Washington in Seattle (Lehmann and Guy). Then

*Radiofrequency Standards,* Edited by B.J.
Klauenberg *et al.,* Plenum Press, New York, 1994

Johnson and his colleagues, Durney and Gandhi, at the University of Utah in Salt Lake City, entered this field and, eventually, at many other places, bioengineering efforts started. This consequent effort, carried out with ever-increasing sophistication, has yielded today a rather detailed insight into the patterns of heat deposition caused by radiofrequency (RF) and MW exposure.

For more than 20 years, most countries used a single field-intensity value to maintain the safety of personnel exposed to radiofrequency radiation (RFR), i.e., the limit of 10 mW/cm$^2$ (100 W/m$^2$), which was first proposed by Schwan in 1953 on the basis of physiological considerations, in particular thermal load and heat balance. During the past 10-15 years, it has become well accepted that the absorption and distribution of RFR in humans are strongly dependent on the frequency of the incident radiation, and consequently new exposure standards have been issued that provide, in general, an added margin of safety over what was previously used.

In this paper an attempt will be made to present the development in time of the NATO standardization agreement (STANAG) on the control and evaluation of exposure of personnel to RF and MW radiation.

## PAST NATO EXPOSURE STANDARD

The original frequency-independent limit of 10 mW/cm$^2$ was tentatively adopted in 1953 on the basis of theoretical considerations by Schwan and his associates. The maximum rate of working that can be sustained over a period of hours requires the body to dissipate about 750 W. Total absorption of 10 mW/cm$^2$ over the cross-sectional body area of about 0.7 m$^2$ would result in an additional heat load of 70 W, which is small compared with that caused by even the lightest manual activity, and is in fact less than the resting metabolic rate. The 10 mW/cm$^2$ level was also lower by a factor of at least 10 than the exposures needed to cause injury to the testes or the eye--the two organs widely regarded as being most sensitive to thermal injury.

The acknowledged assumption in this first standard was that heating of body tissues is the most, if not the only, significant consequence of absorbing RF energy. Consequently, this standard offered the advice that lower levels should be considered in the presence of other heat stresses, while higher levels might be appropriate under conditions of intense cold.

Intensive investigation was subsequently carried out by the US Department of Defense into the biological effects of MW radiation.[6] None of these investigations, however, produced any evidence for a biological effect of levels even approaching the theoretical limit of 100 mW/cm$^2$ that could be considered hazardous for man.

The limit of 10 mW/cm$^2$ for RF and MW exposure was recommended in 1966 by the American National Standards Institute (ANSI C95.1), reaffirmed in 1973, and has been generally accepted by the armed services too.

The ANSI C95.1 standard didn't specify an upper limit of allowable short-time exposure. It specified a safe value based on *average dose* for a period of 0.1 hour (6 min). According to this standard, no individual could be exposed without good reason to a power density in excess of 10 mW/cm$^2$ for longer than 6 minutes. On this basis, it permitted continuous exposure to a level averaging 10 mW/cm$^2$; momentary levels in excess of this value were allowable only if the energy density didn't exceed 1 mW-hr/cm$^2$.

While the limit of 10 mW/cm$^2$ served as a practical exposure level in the US Department of Defense and in NATO nations (STANAG 2345) for several years, it was felt that the duration of exposure was important and that higher levels could be tolerated for shorter periods. Therefore, in 1965 a guide was developed and published by the US Army and US Air Force to permit exposure of personnel within a MW field (300 MHz-300 GHz) only for a specified length of time, determined by the following equation:

$$T_p = 6000/X^2 \tag{1}$$

where $T_p$ = permissible time of exposure in minutes during any 1-hour period, and X = power density (mW/cm$^2$) in the area to be occupied. Because exposures of less than 2 minutes were considered operationally impractical, the use of Equation (1) for power densities above 55 mW/cm$^2$ was contraindicated.

## CURRENT NATO STANAG 2345

Current STANAG 2345 *Control and Evaluation of Exposure of Personnel to Radio-Frequency Radiation* has been promulgated by the Military Agency for Standardization (MAS) in 1982 and is characterized by Annexes A (Definitions), B (Hazard Assessment), C (Limitations of Exposure and Control Measures), and D (Actions to be taken in case of Suspected or Actual Overexposures). The aim was to establish within NATO Forces criteria for the evaluation and control of personnel exposure to radiofrequency radiation within the frequency range 10 kHz to 300 GHz.

Nations that ratified the STANAG agreed to:

i)   Use definitions as listed in Annex A;

ii)  Apply the measures for the protection of exposed personnel from the hazards of RF installations taking into consideration the hazard assessments stated in Annex B;

iii) Apply the principles of the limitations of exposures and radiation protection for exposed personnel to RFR as given in Annex C;

iv)  Take measures as listed in Annex D in the case of suspected or actual overexposure (accident);

v)   Identify areas of biological hazard from RFR by suitable signs.

With an increasing knowledge about RF and MW dosimetry (absorption and distribution of energy in the human body and the differing absorption of the electric and magnetic fields), risk assessment and thus standards are becoming more sophisticated. No longer was a standard merely one power density value over the entire frequency range of interest. Modern standards have frequency ranges that take account of the differences in electromagnetic energy absorption in humans.[7]

In Annex B to STANAG 2345, titled *Hazard Assessment*, a distinction is made between *thermal effects, other biological effects,* and *indirect effects.* Other biological effects are, by definition, those that may also occur from exposures at field or power intensity levels not causing a measurable increase in temperature.

By definition, indirect effects are due to interference with prosthetic medical devices (e.g., cardiac pacemakers) or on flammable materials or on electro-explosive devices.

The current NATO STANAG 2345 deals only with biological effects (thermal and other) on the personnel because indirect effects are taken into account by other specific STANAGs.

To correctly perform a hazard assessment, factors affecting the beam-power density or energy density of the system have to be taken properly into account. Factors listed in the STANAG are:

a) Transmitter characteristics;

b) Pulse width;

c) Pulse repetition frequency;

d) Number of emitters with beams intersecting or illuminating the same area;

e) Distances;
f) Reflecting materials.

Other factors governing the biological effects on personnel, besides the factors mentioned above, are:

a) Exposure duration;
b) Orientation in the field;
c) Environmental conditions;
d) Part of individual exposed.

In the Annex C to STANAG 2345 there is a general statement on the issues that all unnecessary radiation exposure should be avoided and that personnel at risk should be briefed about the hazards and the measures being taken to protect them.

The exposure limits given in Table 1 for frequencies between 10 kHz and 300 GHz are the present NATO STANAG working limits. All limits apply to whole-body exposure from either continuous wave, modulated or pulsed radiation, averaged over any one-second period.

The frequency dependence of power density and electric and magnetic field strength limits recommended in the present STANAG are shown in Figures 1 and 2, respectively.

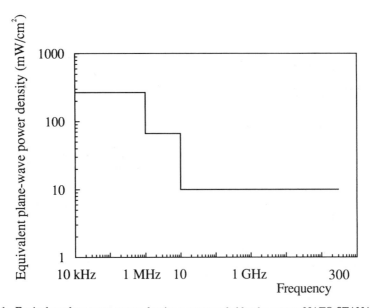

**Figure 1.** Equivalent plane-wave power density recommended by the current NATO STANAG 2345.

Besides the derived limits (root mean square values, rms) shown in Table 1, the following provisions shall be taken into account:

(i)     All exposures are limited to a maximum electric field strength of 100 kV/m in a single pulse;

(ii)    For frequencies less than 30 MHz, both the magnetic field and resultant electric field should be measured; the larger of the calculated power density values will be used to delineate the risk area or to determine the exposure time;

(iii)   The energy density values quoted are the maximum allowed for each 6-minute period.

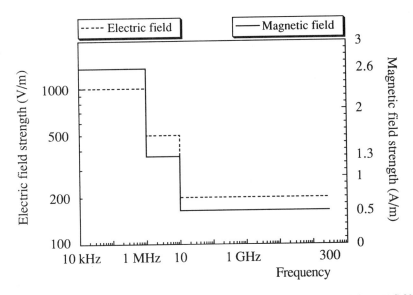

**Figure 2.** Electric and magnetic field strengths recommended by the current NATO STANAG 2345.

For frequencies above 3 kHz, air can be ionized due to voltage breakdown at approximately 1000 kV/m. The resulting electrical discharge could result in hazardous conditions for human exposure. The 100 kV/m limit, as expressed in the item (i), includes thus a safety factor of approximately ten.

**Table 1.** Exposure limits to RF and MW radiation enforced by current NATO STANAG 2345.

| Frequency range (MHz) | Unperturbed rms electric field strength (V/m) | Unperturbed rms magnetic field strength (A/m) | Equivalent plane wave power density (W/m$^2$) | (mW/cm$^2$) |
|---|---|---|---|---|
| 0.01-1 | 1,000 | 2.6 | 2,650 | 265 |
| 1-10 | 500 | 1.3 | 660 | 66 |
| 10-300,000 | 200 | 0.50 | 100 | 10 |

NATO STANAG 2345 also gives advice on how to manage short exposure times, i.e., any six-minute period. Taking into account operational commitments, in any 6-minute period personnel should not be exposed to integrated power density exceeding the values shown in Table 2.

In any period of time $\Delta t$ ranging from 0 to 6 min (expressed in terms of hours, minutes or seconds, respectively), the product of power density P (in mW/cm$^2$) times $\Delta t$ should not exceed the values shown in Table 2.

In order to comply with the STANAG, in case of a suggested or actual overexposure, the occurrence must be investigated and appropriate measures to be taken.

**Table 2**. Integrated power density limits enforced by current NATO STANAG 2345 for short exposure times.

| Frequency range (MHz) | Equivalent plane wave integrated power density | | |
|---|---|---|---|
| | $(mW\text{-}hr/cm^2)$ | $(mW\text{-}min/cm^2)$ | $(mW\text{-}s/cm^2)$ |
| 0.01-1 | 26.5 | 1,590 | 95,400 |
| 1-10 | 6.6 | 396 | 23,760 |
| 10-300,000 | 1 | 60 | 3,600 |

In case of an accident, any overexposure will be reported to the appropriate authorities. The report will contain the following information:

(i)     A detailed account of the circumstances leading to the overexposure;

(ii)    Detail of immediate and subsequent medical findings. EEG and ophthalmic examinations are recommended in due time;

(iii)   Action required to eliminate the hazardous situation or to prevent recurrence.


## PROPOSED REVISION OF CURRENT STANAG 2345

A draft document concerning a proposal for the revision of current standard, i.e., the STANAG 2345 (Edition 2) (1st draft) *Control and Evaluation of Exposure of Personnel to Radio Frequency Radiation* has been, recently, prepared by the General Medical Working Party.[8]

In 1989, nations prepared to ratify the new STANAG were requested to specify a date of implementation, either actual or forecast.

The basic premise of modern risk assessment is that the inherent risk to health from RFR exposures is directly related to the rate of RF energy absorbed and hence the introduction of the concept of specific absorption rate (SAR).

SAR, measured in watts per kilogram, has the advantage of providing a reasonable basis on which to physically scale the exposure parameters that produced results of *in vivo* experiments in various biological systems to the equivalent RF exposure parameters that could be predicted to produce the same effects in humans.

In Annex B to STANAG 2345 (Edition 2), titled *Hazard Assessment*, a distinction is made between *biological effects* and *indirect effects*. Biological effects are furthermore distinguished in *thermal effects* and *specific effects*. Definitions practically remained the same presently adopted in current STANAG 2345.

The proposed NATO STANAG 2345 (Edition 2) makes a distinction between conditions in normal peace time and either in wartime or during exercises. According to the draft document, in peace time personnel should not be exposed to RFR intensities exceeding limits laid down in corresponding national legislations.

The frequency range of greatest concern to present military RFR operations is 3 MHz-30 GHz. In this frequency range, the new STANAG 2345 recommends, as appropriate, safety limits based on whole-body-averaged specific absorption rate (WBA-SAR).

The rationale of the STANAG 2345 (Edition 2) in terms of limitation of exposure for personnel, in wartime and during exercises, is that a WBA-SAR of 4 W/kg has been established as a reasonable value for the adverse effects threshold. Using a safety factor of 10, the RFR protection guides limit the WBA-SAR to 0.4 W/kg or less and the spatial peak SAR (to take in the proper account *hot spots*) to 8 W/kg as averaged over any 1 g of tissue.

The exposure limits given in Table 3 for frequencies between 10 kHz and 300 GHz are the proposed NATO STANAG working limits derived from the WBA-SAR of 0.4 W/kg.

They represent a practical approximation of the incident plane wave power density needed to produce the whole-body-averaged specific absorption rate of 0.4 W/kg. These limits apply to whole-body exposure from either continuous wave or pulsed radiation, averaged over any 6-minute period.

The frequency dependence of power density and electric and magnetic field strength limits recommended in the revised STANAG are shown in Figures 3 and 4, respectively.

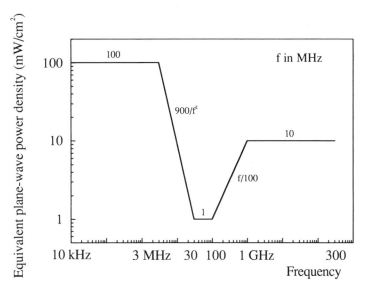

**Figure 3.** Equivalent plane-wave power density recommended by the newly proposed NATO STANAG 2345.

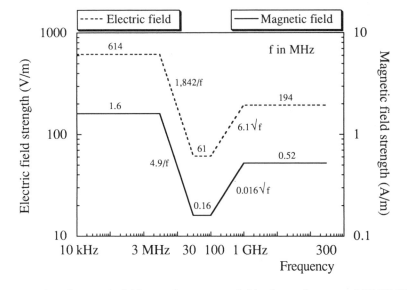

**Figure 4.** Electric and magnetic field strengths recommended by the newly proposed NATO STANAG 2345.

The derived limits (root mean square values, rms) shown in Table 3 may be increased under special circumstances provided that:

(i)     WBA-SAR does not exceed 0.4 W/kg when averaged over any 6-min period;

(ii)    Spatial peak SAR does not exceed 8 W/kg averaged over any 1 gram of tissue;

(iii)   Personnel are adequately protected from electric shock and RF burns through the use of electrical safety matting, safety shoes, or other isolation techniques; and

(iv)    All exposures are limited to a maximum electric field strength of 100 kV/m in a single pulse.

**Table 3**. Exposure limits to RF and MW radiation recommended by the newly proposed NATO STANAG 2345 in wartime and during exercises (f = frequency in megahertz).

| Frequency range (MHz) | Unperturbed rms electric field strength (V/m) | Unperturbed rms magnetic field strength (A/m) | Equivalent plane wave power density (W/m$^2$) | (mW/cm$^2$) |
|---|---|---|---|---|
| 0.01-3 | 614 | 1.6 | 1,000 | 100 |
| 3-30 | 1,842/f | 4.9/f | 9,000/f$^2$ | 900/f$^2$ |
| 30-100 | 61 | 0.16 | 10 | 1 |
| 100-1,000 | 6.1√f | 0.016√f | f/10 | f/100 |
| 1,000-300,000 | 194 | 0.52 | 100 | 10 |

In case of radiating and reactive near-field conditions, mainly in the frequency range 10 kHz-3 MHz, both the electric and magnetic field strength limits must be used to determine compliance.

Exposure to radiofrequency radiation emitted from equipment that radiates at frequencies below 1 GHz and delivers less than 7 W of RF power to the radiating device is considered non-hazardous. These exposures can be excluded from considerations in assessing compliance with the prescribed limits.

**Table 4**. Integrated power density limits recommended by the newly proposed NATO STANAG 2345 for short exposure times (f = frequency in megahertz).

| Frequency range (MHz) | Equivalent plane wave integrated power density (mW-hr/cm$^2$) | (mW-min/cm$^2$) | (mW-s/cm$^2$) |
|---|---|---|---|
| 0.01-3 | 10 | 600 | 36,000 |
| 3-30 | 90/f$^2$ | 5,400/f$^2$ | 324,000/f$^2$ |
| 30-100 | 0.1 | 6 | 360 |
| 100-1,000 | f/1,000 | 6f/100 | 3.6f |
| 1,000-300,000 | 1 | 60 | 3600 |

The NATO STANAG 2345 (Edition 2) also gives advice on how to manage short exposure times, i.e., any six-minute period. Taking into account operational commitments, in any 6-minute period personnel should not be exposed to integrated power density exceeding the values shown in Table 4. In Figure 5 the integrated power density versus time is shown for the frequency range 30-100 MHz. Exposures separated by more than 6 minutes are considered to be separate physiological events.

In order to comply with the STANAG, in case of a suggested or actual overexposure, the occurrence must be investigated and appropriate measures to be taken as already enforced in the present STANAG 2345 (1982).

For the sake of completeness, in Figure 6 the evolution in time of the NATO STANAG 2345 is shown.

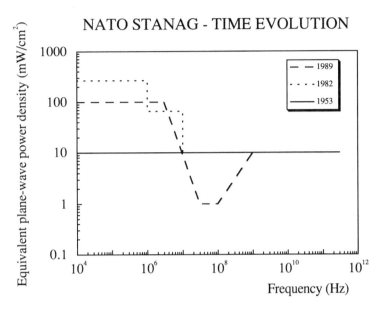

**Figure 5.** Power density limit recommended for short exposure times by the revised STANAG 2345 in the frequency range 30-100 MHz.

## NATO STANAG 2345 Edition 2 - 1st draft

30 - 100 MHz

$$P \ ( \text{mW/cm}^2) \cdot \Delta t \ (\text{min}) = 6$$

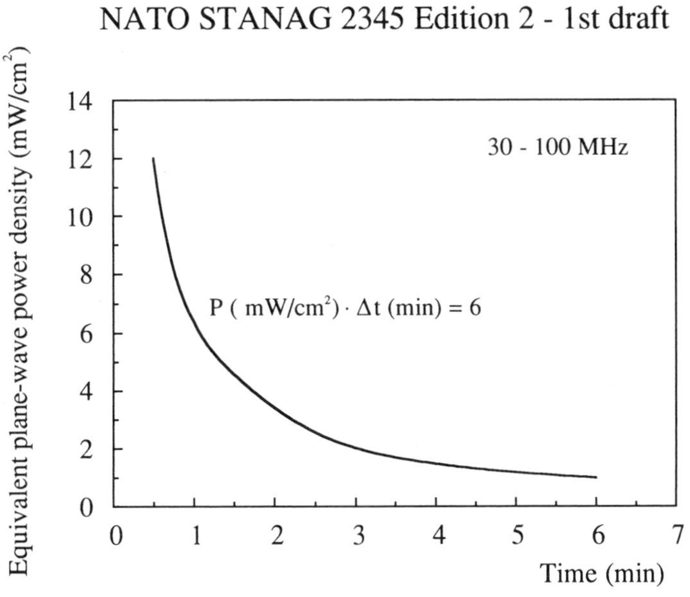

**Figure 6.** Evolution in time of the NATO STANAG.

## FUTURE TRENDS

The acceptance of WBA-SAR in the development of modern frequency-dependent safety standards, as shown in Figure 7, has been a significant improvement. Despite this fact, WBA-SAR seems an inadequate basis for safety guidelines at frequencies greater than 20 GHz and less than about 3 MHz.

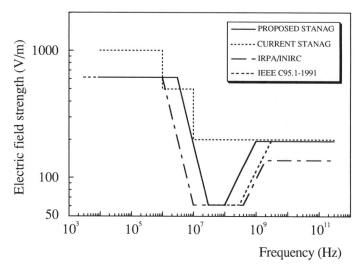

**Figure 7.** Comparison between present and proposed NATO STANAG 2345 and IRPA and IEEE standards.

At frequencies greater than 20 GHz (depth of penetration in tissues of 1.6 mm), microwave energy deposition in biological tissue is very superficial. For instance, at 20 GHz only 1% of the total body mass is actually exposed and absorbs RF energy. Taking into account these considerations, it is advisable to search for some form of localized SAR, instead of WBA-SAR, to serve as a better safety guideline in the frequency range 20-300 GHz. Also, WBA-SAR is not an adequate basis for the safety guideline below about 3 MHz, since absorption in biological tissue decreases as a function of frequency squared below the resonant frequency, becoming very small in this low-frequency range.

Another issue that deserves greater attention in a future NATO STANAG is that exposures should be restricted to limit induced currents in the body that can be associated to contact with objects exposed to RF fields, and potential for RF shocks and burns.

Little information is presently available on the relation of biological effects with peak values of pulsed fields. Thus, under exposure to pulsed fields it is advisable to be conservative in view of some uncertainty about the value of spatial peak SAR, which could be over 20 times the spatially averaged SAR.

Up to now the averaging time, i.e., the appropriate time period over which exposure is averaged for the purpose of determining compliance with the STANAG, always corresponded to a time duration of 6 minutes. In future some refinements will be necessary to permit averaging time to be frequency dependent; this will permit the transition from values of minutes, in the resonant frequency range, to values of seconds necessary to match averaging times suitable at infrared frequencies.

## ACKNOWLEDGMENTS

The author is much indebted to Ms. Alma Paoluzi and Mr. Marco Sabatini for their skilled assistance in preparing this manuscript.

## REFERENCES

1.  M.H. Repacholi, Development of standards - Assessment of health hazards and other factors, *in:* "Biological Effects and Dosimetry of Nonionizing Radiation: Radiofrequency and Microwave Energies," M. Grandolfo, S.M. Michaelson, and A. Rindi, eds., Plenum Press, New York and London (1983).
2.  L. Daily, A clinical study of the results of exposure of laboratory personnel to radar and high frequency radio, *U.S. Nav. Med. Bull.* 41:1052 (1943).
3.  B.I. Lidman and C. Cohn, Effect of radar emanations on the hematopoietic system, *Air Surg. Bull.* 2:448 (1945).
4.  C.I. Barron, A.A. Lowe and A.A. Baraff, Physical evaluation of personnel exposed to microwave emanations, *J. Avia. Med.* 26:442 (1955).
5.  C.I. Barron and A.A.Baraff, Medical considerations of exposure to microwaves (radar), *J.A.H.A.* 168:1194 ( 1958).
6.  S.M.Michaelson, The Tri-service program - a tribute to George M. Knauf, USAF (MC). *IEEE Trans. Microwave Theory and Techniques* MTT19:131 (1971).
7.  M. Grandolfo and K.H. Mild, Worldwide public and occupational radiofrequency safety and protection guides, in: " Electromagnetic Biointeraction. Mechanisms, Safety Standards, Protection Guides," G. Franceschetti, O.P. Gandhi, and M. Grandolfo, eds., Plenum Press, New York and London (1989).
8.  NATO STANAG 2345 (Edition 2, first draft) Control and evaluation of exposure of personnel to radiofrequency radiation, MAS (ARMY) 561-1MED (39th Meeting), dated 19 December 1988 (1989).

# INTERNATIONAL COMMISSION ON NON-IONIZING RADIATION PROTECTION PROGRESS TOWARDS RADIOFREQUENCY FIELD STANDARDS

Michael H. Repacholi

Chairman, ICNIRP
Australian Radiation Laboratory
Victoria, Australia

## ABSTRACT

The International Commission on Non-Ionizing Radiation Protection (ICNIRP) was established in May 1992 by the International Radiation Protection Association (IRPA) as an independent scientific commission to provide advice and guidance on all aspects of protection from NIR exposure to workers, the general public and the environment. ICNIRP's predecessor, the International Non-Ionizing Radiation Committee (INIRC) of the IRPA had developed a substantial reputation in the field of NIR protection. Its procedures for the development of standards will be continued by the ICNIRP and are described here. Current areas of concern and future activities of ICNIRP on protection from radiofrequency (RF) field exposure will also be summarized.

## INTRODUCTION

On 20 May 1992, at its General Assembly in Montreal, International Radiation Protection Association (IRPA) approved a charter establishing an International Commission on Non-Ionizing Radiation Protection (ICNIRP). The Commission is an independent scientific body having a close association with IRPA. The ICNIRP recommends safety guidelines and develops codes of safe practice for the non-ionizing radiations (NIR) in a manner similar to the International Commission on Radiological Protection for ionizing radiations.

In the context of ICNIRP activities, NIR include:
1. electromagnetic radiations and fields having wavelengths of 100 nanometers or greater, or frequencies from zero to 3 PHz ($0\text{-}3.10^{15}$ Hz) and include static electric and magnetic fields, extremely low-frequency fields, radiofrequency fields (including microwaves), infrared, visible and ultraviolet radiation;
2. acoustic fields with frequencies above 20 kHz (ultrasound) and with frequencies below 20 Hz (infrasound).

*Radiofrequency Standards,* Edited by B.J.
Klauenberg *et al.,* Plenum Press, New York, 1994

This paper summarizes the history of activities of the INIRC in the area of RF protection and includes:

- its procedure of in-depth review of the scientific literature in collaboration with the World Health Organization (WHO),
- development of guidelines on RF exposure limits,
- issuing statements to clarify concerns about specific RF emitting devices,
- work with international agencies to develop basic restrictions on which exposure limits can be derived, and
- preparation of codes of safe practice for workers in collaboration with the International Labour Office (ILO).

Current and future activities of the ICNIRP in RF will also be summarized.

---

1. Identification of NIR exposure hazard.
2. Literature review with WHO/other specialized agencies.
   - established health effects
   - gaps in knowledge
   - basis for exposure limits
   - EHC Environmental Health Criteria - WHO/ICNIRP publication
3. Draft exposure limits
   ICNIRP Health Physics publication
4. Codes of practice with ILO/other specialist agencies.
   - ILO/ICNIRP publication
5. Statements on NIR-emitting devices
   - address concerns about specific devices
   - evaluated in relation to standards
   - ICNIRP Health Physics publication

---

**Figure 1.** Steps in development and dissemination of NIR standards.

## DEVELOPMENT OF STANDARDS

Early in its existence, the INIRC evolved a sound procedure for drafting guidelines on human exposure limits (as shown in Figure 1). This involved the conduct of an extensive review of the scientific literature on the effects of NIR on biological systems, including human beings. The review was conducted in collaboration with WHO and was published as an Environmental Health Criteria (EHC) Document. EHC publications included: details on the physical characteristics of the NIR; methods of measurement; interaction mechanisms on biological systems; reviews of biological effects and human studies; summaries of current national standards and the rationales on which they were based; and an overview of protective measures that could be taken to reduce NIR exposure.

One of the primary goals of EHC documents is to distinguish between the established biological effects from those that were hypothesized or preliminary (unconfirmed). This is fundamental to the development of standards because they can only be based on what effects are known and established and lead to possible health hazards. While safety factors can be incorporated into exposure limits to allow for unknowns, to base guidelines on unconfirmed data leads to unnecessarily restrictive limits and increased costs to industry, consumers and the general public, for no known benefit to health.

A second goal of the EHC documents is to summarize the data base of NIR bioeffects with a view to identifying gaps in knowledge. Filling these gaps leads to a better understanding of the health effects of NIR exposure to more clearly defined exposure limits. Thus, EHC documents provide direction for bioeffects research.

More recently, focused, multidisciplinary, large-scale research has been recommended in EHC publications because it was found that many reports were often the result of flawed methodology, lack of quality control and misinterpretation of results. Small experiments generally raised more questions than they answered, were on-off and had the effect of unnecessarily increasing concern about possible health effects. Criteria used to evaluate scientific literature with respect to developing exposure limits have been published by Repacholi and Stolwijk[1] and Repacholi.[2]

With the publication of the essence of a scientific rationale on which exposure limits can be based, the INIRC would prepare guidelines on exposure limits from the conclusions of the EHC documents and more recent publications. Details on what standards should address, whether they should be propagated at all, their benefits and detriments have been outlined by Repacholi.[3] An overview of RF standards has also been summarized in Repacholi.[4]

The primary purpose of the INIRC guidelines documents is to provide a sound scientific rationale on which exposure limits could be based. The Committee recognized that when standards are established, various value judgments are made. The validity of scientific reports has to be considered and extrapolations from animal experiments to effects on humans have to be made. Cost-versus-benefit analyses are necessary, including the economic impact of controls. The limits in the guidelines were based on the scientific data and no consideration was given to economic impact or other nonscientific priorities. However, from available knowledge, the Committee felt that the limits provide a safe, healthy working or living environment from exposure to RF radiation under all normal conditions.

Following the publication of guidelines, the INIRC collaborated with ILO to develop a code of safe practice for workers. In this way, the exposure limits were disseminated to the places where people could use them.


## ACTIVITIES OF THE INIRC/IRPA

### Environmental Health Criteria/Guidelines

The first EHC document on RF was published[5] in 1981 and provided the scientific basis for guidelines published[6] in 1984. Minor adjustments were made to the guidelines in 1988[7] to clarify details of the existing standard. However, since 1981, significant progress was made, predominantly in the area of dosimetry (absorption patterns of RF fields in the human body). Also a large number of experiments on the biological effects of RF were published that necessitated review and evaluation to determine if exposure limits should be altered.

In conjunction with WHO, INIRC prepared an updated RF document that received widespread review and comment. This culminated in a WHO Task Group review in Ottawa in October 1990. The final scientific and language editing has been completed and will be published[8] in early 1993.

This review identified that our knowledge of the effects of pulsed-RF fields was scant. Many new technology devices (e.g., mobile telephones) use pulsed fields, so it is important that these deficiencies be remedied so exposure limits can be reaffirmed or reassessed. The EHC document also concluded that there is no evidence to establish that RF exposure is carcinogenic, although only a few studies have been completed.

**Statements**

Two publications were issued by the INIRC that addressed specific devices seen as a major concern with respect to RF emissions. These were visual display units and magnetic resonance diagnostic devices used in medicine.

**Video display units (VDU).** VDU's emit RF at frequencies around 15-35 kHz from the vertical and horizontal sweep generators. These fields are emitted at very low levels and certainly well below the limits of general public exposure required by national and international standards. The statement concluded that there are no health hazards associated with radiation or fields from VDUs.[9]

**Magnetic resonance.** Concerns about the safety of patients undergoing magnetic resonance (MR) examinations arose with the rapidly increasing popularity of this modality. Of particular concern was the RF energy deposition, especially in the later MR units attempting to reduce imaging times by increasing RF pulse levels.

The INIRC recommended[10] that there should be limitations on the RF to eliminate the elevation of patient body temperature to levels at which local thermal injury or systemic thermal overload may occur. Thus, the primary criterion for radiofrequency exposure was based on elevation of temperature in the skin, body core or in spatially limited volumes of tissues where local temperature increases ("hot spots") may occur under conditions of MR exposure.

It was noted that RF energy at frequencies between 10 and 100 MHz deposited in the body during an MR examination will be converted to heat which will be distributed largely by convective heat transfer through blood flow. As the body temperature increases, there will be an increase in blood flow and cardiac output, as well as an increase in sweat secretion and evaporation.

For whole-body exposures or exposures to the head and trunk, no adverse health effects are expected if the increase in body temperature does not exceed 1°C. In the case of infants, pregnant women and persons with cardiocirculatory impairment, it is desirable to limit temperature increases to 0.5°C. RF specific absorption rates that used these criteria were published in INIRC/IRPA.[10]

**Guidance to International Agencies**

**International Labour Office.** ILO has collaborated with the INIRC almost since the Committee was formed in 1977. This has resulted in a number of publications to inform workers about the hazards of various NIR and what protective measures are necessary to keep NIR exposure below the exposure limits in the IRPA guidelines.

A general document on occupational hazards from NIR was published in 1985[11] and a more specific technical review on protection of workers against RF in 1986.[12] The latter document gave a brief overview of RF bioeffects and health hazards in a language understandable by workers and informed them of the IRPA RF guidelines and how to comply with them.

Video display units and possible radiation hazards to workers were also of interest to ILO. A detailed description of VDU operation, NIR emissions, operator accessories, assessment of bioeffects literature and possible health effects were published.[13] With respect to RF emissions the INIRC/IRPA[9] statement was repeated.

**Commission of the European Communities.** Since 1970 the CEC has been concerned with the influence of non-ionizing radiation on human health. In its Action Programme relating to the implementation of the Community Charter of Basic Social

Rights for Workers, the Commission has included the presentation of a Proposal for a Council Directive on the minimum safety and health requirements regarding the exposure of workers to the risks caused by physical agents.

Concerning non-ionizing electromagnetic radiation, the Commission invited experts, who are involved in each of their countries in protection against non-ionizing radiation, to formulate proposals for basic restrictions for protection against occupational exposure. These proposals were to be based where appropriate on the activities of the IRPA/INIRC that represents the opinions of the scientific community at large and ensures liaison with other relevant international organizations such at the ILO and the WHO.

For RF it was noted[14] that:

1. Associations between cancer incidence and exposure to electromagnetic fields have been reported in epidemiological studies. However, the data do not establish a causal relationship nor do they provide a basis for risk assessment.

2. Electric and magnetic fields at frequencies up to about 100 kHz coupling to conducting objects may cause electric currents to pass through the human body in contact with the object. The magnitude and spatial distribution of such currents depend on frequency, quality of field coupling, size, age and sex of the person, and area of contact.

   Transient discharges (sparks) can occur when people and conducting objects exposed to strong fields come into close proximity. The fields can interfere with medical implants such as cardiac pacemakers and cause malfunction of the device. The sensitivity of medical implants to electromagnetic interference is in general not known. Cardiac pacemakers may show different performance when exposed to electromagnetic fields.

3. For electric and magnetic fields up to 100 kHz, the following can be stated:

   Electric fields induce a surface charge on an exposed body. This results in a current distribution inside the body, depending on exposure conditions, size, shape and position of the exposed body in the field. Magnetic fields induce electric fields and currents in the body depending on exposure conditions and body dimension. For these fields, the basic restrictions recommended were:

   (a)  current densities induced in the whole body and for the working day not to exceed $f/100$ mA/m$^2$ where $f$ = frequency in Hz (in range 400 Hz - 100 kHz), and current densities must be averaged over periods shorter than 1 s and over cross sections of 1 cm$^2$, perpendicular to the current direction.

   (b)  contact currents for any one foot or hand in contact with conducting objects must not exceed 1.5 mA for frequencies less than 3 kHz or $0.5f$ mA for frequencies in range 3 kHz - 100 kHz.

4.  For RF fields 100 kHz - 300 GHz the restrictions are designed to prevent heating, electric shocks and RF burns. From a literature review a threshold for effects considered to be detrimental to health is observed at 4 W/kg. Table 1 gives the basic restrictions for workers.

## FUTURE ACTIVITIES OF ICNIRP

At its final meeting in Vancouver in 1992, the INIRC drafted a four-year work plan of activities for the new ICNIRP. These included development of the following documents:

a. Guidelines on limits of exposure to electromagnetic fields 0-100 kHz.
b. Guidelines on limits of exposure to pulsed RF fields.
c. Statement on police radars.
d. Statement on mobile telephones.
e. Practical guides for protection of workers using RF heaters (ICNIRP/ILO).

f. Manual on medical handling of accidental overexposure to NIR (ICNIRP/ILO/WHO).

g. Manual on protection of health care workers against NIR. (ICNIRP/ILO/WHO).

All of the above documents were discussed further at the first ICNIRP meeting in Munich, May 7-12, 1993.

**TABLE 1.** Basic restrictions for workers in the frequency range 100 kHz-300 GHz.[14]

| Frequency | Induced current density | Contact current (mA) | Whole body average SAR | Local peak SAR (in the limbs) | Local peak SAR (head and trunk) |
|---|---|---|---|---|---|
| 100 kHz-10 MHz[a] | $f/100$ A/m$^2$ | 50 | 0.4 W/kg | 2 W/(0.1 kg) | 1 W/(0.1 kg) |
| 10 MHz-300 GHz[b] | | 50[c] | 0.4 W/kg | 2 W/(0.1 kg) | 1 W/(0.1 kg) |

f    frequency in kHz

(a)   In this frequency range both the induced current density and SAR limits should not be exceeded.

(b)   In order to avoid auditory effects associated with short (< 30 ls) pulses at frequencies above 300 MHz the specific absorption should not exceed 10 mJ/kg.

(c)   This value applies only up to 100 MHz.

## REFERENCES

1. M.H. Repacholi and J.A.J. Stolwijk, Criteria for evaluating scientific literature and developing exposure limits. Radiation Protection in Australia 9(3): 79-84 (1991).

2. M.H. Repacholi, Development of standards-assessment of health hazards and other factors, *in:* Biological Effects and Dosimetry of Non-ionizing Radiation, M. Grandolfo, S.M. Michaelson and A. Rindi, eds., Plenum Press, New York and London, pp 611-625 (1983).

3. M.H. Repacholi, Guidelines and standards, *in:* Non-Ionizing Radiation, M.W. Green, ed., University of British Columbia Press, Vancouver, pp 465-482 (1992).

4. M.H. Repacholi, Radiofrequency field exposure standards: Current limits and the relevant bioeffects data, *in:* Biological Effects and Medical Applications of Electromagnetic Energy, O.P. Gandhi, ed., Prentice Hall, New Jersey, pp 9-27 (1990).

5. UNEP/WHO/IRPA, United Nations Environment Programme, World Health Organisation, International Radiation Protection Association, Radiofrequency and Microwaves, Environmental Health Criteria 16. WHO, Geneva (1981).

6. International Radiation Protection, Association/International Radiation Protection Association (INIRC/IRPA). Interim guidelines on limits of human exposure to radiofrequency electromagnetic fields in the frequency range from 100 kHz to 300 GHz, *Health Physics* 46: 975-984 (1984).

7. International Radiation Protection Association/International Radiation Protection Association (INIRC/IRPA), Guidelines on limits of exposure to radiofrequency electromagnetic fields in the frequency range from 100 kHz to 300 GHz, *Health Physics* 54(1): 115-123 (1988).

8. UNEP/WHO/IRPA, United Nations Environment Programme/World Health Organisation/International Radiation Protection Association, Electromagnetic fields 300 Hz to 300 GHz, Environmental Health Criteria 137. WHO, Geneva (in press) (1992).

9. International Non-Ionizing Radiation Committee/International Radiation Protection Association (INIRC/IRPA), Alleged radiation risks from visual display units. International Non-Ionizing Radiation Committee of IRPA, *Health Physics* 54(2): 231-232 (1988).

10. International Non-Ionizing Radiation Committee/International Radiation Protection Association (INIRC/IRPA), Protection of the patient undergoing a magnetic resonance examination, *Health Physics* 61(6): 923-928 (1991).

11. International Labour Office (ILO), Occupational hazards from non-ionizing electromagnetic radiation. Occupational Safety and Health Series No. 53, ILO, Geneva (1985).

12. International Labour Office (ILO), Protection of workers against radiofrequency and microwave radiation: a technical review. Occupational Safety and Health Series No. 57 (in English and French), ILO, Geneva (1986).

13. International Labour Office (ILO), Video display units - radiation protection guidance, Occupational Health and Safety Series No. XX, ILO, Geneva (in press) (1992).

14. S.G. Allen, J.H. Bernhardt, C.M.H. Driscoll, M. Grandolfo, G.F. Mariutti, R. Mathes, A.F. McKinlay, M. Steinmetz, P. Vecchia, and M. Willock, Proposals for basic restrictions for protection against occupational exposure to electromagnetic non-ionizing radiations. Recommendations of an international working group set up under the auspices of the Commission of the European Communities, *Physica Medica* 7(2): 77-89 (1991).

# EUROPEAN COMMUNITIES PROGRESS TOWARDS ELECTROMAGNETIC FIELDS EXPOSURE STANDARDS IN THE WORKPLACE

Gianni F. Mariutti

Physics Laboratory
National Institute of Health
Rome, Italy

## INTRODUCTION

The regulatory activity of the European Community (EC) on protecting humans from non-ionizing radiation dates back to 1970. That year the Commission's Directorate General for Employment, Social Affairs and Education was given the responsibility for protecting human health. A Working Group of experts was established and a technical document containing the basic restrictions for health protection from microwaves was prepared. This document was converted into a proposal for a directive to be submitted to the Council for · approval. This proposal, aimed to protect workers and the general public from microwave radiation hazards, was published in the Official Journal of the European Communities in 1980. In 1981 the European Parliament and the Economic Social Committee made a positive evaluation, but, in spite of that, the proposed Directive has never been approved by the Council of Ministers. From 1983 to 1984 the proposal on microwaves was modeled and aligned according to the framework of other Council Directives on the protection of workers from hazards due to exposure to other chemical, biological and physical agents. The introduction of basic limits defined in terms of Specific Absorption Rate (SAR) was the peculiar improvement of workers protection philosophy adopted by the EC at that time. The basic restriction for SAR was defined at the value of 0.4 W/kg averaged over any period of six minutes and over the whole body mass and at the value of 4 W/kg for one gram of tissue. However, no agreement was reached on the derived limits expressed in terms of electric and magnetic field intensity or power density.

In conclusion, ten years' discussions have not been sufficient to reach any practical conclusion. This situation remained unchanged until 1989. The approval of the Council Directive 89/391, whose Article 16 decreed the introduction of measures to encourage improvements in both health and safety of workers in the workplace, hastened the technical and legal action of the Commission on protection from health hazards arising from exposure to electromagnetic fields and other physical agents. Two other political decisions were relevant to the Commission's decision to propose new protective provisions: the approval in 1989 of the Community Charter of Fundamental Social Rights of Workers; and

*Radiofrequency Standards*, Edited by B.J.
Klauenberg *et al.*, Plenum Press, New York, 1994

the Resolution of European Parliament inviting the Commission to draw up a proposal of Directive against risks due to exposure to physical agents at work, in June 1992.

The preparation of the proposal necessitated a critical analysis of existing knowledge on the health effects of the physical agents, with the exception of those covered by the European Atomic Energy Commission Treaty (ionizing radiation), which include noise; mechanical vibrations; static and low-frequency electric and magnetic fields, radiofrequencies and microwaves; and optical radiation from coherent and incoherent sources. The Commission invited experts from three European institutions who are involved in their countries in protecting against non-ionizing radiation, namely, the Federal Office of Radiological Protection (Neuherberg, Germany), the National Institute of Health (Rome, Italy) and the National Radiological Protection Board (Chilton, UK), to evaluate the technical aspects of protection and formulate a proposal. This Working Group, financially supported by the EC Commission, prepared a document that defines the risk, the threshold values for effects considered adverse to human health, and a set of basic restrictions, intervention levels, and minimum health and safety requirements to protect workers in the workplace. The final version of the Working Group technical report was published in June 1991.[1] The Working Group's proposal is in good agreement with the recommendations issued by the International Non-Ionizing Radiation Committee of the International Radiation Protection Association (IRPA-INIRC),[2] that represents the opinions of the scientific community at large and ensures liaisons with other relevant international organizations, such as the International Labour Office (ILO) and the World Health Organization (WHO). Table 1 shows, for example, the basic restrictions for workers in the frequency range 100 kHz - 300 GHz proposed by the Working Group.

**Table 1.** Basic restrictions for workers in the frequency range 100 kHz-300 GHz.[1]

| Frequency | Induced current density | Contact current (mA) | Whole body average SAR | Local peak SAR (in the limbs) | Local peak SAR (head and trunk) |
|---|---|---|---|---|---|
| 100 kHz-10 MHz[a] | $f/100$ A/m$^2$ | 50 | 0.4 W/kg | 2W/(0.1 kg) | 1 W/(0.1 kg) |
| 10 MHz-300 GHz[b] | | 50[c] | 0.4 W/kg | 2W/(0.1 kg) | 1 W/(0.1 kg) |

f = frequency in kHz
a) In this frequency range both the induced current density and SAR limits should not be exceeded.
b) In order to avoid auditory effects associated with short (<30 μs) pulses at frequencies above 300 MHz the specific absorption should not exceed 10 mJ/kg.
c) This value applies only up to 100 MHz.

During 1991-1992 several technical meetings of experts of all Member States were held at Luxembourg to discuss and amend the Commission's draft. Finally, a proposal for a Council Directive on the minimum health and safety requirements regarding the exposure of workers to the risk arising from physical agents was officially issued on December 1992.[3]

## AIM, SCOPE AND STRUCTURE OF THE COMMISSION'S PROPOSAL FOR A COUNCIL DIRECTIVE

The main aim of this proposal for a Council Directive on the minimum health and safety requirements for workers' exposure to risks arising from noise, mechanical vibrations, optical radiation, and static and time-varying electric and magnetic fields up to 300 GHz, is both to improve gradually the level of protection ensured to workers and to harmonize the minimum safety and health requirements with respect to their social and economic impact at Community level. Given the considerable differences existing between national provisions within the Community (Table 2), that in turn results in considerably different levels of protection afforded by workers and distortion of competition, the harmonizing process represents a substantial progress and in itself amply justifies the present legislative action of EC authorities. All physical agents that are considered in the proposal for a Council Directive are largely present in many workplaces, but the exposure can induce adverse health effects only when some values are exceeded.

**Table 2**. Comparative table of existing national regulations on non-ionizing radiation: microwaves, radiofrequency, ELF in EC Member States.[3]

| | Specific legislation | Ref. to standards in general legislation | Recommendations Guidelines | Recommended exposure limits |
|---|---|---|---|---|
| Belgium | | x | Medical exams | |
| Denmark | | x | (x) | ANSI, IRPA |
| France | | | x | ANSI, IRPA |
| Germany | x | | x | Spec.applic. |
| Greece | | x | | ACGIH |
| Ireland | | | | |
| Italy | | | | |
| Luxembourg | | | | |
| Netherlands | (x) | x | | ACGIH |
| Portugal | | x | | |
| Spain | | x | | ANSI, ACGIH, IRPA |
| United Kingdom | | x | x | NRPB (IRPA) |

The Commission has proposed a single document to control hazards due to exposure of workers to all the above-mentioned physical agents rather than four different directives, each one covering the risk due to a single agent.

According to the Commission's views, this structure is largely justified by the observation that a common strategy can be adopted to cope with hazards involved, despite the fact that effects on human health from one agent to the other can differ. In short, the main body of the proposed Directive consists of 19 articles that define the general provisions that apply to all agents, and four distinct Annexes, which contain specific guidance for noise, mechanical vibrations, optical radiation, and static and time-varying electric and magnetic fields up to 300 GHz. For each agent the corresponding Annex defines, in qualitative and quantitative terms, the specific hazard and provisions respectively.

## GENERAL PROVISIONS

To achieve its targets in terms of the level of protection, the Commission has proposed a legal text flexible and clear enough to have the same meaning in all Member States, so that provisions can be managed properly into practical instructions to achieve protection targets and easily updated to technical progress.

The main body of the Directive contains definitions, terms, minimum requirements, and provisions aimed to protect workers from risk arising or likely to arise from exposure to the mentioned physical agents. The most relevant aspects covered by the common part of the Commission's proposal are: a) the assessment and measurement of each physical agent to control the exposure and to reduce the risk; b) worker's information training and consultation; c) worker's health surveillance; d) personal protective equipment, working equipment and working methods to prevent a risk of overexposure; e) particularly dangerous activities, extension of exposure, interference and indirect effects. Levels of exposure are established on the ground of an overall strategy aimed to reduce the hazards to the lowest reasonable level by means of a balanced process that necessarily must consider health effects, technical and economical aspects, and conflicting requirements as well. Three reference levels of exposure are defined, which in turn correspond to three different levels of risk. Ceiling level defines the exposure value that, for an unprotected person, give rise to hazards considered unacceptable by the Commission. This level of exposure is the maximum permissible value and cannot be exceeded. All the provisions of the Directive must be implemented to exclude any possibility of such a risk occurring. Threshold level expresses in quantitative terms an exposure value below that, according to current scientific knowledge, a health risk is no longer present. This value is the goal towards which the employer shall address an implementation program of Directive provisions.

No specific actions to protect workers are imposed on industrial activity when exposure is below the threshold level. The action level is defined as a value, placed between the threshold and the ceiling levels, corresponding to a level of risk above which the exposure may become a cause of concern. In this specific situation an action of protection must be undertaken, the exposure should be carefully evaluated, and provisions must be implemented to control the hazard.

The Directive states the protection targets in terms of practical results to be achieved, provides a response to practical needs, and makes use of technical standards to comply with its provisions.

However, the Commission is convinced that its option to propose a hard and fast set of rules to be followed by all the Member States in defining equivalent practical results is not sufficiently detailed to respond to new future situations. Therefore, to give a practical response to these needs, the Commission is aware that additional documents, within the framework of the Directive, should be proposed and adopted to better explain certain concepts, basic principles, methodologies, management of technical progress, assessment of the risk, measurements, special risk groups, and adaptation of protection to technical progress.


## THE PROTECTION OF WORKERS
## FROM EXPOSURE TO FIELDS AND WAVES

Annex 4 of the proposed EC Directive lays down a set of definitions, provisions and obligations specifically addressed to protect workers from the risk arising or likely to arise from exposure to static and time-varying electric and magnetic fields with a frequency up to 300 GHz. Three relevant biological effects are taken into consideration to establish the risk to health and safety of exposed workers: 1) heat due to energy absorption that results in an additional thermal load with respect to heat production of body metabolism; 2) the circulation of induced currents whose biological effect arises not only from tissue heating but also from stimulation of excitable tissues, such as nerve and muscles; or 3) effects on vision and shock and burns due to transient discharges or contact currents. This last effect depends strongly on the voltage of the charged objects, on contact area, on the sensitivity of the involved part of the body, and on impedance of the current path. The quantities used as predictors of the hazard and to establish basic restrictions are the induced current density; the contact current; the time rate of electromagnetic energy absorption by unit of mass in the biological system, defined as Specific Absorption Rate (SAR); and the energy absorption expressed in terms of Specific Absorption (SA).

### Ceiling levels

Ceiling levels to protect workers in the workplace established in the Commission's proposal for a EC Directive are reported in Table 3.

The levels are expressed in terms of basic restrictions and reflect the INIRC-IRPA recommendations, with the exception of the 4 Hz - 1 kHz frequency range where the Commission considered the prevailing need to limit the constraints imposed on industrial activity with respect to possible unpleasant field perception. However, a specific article of the Annex imposes an obligation to inform workers exposed to an electric field above 5 kVm$^{-1}$ that apparently harmless perception effects on the surface of the body may occur.

**Table 3.** Ceiling levels.[3]

| Frequency | Induced current density in head and trunk | Contact current | SAR | | Local peak in the head and trunk |
| | | | Whole body average | Local peak in the limbs | |
| | [A m$^{-2}$] | [mA] | [W kg$^{-1}$] | [W(0.1 kg)$^{-1}$] | [W(0.1kg)$^{-1}$] |
|---|---|---|---|---|---|
| 0 - 1 Hz | 0.04 | 1.5 | * | * | * |
| 1 - 4 Hz | 4 x 10$^{-5}$/f | 1.5 | * | * | * |
| 4 Hz-1 kHz | 0.010 | 1.5 | * | * | * |
| 1 - 3 kHz | f/100 | 1.5 | * | * | * |
| 3 - 100 kHz | f/100 | f/2 | * | * | * |
| 100 kHz-10 MHz | f/100 | 50 | 0.4 | 2 | 1 |
| 10 - 100 MHz | * | 50 | 0.4 | 2 | 1 |
| 100 MHz-300 GHz | * | * | 0.4 | 2 | 1 |

* not relevant at these frequencies; f in kHz

For static, extremely slow time-varying electric and magnetic fields, and time-varying electric and magnetic fields at frequencies up to 100 kHz, the ceiling levels are established and expressed in terms of: a) induced current density A/m$^2$ in the head and in the trunk due to direct coupling between electric and magnetic fields with the human body, and b) contact currents passing through the human body in contact with conducting objects coupled to electric and magnetic fields. The ceiling level of 40 mA/m$^2$ at 1 Hz relates to a flux density of 200 mT to induce this current density in a standard trunk of radius 30 cm. The ceiling level for the contact current defines its average value over a period of one second. Within this time interval, the peak value cannot exceed ten times the average value. For frequencies higher than 10 MHz, the ceiling level is defined by the maximum permissible value of the SAR, the most appropriate dosimetric quantity. The value of SAR should not exceed 0.4 W/kg when averaged over any 6-minute period and over the whole body mass. However, when this condition is satisfied, the local peak SAR averaged over any 6 minutes and 0.1 kg mass in the extremities and in the remaining parts of the body can rise up to 2 W/kg and 1 W/kg respectively. For exposure to pulsed microwaves of less than 30 µs duration at frequencies above 300 MHz, the ceiling level to protect workers from auditory effects is defined by a value limit (10 mJ/kg) on the energy absorption (Specific Absorption).

**Threshold levels**

Threshold levels for occupational exposure are also defined in terms of basic limits and expressed by means of the same quantities used to set ceiling levels of Table 2. They are established quantitatively at 1/5 of the ceiling levels. By way of example, the threshold level for whole-body-averaged SAR in the frequency range 10 MHz-300 GHz is 0.08 W/kg when averaged over any 6-minute period. The Commission, therefore, is proposing a threshold value that corresponds to that recommended for protection of general population in the INIRC-IRPA guidelines. In the proposed Directive, the supplementary safety factor introduced in threshold levels should be considered in the light of worker's protection for long term exposure. However, this approach seems quite interesting with respect to future activities of EC, specifically addressed to protection of general population.

**Action levels**

Action levels for the whole frequency range 0-300 GHz are reported in Table 4.

**Table 4.** Action levels.[3]

| Frequency | H $[A \cdot m^{-1}]^{\Delta \circ}$ | B $[\mu T]^{\Delta}$ | E $[V \cdot m^{-1}]^{\Delta}$ | P $[W \cdot m^{-2}]^{\Delta}$ |
|---|---|---|---|---|
| < 1 Hz | $1.63 \times 10^5$ | $2 \times 10^5$ | $6.14 \times 10^4$ | * |
| 1 Hz - 10 Hz | $0.163/f^2$ | $0.2/f^2$ | $6.14 \times 10^4$ | * |
| 10 Hz - 1 kHz | $16.3/f$ | $20/f$ | $614/f$ | * |
| 1 kHz - 300 kHz | $16.3$ | $20$ | $614$ | * |
| 300 kHz - 1 MHz | $4.9 \times 10^3/f$ | $6 \times 10^3/f$ | $614$ | * |
| 1 MHz - 10 MHz | $4.9 \times 10^3/f$ | $6 \times 10^3/f$ | $6.14 \times 10^5$ | * |
| 10 MHz - 30 MHz | $4.9 \times 10^3/f$ | $6 \times 10^3/f$ | $61.4$ | 10 |
| 30 MHz - 400 MHz | $0.163$ | $0.2$ | $61.4$ | 10 |
| 400 MHz - 2 GHz | $2.58 \times 10^{-4} f^{0.5}$ | $3.16 \times 10^{-4} f^{0.5}$ | $9.7 \times 10^{-2} f^{0.5}$ | $2.5 \times 10^{-5} f$ |
| 2 GHz - 150 GHz | $0.364$ | $0.45$ | $137$ | $50$ |
| 150 GHz - 300 GHz | $2.96 \times 10^{-4} f^{0.5}$ | $3.7 \times 10^{-5} f^{0.5}$ | $1.12 \times 10^{-2} f^{0.5}$ | $3.33 \times 10^{-7} f$ |

* not relevant at these frequencies
$^{\Delta}$ f in kHz
$^{\circ}$At frequencies of 10 MHz or greater, the value of H may be increased to that calculated from the formula:
$$5/6 \, (Em^2 / 377) + 1/6 \, (377 \, H^2) \leq P$$
where Em is the measured electric field strength $[V \cdot m^{-1}]$; H and P are the values in the table at the frequency considered.

They are expressed by using the quantity's magnetic field strength H (ampere/m), magnetic flux density B (tesla), electric field strength E (volt/m), and power density in free space and far-field conditions (watt/m²), which can be obtained directly by means of a measuring instrument. Action levels are derived from basic threshold levels reported on Table 3 but do not represent limit values to be respected in any case because these levels fully comply with ceiling levels. Action levels provide a response to practical needs to get the results prescribed by the Directive, however their values should be understood in terms of an action to be undertaken when workers are likely to be exposed to these levels. This action, mentioned in the main body of the proposed Directive and in the Annex on fields and waves, provides: worker's information concerning health and safety relating to exposure and its potential risk; training covering, in particular, the implementation of measures pursuant the Directive; the obligation to comply with protective and preventive measures in accordance with national legislation; wearing personal protective equipment and the information on fields and waves produced by work equipment likely to result in exposure to such values.

An action that includes a program of technical and/or work organization, measures aimed to reduce exposure, delimiting of areas and the restriction of access and training of operator is triggered by values of H, B, or E 1.6 times those reported in Table 4.

# CONCLUSIONS

The proposed Directive on worker's protection from physical agents will now be submitted to legal procedure that leads to its formal approval by the Council of Minister. How long this process will last and possible substantial modifications of the Directive's content cannot be predicted at the moment. The Commission proposal, however, represents an appreciable effort to put into a logical framework a set of clear terms, concepts, and provisions that should be understood by legislators of all Member States and easily updated respect to future needs. Its possible approval will change deeply the level of protection afforded to workers exposed to physical agents in many Member States, in particular, those (Table 2) where there is no national legislation in this field. As far as protection from electric and magnetic fields and waves is concerned, the Directive's rationale is based on well-established biological effects and does not consider subtle and still unclear effects mainly related to long-term low-level exposures. Figures 1 and 2 are the graphic representations of the action values of the electric and magnetic field intensities and power density. They visually outline the philosophy followed by the Commission. In the frequency range encompassing radiofrequencies and microwaves that include the subresonance range, the resonance range, and the hot spot range, the significant dosimetric quantity, which is kept constant and represents the basic limit, is SAR. In practice, to comply with this limit it is necessary to define derived limits dependent on frequency according to both energy absorption and its distribution within the exposed body which also varies with frequency. With regard to mobile radio equipment, the Directive states that: the particularities of conditions of use and the rapidity of technological changes require that the ceiling levels be determined following further consideration. The management of Directive's provisions, will require additional documents, particularly on how measurement should be carried out. This result could be achieved through cooperation between EC and the European Committee for Electrotechnical Standardization (CENELEC), which includes 18 EC and European Free Trade Association countries. CENELEC is now preparing a draft for a standard on human exposure to electromagnetic energy in the frequency range 10 kHz - 300 GHz.

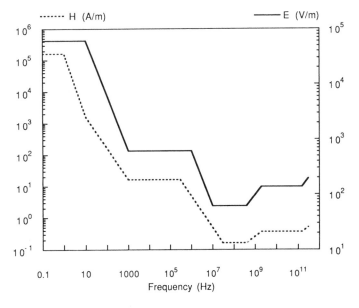

**Figure 1.** Graphic representation of the action values of the electric and magnetic fields proposed by EC Commission for workers' protection in the frequency range 0-300 GHz.

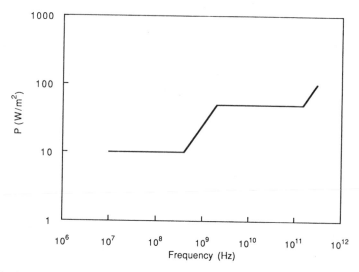

**Figure 2.** Graphic representation of the action values of the electromagnetic power density proposed by EC Commission for workers' protection in the frequency range 0-300 GHz.

## ACKNOWLEDGEMENTS

The author wishes to thank Ms Paola Di Ciaccio for her technical support.

## REFERENCES

1. S.G. Allen, J.H. Bernhardt, C.M.H. Driscoll, M. Grandolfo, G.F. Mariutti, R. Matthes, A.F. McKinlay, M. Steinmetz, P. Vecchia, M. Whillock. "Proposals for basic restrictions for protection against occupational exposure to electromagnetic non-ionizing radiations. Recommendations of an international working group set up under the auspices of the Commission of the European Communities", *Physica Medica*, vol.VII (2), (April-June 1991).

2. INIRC-IRPA, Guidelines on limits of exposure to radiofrequency electromagnetic fields in the frequency range from 100 kHz to 300 GHz, *Health Physics*, 54 (1): 115-123 (1988).

3. CEC Proposal for a Council Directive on the minimum health and safety requirements regarding the exposure of workers to the risks arising from physical agents, COM (92) 560 final - SYN 449.

# ANSI/IEEE EXPOSURE STANDARDS FOR RADIOFREQUENCY FIELDS

James C. Lin

University of Illinois at Chicago
1030 SEO (M/C 154)
851 S. Morgan Street
Chicago, IL 60607-7053

## INTRODUCTION

The fact that exposure to continous, sinusoidal, time-varying, microwave and radiofrequency fields can produce untoward biological effects in human beings has been recognized for some time. Indeed, a sizable volume of studies has suggested that at both low- and high-power levels, exposure to microwave and radiofrequency fields could produce heating of tissues in the body. The thermalization may or may not be detectable using available temperature sensing devices as temperature elevations. Nevertheless, the absorbed energy initiates vibration and rotation of polar molecules in tissue water that would ultimately dissipate in heat.[1] This information was instrumental in the recommendation by the United States of American Standards Institute (USASI) to establish in 1966, a power density of 10 mW/cm$^2$ as the standard for human exposure to microwave and radiofrequency fields over any 6-min period.[2]

In October 1968, the U.S. Congress adopted the "Radiation Control for Health and Safety Act (PL 90-602)" to protect the public from unnecessary exposure to potentially harmful radiation, including microwaves emitted by electronic products. Together with the former Soviet Union's far more conservative standards for long-term exposure, this Act posed new questions on the adequacy of the existing knowledge of biological effects of electromagnetic fields and of the protection afforded the public from the harmful effects of these fields.

The resulting research has produced a large literature base, which was used by several national and international organizations to develop safety standards and exposure guidelines. However, this discussion shall cover only the standard recently published by the Institute of Electrical and Electronic Engineers (IEEE) and recognized by American National Standards Institute (ANSI) as an American national standard.[3] Until recently, the U.S. standard was under the auspices of ANSI. Since 1960, the development of a revision of the 1982 standard came under the sponsorship of the IEEE Standards Board.

*Radiofrequency Standards*, Edited by B.J.
Klauenberg *et al.*, Plenum Press, New York, 1994

## THE ANSI/IEEE STANDARD

The American National Standards Institute/Institute of Electrical and Electronics Engineers (ANSI/IEEE) Standard for safe human exposure to 3 kHz-300 GHz radiofrequency fields[3] includes extensive revisions of earlier versions of these standards. A capsule guide to the Standard is given in Figure 1. It presents recommendations for prevention of adverse or harmful effects in humans. However, it is a voluntary standard and thus provides only guidelines for maximum permissible exposure (MPE) limits on electric and magnetic fields or current and power densities under normal environmental conditions.

## NEW REVISIONS

At the time the standard was promulgated, the subcommittee responsible for the recommendation consisted of 122 members. In addition, the balloting committee (IEEE Standard Coordinating Committee 28) that approved the guidelines for submission to the IEEE Standards Board had 46 members. The entire effort spanned nearly 10 years. The Standard relied on the finest interpretation of a list of 321 papers selected from the peer-reviewed literature and later was supplemented by an additional 60 bibliographical entries that were deemed to have biological, engineering, and statistical validity.

Among the various new provisions are:

(1) The expansion of frequency range from 3 kHz to 300 GHz to include quasi-static and quasi-optic fields. It is essential to incorporate parameters other than the average specific absorption rate (SAR) to meaningfully delineate the influence of different regions of the spectrum. Above 6.0 MHz, the pertinent metric becomes power density. Below 0.1 MHz, stimulation of excitable biological tissues plays a dominant role and the primary metric is induced current density.

(2) Contact and induced current limits to deter shock and burn hazard. Current limits (rms values averaged over any 1 second) are favored over electric field limits because induced currents are functions of the exposed subject and the exposure environment. They do not provide protection against transient spark discharges when making or breaking contact with charged objects in radiofrequency (RF) fields.

(3) Two-tiered MPE to distinguish between occupational vs. general public exposures. An extra safety factor in adopted for the uncontrolled environments that include environments where individuals are not specifically involved in the operation or use of equipment that radiates significant electromagnetic energy.

(4) Peak power limits to prevent unintentional high power exposure and to preclude high specific absorption for decreasingly short pulses. For pulses shorter than 100 ms, the permissible limit is 100 kV/m.

(5) Relaxation of MPE limits for partial body exposures for all parts of the body except the eyes and the testes. For example, in narrow cross sections as the wrists and ankles, higher absorption rates (up to 20 W/kg) are allowed.

(6) Relaxation of MPE limits for low frequency magnetic fields to better correspond to whole body SAR limits at frequencies below 3 MHz. The new limits will result in an SAR of 0.08 W/kg or less.

(7) Frequency-dependent averaging times, from 30 min down to 10 sec, these limits will mitigate against shock or burn upon grasping contact with objects in an RF environment. At higher frequencies, these limits agree with averaging times derived from infrared considerations.

## MPE OBJECTIVES

As noted previously, the metric used for each frequency region covered by the Standard differs. At low frequencies, the MPEs, given in field parameters, are motivated by limiting contact and induced currents to guard against RF shocks and burns. At high frequencies, the MPEs, specified in power densities, are primarily governed by the desire to protect against skin burns. The inversely, frequency dependent averaging time prevents high SARs that otherwise could result in skin burns. In the resonance region, the concept of SAR and its frequency-dependent connection to fields and power density formed the basis for the MPEs. An SAR of 4 W/kg, temporally and spatially averaged over the whole body mass was adopted as the working threshold for adverse biological effects in humans. Above this threshold, disruption of work schedules in trained rodents and primates has been demonstrated. It had been noted that a metabolic heat production rate of 4 W/kg falls well within the normal range of human thermoregulatory capacity. Clearly, the ANSI/IEEE C95.1-1992 Standard provides recommendations to prevent adverse thermal effects on the functioning of human body, although the assessment criteria for reports of biological effects were without regard to mechanisms of interaction.

## ACKNOWLEDGMENT

This work was supported in part by the National Science Foundation under Grant No. BCS-9110184.

## REFERENCES

1.  S.M. Michaelson and J.C. Lin. "Biological Effects and Health Implications of Radiofrequency Radiation," Plenum Press., New York (1987).
2.  USASI, Safety Levels of Electromagnetic Radation with Respect to Personnel, C95.1., USA Standards Institute, New York (1966).
3.  ANSI/IEEE C95.1 IEEE Standard for Safety Levels with Respect to Human Exposure to Radiofrequency Electromagnetic Fields, 3 kHz to 300 GHz, *IEEE, New York* (1992).

# RADIOFREQUENCY RADIATION SAFETY GUIDELINES
# IN THE FEDERAL REPUBLIC OF GERMANY

Klaus W. Hofmann

Forschungsinstitut für Hochfrequenzphysik (FHP)
Forschungsgesellschaft für Angewandte Naturwissenschaften e.V. (FGAN)
Bad Neuenahrer Str. 20
D 53343 Wachtberg-Werthhoven, Germany

## PRELIMINARY REMARK

In the Federal Republic of Germany, there are no regulations in the field of Radiofrequency Radiation Safety set by law. The guidelines have been worked out under the responsibility of Verband Deutscher Elektrotechniker/Association of German Electrical Engineers (VDE). Their use is wholly voluntary. But as a rule, they serve as "state of the art" and are cited in court as "anticipated expertises," i.e., already existing ones.

## THE STANDARD SETTING ORGANIZATION

The VDE and Deutsches Institut fuer Normung/German Institute for Standardization (DIN) constitute the Deutsche Elektrotechnische Kommission/German Electrotechnical Commission (DKE) as a joint organization, the juridical responsibility for running the DKE lying in the hands of the VDE.

The DKE is the national organization responsible for the elaboration of standards and safety specifications covering the whole area of electrical engineering in Germany.

The DKE, as a member of international electrotechnical standardization organizations, represents and safeguards German interests within such bodies as:
- International Electrotechnical Commission (IEC) and its associated bodies
- European Committee for Electrotechnical Standardization (CENELEC)
- and others.

The results of DKE-work are issued in the form of, for example:
- DIN/VDE - Standards
- DIN/EN - Standards (European Standard)
- DIN/IEC - Standards (worldwide Standards).

DKE work is done by different committees, subcommittees and working groups consisting of delegates from associations, institutions or firms that are interested in electrotechnical standardization. Members of committees and subgroups serve voluntarily and without compensation.

## THE GERMAN STANDARD DIN/VDE 0848

One of the most important requirements in setting standards by DKE is their general acceptance by the public. Therefore, after elaboration within the DKE, the standard is published as a draft on yellow paper (the so-called "Yellow-Print"). Comments and proposals for significant changes are then expected from any interested party within a certain period set for filing an objection. These objections are discussed with the delegates of the objecting party or with individual objectors. If the changes derived from these discussions are very serious, a new Yellow- Print will be published. Objections are possible again and discussion starts anew. (In the case of European standards/international standards, the draft is published on pink paper, the "Pink-Print". This differentiation is very useful, because in these days of harmonization within the European Community it is then very easy to identify the existence of an international activity on standardization in this field). After this procedure, the draft standard may become valid, documented by the color of the paper on which it is to be printed: it will become a "White-Print" supplied with the date of validity. Each DIN/VDE - standard has to undergo the procedure described. If it consists of several parts (as in the case of the standard to be presented here) each part is treated separately.

The actual German Standard DIN/VDE 0848 dealing with "Safety in Electromagnetic Fields" consists of five parts that have presently reached different stages, or colorprints.

The five parts of the standard are listed as:

**Part 1:** "Measuring and calculation procedures in the frequency range from 0 Hz to 300GHz."

Date of the standard before revision was February 1982. The draft of June 1992 has been approved and is being published and issued as a "White-Print" with date of validity February 1993. There is also a "Pink-Print," because part 1 of DIN/VDE 0848 is submitted to CENELEC Technical Committee (TC 111) for discussion.

The purpose of this standard is to compare measured or calculated physical quantities with limiting values. Its importance to DIN/VDE 0848 as a whole lies in paragraph 2 "Definitions," especially, where the relevant terms are listed.

**Part 2:** "Protection of persons in the frequency range from 30 kHz to 300 GHz."

(In comparison to part 1, this part covers another frequency range due to the fact that different basic effects result in splitting the frequency range. The range from 0 Hz to 30 kHz is covered by Part 4).

Part 2 partly replaces the 1984 standard and was published in October 1991 as a "Yellow-Print." It has been submitted to CENELEC TC 111 for discussion. In accordance with the CENELEC rules for European standardization, this draft could not have been implemented to give it the status of a National Standard due to the fact that European standardization concerning the safety of electromagnetic fields had already been initiated (stand-still agreement). But as progress is very slow in European Standardization, an application for exception has been made and positively decided by CENELEC. Thus, this draft ("Yellow-Print") of October 1991 can become a valid national standard, in principle.

Part 2 of DIN/VDE 0848 will be discussed in more detail in the following chapter.

**Part 3:** "Protection against explosives in the frequency range from 10 Hz to 30 GHz"

This standard provides guidance to the prevention of inadvertent ignition of flammable atmospheres in the presence of electromagnetic fields, as emitted by radiating elements such as antennas.

There had been an edition of March 1985; a revision has become necessary. The draft of February 1993 is in preparation to be published as "Yellow-Print." This can be done because there are no international activities on standardization at the moment. Nevertheless, this part of DIN/VDE 0848 has widely been matched with the contents of future BSI Standards (BSI = British Standards Institution) to get a well-based input for a future European Standard.

**Part 4:** "Limits of field strengths for the protection of persons in the frequency range from 0 to 30 kHz."

This part of DIN/VDE 0848 is valid since October 1989 ("White-Print") and covers the range from static to low-frequency fields. Formerly, Part 2 of DIN/VDE 0848 covered the whole frequency range from 0 Hz up to 3000 GHz (at that time), but different basic concepts for the limiting values of incident electromagnetic energy led to splitting the frequency range and the working group, too. Thus, a separate working group was founded for the lower frequency range. This group decided not to follow the proposals made by the group responsible for the levels of higher frequencies. Therefore, a significant difference can be recognized in the philosophy of setting standards. For example, the low frequency standard has not adopted the splitting of the limiting values of exposure in view of "controlled/uncontrolled areas," otherwise "workers/population."

**Part 5:** "Protection against inadvertent initiation of electroexplosive devices (bridge-wire) in the frequency range from 10 kHz to 1 GHz (frequency range > 1 GHz in preparation)."

This part of DIN/VDE 0848, in a certain manner, is comparable to Part 3 (flammable atmospheres). Subject here is the prevention from inadvertent initiation of electroexplosive devices by electromagnetic fields as generated by technical installations. This standard has not yet left the stage of a first draft. It is still under consideration, although already matched with proposals from BSI Panel EEL/23/-/1 in a first step.

For the discussion of human exposure to electromagnetic fields in the upper frequency range, only Parts 1 and 2 of the 5 parts of the DIN/VDE 0848 are of importance.


## DIN/VDE 0848/ PART 2

The actual Standard DIN/VDE 0848 /Part 2 had its predecessor in the edition of July 1984, from which it differs significantly. The first draft of this edition was published in 1978 and was the result of the work of a working group installed by Committee 764 of the German Electrotechnical Commission in 1975. Before this period there had only been a guideline edited by the German Association for Radar and Navigation/Deutsche Gesellschaft für Ortung und Navigation (DGON), which limited exposure to a frequency-independent level of 10 mW/cm$^2$, as postulated by Hermann Schwan in the early fifties. It should be mentioned that the work of Research Study Group 2 of NATO Panel VIII on Standardization Agreement (STANAG 2345) was very helpful for the work in the (former) Federal Republic of Germany to achieve a considerable step forward at that time.

Part 2 of DIN/VDE 0848 is not as long as the NRPB draft of June 1992 or IEEE C.95.1-1991. Only the actual STANAG 2345 may yet be less voluminous. This lies in the philosophy of DKE to reduce text to a minimum and to direct attention to relevant contents. It may be that explanation and comment are not as detailed as sometimes desired. The standard, following this philosophy, is therefore divided into two parts - a normative one and an informative one. The normative part contains definitions, basic requirements, tables and figures, whereas the informative part gives some explanations necessary for understanding the normative part.

An essential part of any regulation is usually the scope, which is cited here for the German standard:

- "This standard applies to the protection of persons from direct and indirect effects of electric, magnetic and electromagnetic fields in the frequency range from 30 kHz to 300GHz.

It is not applicable to patients in the case of medical applications of said fields."

In this scope a difference is made between direct and indirect effects. The greater part of this standard concerns the direct effects of exposure to electric, magnetic and/or electromagnetic fields. The smaller part deals with indirect effects, as occur from touch voltages (which is mainly laid down in another German standard DIN/VDE 0800) or those that may affect pacemakers. Reference is made to a European Draft Standard EN 50061:1988/A1:1991 or DIN/VDE 0750/Part 9A1. Indirect effects, caused by currents in the body, are included in the limiting values of this standard, of course.

Basic considerations led to a splitting of the limiting values over the whole frequency range (see Figure 1). Exposure area 1 applies to controlled areas where general access is possible, if it is ensured that exposure is only of short duration. This exposure area is derived from the so-called safety concept, which ensures avoidance of hazards provided an equivalent safety factor is adhered to. This concept includes effects such as unnatural stimulation of sensory receptors, nerve and muscle cells, disturbance of heart action, and hazards due to the occurrence of temperature rise.

Exposure area 2 remains for all others. The limiting values here follow the precautionary concept and are given by an additional safety factor. (In the case of power flux densities, this factor is 5). This second limiting value shall consider the special need for protection of groups of sensitive persons, the possibility of permanent field effects as well as the involuntary or unintentional exposure of persons, and shall avoid considerable annoyances due to field effects.

The limiting values themselves result from considerations of two basic concepts: current densities in the human body and the specific absorption rate (SAR). The first concept applies predominantly to the lower frequency range, whereas the latter refers mainly to the higher frequency range. These well-established fundamental concepts lead to base limit values as listed in paragraph 4.1.1. of the German standard (Figure 2).

What is called base limit values in the German standard corresponds to the definition of "basic restrictions" in the NRPB draft, for instance. These are: body current, current density, specific absorption rate (SAR), and specific absorption (SA). But in practical work, these dosimetric quantities cannot be obtained by means of measurements. Therefore, derived limit values (or reference levels) are given to make the standard practicable. These quantities are electric and magnetic field strength and power flux density, which are then specified in such a manner that the fields of basic restrictions are not exceeded even if the most unfavorable conditions of effects are taken as a basis.

Important other differentiations are made in the standard concerning the exposure time. The basic quantity here is the endogenous thermoregulation with an assumed time constant of 6 minutes. This leads to different tables of the standard, for which the reference levels are given in r.m.s. values or peak limit values, respectively.

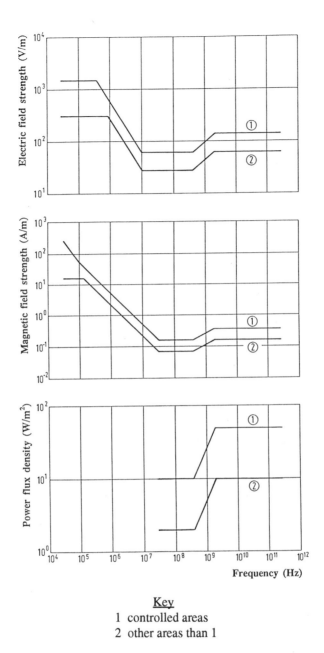

**Figure 1.** Maximum permissible exposure levels, exposure duration ≥ 6 min.

In concluding this chapter, it should be mentioned that the German Standard DIN/VDE 0848/Part 2 has been adopted by the German Armed Forces, with some special additions for the practice in military installations.

## FINAL REMARK

A standard, regulation or guideline does not attain its validity only from its factual contents. A very important aspect in the assessment of its effectiveness is its acceptance by the affected public. Of course, it is very easy to set limiting values low enough to be

| Frequency Range | Effect | Exposure Area 1 | Exposure Area 2 |
|---|---|---|---|
| 30 kHz to 100 kHz | locally limited areas | $\dfrac{1}{12}\,f\,\dfrac{mA}{m^2 Hz}$ *) | $\dfrac{1}{30}\,f\,\dfrac{mA}{m^2 Hz}$ *) |
| > 100kHz to 1MHz | whole body | 0, 4 W/kg **) | 0, 08 W/kg **) |
| | locally limited areas | $\dfrac{1}{12}\,f\,\dfrac{mA}{m^2 Hz}$ *) | $\dfrac{1}{30}\,f\,\dfrac{mA}{m^2 Hz}$ *) |
| > 1 MHz to 300 GHz | whole body | 0, 4 W/kg **) | 0, 08 W/kg **) |
| | locally limited areas | 10 W/kg ***)<br>10 mJ/kg ****) | 2 W/kg ***)<br>2 mJ/kg ****) |
| | hand, wrist, ankle | 20 W//kg ***) | 4 W/kg***) |

| | |
|---|---|
| *) | The current density values shall be averaged over surface elements of 1 cm$^2$ at right angles to the current direction as well as over time intervals of 1 s. |
| **) | Averaged over the complete body and over each 6 min interval. |
| ***) | Averaged over each 100g of tissue mass and over each 6 min interval. |
| ****) | Peak value of the specific absorption averaged over each 10 g tissue mass. |

**Figure 2.** Base limits for electric current density, the specific absorption and specific absorption rate (excerpt from DIN/VDE 0848 Part 2).

favored by broadest parts of the population. But this will, at once, provoke the disapproval of producers and users of equipment to be controlled by the standard. These parties would prefer less stringent limiting values, because development and costs in production and operation would be much more modest. Somewhere between these limits a compromise has to be worked out. And a compromise - however well-balanced and proven it may be - will always bear some quarrel in itself.

This applies to the German Standard DIN/VDE 0848, too. Again and again action groups in the public claim that levels would be too high and athermic effects would not have been taken into account. Eastern standards are cited here. Also, the structure of working groups and committees is criticized. They would consist only of lobbyists of producers and users.

From such statements it can easily be derived that acceptance of standards, elaborated by such groups, will very often be poor. But the question is always from where the experts should come, the specialists who know matter and facts, if not from the party of producers and users, from universities and proper authorities! In so far, especially Part 2 of the German Standard which should be presented here in an overview was elaborated by the members of the working group with a high degree of responsibility. It is wholly another question if more international cooperation can lead to more independence of working groups and, thus, can strengthen the trust of the general public in the results of such work for setting standards.

This should be one of the goals in the development of a new Standardization Agreement STANAG 2345.

**Session B: Considerations for NATO Standardization
Agreement**

**Chair: G. Mariutti**


New Technologies for Dosimetry: Slow Luminescence
*Johnathan L. Kiel, John G. Bruno, and William D. Hurt*


Biological Effects of High-Peak-Power Microwave Energy
*Edward C. Elson*


Some Recent Applications of FDTD for EM Dosimetry:
ELF to Microwave Frequencies
*Om P. Gandhi*


Microwave Exposure Limits for the Eye: Applying Infrared
Laser Threshold Data
*David H. Sliney and Bruce E. Stuck*


New IEEE Standards on Measurement of Potentially Hazardous
Radiofrequency/Microwave Electromagnetic Fields
*Ron C. Petersen*


Risk Assessment of Human Exposure to Low Frequency Fields
*Jürgen H. Bernhardt*

# NEW TECHNOLOGIES FOR DOSIMETRY:
# SLOW LUMINESCENCE

Johnathan L. Kiel, John G. Bruno, and William D. Hurt[*]

USAF Armstrong Laboratory
Brooks AFB, TX 78235-5102

## INTRODUCTION

Measurement of nonionizing electromagnetic radiation absorption (dosimetry) involves one of two basic approaches: invasive measurement with temperature sensing probes or noninvasive measurement of re-irradiated energy. The latter approach involves millimeter radiometry,[1,2] calorimetry,[2] or thermography.[2] Although the former methods have better point resolution, depth of measurement, and can be used in real time, the latter methods, with the exception of calorimetry, can give continuous spatial (surface) and temporal measurement.[1,2,3] The invasive thermal probes may miss local hot spots or disturb the measurement.

Modelling methods validated by experimental measurements are the least invasive and potentially have the best qualities of both actual measurement techniques.[2,4,5,6,7] However, they are limited in resolution and accuracy by the assumptions on which they are based.[8]

This paper begins with thermochemiluminescent microdosimetry (TCM),[9] which uses molecular temperature probes, and ends with generalization of the mechanism of the method to other potential dosimeters, biochemical and artificial. TCM can potentially provide continuous spatial and temporal resolution of temperature and specific energy absorption to the microscopic (cellular and subcellular) level.[9,10] Furthermore, it can theoretically measure interactions of objects with electromagnetic fields (EMF) with minimal perturbation of the incident fields.

All dosimetry techniques, to be useful in assessing safety of emitters or predicting the magnitude of interactions with the radiation, should measure properties of the radiation relevant to the biological effects. Temperature increases and electric and magnetic field interactions are the most likely generators of biological effects. Slow luminescence, the underlying mechanism of TCM, is capable of measuring all three types of interaction.

---

[*] These views and opinions are those of the author and do not necessarily state or reflect those of the U.S. Government.

*Radiofrequency Standards*, Edited by B.J.
Klauenberg *et al.*, Plenum Press, New York, 1994

## THEORY, MATERIALS, AND METHODS

Thermochemiluminescence was first associated with diazoluminomelanin (DALM), a biochemical that has been chemically and biologically synthesized.[11,12] Under peroxidizing conditions, DALM yields steady-state luminescence for long periods that assumes a level based upon the temperature of the solution in which it is dissolved.[9] The solutions can be very dilute (.05%),[11] providing for dilution in tissue-equivalent solvents (saline and glycerin solvents) without disturbing the tissue-equivalent dielectric properties. Although in high concentrations (5%), DALM solutions show high ionic conductivity at low frequencies,[9] Figure 1 shows that at low concentrations, biosynthesized DALM solution behaves dielectrically like water in which it is dissolved. The permittivity properties of the bacterially biosynthesized DALM were measured using an open-ended coaxial-line sensor and a computer-controlled automatic network analyzer.[13] The system was calibrated with the sensor open-circuited, and immersed in methanol and a saline solution to minimize the errors related to the system imperfections.[7] The sensor was then immersed into the material to be tested.

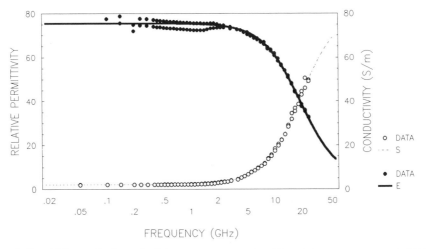

**Figure 1.** Permitivitty and ionic conductivity measurements made of bacterial suspension grown in 3AT medium.[10] See the Theory, Materials, and Methods for procedural details. The dielectric properties are what would be expected from water with the same ionic content, even though the bacteria in the suspension have synthesized diazoluminomelanin.

In simple theoretical terms, the slow luminescence of DALM can be explained based on slow fluorescence[14,15,16,17,18] and a pseudoequilibrium between activated states established during the steady-state condition. The equilibrium of activated states, singlet ($S_1$) and triplet ($T_1$), can be represented as follows:

$$S_1 \longleftrightarrow T_1(\pm)$$
$$T_1(\pm) \longleftrightarrow T_1 + kT$$

The pseudoequilibrium constant Q, or ratio of the concentrations of the activated states, also represents the steady-state chemiluminescence (or fluorescence):

$$Q = [S_1]/[T_1(\pm)]$$
$$S_1 \longleftrightarrow S_0 + Q$$

where: $S_0$ is the ground state. $T_1$ is a metastable state with a lifetime estimated to be in milliseconds to seconds and that decays to ground state by nonradiative transfer (breaking of bonds, forming free radicals), vibrational cascade (release of heat), or emission of red light. Since the intersystem crossing between the $S_1$ state and the charged triplet is "lossless" and the transition between the charged and uncharged triplet states is within kT, the luminescence correlates very well with absolute temperature and the specific absorption rate (SAR) of energy into the system. This relationship of temperature and energy in the steady-state system is best illustrated by the following combined chemical equation:

$$S_0 + Q <---> S_1 <---> T_1(\pm) <---> T_1 + kT$$

Therefore, this pseudoequilibrium can be disturbed to yield more or less heat and light, but the two are always correlated. The excitation for the ground state is provided endogenously by chemical bond breaking and making (chemiluminescence) or by photoexcitation (as in slow fluorescence). This pseudoequilibrium can be converted into an energy content for the system using the following equation:

$$\pm E = RTlnQ$$

where: E is the free energy entering or leaving the local environment, R is the molar gas constant, T is the absolute temperature, and lnQ is the natural log of the steady-state thermochemiluminescence. The intensity of the slow luminescence (I) is related to the energy gap ($E_n$) between the $S_1$ and $T_1$ states. This relationship can be expressed by the following simple exponential equation:

$$I = Ae^{-En/kT}$$

Where: A is the maximum slow luminescence (or fluorescence) intensity. To distinguish the contributions of kT and incident radiofrequency radiation (RFR), a baseline measurement of temperature and thermochemiluminescence is made ($T_1$ and $Q_1$) and then the steady-state thermochemiluminescence is measured during exposure to the radiation ($T_2$ and $Q_2$). The absolute temperature in both cases can be determined from calibration curves of temperature vs. lnQ, since they have a high correlation coefficient.[9] The SAR would then be given (normalized to moles of luminescent compound) by the following equation:

$$E_2 - E_1 = RT_2lnQ_2 - RT_1lnQ_1$$

Besides chemically synthesizing DALM and biosynthesizing it in bacteria, we have biosynthesized it in human leukemia cells (HL-60). HL-60 cells were grown to about $10^6$ cells per ml in RPMI-1640 medium containing 10% heat-inactivated fetal bovine serum and incubated for 2 days in the medium supplemented with 400 mg/ml of 3-amino-L-tyrosine hydrochloride (3AT). The pH of the medium was adjusted to 7.2 to 7.4 following addition of 3AT because it is an acid salt and the phenol red was deleted from the medium to prevent chemical reaction with oxidizing 3AT. After growth in the 3AT medium the cells were washed twice in pH 7.4 phosphate buffered saline (PBS) and resuspended at $10^6$/ml in 5 ml of 0.1 M $NaNO_2$ in PBS adjusted to pH 4.0 for diazotization of the aminomelanin formed in the leukemia cells. Subsequently, 50 ml of 10 g/ml luminol in DMSO were added and the cells were incubated in the dark at 37°C with 5% $CO_2$ for 24 h. Two ml of cell lysate and 40 ml of 3% hydrogen peroxide (0.06% final concentration) were added together to form the reaction mixture in a polystyrene cuvette. The cuvette was placed in the Quantitative Luminescence Imaging System (QLIS)[19] and irradiated with

2450 MHz radiation (continuous wave) from 0 to 20 W input power. Temperature was recorded simultaneously by means of a Thermal Control System[20] run by a separate IBM-compatible model 386 personal computer with custom designed software (Questech, Inc., San Antonio, Texas).

## RESULTS AND DISCUSSION

Thermochemiluminescent microdosimetry promises to address two major concerns of RFR investigators: temporal and spatial resolution of radiation absorption. The rapid response of the thermochemiluminescence prevents interference from thermal diffusion and the use of dilute solutions of a molecular probe prevents probe perturbation of the fields. However, there are interactions between electromagnetic fields and molecular probes that operate through the slow luminescence mechanism that must be considered. Figure 2 shows that adding material that enhances heating of the dosimetric solutions (such as magnetite, a conductive salt that also interacts with magnetic field components) does not necessarily result in enhanced chemiluminescent response. Also, not all luminescent materials (such as luminol in steady-state luminescence) are suitable as a dosimeter.[21] Figure 2 shows that luminol had a linear response to microwave energy (2450 MHz) input until it reached 30°C, then it increased in luminescent response with time, although the temperature remained constant (see Figure 3). In Figure 3, the magnetite particles coated with DALM showed enhanced thermalization as expected even though the thermochemiluminescence was decreased for each given temperature (Figure 2). Figure 4 shows that the DALM made in HL-60 leukemia cells behaved like chemically-synthesized DALM. The response to microwave energy input was nearly linear at lower energy because the output of activated states of DALM was low, but became nonlinear (exponential) at higher energies, when the concentration of activated states rapidly rose.

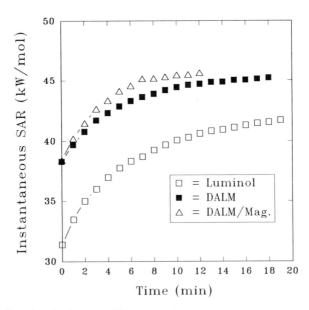

**Figure 2.** Comparison of the luminescence (Q in photons/sec) of peroxidizing luminol, diazoluminomelanin (DALM), and diazoluminomelanin-coated microscopic magnetite particles at different thermal energies (RT; where R = the molar gas constant and T = the absolute temperature in degree Kelvin) in a 2450 MHz, 25 W incident power microwave field. The details of the experiments are found in Reference 11.

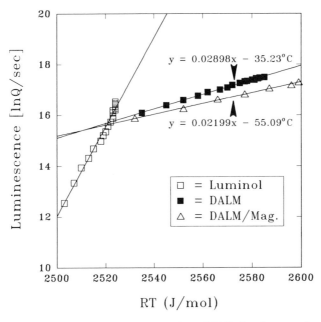

**Figure 3.** The same data as in Figure 2 were used to plot the total SAR of environmental energy (from microwave radiation and kT) over time for luminol, diazoluminomelanin (DALM), and DALM-coated magnetite particles in a 25 W, 2450 MHz microwave field. See the text for calculation of SAR and Reference 11 for details of the experiments.

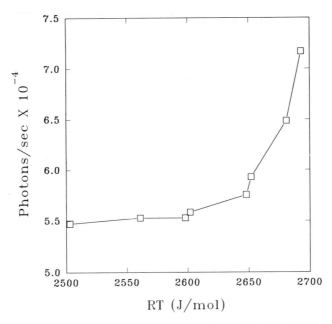

**Figure 4.** Luminescence (photons/sec) of peroxidizing diazoluminomelanin, biosynthesized in HL-60 human leukemia cells, at different thermal energies generated by exposure to 2450 MHz radiation. See Theory, Materials, and Methods for details.

What is the explanation for this unexpected behavior? It is found in the slow luminescence mechanism:

$$S_0 + Q <---> S_1 <\text{--Magnetic field/Electric field--}> T_1(\pm) <---> T_1 + kT$$

By imposing a selection factor, the external magnetic field component, on $T_1$ which exists in x, y, and z orientations, one such orientation is favored: the one aligned with the external magnetic field. Therefore, the equilibrium shifts to $S_1$ because of the nondegeneracy of the unaligned magnetic dipole states. The result is that immediate luminescence is increased, but slow luminescence is decreased. Currently, we have not tested this effect. However, if the electric field is used as the selection factor, then the charged triplet state is favored over the uncharged $S_1$ state and heating of the solution is seen with a decrease in luminescence sensitivity to heating. This result was observed (Figures 3 and 4) with the DALM-coated magnetite.

How can this correlation of changes in thermochemiluminescence and enhanced heating be corrected for in making dosimetric measurements, including temperature measurements with thermochemiluminescence, in EMFs? There are at least two ways to accomplish this: first, find the maximum output (A) of photons/sec of a given preparation of the slow luminescer; and second, prepare calibration plots of RT vs. luminescence (lnQ). Maximum output per second of photons (A) may be experimentally determined by heating the preparations to a temperature where the steady-state condition is destroyed and recording the peak output in photons/sec or by running a reaction at a fixed temperature until the steady-state condition is exhausted[12] and integrating the photonic output and treating it as if it were compressed into a second. This total photonic output represents the total amount (one photon representing one activated-state molecule) of available activated-state molecules (moles) that can emit light. The rising and falling tails of a DALM reaction can be of considerable length (microgram solutions of DALM may luminesce in excess of 52 h).[12] Because absolute temperature and luminescence correlate, they correct for changes in the thermochemiluminescent response in RFR fields, if the calibration plots of the thermochemiluminescent response (lnQ) are based on temperatures (RT) physically measured outside of an EMF. The equations inset in Figure 2 are the calibration equations for the specific preparations of DALM measured within the EMF. The formulas should be of the nature of:

$$y = mx + b$$

where: y is the luminescence (lnQ) and m is the RFR sensitivity, and b should equal the $0^0$ point for thermochemiluminescence (Q = 1). The temperatures given for b in Figure 2 are extrapolated temperatures at which Q < 1. The preparations did, in fact, visibly luminesce in the frozen state. Theoretically, because of quantum effects, Q = 1 should be at $0^0$ K. However, the practical limit of the QLIS is about 10,000 photons/sec because it samples for 33 ms per frame (video rate) and has a physical limit on the photosensitivity of its microchannel plates and low-light level camera.

The slow luminescence mechanism has biological relevance as a dosimetric technique. Magnetic fields in the kilogauss range inhibit slow fluorescence and enhance fast fluorescence in photosynthetic pathways,[15,22] microwave radiation inhibits autoxidation of hemoglobin mediated by green hemoprotein[23] and enhances extracellular thiol production by macrophages in vitro (a stress response).[24] The pathophysiological consequences of these effects remain to be determined.

# SUMMARY

Thermochemiluminescent microdosimetry promises to provide high temporal and spatial resolution of radiofrequency radiation absorption measurements in phantom material. The method has the potential for distinguishing interactions based on thermalization from those related to direct magnetic or electric field coupling to biochemical reactions. The mechanism upon which it is based, slow luminescence, is operational in nature in photosynthesis, autoxidations of hemoproteins, and other redox pathways. Therefore, this dosimetric method is likely to be highly relevant to biological effects.

# ACKNOWLEDGMENTS

This research was supported in part by the United States Air Force Office of Scientific Research.

# REFERENCES

1. M. Gautherie, J. Edrich, R. Zimmer, J.L. Guerguin-Kern, and J. Robert , Millimeter-wave thermography -- Application to breast cancer, *Journal of Microwave Power* 14(2): 123-129 (1979).
2. W.D. Hurt. Specific absorption rate measurement techniques, *in* "Proceedings of 22nd Topical Meeting on Instrumentation," 4-8 Dec 1988, South Texas Chapter of the Health Physics Society, San Antonio, Texas, pp. 139-151 (1988).
3. T.P. Ryan, R.R. Wikoff, and P.J. Hoopes, Design of an automated temperature mapping system for ultrasound or microwave hyperthermia, *Journal of Biomedical Engineering* 13: 348-354 (1991).
4. J.-Y. Chen and O.P. Gandhi, RF Currents induced in an anatomically-based model of a human for plane-wave exposures (20 - 100 MHz), *Health Physics* 57: 89-98 (1989).
5. M. Grandolfo, P. Vecchia, and O.P. Gandhi, Magnetic resonance imaging: Calculation of rates of energy absorption by a human-torso model, *Bioelectromagnetics* 11: 117-128 (1990).
6. J.-Y. Chen and O.P. Gandhi, Currents induced in an anatomically based model of a human for exposure to vertically polarized electromagnetic pulses, *IEEE Transactions on Microwave Theory and Techniques* 39(1): 31-39 (1991).
7. A. Kraszewski, M.A. Stuchly and S.S. Stuchly, ANA Calibration method for measurements of dielectric properties, *IEEE Transactions on Instruments and Measurements* IM-32: 385-387 (1983).
8. O.P. Gandhi, Automated Radiofrequency Radiation Dosimetry, USAFSAM-TR-90-37 (1990).
9. J.L. Kiel, C. Gabriel, D.M. Simmons and E.H. Grant, Diazoluminomelanin: A conductive luminescent polymer with microwave and radiowave absorptive properties, *in* "Proceedings of the Twelfth Annual International Conference of the IEEE Engineering in Medicine and Biology Society," P.C. Pedersen and B. Onaral, eds., vol. 12(4), IEEE, Philadelphia, pp. 1689-1690 (1990).
10. J.L. Kiel, J.E. Parker, J.L. Alls, and R.A. Weber, Self-labeling of bacteria with a luminescent polymer, *in* "Proceedings of the Thirteenth Annual Conference of the IEEE Engineering in Medicine and Biology Society," J.H. Nagel and W.M. Smith, eds., vol. 13(4), IEEE, Philadelphia, pp. 1605-1606 (1991).
11. J.L. Kiel, G.J. O'Brien, D.M. Simmons, and D.N. Erwin, Diazoluminomelanin: A synthetic electron and nonradiative transfer biopolymer, *in* "Charge and Field Effects in Biosystems-2," M.J. Allen, S.F. Cleary, and F.M. Hawkridge, eds., Plenum Press, New York, pp. 293-300 (1989).
12. J.L. Kiel, G.J. O'Brien, J. Dillon and J.R. Wright, Diazoluminomelanin: A synthetic luminescent biopolymer, *Free Radical Research Communications* 8:115-121 (1990).
13. E.C. Burdette, F.L. Cain and J. Seals, *In vivo* probe measurement technique for determining dielectric properties at UHF through microwave frequencies, *IEEE Transactions on Microwave Theory and Techniques* MTT-28: 414-427 (1980).
14. R.E. Blankenship, T.J. Schaafsma and W.W. Parson, Magnetic field effects on radical pair intermediates in bacterial photosynthesis, *Biochimica et Biophysica Acta* 461: 297-305 (1977).

15. S.G. Boxer, C.E.D. Chidsey and M.G. Roelofs, Magnetic field effects on reaction yields in the solid state: An example from photosynthetic reaction centers, *Annual Review of Physical Chemistry* 34: 389-417 (1983).

16. R.E. Blankenship, T.J. Schaafsma and W.W. Parson, Radical-pair decay kinetics, triplet yields and delayed fluorescence from bacterial reaction centers, *Biochimica et Biophysica Acta* 680: 44-59 (1982).

17. R.A. Goldstein and S.G. Boxer, The effects of very high magnetic fields on the delayed fluorescence from oriented bacterial reaction centers, *Biochimica et Biophysica Acta* 977: 70-77 (1989).

18. Y.-P. Sun, D.F. Sears, Jr. and J. Saltiel, Resolution of benzophenone delayed fluorescence and phosphorescence spectra. Evidence of vibrationally unrelaxed prompt benzophenone fluorescence, *Journal of the American Chemical Society* 111: 706-711 (1989).

19. C.R. Batishko, K.A. Stahl, D.N. Erwin and J.L. Kiel, A quantitative luminescence imaging system for biochemical diagnostics, *Review of Scientific Instrumentation* 61(9): 2289-2295 (1990).

20. J.L. Kiel, D.N. Erwin, and D.M. Simmons, Flow-Through Cell Cultivation System. US Patent 5,028,541, July 2, 1991.

21. T. Smith, A. Mackie, S. Van Waggoner, G. Gandy, M. Washburn, R. Self, A. Horn, J. Kiel, and J. Wright, Chemiluminescent microwave/thermal dosimetry based on luminol and metal oxide catalysts, *Microchemical Journal*, in press (1992).

22. A. Sonneveld, L.N.M. Duysens and A. Moerdijk, Sub-microsecond chlorophyll a delayed fluorescence from photosystem I: Magnetic field-induced increase of the emission yield, *Biochimica et Biophysica Acta* 636:39-49 (1981).

23. J.L. Kiel, C. McQueen, and D.N. Erwin, Green hemoprotein of erythrocytes: Methemoglobin superoxide transferase, *Physiological Chemistry and Physics and Medical NMR* 20:123-128 (1988).

24. J.L. Kiel, J.E. Parker, J.L. Alls, and S.B. Pruett, The cellular stress transponder: Mediator of electromagnetic effects or artifacts? *Nanobiology* 1(4):491-503 (1992).

# BIOLOGICAL EFFECTS OF HIGH-PEAK-POWER MICROWAVE ENERGY

Edward C. Elson[*]

Walter Reed Army Institute of Research
Department of Microwave Research
Washington, D.C. 20307-5100

## INTRODUCTION

At the Walter Reed Army Institute of Research, a research program designed to study the biological effects of non-ionizing radiation has focused on the effects of very short but high peak-power bursts of microwave energy. Cardiovascular, ocular and behavioral effects have been studied. This paper presents some of the behavioral effects that have been observed.

The Department of Defense has supported such research in conjunction with programs studying possible incapacitation of enemy force structure through defeat or killing of electronic assets on a future battlefield. The radiofrequency, mostly microwave transmissions that would be employed, cannot be directed or controlled completely and personnel exposure could be a problem. For such short-burst, high peak-power, "low" repetition rate transmissions, scant biomedical data are available.

The Walter Reed program uses transmitters similar to those likely to be used in a putative, directed-energy battle scenario. The fluence on a biological target, the peak power, repetition rate, and average power are systematically varied, and the "dose" of electromagnetic energy related to a number of biological effects or endpoints. For behavioral studies, the endpoints are similar to those employed in testing the toxicity of drugs.

## BACKGROUND

The American National Standards Institute (ANSI) C95.1-1992 safety guide for human exposure to radiofrequency electromagnetic fields, a voluntary guide widely used in the U.S., was derived mostly from investigations of continuous-wave rather than pulsed fields. A small amount of pulsed data was used to formulate limits on peak power in the 1992

---

[*] These views and opinions are those of the author and do not necessarily state or reflect those of the U.S. Government.

*Radiofrequency Standards,* Edited by B.J.
Klauenberg *et al.,* Plenum Press, New York, 1994

iteration. More recent experimental evidence suggests that the guidelines may need modification. The experiments described below attempt to address the adequacy of the guidelines. A central question is whether energy delivered in short but intense pulses produces biological effects of any kind, as compared to energy delivered in continuous wave form at exactly the same average power over time, especially for average powers within the ANSI permissible exposure limits. The answer obviously affects the issue of whether the existing radiofrequency protection guidelines can be applied to the directed-energy environment of the future.

## ANIMAL EXPERIMENTS

The only practical way of answering the question is to perform controlled experiments in animals using well-defined, reproducible, biological endpoints. The energy dose required to produce such endpoints must also be controlled and measured. One seeks to identify energy thresholds for such effects. Statistical methods of analysis, requiring an appropriate minimum number of animals, are always required. It is essential to understand and control a large number of potentially confounding variables. Rats were used in all cases. Issues of biological and electromagnetic scaling to human beings must then be confronted. Those issues will not be considered here.

Four different sets of experiments were performed:
1. Evaluation of memory retention.
2. Evaluation of treadmill performance.
3. Evaluation of time perception.
4. Effect on circadian rhythms.

Such tests evaluate very different psycho-physiologic processes and thereby provide information on the range or extent of functional alterations.

The parameters of microwave exposure were identical in all experiments and are tabulated as follows:

Frequency: 3 GHz
Pulse width: 80 ns
Pulse repetition rate: 0.125 Hz
Maximum transmitter peak output power: 700 MW
Average power of pulse: 200 MW
Average transmitter output power: 2 W
Peak power density: 700 kW/cm$^2$
Average power density (over 80 ns): 200 kW/cm$^2$
Average power density: 0.2 mW/cm$^2$
Whole-body average SAR: 0.07 W/kg

As shown, the average power density was 0.2 milliWatts/cm$^2$ over the approximately 26 minute period of each episode of exposure. The ANSI radiofrequency protection guideline allows a maximum exposure of 10 milliWatts/cm$^2$ in any six-minute period at a frequency of 3 GHz. The experiments were designed to ascertain if any effects might be found under pulsed conditions at the average power of 0.2 milliWatts/cm$^2$. It should be noted that the C95.1-1992 iteration of the standard imposed a peak-field limit of 100 kV/m throughout the radiofrequency spectrum. This is approximately 3 kW/cm$^2$. In view of this restriction, human exposures at levels described in this paper would not be permitted.

1. **Memory retention.** In rats (and man) the acquisition of information is initially committed to short-term memory. A consolidation process commits the information to

long-term memory in about 30 minutes in rats. The information is then stored in a more stable way. A perturbation of this process within 30 minutes following information acquisition can disrupt consolidation to long-term memory. The same perturbation coming after the consolidation process has no impact on memory. The process can be measured by placing water-deprived rats in a Y maze and permitting "one trial learning" to find the water, exposing to microwaves in less than one half hour, and retesting Y maze performance 24 hours later, recording the number of errors in the maze. This experiment was performed and resulted in a significant increase in the number of errors made by exposed animals, and also animals exposed to minus 30 db power, compared to cage controls.[1] A mental process is affected at low, "non-thermal" levels permitted by average power standards.

2. **Treadmill performance.** Rats were trained 5 minutes per day for 8 days, on a treadmill with an 8° incline, running at 10 meters per minute. Animals failing to stay on the treadmill experienced 0.5 milliAmp shocks provided for motivation. Sham-exposed and cage controls were compared to exposed animals. Exposed animals were tested immediately following exposure. The running time of exposed animals was significantly decreased.[2] The conclusion is that low-level "non-thermal" energy delivered in pulses decreased the endurance of rats.

3. **Time perception.** Animals were trained and became proficient in discriminating between light pulses of 0.5 and 5.0 seconds duration. A food reward reinforced the correct selection of the appropriate lever. The trained animals were then subjected to pulses of varying intermediate duration and patterns of response recorded. The pulse duration for which the percentage of responses is 50% long (or short) is identified as the bisection point. Changes in the pattern of response following exposure to microwave radiation suggest an effect of the radiation on the nervous system. The number of null responses, that is, a failure to respond by lever press after 10 seconds, was found to increase significantly and proportionate to power delivered. The maximum effect was seen at all power levels at the bisection point, the point at which the animal would logically be the most indecisive.[3] The conclusion is that the confidence or degree of decisiveness of the animal is significantly decreased in response to "non-thermal" levels of microwave power.

4. **Effect on circadian rhythms.** Rats were maintained in a controlled alternating light and dark environment (24-hour cycle). The nocturnal habits of rats normally result in 90% of feeding occurring during the dark cycle. Following exposure to microwave energy rats were placed in home cages equipped with response levers and pellet dispensers. Controls were similarly placed. The experiment demonstrated that patterns of food intake were significantly affected suggesting disrupted rest-activity cycles.[4]

## CONCLUSIONS

It is technically difficult to measure functional changes in animals during microwave exposures, and therefore these experiments focused on immediate post-exposure effects. There was a degradation in the quality of response in several different testing situations. The duration and reversibility of such effects could not be evaluated. Additional studies would be required to address such issues. The important question of extrapolation to the human condition was not addressed. The relevance of the particular nature of these electromagnetic exposures to a realistic directed-energy environment awaits further evolution in the systems development community. The data are useful in assessing the dimensions of the problem and can be employed to guide further research.

On the important issues of whether any effects can be identified in animals at average energies within the existing safety standards and not associated with conventional heating, the verdict is emphatically that such effects can be identified, and although not apparently catastrophic, could interfere in a deleterious manner with human task performance.

## REFERENCES

1. T.G. Raslear, Y. Akyel, R. Serafini, F. Bates, and M. Belt. Memory consolidation in the rat following high-peak power pulsed microwave irradiation, *Annual International Conference of the IEEE Engineering in Medicine and Biology Society,* 13(2):958 (1991).
2. Y. Akyel and M. Belt. The effects of high-peak power microwaves on treadmill performance in the rat, in *Electricity and Magnetism in Biology and Medicine,* Blank, M., Ed., 668 (1993).
3. T.G. Raslear, Y. Akyel, F. Bates, M. Belt, S.T. Lu. Temporal bisection in rats: the effects of high-peak-power pulsed microwave irradiation, *Bioelectromagnetics,* 14:459 (1993).
4. T.G. Raslear, Y. Akyel, R. Serafini, F. Bates, and M. Belt. Food demand and circadian rhythmicity following high-peak power pulsed microwave irradiation, *Annual International Conference of the IEEE Engineering in Medicine and Biology Society,* 13(2):962 (1991).

# SOME RECENT APPLICATIONS OF FDTD FOR EM DOSIMETRY: ELF TO MICROWAVE FREQUENCIES

Om P. Gandhi

Department of Electrical Engineering
University of Utah
Salt Lake City, Utah 84112

## ABSTRACT

In this paper, we describe some of the recent applications of the finite-difference time-domain (FDTD) method for a number of problems in bioelectromagnetics. This method, used in the past for whole-body or partial-body exposures due to spatially uniform or nonuniform (far-field or near-field) sinusoidally varying electromagnetic fields, and for low-frequency transient fields, such as those for an electromagnetic pulse, has now been modified and used for the following new applications:

1. For short nanosecond pulses with ultrawide bandwidths, a frequency-dependent FDTD has been formulated, which uses frequency-variable properties of the various tissues using the best-fit two-relaxation-constant Debye equations.
2. The FDTD code has been modified and used for specific absorption rate (SAR) calculations for radiofrequency (RF) magnetic fields typical of new and emerging magnetic resonance imaging (MRI) techniques.
3. Using scaled higher quasi-static frequencies, the FDTD method has been used for calculations of internal fields and induced current densities in an anatomically based model of the human body for electric, magnetic, or combined electromagnetic (EM) fields at power-line frequencies.

We also describe a new convolution method to alleviate the problem of having to run computer-memory-intensive anatomically based models repeatedly as the incident fields are varied in time and/or space domains. In this method, the impulse response of the heterogeneous model in time and space domains is obtained and stored. This may then be convolved with the prescribed time and/or space variations of the incident fields. Since convolution integrals are relatively easy to calculate and do not need a large computer memory, high-resolution dosimetric calculations should be possible using smaller computers or PCs.

*Radiofrequency Standards*, Edited by B.J.
Klauenberg *et al.*, Plenum Press, New York, 1994

## INTRODUCTION

Great strides have been made in the last few years in the area of numerical dosimetry using anatomically based models of the human body. Perhaps the most successful and the most promising of the numerical methods for SAR calculations at the present time is the finite-difference time-domain (FDTD) method.[1-7] For numerical calculations this method requires a computer memory and central processing unit (cpu) time proportional to N as opposed to $N^3$ for the competing Method of Moments (MOM), where N is the number of cells into which an absorbing body is divided. For bioelectromagnetic applications, use of the FDTD method has, therefore, allowed us to obtain SAR distributions for an anatomically based 45,024-cell model of the human body as opposed to the 100-500-cell models that were possible with MOM, permitting a degree of resolution that would have been unthinkable just four or five years ago.

The FDTD method has been found to be extremely versatile and has been used for whole-body or partial-body exposures due to spatially uniform or nonuniform (far-field or near-field) sinusoidally varying electromagnetic fields, and for low-frequency transient fields such as those for an electromagnetic pulse (EMP). We have also recently extended the use of the FDTD method to obtain SAR and induced current distributions in a 45,024-cell sixteen-tissue, anatomically based model of the human body to frequencies as high as 915 MHz.[8] This work has allowed us to obtain SARs for the various organs (brain, eyes, heart, lungs, liver, kidneys, and intestines) as well as the various parts of the body (head and neck, torso, legs, and arms) as a function of frequency in the band 100-915 MHz. Special attention was given to the frequency band 110-200 MHz to see if we could predict the high SARs that had been observed for the neck by Stuchly et al.[9] at 160 MHz using a four-tissue (skeleton, brain, lungs, and muscle) heterogeneous model of a human. In this work, only three frequencies, 160, 350, and 915 MHz, and an isolated human model had been studied. By calculating SAR distributions at frequency intervals of 10 MHz, we found that the highest layer-averaged SAR for the neck section was obtained not at 160 MHz, but rather at 200 MHz, where the highest local SAR of 3.32 W/kg was calculated for the center of the neck.

The FDTD algorithm has recently been modified and used for the following newly emerging applications:

1. For ultrashort pulses where extremely wide bandwidths (on the order of 1 GHz) are involved, the code has been modified to include frequency-dependent properties of the tissues using the best-fit two relaxation Debye equations.[10,11] This frequency-dependent finite-difference time-domain ((FD)$^2$TD) method can also be used to determine SAR distributions at multiple frequencies based on a single numerical run with an impulse using the Fourier components at various frequencies.

2. The FDTD code has also been used to calculate SARs in the exposed parts of the human body due to RF magnetic fields that are used in magnetic resonance imaging (MRI). Newer techniques for MRI are leading to the use of higher static magnetic fields and correspondingly higher RF magnetic fields. Based on the solution of a complete set of Maxwell's equations, the FDTD method is usable at any frequency, including RF magnetic-field frequencies as high as about 200 MHz that are proposed for the MRIs of the future.[12]

3. For one of the recent applications, the FDTD code has been adapted for calculations of internal **E** and **H** fields and induced current densities for human exposure to electromagnetic fields (EMFs) at power-line frequencies.[7] For these calculations, both the magnetic and electric incident fields have been considered and use is made of a scaled higher quasi-static frequency[13] on the order of 5-10 MHz to help reduce the computation time by over five orders of magnitude.

4. To alleviate the problem of having to run the computer-memory-intensive anatomically based model of the human body repeatedly as the incident fields are varied in time and/or space domains, we are developing simple and efficient techniques based on the convolution theory.[14] In this approach, the impulse response of the heterogeneous model is obtained and stored using an impulse in the time domain or space domain (spatially localized incident fields) or both, and convolved with the prescribed variations of the incident fields in time and space domains. Since convolution integrals are relatively easy to calculate and do not need a large computer memory, high-resolution dosimetric calculations should therefore be possible using a small computer or a personal computer (PC), as long as the direction of incidence on the body is maintained. This method inherently assumes that coupling back to the source is negligible, which is true for all far-field sources and for several near-field sources such as leakage fields of RF sealers, microwave ovens, etc.

## THE FINITE-DIFFERENCE TIME-DOMAIN METHOD

The FDTD method for electromagnetic calculations has been described in a number of publications (see, e.g., Taflove and Umashankar[15] and Gandhi[16] for a review of the procedure). We have used the method for calculations of the distributions of electromagnetic (EM) fields and SARs in sixteen-tissue anatomically based models of the human body for whole-body or partial-body exposures due to far-field (plane-wave) or near-field irradiation conditions.[1-8] In this method, the time-dependent Maxwell's curl equations

$$\nabla \times \mathbf{E} = -\mu \frac{\partial \mathbf{H}}{\partial t} \; , \quad \nabla \times \mathbf{H} = \sigma \mathbf{E} + \varepsilon \frac{\partial \mathbf{E}}{\partial t} \tag{1}$$

are implemented for a lattice of subvolumes or "cells" that may be cubical or parallelepiped with different dimensions $\delta_x$, $\delta_y$, and $\delta_z$ in x-, y-, or z-directions, respectively. The components of $\mathbf{E}$ and $\mathbf{H}$ are positioned about each of the cells as shown in x and calculated alternately with half-time-steps where the time step $\delta t = \delta/2c$, where $\delta$ is the smallest of the dimensions used for each of the cells and c is the maximum phase velocity of the fields in modeled space. Since some of the modeled volume is air, c corresponds to the velocity of EM waves in air.

In the FDTD method, it is necessary to represent not only the scatterer/absorber such as the human body or a part thereof, but also the EM source(s), including their shapes, excitations, etc., if these sources are in the near-field region.[4,5] The far-field sources, on the other hand, are described by means of incident plane-wave fields prescribed for a "source" plane,[1-3, 6-8] typically 6-10 cells away from the exposed body. The source-body interaction volume is subdivided into cells of the type shown in Fig. 1. The interaction space consisting of several hundred thousand to a few million cells is truncated by means of absorbing boundaries. The prescribed incident fields are tracked in time for all cells of the interaction space. The problem is considered completed when either the fields have died off or, for sinusoidal excitation, when a sinusoidal steady-state behavior for $\mathbf{E}$ and $\mathbf{H}$ is observed for the interaction space.

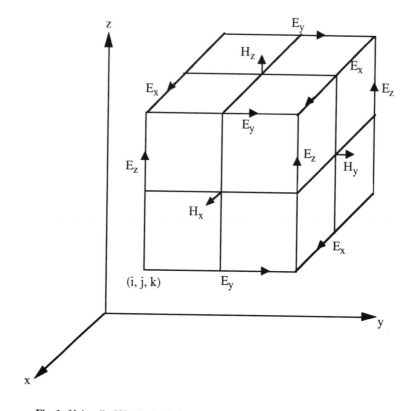

**Fig. 1.** Unit cell of Yee lattice indicating positions for various field components.

## THE FREQUENCY-DEPENDENT FDTD METHOD

As previously mentioned, the frequency-dependent FDTD or (FD)$^2$TD method is needed for short pulses where wide bandwidths are generally involved. Two general approaches have been used for the (FD)$^2$TD method. One approach is to convert the complex permittivity from the frequency domain to the time domain and convolve this with the time-domain electric fields to obtain time-domain fields for dispersive material. This discrete time-domain method may be updated recursively for some rational forms of complex permittivity, which removes the need to store the time history of the fields and makes the method feasible. This method has been applied to materials described by a first-order Debye relaxation equation,[17-19] a second-order Lorentz equation with multiple poles,[20] and to a gaseous plasma.[21]

A second approach is to add a differential equation relating the electric flux density **D** to the electric field **E** and solve this new equation simultaneously with the standard FDTD equations. This method has been applied to 1D and 2D examples with materials described by a first-order Debye equation or second-order single-pole Lorentz equation,[22] and to 3D sphere and homogeneous two-thirds muscle-equivalent man model with properties described by a second-order Debye equation.[10,23] In the following we describe this differential equation approach, which has now been used for induced current and SAR calculations for a heterogeneous model of the human body.[11]

The time-dependent Maxwell's curl equations used for the FDTD method are:

$$\nabla \times \mathbf{E} = -\frac{\partial \mathbf{B}}{\partial t} = -\mu \frac{\partial \mathbf{H}}{\partial t} \tag{2}$$

$$\nabla \times \mathbf{H} = \frac{\partial \mathbf{D}}{\partial t} \tag{3}$$

where the flux density vector $\mathbf{D}$ is related to the electric field through the complex permittivity $\varepsilon^*(\omega)$ of the local tissue by the following equation:

$$\mathbf{D} = \varepsilon^*(\omega) \mathbf{E} \tag{4}$$

Since Eqs. 2 and 3 are solved iteratively in the time domain, Eq. 4 must also be expressed in the time domain. This may be done by choosing a rational function for $\varepsilon^*(\omega)$ such as the Debye equation with two relaxation constants:

$$\varepsilon^*(\omega) = \varepsilon_0 \left[ \varepsilon_\infty + \frac{\varepsilon_{s1} - \varepsilon_\infty}{1 + j\omega\tau_1} + \frac{\varepsilon_{s2} - \varepsilon_\infty}{1 + j\omega\tau_2} \right] \tag{5}$$

Rearranging Eq. 5 and substituting in Eq. 4 gives

$$\mathbf{D}(\omega) = \varepsilon^*(\omega)\,\mathbf{E}(\omega) = \varepsilon_0 \frac{\varepsilon_s + j\omega\,(\varepsilon_{s1}\,\tau_2 + \varepsilon_{s2}\,\tau_1) - \omega^2\,\tau_1\,\tau_2\,\varepsilon_\infty}{1 + j\omega\,(\tau_1 + \tau_2) - \omega^2\,\tau_1\,\tau_2}\,\mathbf{E}(\omega) \tag{6}$$

where the dc (zero frequency) dielectric constant is given by

$$\varepsilon_s = \varepsilon_{s1} + \varepsilon_{s2} - \varepsilon_\infty \tag{7}$$

Assuming $e^{j\omega t}$ time dependence, we can write Eq. 6 as a differential equation in the time domain

$$\tau_1\tau_2 \frac{\partial^2 \mathbf{D}}{\partial t^2} + (\tau_1 + \tau_2) \frac{\partial \mathbf{D}}{\partial t} + \mathbf{D} =$$

$$\varepsilon_0 \left[ \varepsilon_s\,\mathbf{E} + (\varepsilon_{s1}\,\tau_2 + \varepsilon_{s2}\,\tau_1) \frac{\partial \mathbf{E}}{\partial t} + \varepsilon_\infty\,\tau_1\,\tau_2 \frac{\partial^2 \mathbf{E}}{\partial t^2} \right] \tag{8}$$

For the (FD)$^2$TD method, we need to solve Eqs. 2 and 3 subject to Eq. 8. As in Gandhi et al.,[10,23] we write these equations in the difference form, solve Eq. 8 to find $\mathbf{E}$, Eq. 2 to find $\mathbf{H}$, and Eq. 3 to find $\mathbf{D}$ at each cell location. The $\mathbf{E} \to \mathbf{H} \to \mathbf{D}$ loop is then repeated until the pulse has died off.

# ANATOMICALLY BASED MODELS OF HUMANS

For most of the calculations given in this paper, the anatomically based model described in our earlier publications[2-7] has been used. This model is based on the cross-sectional diagrams of the human body given in the book by Eycleshymer and Schoemaker.[24] This book contains cross-sectional diagrams of the human body that were obtained by making cross-sectional cuts at spacings of about one inch in human cadavers. The process for creating the data base of the man model was the following: first of all, a quarter-inch grid was taken for each single cross-sectional diagram and each cell on the grid was assigned a number corresponding to one of the sixteen tissue types (muscle, fat, bone, blood, intestine, cartilage, liver, kidney, pancreas, spleen, lung, heart, nerve, brain, skin, eye) or air. Thus the data associated with a particular layer consisted of three numbers for each square cell: x and y positions relative to some anatomical reference point in this layer, usually the center of the spinal cord; and an integer indicating which tissue that cell contained. Since the cross-sectional diagrams available in Eycleshymer and Schoemaker[24] are for somewhat variable separations, typically 2.3-2.7 cm, a new set of equispaced layers was defined at 1/4-in (0.635 cm) intervals by interpolating the data onto these layers. Because the 1/4-in cell size was too small for the memory space of readily accessible computers at that time, the proportion of each tissue type was calculated next for somewhat larger cells of size 1/2 in (1.27 cm) combining the data for $2 \times 2 \times 2 = 8$ cells of the smaller dimension. Without changes in the anatomy, this process allows some variability in the height and weight of the body. We have taken the final cell size of 1.31 cm (rather than 1/2 in) to obtain the total height and body weight of 175.5 cm and 69.6 kg, respectively. With a resolution of 1.31 cm, the model consists of a total of 45,024 cells that are either totally or partially within the human body. If the shoe-wearing condition of the model is needed, this is simulated by using a separation layer of rubber ($\varepsilon_r = 4.0$) that is 2 layers or 2.62 cm thick between the feet and the ground plane.

# MODELING OF THE TISSUE PROPERTIES WITH THE DEBYE EQUATION

For ultrawideband calculations using the (FD)$^2$TD method, the measured properties for the various tissues have been fitted to the Debye equation 5 with two relaxation constants.[10,11,23] The measured properties of biological tissues (muscle, fat, bone, blood, intestine, cartilage, lung, kidney, pancreas, spleen, lung, heart, brain/nerve, skin, and eye) were obtained from.[25] Optimum values for $\varepsilon_{s1}$, $\varepsilon_{s2}$, $\varepsilon_\infty$, $\tau_1$, and $\tau_2$ in Eq. 5 were obtained by nonlinear least squares matching to the measured data for fat and muscle. All other tissues have properties falling roughly between these two. Optimum values shown in Table 1 for $\varepsilon_{s1}$, $\varepsilon_{s2}$, and $\varepsilon_\infty$ for all tissues were then obtained with $\tau_1$ and $\tau_2$ being the average of the optimized values for fat and muscle. This was done to facilitate volume averaging of the tissue properties in cells of the heterogeneous man model. Having $\tau_1$ and $\tau_2$ constant for all tissues allowed linear (volume) averaging of the $\varepsilon$ values for each tissue in a given cell to calculate $\varepsilon$ values for that cell. The measured tissue properties and those computed from the Debye equation with $\tau_1$ and $\tau_2$ being the average of fat muscle are shown in Fig. 2 for fat and muscle. Similar comparisons were also obtained for the other tissue types.

# COUPLING OF AN ULTRAWIDEBAND PULSE TO THE HUMAN BODY

The (FD)$^2$TD method has been used to calculate coupling of an ultrashort pulse to the heterogeneous model of the human body. From the calculated internal fields we calculated the vertical currents passing through the various layers of the body by using the following equation:

$$I_z(t) = \delta^2 \sum_{i,j} \frac{\partial D_z}{\partial t}$$

(9)

where $\delta$ is the cell size (= 1.31 cm), and the summation is carried out for all cells in a given layer. We also calculated the layer-averaged absorbed energy density or specific absorption (SA) and the total energy W absorbed by the whole body using the following relationships:

$$S A \Big|_{\text{layer } k} = \frac{\delta t}{N_k} \sum_t \frac{E(i, j, k, t)}{\rho(i, j, k)} \cdot \frac{\partial D(i, j, k, t)}{\partial t}$$

(10)

$$W = \delta t \cdot \delta^3 \sum_t E(i, j, k, t) \cdot \frac{\partial D(i, j, k, t)}{\partial t}$$

(11)

In Eqs. 10 and 11, $\delta t$ is the time step (= $\delta/2c$ = 0.02813 ns) used for the time-domain calculations, $N_k$ is the number of cells in layer k of the body, and $\rho(i, j, k)$ is the mass density in kg/m$^3$ for each of the cells in the corresponding layers.

**Table 1.** Debye constants for tissues.

$$\tau_1 = 46.2 \times 10^{-9} \text{ s}$$
$$\tau_2 = 0.91 \times 10^{-10} \text{ s}$$

(average of optimum for fat and muscle)

| Tissue | $\varepsilon_\infty$ | $\varepsilon_{s1}$ | $\varepsilon_{s2}$ |
|---|---|---|---|
| Muscle | 40.0 | 3948. | 59.09 |
| Bone/Cartilage | 3.4 | 312.8 | 7.11 |
| Blood | 35.0 | 3563. | 66.43 |
| Intestine | 39.0 | 4724. | 66.09 |
| Liver | 36.3 | 2864. | 57.12 |
| Kidney | 35.0 | 3332. | 67.12 |
| Pancreas/Spleen | 10.0 | 3793. | 73.91 |
| 1/3 Lung | 10.0 | 1224. | 13.06 |
| Heart | 38.5 | 4309. | 54.58 |
| Brain/Nerve | 32.5 | 2064. | 56.86 |
| Skin | 23.0 | 3399. | 55.59 |
| Eye | 40.0 | 2191. | 56.99 |

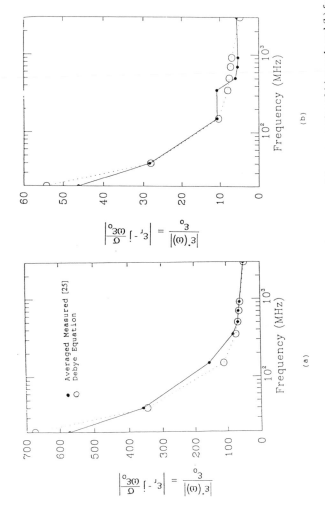

**Fig. 2.** Fit of Debye equations with two relaxation constants (5) to measured tissue properties of (a) muscle and (b) fat.

For the various calculations, we have used both the isolated model of the human body as well as the model standing vertically on a conducting ground plane.

A typical ultrawideband pulse with a prescribed peak amplitude of 1.1 V/m is shown in Fig. 3 in the time domain. The pulse shape was provided on a diskette by Jim O'Loughlin of Kirtland Air Force Base, New Mexico, courtesy of Dr. David N. Erwin of Armstrong Laboratory, Brooks Air Force Base, Texas. It is interesting to note that the pulse has a rise time of about 0.2 ns and a total time domain of about 7-8 ns. We have calculated the Fourier spectrum of the prescribed pulse that is shown in Fig. 4. Most of the energy in the pulse is concentrated in the 200-900 MHz band with the peak of the energy being at about 500 MHz.

We assumed the incident fields to be vertically polarized, since this polarization is known to result in the strongest coupling for standing individuals. Using the $(FD)^2TD$ procedure described earlier, we have calculated the temporal variations of total vertical currents for the various sections of the body, for both the shoe-wearing grounded, and ungrounded exposure conditions of the model, respectively. The current variations for a couple of representative sections, such as those through the eyes and the bladder, are given in Figs. 5a and 5b, respectively. The calculated peak currents for the various sections are on the order of 1.1 to 3.2 mA/(V/m). It is interesting to note that there is very little difference in the induced currents whether the model is grounded or not. This is because most of the energy in the pulse is at frequencies in excess of 300 MHz, where the effect of the ground plane on the induced currents or the SARs is minimal.[8]

In Fig. 6 we have plotted the peak current for each section of the body with a resolution of 1.31 cm. The maximum peak current of 3.5 mA, which is 3.2 mA/(V/m), occurs at a height of 96.3 cm above the bottom of the feet. A very similar result had previously been observed for calculations using isolated and grounded models of the human body for plane-wave exposures at frequencies of 350-700 MHz, where the highest induced currents on the order of 3.0-3.2 mA/(V/m) were calculated for sections of the body that are at heights of 85-100 cm relative to the feet.[8]

Using Eqs. 10 and 11, we have also calculated the specific absorptions (SA) and the total absorbed energy for exposure to the ultrawideband pulse of Fig. 3. The specific absorptions are plotted in Fig. 7 as a function of height above the feet of the various sections of the body for isolated and shoe-wearing conditions. Note that because of the very limited time duration of the pulse (7-8 ns), the specific absorptions are on the order of 0.02 to 0.20 pJ/kg. Using Eq. 11, the total energy absorbed by the body as a function of time has been calculated and is shown in Fig. 8. The energy is virtually all absorbed in the first 6 to 8 ns. The total energy absorbed by the body exposed to a single pulse is calculated to be 2.0 and 1.91 pico Joules for isolated and shoe-wearing grounded conditions, respectively.

## SAR DISTRIBUTIONS DUE TO RF MAGNETIC FIELDS OF MRI

MRI is becoming an increasingly important tool for medical diagnostic applications. New and emerging technologies are leading to use of higher static magnetic fields (up to 4T) and associated higher RF frequencies (up to 200 MHz).[26] Recently suggested safety guidelines by the USFDA[27] and the NRPB (UK)[28] limit the SARs within the body for exposure to the RF magnetic fields used for the MRI. We have, in the past, used the impedance method to calculate the SAR distributions in the 1.31-cm-resolution anatomically based model of the human body.[29] Because of the quasi-static approximation, the impedance method cannot, however, be used for frequencies in excess of 30-40 MHz. To alleviate this frequency limitation, we have been investigating the applicability of

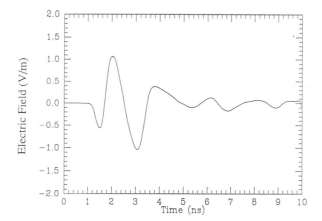

**Fig. 3.** The prescribed electromagnetic pulse. Peak incident field = 1.1 V/m.

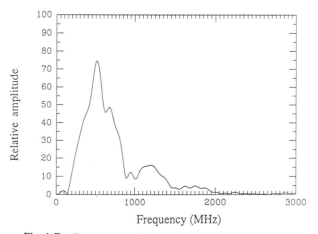

**Fig. 4.** Fourier spectrum of the electromagnetic pulse of Fig. 3.

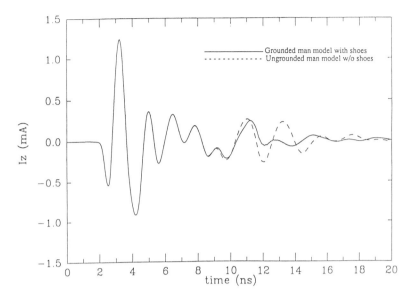

5a. Section through the eyes (height above the bottom of the feet = 168.3 cm).

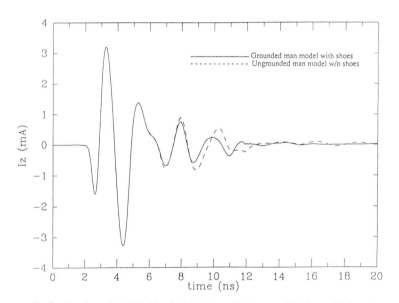

5b. Section through the bladder (height above the bottom of the feet = 91.0 cm).

**Fig. 5.** Currents induced for the various sections of the body for shoe-wearing grounded and ungrounded conditions of exposure. $E_{peak} = 1.1$ V/m.

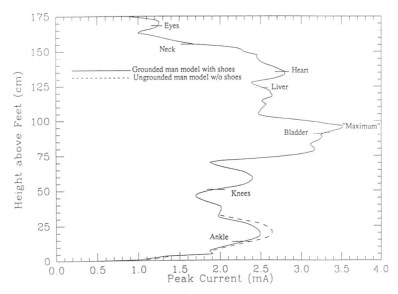

**Fig. 6.** Peak currents induced for the various sections of the body for shoe-wearing grounded and ungrounded conditions of the model. $E_{peak}$ = 1.1 V/m.

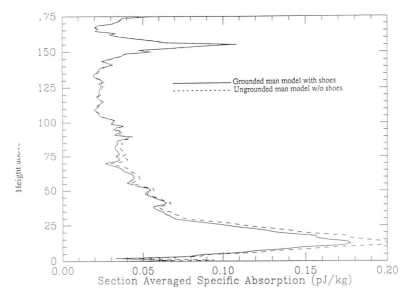

**Fig. 7.** Variation of section or layer-averaged specific absorption for the ultrawideband pulse of Fig. 3. $E_{peak}$ = 1.1 V/m.

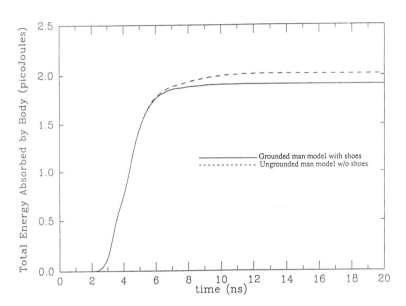

**Fig. 8.** Total energy absorbed by the body as a function of time.

FTDT for SAR calculations for parts of the body that are exposed to RF magnetic fields of magnetic resonance imagers. For this application, we have modified the FDTD algorithm so that a sinusoidal uniform equiphase magnetic field, either linearly or circularly polarized, is postulated for each of the cells for the exposed parts of the body as the initial excitation. Solution of the Maxwell's equations for subsequent times give rise to the spatial nonuniformities and creation of internal (and external) electric fields.

To test the feasibility of using the FDTD method for such calculations, we have used a test case of a lossy sphere of radius 20 cm. We initially used a lower frequency of 5 MHz, where the results could be compared with the analytical solution that can be obtained from Faraday's law of induction.

$$\oint \mathbf{E} \cdot d\ell = -\frac{\partial}{\partial t} \int \mathbf{B} \cdot d\mathbf{S} \tag{12}$$

For a circularly symmetric model such as a sphere

$$\mathbf{E} = E_\phi \hat{\phi} = -\frac{j \omega \mu_o H r}{2} \hat{\phi} \tag{13}$$

In Fig. 9, curve a shows the plot of magnitude of $E_z$ $(= |E_\phi|)$ along the central y axis of the sphere obtained from the analytical Eq. 13 and the results obtained by using the modified FDTD algorithm. For the calculations, properties assumed for the sphere were $\varepsilon_r$ = 340, $\sigma$ = 0.5 S/m, corresponding to the properties for high-water-content tissues such as muscle at 5 MHz.[30] Agreement of the FDTD-calculated results with the analytical results (curve a) is excellent.

After verifying the feasibility of using the FDTD method, we next applied it for calculations of induced RF electric fields for an applied $\mathbf{H} = H_x \hat{x} = 1$ A/m at 200 MHz. For these calculations, we assumed $\varepsilon_r$ = 37.0 and $\sigma$ = 0.85 S/m, corresponding to the properties of 2/3 muscle at 200 MHz.[30] The calculated variation of $|E_z|$ along the central y axis is shown as curve b in Fig. 9. For comparison, we have also plotted the linear variation given in by Eq. 13 as curve c in Fig. 9. It is interesting to note that the induced E fields (and, hence, SARs) based on the solution of the complete set of Maxwell's equations (the FDTD method) are considerably lower than those that will be given by solving just the Faraday's law of induction Eq. 13. This is not surprising, since the latter procedure is valid only for low frequencies where quasi-static approximations can be made.

We have also carried out calculations for another test case for which a homogeneous muscle-equivalent cylinder of dimensions 20 cm was used for calculations at 10, 30, and 100 MHz. The calculated SAR variations for this test case were found to be similar to those given in a paper by Bottomley and Andrew.[31] It was interesting to note both in the test case for the sphere (Fig. 9) and the cylinder that the SARs at higher frequencies do not increase as rapidly as $\omega^2$. We are presently in the process of extending this procedure for calculations of SARs for the anatomically based model of the human body.

## NUMERICAL DOSIMETRY AT POWER-LINE FREQUENCIES

As described in Gandhi and Chen,[7] the FDTD method has also been adapted for calculations of internal **E** and **H** fields and induced current densities for exposure to purely electric, purely magnetic, or combined electric and magnetic fields (EMFs) at power-line frequencies. It is recognized that the conductivities of several tissues (skeletal muscle, bone, etc.) are highly anisotropic for power-line frequencies. This has, however, been neglected in the first instance and will be included in the future when models based on separate identification of these tissues are available.

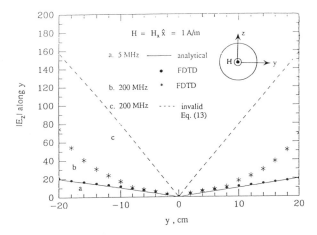

**Fig. 9.** FDTD-calculated and analytical variations of the magnitude of the induced **E** field for the central y axis of a homogeneous lossy sphere used as a test case. For a lossy sphere, radius = 20 cm; $\varepsilon_r$ = 340, $\sigma$ = 0.5 S/m at 5 MHz; $\varepsilon_r$ = 37.0, $\sigma$ = 0.85 S/m at 100 MHz. Note the considerably lower **E** fields (and, hence, SARs) than those projected by quasi-static analytical Eq. 13 for higher-frequency RF magnetic fields.

Both sinusoidal and prescribed time-varying incident fields can be used with the FDTD procedure -- hence the method is well suited also for transient exposures that are often of interest at power-line-related frequencies. For sinusoidally varying fields, the solution is completed when a sinusoidal steady-state behavior for **E** and **H** fields is observed for each of the cells. For lossy biological bodies this typically takes a stepped time on the order of 3 to 4 time periods of oscillation. Since $\delta t$ is fixed for a given cell size, a larger number of iterations is therefore needed at lower frequencies. Because of the horrendous number of iterations, the FDTD procedure would be clearly inapplicable for calculations at power-line frequencies were it not for the quasi-static nature of the coupling as previously pointed out by Kaune and Gillis[32] and Guy et al.[13] Using a logic similar to these authors, the fields outside the body depend not on the internal tissue properties, but only on the shape of the body as long as the quasi-static approximation is valid, i.e., the size of the body is a factor of 10 or more smaller than the wavelength, and $|\sigma + j\omega\varepsilon| >> \omega\varepsilon_o$ where $\sigma$ and $\varepsilon$ are the conductivity and the permittivity of the tissues, respectively, $\omega = 2\pi f$ is the radian frequency, and $\varepsilon_o$ is the permittivity of the free space outside the body. Under these conditions, the electric fields in air are normal to the body surface and the internal tissue electric fields are given from the boundary conditions in terms of the fields outside

$$j\omega\varepsilon_o\hat{n} \cdot \mathbf{E}_{air} = (\sigma + j\omega\varepsilon)\hat{n} \cdot \mathbf{E}_{tissue} \qquad (14)$$

A higher quasi-static frequency f′ may therefore be used for irradiation of the model and the internal fields E′ thus calculated may be scaled back to frequency f of interest, e.g., 60 Hz. From Eq. 14 we can write

$$E_{tissue}(f) = \frac{\omega}{\omega'} \frac{(\sigma' + j\omega\varepsilon')}{(\sigma + j\omega\varepsilon)} E'_{tissue}(f')$$

$$\simeq \frac{f \sigma'}{f' \sigma} E'_{tissue}(f') \tag{15}$$

assuming that $\sigma + j\omega\varepsilon \simeq \sigma$ at both f′ and f.

For our calculations, we have used a full-scale anatomically based model of the human body and a frequency f′ of 5-10 MHz to reduce the computation time by orders of magnitude. Since in the FDTD method one needs to calculate in the time domain until convergence is obtained (typically 3-4 time periods), this frequency scaling to 5-10 MHz for f′ reduces the needed number of iterations by over 5 orders of magnitude. At the higher irradiation frequency f′, we have taken $\sigma' = \sigma$, i.e., conductivities of the various tissues at 60 Hz. Furthermore, we have taken the incident E field $E_i(f') = 60\ E_i(f)/f'$ to obtain $E_{tissue}(f)$ at say $E_i(f) = 10$ kV/m. The incident magnetic field $H_i(f')$ has similarly been taken to be considerably lower (= 60 $H_i(f)/f'$) to account for the fact that the induced current densities and internal electric fields are proportional to the frequency of the incident fields and would therefore be higher at the assumed frequency f′.

After verifying the accuracy of this approach with test cases using homogeneous and layered spheres[7] where the results for the internal field variations could be compared with those using the analytic Mie series solutions,[33] we also checked the results against the experimental data on a mannequin given by Deno[34] and the variations of the induced currents calculated along the height of the body by DiPlacido et al.[35] The agreement with the results of these two authors who had used a vertical electric field, such as that under a high-voltage power line, was found to be very good.

Shown in Fig. 10 are the calculated results using the anatomically based model where the conductivities used for the various tissues are as given in Table 2. Recognizing the anisotropy in the conductivity of the skeletal muscle, two different values of muscle conductivities are taken for curves (1) and (2). For these curves a higher conductivity of 0.52 S/m is taken for the skeletal muscle and an average value of 0.11 S/m is taken for the muscle in the interior of the body. For curves (3) and (4), however, a lower conductivity of 0.11 S/m is taken for all of the muscle, interior or skeletal. The results shown in Fig. 10 curves (1), (3), and (4) are for $E_{inc} = 10$ kV/m (vertical) and $H_{inc} = 26.5$ A/m from side to side of the model. To point out the preponderance of the induced currents due to incident electric field, $H_{inc} = 0$ is assumed for the calculations shown in curve (2). It is interesting to note that the layer currents due to E-field exposure alone are almost 98-99 percent of the currents calculated for the combined electric and magnetic fields. It is also interesting to note that the calculated foot currents of 155-160 μA are in excellent agreement with 165 μA that would be projected from the measurements of Deno[34] for the human.

For the combined electric- and magnetic-field exposure condition of curve (1), the current density distributions for three representative sections through the brain, the heart, and the ankles are shown in Figs. 11a-c, respectively. For each of the sections the letter F denotes the front of the body. For comparison the current density distributions for the same sections due to incident magnetic field alone ($H_{inc}$ = 26.5 A/m or B = 33 µT) are shown in Figs. 11d-f. It is interesting to note that considerably smaller current densities are obtained due to magnetic fields alone, as compared to the combined EMFs.

## A NEW CONVOLUTION PROCEDURE FOR NUMERICAL DOSIMETRY

To alleviate the problem of having to run computer-memory-intensive anatomically based models of the human body repeatedly, we are developing simple and efficient techniques based on convolution theory, where the impulse response of the heterogeneous

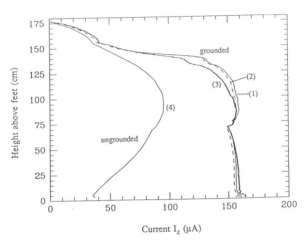

**Fig. 10.** Calculated layer currents for anatomically based grounded and ungrounded models exposed to EMFs at 60 Hz. For curves (1) and (2) σ = 0.11 S/m for the interior muscle. For curves (3) and (4) σ = 0.11 S/m for all of the muscle. E = 10 kVm (vertical), H = 26.5 A/m from side to side for all of the curves except for (2) for which only E-field exposure is assumed.

model would be obtained and stored using an impulse in the time domain or space domain (spatially localized incident fields) or both, and convolved with the prescribed variation of the incident fields in time and space domains. We illustrate this procedure in the following with the example of the response of the model for exposure to pulses or transients of prescribed shapes.[36] For impulse response calculations, a narrow unit magnitude flat-top "impulse" of time duration T´ = 5 to 10 δt is assumed. The impulse-induced current response $H_i(t)$ for the various sections of the body (such as sections through the eyes, neck, heart, liver, etc.) is calculated and stored. Induced currents $I_i(t)$ for these sections for a prescribed incident pulse E(τ) can then be calculated from the convolution integral

71

$$I_i(t) = \int_0^T E(\tau)\, H_i(t - \tau)\, d\tau \tag{16}$$

where the integration time T is a period in excess of the time duration of the incident pulse. An alternate procedure that has also been found to be equally applicable is to work with Eq. 16 in terms of the Fourier transforms $\mathcal{F}\{H_i(t)\}$, $\mathcal{F}\{E^{im}(t)\}$, and $\mathcal{F}\{E(t)\}$ of the induced currents $H_i(t)$, the broadband initial impulse $E^{im}(t)$, and the prescribed pulse E(t) for which the induced currents $I_i(t)$ are desired. In terms of the Fourier transforms Eq. 16 can be written as

$$I_i(t) = \mathcal{F}^{-1}\left\{\frac{F\{E(t)\}\; F\{H_i(t)\}}{F\{E^{im}(t)\}}\right\} \tag{17}$$

where the discrete Fourier transforms needed for Eq. 17 can be efficiently calculated using the Fast Fourier Transform (FFT) algorithm. Both of the above approaches are relatively simple and do not need a large computer memory, making it possible to use a small computer or PC for calculating the response of the human body to any desired incident pulse. We show in Figs. 12a and 12b the comparison of the vertical currents induced for a couple of representative sections of the body for the ultrawideband pulse of Fig. 3. In both Figs. 12a and 12b the time-domain variations of the currents calculated with the simpler impulse response and the convolution method are in excellent agreement with the results obtained from the exact simulation with the conventional $(FD)^2TD$ method.

A similar procedure may be also be usable for prescribed spatial variations of the incident fields.

**Table 2.** Tissue conductivities used for calculations at the power-line frequency of 60 Hz..

| TISSUE TYPE | $\sigma$ S/m |
|---|---|
| Air | 0 |
| Muscle | 0.52 or 0.11 |
| Fat, Bone | 0.04 |
| Blood | 0.6 |
| Intestine | 0.11 |
| Cartilage | 0.04 |
| Liver | 0.13 |
| Kidney | 0.16 |
| Pancreas | 0.11 |
| Spleen | 0.18 |
| Lung* | 0.04 |
| Heart | 0.11 |
| Nerve,Brain | 0.12 |
| Skin | 0.11 |
| Eye | 0.11 |

*We have used 33 percent lung tissue and 67 percent air for the dielectric properties of the lung.

```
                              63   65   21
                         50   68   63   92   76   65   61   53   20
              30    73   68   72   82   88  156  201  244  145  151   31   25
         59   97    63  117  114  117  106  109  108   81  254  285   75   31
        117   96   178  126  114  125   97   98  119  120   84   37   87   92   17
   102  100  181   147  121  103  105   98  101   94   93  114  147   55   58   10
    95  125  183   134  118  112  104  113  112   64   51   41   37   34   17   34
    91   72  188   132  127  128  104  104  102   55   45   39   48   45   46   57   44       F
   102  130  184   134  117  109  104  113  113   65   52   40   37   35   16   46
   125  105  181   147  120  109  105   95  100   93   96  124  162   57   55    8
        123   94   177  124  114  126   96   96  116  122   86   34   92   79   17
         43   91    59  116  111  105  100  108  109   88  208  264   70   31
              29    61   60   72   82   90  133  146  216  149  147   33   29
                    30   45   45   54   67   66   54   39   18
                              38   36   12
```

a.  Section through the brain (height above ground = 167 cm).

```
                         15   48   50   34
                    26  149  281  192  121   82   13
               27  226  365  380  325  354  351   99   12
          29   81  356  370  379   45  108  425  381  181
          27   85  359  371  381  217  211  424  428  243
          20   78  351  360  374  344  360  388  394  178
               24  194  247  248  162  227  230  175   24
                    13   26   32   33   34   37   39

          17   32  148  161  175  178  173  162  152   31   30
     21   47  273  429  452  403  462  288  307  433  201  274  179  102   17
     31  148  442  396  266  177  110   51   53  109   46   53  300  411  190   48
     26  301  336  372   49  191   56   64   97   61   58   45   53  143  221  149   54
     24  290  112  107   60   57   58  105  107   99   96   93   59   53   78  238  151   16
     23  295   97   49   55   55   59  114  104  100   93  108  108   55   91  179  268   61
     23  308  223  121   49   53   54   63  110  104   90   93  108   99  128  118  271  120
     23  320  202  243   50   56   51   48   48   77  121   94   96  103  113  108  169  224   10
     23  333  104  198   40   48   61  174   48   61  220   97   96   96  103  104  265  190   37
     23  326  368   40   42   50   55  198   64   74  151  103   89  120  159  199  122  177   34   F
     23  310  407   41   47   42   75  192   83  109  176  106  133  316  346  238   57   90
     22  286  245   71   71   39   73  365   83  139  319   72  299  360  356  190   63   98
     23  319  383   55   62   41   45   46   78  107  398  252  394  332  301   73   78  105
     23  325  430   42   46   43   49   58   67  126  493  378  405  248   54   80  148  146
     23  328  278   41   42  145  135   79   65   65  200  184  251  237   78  135  296  219
     23  315  216   46   48  184  150   73  138   49   94  297   91   91  168  128  269  194
     23  278  190   41   46   51   56   66  120   56  133  191  173   91   95  154  359   96
     24  270   80   42   51   55   55   57   50   47   48   40   77   92  179  211  367   46
     25  270   93   46   50   51  117  141   87   44   57   59  102  108  125  298  207   11
     28  281  268  185   51   51  111   62  177   42   97  109  116  138  166  128   43
     42  109  332  478  192   52   54   53   50   90  124  126  238  362  169  100
     34   77  381  444  416  385  375  157  169  380  340  160  353  164   40   12
          23  116  265  356  364  364  347  339  247  154  136   22   10
          14   14   22   32   33   34   34   34   35   34

                    2   65  100  100   76   91   95   54    8
               55  315  352  360  251  344  336  312   78
          17   78  348  361  372  339  293  411  417  233
          27   81  339  357  372   94  120  413  398  206
          48   66  297  342  367  140  220  374  216   66
               32   88  259  314  284  246  172   28
                    35   28  102   65   27   17
```

b.  Section through the heart (height above ground = 133 cm).

```
              215   616   982   807
         208  2610  4074  3317  5285  1652
         589  2629  6797  1216  1696  2879   316
         605   646  1651  1274   633  1693   153
              648   843   884   900
```

F

```
              678   884   906   912
         624   683  1702  1311   656  1661   150
         595  2610  6787  1213  1698  2869   315
         215  2606  4079  3308  5288  1658
              215   614   977   811
```

c.  Section through the ankles (height above ground = 15.1 cm).

**Fig. 11.** Calculated total current densities in nA/cm$^2$ for the various cells ($\delta = 1.31$ cm) of a grounded anatomically based human model. Figures a-c are for combined E and H fields, E = 10 kV/m (vertical), H = 26.5 A/m (from side to side of the body), while Figs. d-f are for the magnetic fields alone.

```
                                        2    2    1
                          1    2    4    4    4    4    3    1
              1    1    1  1    2    3    4    9   13   17   12   14    3    1
         1    1    1    5  3    4    5    6    7    7    5   21   25    7    3
         2    1    5    4  4    4    5    5    6    8    9    7    3   10   11    1
    2    1    3    4    4  4    5    5    7    6    7   10   14    6    8    1
1    2    4    3    4    4  5    5    6    7    4    4    3    3    4    2    7
1    1    5    3    4    5  5    5    6    3    3    3    5    5    7    9    5       F
2    2    5    3    4    4  5    6    7    4    4    3    3    4    2    7
3    2    4    4    4    4  5    5    6    6    7   10   15    6    8    1
    3    2    6    4    4  6    5    6    8    9    7    3   10   10    1
    2    2    2    4    4  5    5    7    8    6   18   24    7    3
         1    1    2    3  3    4    8   11   16   12   14    3    1
              1    1         2    3    4    4    4    3    1
                             2    1    1
```

d.  Section through the brain (height above ground = 167 cm).

```
                   5    6    1              1
    1   11   13    8    2         1    7    3
    4   18   13    7              7   13    9
    5   18   13    7    1    1    7   15   12
    4   17   12    7    2    2    7   14    9
    1   10    9    5    1    1    4    6    1


         1    5    3    2    2    4    5    6    1    1
    3   14   16   10    4    6    6    9   16   10   16   12    8    1
1  10   23   15    6    2    1    1    1    4    2    3   20   33   17    4
   20   17   13    1    2         1    2    2    2    2    3   11   20   15    6
   19    5    3    1              2    3    3    4    5    4    4    7   24   17    1
   18    4    1    1    1    1    2    2    3    4    6    7    4    8   18   30    7
   19   11    5    1    1    1    1    2    3    4    5    7    8   11   11   31   15
   20   10   10    1    1              1    2    5    5    6    8   10   11   19   28       1
   21    5    8    1    1    1    2    1    2    9    5    6    8   10   11   31   24       4
   21   19    1    1    1    3    1    2    6    5    6    6   10   16   23   14   23   4   F
   20   21    1    1    1    3    1    3    7    5    9   27   36   28    7   12
   19   13    3    2    1    1    5    2    4   12    3   20   30   37   23    7   13
   21   20    2    2    1              1    3   15   12   26   28   31    9    9   14
   21   22    2    1    1         1    1    3   19   19   26   21    6    8   18   20
   21   14    1    1    3    2    1    1    2    7    9   17   20    8   14   35   29
   20   11    1    2    4    2    1    3    1    4   15   10    6    8   16   13   32   26
   18    9    1    1    1    0    1    3    2    5   10   11    8    9   15   42   12
   18    3    1    1    1    1    1    1    1    1    2    5    7   16    9   21   42    5
   19    5    2    1    1    2    3    2    1    2    3    6    8   11   30   23    1
1  19   14    8    1    1    2    1    5    1    4    6    7   10   14   13    4
1   7   18   19    4         1    1    1    3    6    7   15   27   15   10
    5   20   17   10    5    5    3    4   13   15    9   22   12    3    1
    1    6   10    9    5    4    7   10    9    7    7    1


              4    4    2    1    1    2    2
    3   17   13    8    2    2    7   12    4
    5   19   13    8    3    1    7   15   12
    5   18   13    8              7   14   11
1   3   16   13    8    1    1    6    8    3
    4   10    8    3    2    3    1
    1    2
```

e.  Section through the heart (height above ground = 133 cm).

```
              3   11   18   15
    3   46   75   63  104   33
   10   46  125   23   33   58    6
   10   11   30   24   12   34    3
   11   15   16   17
```

F

```
             12   16   17   18
   10   12   31   25   13   34    3
   10   46  126   23   33   59    6
    3   46   75   63  105   34
    3   11   18   16
```

f.  Section through the ankles (height above ground = 15.1 cm).

**Fig. 11.** (continued)

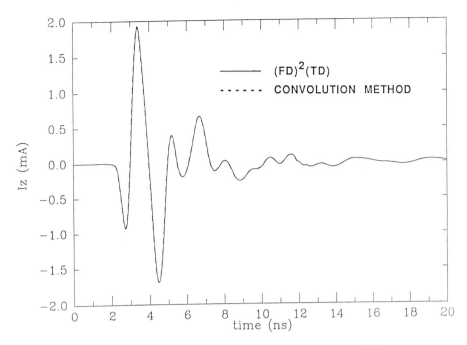

a. Section through the liver (height above the bottom of the feet = 123.8 cm).

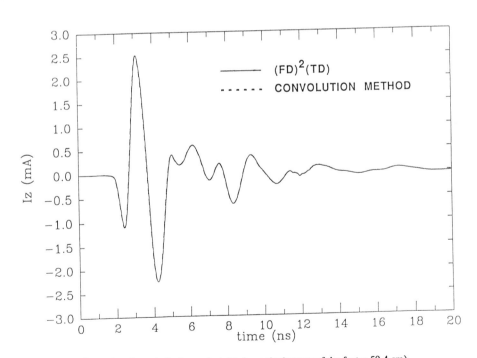

b. section through the knees (height above the bottom of the feet = 50.4 cm)

**Fig. 12.** Comparison of the induced currents calculated using the convolution method with the conventional (FD)2 TD method for two representative sections of the body. Shoe-wearing grounded model of the body was used for the calculations.

## CONCLUDING REMARKS

From the foregoing it can be seen that the FDTD method is very versatile and has been used for a number of important and meaningful problems in the field of bioelectromagnetics. Some of the future applications involve improving the efficiency of the code by techniques such as use of the expanding grid rather than the regular grid, elimination of the interior of the body at higher frequencies, since this region is relatively shielded from incident fields, and use of truncated models of the body at microwave frequencies where there is a lack of coupling between the various regions of the body. We also plan to develop the new convolution method for use with prescribed variations of the incident fields in time and space domains. In this method, high-resolution impulse response of the model, in time and space domains, would be stored and used for subsequent calculations. Since convolution integrals are relatively easy to calculate and do not need a large computer memory, high-resolution dosimetric calculations should be possible using smaller computers or PCs. We also plan to focus on computer graphical displays for visualization of the calculated data.

## REFERENCES

1.  D.M. Sullivan, D.T. Borup and O.P. Gandhi, Use of the finite-difference time-domain method in calculating EM absorption in human tissues, *IEEE Trans. Biomed. Engg.* BME-34:148-150 (1987).
2.  D.M. Sullivan, O.P. Gandhi and A. Taflove, Use of the finite-Difference time-domain method in calculating EM absorption in man models, *IEEE Trans. Biomed. Engg.* BME-35:179-186 (1988).
3.  J.Y. Chen and O.P. Gandhi, RF currents induced in an anatomically based model of a human for plane-wave exposures 20-100 MHz, *Health Physics*, 57:89-98 (1989).
4.  J.Y. Chen and O.P. Gandhi, Electromagnetic deposition in an anatomically based model of man for leakage fields of a parallel-plate dielectric heater," *IEEE Transactions on Microwave Theory and Techniques*, MTT-37:174-180 (1989).
5.  C.Q. Wang and O.P. Gandhi, Numerical simulation of annular-phased arrays for anatomically based models using the FDTD method, *IEEE Transactions on Microwave Theory and Techniques*, MTT-37:118-126 (1989).
6.  J.Y. Chen and O.P. Gandhi, Currents induced in an anatomically based model of a human for exposure to vertically polarized EMP, *IEEE Transactions on Microwave Theory and Techniques*, MTT-39:31-39 (1991).
7.  O.P. Gandhi and J.Y. Chen, Numerical dosimetry at power-line frequencies using anatomically based models, *Bioelectromagnetics*, Suppl 1, 43-60 (1992).
8.  O.P. Gandhi, Y.G. Gu, J.Y. Chen and H.I. Bassen, Specific absorption rates and induced current distributions in an anatomically based human model for plane-wave exposures (at frequencies 20-915 MHz), *Health Physics*, 63 (3):281-290 (1992).
9.  S.S. Stuchly, A. Kraszewski, M.A. Stuchly, G.W. Hartstgrove and R.J. Spiegel, RF energy deposition in a heterogeneous model of man: far-field exposures, *IEEE Trans. Biomed. Engg*, BME-34:951-957 (1987).
10. O.P. Gandhi, B.Q. Gao and J.Y. Chen, A frequency-dependent finite-difference time-domain formulation for induced current calculations in human beings, *Bioelectromagnetics*, 13 (6):543-555 (1992).
11. C.M. Furse, J.Y. Chen and O.P. Gandhi, A frequency-dependent finite-difference time-domain method for induced current and SAR calculations for a heterogeneous model of the human body, accepted for publication in *IEEE Transactions on Electromagnetic Compatibility*.
12. O.P. Gandhi and J.Y. Chen, Absorption and distribution patterns of RF magnetic fields used or planned for MRI with frequencies up to 200 MHz, paper presented at the Meeting of the Bioelectromagnetics Society, Los Angeles, California, June 13-17 (1993).
13. A.W. Guy, S. Davidow, G.Y. Yang and C.K. Chou, Determination of electric current distributions in animals and humans exposed to a uniform 60-Hz high-intensity electric field, *Bioelectromagnetics*, 3:47-71 (1982).

14. O.P. Gandhi, J.Y. Chen, C.M. Furse and Y. Cui, A simple convolution procedure for calculating coupling to the human body for EM fields of prescribed time and space variations, paper presented at the Meeting of the Bioelectromagnetics Society, Los Angeles, California, June 13-17 (1993).

15. A. Taflove and K.R. Umashankar, "The Finite-Difference Time-Domain Method for Numerical Modeling of Electromagnetic Wave Interactions with Arbitrary Structures," Chapter 8 *in:* "Progress in Electromagnetics Research PIER 2," Michael A. Morgan, Editor, Elsevier Science Publishing Company, New York, New York (1990).

16. O.P. Gandhi, "Numerical Methods for Specific Absorption Rate Calculations," Chapter 6 *in:* "Biological Effects and Medical Applications of Electromagnetic Energy," Om P. Gandhi, Editor, Prentice Hall, Englewood Cliffs, New Jersey (1990).

17. R.J. Luebbers, F.P. Hunsberger, K.S. Kunz, R.B. Standler and M. Schneider, A frequency-dependent finite-difference time-domain formulation for dispersive materials, *IEEE Transactions on Electromagnetic Compatibility*, EMC-32:222-227 (1990).

18. M.D. Bui, S.S. Stuchly and G.I. Costache, Propagation of transients in dispersive dielectric media, *IEEE Transactions on Microwave Theory and Techniques*, MTT-39:1165-1172 (1991).

19. D.M. Sullivan, A frequency-dependent FDTD method for biological applications, *IEEE Transactions on Microwave Theory and Techniques*, MTT-40:532-539 (1992).

20. R. Luebbers, F. Hunsberger and K. Kunz, FDTD for nth order dispersive media, submitted to *IEEE Transactions on Antennas and Propagation*, (1992).

21. R.J. Luebbers, F. Hunsberger and K.S. Kunz, A frequency-dependent finite-difference time-domain Formulation for Transient Propagation in Plasma, *IEEE Transactions on Antennas and Propagation*, Vol. AP-39:29-34 (1991).

22. R.M. Joseph, S.C. Hagness and A. Taflove, Direct time integration of Maxwell's equations in linear dispersive media with absorption for scattering and propagation of femtosecond electromagnetic pulses, *Optics Letters*, Vol. 16 (18):1412-1414 (1991).

23. O.P. Gandhi, B.Q. Gao and J.Y. Chen, A frequency-dependent finite-difference time-domain formulation for general dispersive media, *IEEE Transactions on Microwave Theory and Techniques*, Vol. 41, April (1993).

24. A.C. Eycleshymer and D.M. Schoemaker, *A Cross-Section Anatomy*, Appleton-Century-Crofts, New York (1970).

25. C.H. Durney et al., *Radiofrequency Radiation Dosimetry Handbook*, Second Edition, June 1978.

26. R.L. Magin, R.P. Liburdy and B. Persson, Editors, Biological Effects and Safety Aspects of Nuclear Magnetic Resonance Imaging and Spectroscopy, *Annals of the New York Academy of Sciences*, Vol. 649 (1992).

27. U. S. Food and Drug Administration: Magnetic Resonance Diagnostic Device: Panel Recommendation and Report on Petitions for MR Reclassification. Fed. Reg. 53:7575, (1988).

28. National Radiological Protection Board, U. K., Revised Advice on Acceptable Restrictions of Patient and Volunteer Exposure During Clinical Magnetic Resonance Diagnostic Procedures (1990).

29. N. Orcutt and O.P. Gandhi, A 3-D impedance method to calculate power deposition in biological bodies subjected to time-varying magnetic fields, *IEEE Transactions on Biomedical Engineering*, 35:577-583 (1988).

30. C.C. Johnson and A.W. Guy, Nonionizing electromagnetic wave effects in biological materials and systems, *Proceedings of the IEEE*, 60:692-718 (1972).

31. P.A. Bottomley and E.R. Andrew, RF magnetic field penetration, phase shift, and power dissipation in biological tissue: implications for NMR imaging, *Physics in Medicine and Biology*, 23:630-643 (1978).

32. W.T. Kaune and M.F. Gillis, General properties of the interaction between animals and ELF electric fields, *Bioelectromagnetics*, 2:1-11 (1981).

33. J.R. Mautz, Mie series solution for a sphere, *IEEE Transactions on Microwave Theory and Techniques*, MTT-26:375 (1978).

34. D.W. Deno, Currents induced in the human body by high voltage transmission line electric field -- measurement and calculation of distribution and dose, *IEEE Transactions on Power Apparatus and Systems*, PAS-96:1517-1527 (1977).

35. J. DiPlacido, C.H. Shih and B.J. Ware, Analysis of the proximity effects in electric field measurements," *IEEE Transactions on Power Apparatus and Systems*, PAS-97:2167-2177 (1978).

36. J.Y. Chen, C.M. Furse and O.P. Gandhi, A simple convolution procedure for calculating currents induced in the human body for exposure to electromagnetic pulses, accepted for publication in *IEEE Transactions on Microwave Theory and Techniques*.

# MICROWAVE EXPOSURE LIMITS FOR THE EYE: APPLYING INFRARED LASER THRESHOLD DATA

David H. Sliney[†] and Bruce E. Stuck[‡*]

[†]Laser Microwave Division
US Army Environmental Hygiene Agency,
Aberdeen Proving Ground, MD 21010-5422 USA

[‡]US Army Medical Research Detachment,
Brooks AFB, TX 78235-5138 USA

## INTRODUCTION

A large body of knowledge exists on the adverse effects of intense infrared radiant energy upon the eye. Infrared heat cataracts were common in industry at the turn of the century, but are now rare. Epidemiological and animal studies permitted the establishment of occupational exposure limits to infrared radiation. More recently, infrared laser bioeffects studies have permitted an understanding of both single-pulse and repetitive-pulse infrared laser injury to the cornea. From a knowledge of penetration depth at different laser wavelengths, one should be able to predict the adverse effects of higher frequency pulsed microwaves upon the eye. These predictions are compared with published data of microwave effects upon the cornea.

## BACKGROUND

Since the early 1950s it has been recognized that because of poor heat dissipation in the lens of the eye, ocular thermal injury from microwave radiofrequency (RF) exposure may be of particular interest. Schwann, when queried by the Navy in 1953 on this subject, singled out the eye,[1,2] and there were early reports of cataracts produced by diathermy of the head. However, when biological laboratory studies of animals were performed, the eye was shown to be remarkably resilient to thermal injury.[3-6] In order to achieve a microwave cataract in rabbits, the animal must be placed in a cooling blanket to avoid death from whole-body heating. Threshold for acute cataractogenesis in the rabbit is of the order of 150 mW/cm$^2$ and in the primate, still higher.[3-6] It is not surprising, then, that in cases

---

* These views and opinions are those of the authors and do not necessarily state or reflect those of the U.S. Government.

where cataracts in humans have been generally accepted as related to microwave exposure, local irradiation of the eye took place. Early radar workers who would peer down an open waveguide to make adjustments, certainly could receive very heavy ocular exposures, and may not have experienced a very strong aversion from facial heating to limit their exposures.[7] Likewise, because of the small beam diameter of lasers, eye injuries from infrared lasers have taken place--although normally from single-pulse sources, where the exposure was not determined by human behavior, but by the duration of the pulse event.[6,19] Because of the similarities of infrared penetration depths with those from higher microwave frequencies, there should be value in drawing comparisons between the two spectral regions. This may be particularly valuable with regard to short-pulse exposures, where there is a considerable laser biological data base upon which to draw.

## CIE Spectral Bands

The International Commission on Illumination (CIE) has designated seven spectral bands useful in the discussion of biological effects of optical radiation: UV-A, UV-B, UV-C, Visible (light), IR-A, IR-B, and IR-C. The infrared bands are based upon absorption properties of water: At IR-A (760-780 to 1400 nm) water has significant transmission; at UV-B (1400 to 3000 nm, i.e., 1.4 to 3.0 μm) water has a penetration depth of the order of only 1 mm; and at IR-C (3.0 μm to 1 mm, i.e., 100 THz down to 300 GHz) only very superficial surface heating can take place in water and biological tissue.

## DOSIMETRIC CONCEPTS FOR INFRARED RADIANT ENERGY

### Quantities and Units

Two sets of optical-measurement quantities and units are useful in defining light exposure of the eye and skin: radiometric and photometric. Radiometric quantities, such as radiance-used to describe the "brightness" of a source [in $W/cm^2/sr$], and irradiance--used to describe the irradiance level on a surface [in $W/cm^2$], are particularly useful for hazard analysis. Radiance and luminance are particularly valuable because these quantities describe the source and do not vary with distance. Photometric quantities, such as luminance (brightness in $cd/cm^2$ as perceived by a human "standard observer") and illuminance in lux (the "light" falling on a surface) indicate light levels spectrally weighted by the standard photometric visibility curve, that peaks at 550 nm for the human eye, and are therefore not useful in discussions of infrared hazards. When photobiologists quantify a photochemical effect, they recognize that it is not sufficient to specify the number of photons-per-square-centimeter (photon flux density) or the irradiance ($W/cm^2$) since the efficiency of the effect will be highly dependent on wavelength. Generally, shorter-wavelength, higher-energy photons are more efficient, and it has been generally accepted that photochemical effects are almost non-existent in the infrared. Hence we shall consider thermal effects and must describe both the irradiance and duration of exposure. Reciprocity does not exist because of heat flow, and the time-temperature dependence of thermal injury requires us to consider the area of exposure. Radiant exposure in $J/cm^2$ is the product of the irradiance in $W/cm^2$ and the exposure duration in s. Radiant exposure is the "dose" in photobiology, and it is used with laser bioeffects studies of short exposure durations, where heat-flow does not greatly minimize the temperature rise (Figure 1).

### The Importance of Geometry in Retinal Exposure

Because of the imaging properties of the human eye, one must employ the concept of radiance to determine retinal exposure dose in the wavelength region of 400 nm to 1400 nm (the visible and near-infrared bands). From knowledge of the optical parameters of the human eye and from radiometric parameters of a light source, it is possible to calculate irradiances (dose rates) at the retina. Exposure of the anterior structures of the human eye to near infrared radiant energy is also of interest; and the relative position of the light source and the degree of lid closure can greatly affect the proper calculation of this ultraviolet exposure dose.

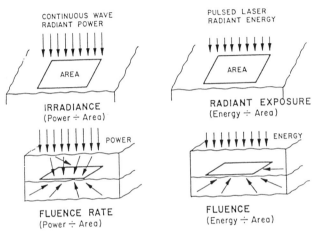

**Figure 1.** Radiometric concepts used in laser bioeffects studies.[8]

## OPTICAL RADIATION AND THE EYE

The eye is well adapted to protect itself against optical radiation (ultraviolet, visible and infrared radiant energy) from the natural environment, and mankind has learned to use protective measures, such as clothing to shield against the harmful effects upon the skin from the ultraviolet radiation (UVR) present in sunlight. The eye is protected against bright light by the natural aversion response to viewing bright light sources. The aversion response normally protects the eye against injury from viewing bright light sources such as the sun, arc lamps and welding arcs, since this aversion limits the duration of exposure to a fraction of a second (about 0.25 s). When a source of radiant energy has no significant visible component, this rapid natural aversion response is not present, e.g., exposure from a microwave or purely infrared source. However, the cornea is also sensitive to heating, and an aversion response also exists to limit lengthy exposures lasting several seconds or longer; but, this latter response has not been carefully studied.

Three separate types of hazards to the eye and skin from infrared optical sources are evaluated separately when one is dealing with lasers:

(1) Near-infrared thermal hazards to the lens (approximately 800 nm to 3000 nm: i.e., 0.8 to 3.0 μm).

(2) Near-infrared thermal injury to the retina (approximately 770 nm to 1400 nm; i.e., 0.77 μm to 1.4 μm).

(3) Thermal injury (burns) of the skin (approximately 770 nm to 1 mm; i.e., 0.77 to 1,000 μm) and of the cornea of the eye (approximately 1400 nm to 1 mm; i.e., 1.4 μm to 1,000 μm).

## Thermal Injury

Studies of potential thermal hazards to the eye typically include a consideration of the contributions of IR-A (700-1400 nm) and IR-B (1.4 μm - 3.0 μm). In contrast to blue light, IR-A is very ineffective in producing retinal injuries.[20,21] While skin pigmentation and ambient tissue temperature will both affect the threshold for thermal injury to the skin, these factors play a relatively minor role. Thermal injury will occur when tissue such as the skin or cornea experiences a temperature elevation T of about 10-20°C (to at least 45-47°C). Greater temperature elevations are required to produce coagulation (a thermal burn) when the duration of exposure is shortest. Melanin granules in the retina actually have been shown to reach incandescence for pulsed laser injury thresholds in the nanosecond regime, although surrounding tissue is still damaged at a ΔT of about 20°C. With this temporal dependence, thermal injury is referred to as a rate process; the time-temperature history of the exposed tissue determines whether microscopically visible injury occurs.[6,12] Acute thermal injury in the human's natural environment and work environment is rarely the result of exposure to optical radiation. Thermal injury of the skin can result from contact with a hot object or exposure to a flame. One's natural aversion response to high heat (i.e., to ΔT > 10°C) normally occurs sufficiently fast that one is not burned when approaching an intense lamp or other optical source. Only with pulsed lasers is this exposure situation likely to occur.

## Retinal Injury

Retinal injury has been studied in great detail and mathematical models of heat flow that incorporate the rate-process of thermal injury are successful in predicting injury from pulsed and CW laser radiation. In the visible spectrum, however, thermal injury must be distinguished from photochemical injury mechanisms. The principal retinal hazard resulting from viewing bright light sources is *photoretinitis*; e.g., solar retinitis with an accompanying scotoma that results from staring at the sun. Solar retinitis was once referred to as "eclipse blindness" and associated "retinal burn." Only in recent years has it become clear that photoretinitis results from a photochemical injury mechanism following exposure of the retina to shorter wavelengths in the visible spectrum, i.e., violet and blue light.[20,21] Although solar retinitis does not result from a thermal injury mechanism, intense artificial, near-infrared source, such as a filtered arc-lamp with the visible component removed, or a near infrared laser can produce a sufficient retinal irradiance to produce a purely thermal retinal injury. In most of the visible and near infrared the absorbing chromophore responsible for the highest temperature rise is melanin, and it is this that produces retinal thermal burns at temperature rises of 10 to 20°C for exposure times ranging from 1 ms to 10 s, respectively. In the spectral range from 1.2 to 1.4 μm only about 1 to 5 % of the infrared energy reaches the retina and melanin no longer contributes to a significant temperature rise. Thus, by the appropriate choice of wavelength and tissue, it is possible to provide examples of ocular thermal injury for penetration depths ranging from about 1 μm to several mm, and for exposure durations ranging from a few nanoseconds to several minutes.

## Thermal vs. Photochemical Injury Mechanisms

Optical radiation effects generally result from either thermal or photochemical injury mechanisms. Although photochemical injury is not known to occur in the microwave region, it is still relevant to review key distinguishing factors that relate to each of these mechanisms. For photochemical injury, the product of the dose-rate and the exposure duration always must result in the same exposure dose (a radiant exposure dose in $J/cm^2$) to produce a threshold injury. For example, a photochemical, "blue-light" retinal injury (photoretinitis) can result from viewing either an extremely bright light for a short time, or a less bright light for longer exposure periods. This characteristic of photochemical injury mechanisms is termed reciprocity. It helps to distinguish these effects from thermal burns, where heat conduction requires a very intense exposure within seconds to cause a retinal coagulation; otherwise, surrounding tissue conducts the heat away from the retinal image. Furthermore, the size of a photochemical lesion generally corresponds exactly to the exposure area; whereas, because of heat conduction away from the site of exposure, a thermal lesion can be either smaller or larger than the exposure area.

Occupational safety limits for exposure to bright light and infrared radiation are based upon both our knowledge of injury thresholds *and mechanism*, and have been corroborated for the skin and eye from human accident data. For any photochemical injury mechanism, one must consider the *action spectrum*, which describes the relative effectiveness of different wavelengths in causing a photobiological effect. For example, the action spectrum for photochemical retinal injury peaks at approximately 440 nm and is dominant over a spectral band of only about 100 nm. Although thermal injury is not nearly as dependent upon wavelength as photochemical injury mechanisms in the ultraviolet and short-wavelength visible bands, the spectral distribution of the infrared source can also be important to determine how much energy reaches the deeper cornea, the lens and finally to the retina (if at all).

## HUMAN EXPOSURE LIMITS

Internationally, a general consensus exists on limits of exposure to both ultraviolet and laser radiation. The International Commission on Non-ionizing Radiation Protection (ICNIRP) (formerly the International Non-ionizing Radiation Committee [INIRC] of the International Radiation Protection Association [IRPA]), published Guidelines on Limits of Exposure to Laser Radiation in 1985[15] and revised them in 1988. ICNIRP guidelines are developed through collaboration with the World Health Organization (WHO) by jointly publishing criteria documents that provide the scientific data base for the exposure limits.[16] The International Electrotechnical Commission (IEC) MPEs are essentially the same as those of ICNIRP.

Fewer organizations have published exposure limits (ELs) for non-coherent optical radiation. The American Conference of Governmental Hygienists (ACGIH). The ACGIH refers to its ELs as "Threshold Limit Values," or TLVs.[13-14] The current ACGIH TLV's for infrared exposure are based in large part on ocular injury data from animal studies and from data from human retinal injuries resulting from viewing the sun, lasers and welding arcs. The TLVs also have an underlying assumption that outdoor environmental exposures to visible radiant energy are normally not hazardous to the eye except in very unusual environments such as snow fields and deserts.

The data that have been used as the basis of an exposure limit for chronic exposure of the anterior of the eye to infrared radiation are very limited.[1,6,16,18,19] Sliney and Freasier[10] stated that the average corneal exposure from infrared radiation in sunlight was of the order of 1 mW/cm². Glass and steel workers exposed to infrared irradiances of the order of 80-400 mW/cm$^2$ daily for 10-15 years have reportedly developed lenticular opacities.[10]

The ACGIH guideline for IR-A exposure of the anterior of the eye is designed to avoid thermal injury of the cornea and possible delayed effects upon the lens of the eye (cataractogenesis). The infrared radiation (770 nm < λ< 3 μm) is limited for lengthy (t > 1,000 s, i.e., about 16.7 min.) to 10 mW/cm$^2$, and to 1.8 $t^{-3/4}$ W/cm$^2$ for shorter exposure durations.

Pitts and Cullen[11] showed that the threshold radiant exposures causing lenticular changes from IR-A were of the order of 5000 J/cm$^2$. Threshold damage irradiances were at least 4 W/cm$^2$. There is also a second ACGIH criterion to protect the retina against thermal injury from viewing specialized infrared illuminators that have visible light filtered out so that the aversion response stimulus is not present.[13] To protect the cornea from pulsed infrared laser sources for durations less than 10 s the ACGIH and ANSI and IEC have always limited far-infrared radiant exposures (μ > 2.6 μm, where all of the energy is absorbed in the most superficial layer of the cornea) to:

$$H_{EL} = 0.56 \, t^{1/4} \text{ J/cm}^2 \tag{1}$$

## NEW INFRARED EL'S FOR LASERS

The new Maximum Permissible Exposure (MPE) limits in the American National Standard, ANSI Z136.1-1993 laser safety standard,[17] the new TLVs of the ACGIH, the Exposure Limits (ELs) of the IEC and ELs of ICNIRP are all changing in the infrared spectrum.[13-17] The limits are being readjusted in the spectral region from 1.1 μm to 1.4 μm for all exposure durations. But most importantly, in the spectral region between 1.4 μm and 2.8 μm, the limits are increased for pulsed lasers (Figure 2). Further biological research

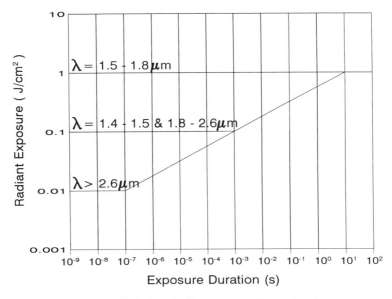

**Figure 2.** Revised laser emission limits for pulsed laser exposure for wavelengths greater than 1.4μm.[17]

made these changes possible, and the increasing use of diode lasers at 1.3 μm and 1.55 μm made more realistic MPEs urgently needed. The studies and analysis of data that were made during this revision effort makes possible an improved picture of thermal injury to the anterior segment of the eye. The penetration depth as shown in Figure 3 illustrates the reason that for pulsed exposure, at wavelengths where great penetration takes place, the thresholds for single-pulse thermal injury increases.[9] This data base should provide as well a better insight into potential thermal injury of the eye from microwave radiation.

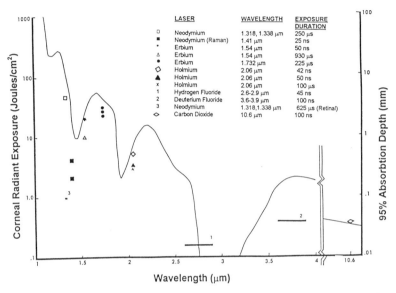

**Figure 3.** Thresholds for laser injury to the cornea for infrared laser wavelengths as a function of wavelength from 1-10 μm. The solid curve is the depth (ordinate on the right side) at which 95% of the incident energy is absorbed in physiological saline (the absorption properties of saline approximate those of the outer ocular media.[9]

## RECENT STUDIES OF PULSED MICROWAVE EXPOSURES OF THE EYE

Recently, Kues, et al.,[22] have reported findings of both corneal endothelial abnormalities and retinal changes in primates exposed to 2.54 GHz radiation, characterized by very high peak power, repetitive pulsed exposures for 4-hour periods on a daily or weekly basis extending over a week or more.[22] The average power density was only 15 mW/cm$^2$ and the SAR was 3.9 W/kg, but since the pulse width was 10 μs and the pulse repetition frequency (PRF) was 100 Hz, the duty cycle was only 0.001, therefore the peak power density was of the order of 15 W/cm$^2$. When one calculates these as radiant exposures, the threshold per pulse is: 0.15 mJ/cm$^2$ per pulse. The laser exposure limit for a penetration depth corresponding to 1.54 μm is 1.0 J/cm$^2$ for all exposures within 1 s; hence, at 100 Hz, the per-pulse laser EL would be 10 mJ/cm$^2$ per pulse, or 100 times the Kues value. However, for a very superficial penetration depth corresponding to a $CO_2$ laser (10.6 μm), the EL is:

$$H_{EL} = 0.56 \, C_p \, t^{1/4} \text{ J/cm2} \tag{2}$$

85

where $C_p$ is a pulse correction factor equal to the number of pulses N in an exposure raised to the minus 0.25 power:

$$C_p = N^{-1/4} \qquad (3)$$

This important pulse-additivity rule is based upon a great deal of empirical data, although there have been efforts to derive some rate constants for thermal models that fit this rule.[23-24] Therefore, a 100 s exposure with $10^4$ pulses would lead to a $C_p = 0.1$. The other expression for a pulse width of 10 μs would lead to 31.5 mJ/cm$^2$ for a single pulse, or 3.15 mJ/cm$^2$ for $10^4$ pulses (over 10 s). This simply illustrates how the laser ELs are applied. Now, to compare this with a 4-hour (1.5 X $10^4$ s) exposure used in of Kues' experiments, N would have a value of 1.5 X $10^6$ pulses, and the laser EL would be reduced to: (31.5 mJ/cm$^2$)(0.0286) = 0.9 mJ/cm$^2$, a value that is still a factor of 6 above the Kues data point. This is puzzling, since there can be no thermal explanation for this effect if there are not artifacts in the experimental procedure. The peak power is insufficient for breakdown of membranes, or even for stimulation of nerve and muscle cells based upon the data of Foster and Schwann.[25] One is inclined to look at the geometry of exposure and possible artifacts.

## CONCLUSIONS

Based upon current knowledge of microwave biological effects, the current laser ELs provide a basis for recommending their use in the microwave frequency domain for pulsed exposure, depending upon the penetration depth of the electromagnetic fields. However, it is also shown that there appear to be some biological data for high peak power, microsecond pulsed exposure for very long exposure that cannot be explained by current thermal injury theory. Without a dose response curve, it is impossible at this stage to provide any suggested revisions of guidelines based upon this one set of data.

## REFERENCES

1. D.H. Sliney. Non-ionizing radiation, in (L.V. Cralley, Ed.), Industrial Environmental Health, the Worker and the Community, New York, Academic Press, pp. 171-241 (1972).
2. W.B. Deichmann and F.H. Stephens. *Industr. Med. Surg.*, 30:221 (1961).
3. R.L. Carpenter, Ocular effects of microwave radiation, *Bull. N.Y. Acad. Med*, 55:1048-1057 (1979)
4. E. Lydahl, L.E.Paulsson and B. Philipson, Effects of microwaves, separately and in combination with galactose, on the eye, *in:* Current Concepts in Ergophthalmology, Societas Ergophthalmologica Internationalis, B.T. Tengroth and D. Epstein, eds.,Stockholm (1978).
5. P. Kramar, C. Harris, A.F. Emery, and A.W. Guy, Acute microwave irradiation and cataract formation in rabbits and monkeys, *J. Microwave Power*, 13:239-249 (1978).
6. B. Appleton, S.E. Hirsch and P.V.K. Brown, Investigation of single-exposure microwave ocular effects at 3000 MHz, *Ann. NY Acad. Sci.,* 247:125-134 (1978).
7. E. Aurell, B. Tengroth, Lenticular and retinal changes secondary to microwave exposure, *Acta Ophthal* 51:764-771, (1973).
8. D.H. Sliney and M.L. Wolbarsht. Safety with Lasers and Other Optical Sources. New York: Plenum Publishing Corp (1980).
9. B. Stuck, D.J. Lund and E.S. Beatrice, "Ocular Effects of Laser Radiation from 1.06 to 2.06 μm," SPIE Vol 229, Ocular Effects of Non-Ionizing Radiation, pp 115-120, (1980).
10. D.H. Sliney and B.C. Freasier, The evaluation of optical radiation hazards, *Applied Opt*, 12(1):1-24 (1973) .
11. D.G. Pitts and A.P. Cullen, Determination of infrared radiation levels for acute ocular cataractogenesis, *Albrecht von Graefes Arch Klin Ophthalmol*, 217:285-297 (1981).

12. F. Hillenkamp. Laser Interactions with Biological Tissue, *in:* Lasers in Biology and Medicine, F. Hillenkamp, C. A. Sacchi, and T. Arrechi, eds., Plenum Press, New York (1980).

13. American Conference of Governmental Industrial Hygienists (ACGIH) (1993), TLV's, Threshold Limit Values and Biological Exposure Indices for 1993-1994, American Conference of Governmental Industrial Hygienists, Cincinnati, OH.

14. ACGIH Documentation for the Threshold Limit Values, 4th Edn., American Conference of Governmental Industrial Hygienists, Cincinnati, OH (1991).

15. IRPA, International Non-Ionizing Radiation Committee, Guidelines for Limits of Human Exposure to Non-Ionizing Radiation, New York, MacMillan (1991).

16. World Health Organization [WHO], Environmental Health Criteria No. 23, Lasers and Optical Radiation, joint publication of the United Nations Environmental Program, the International Radiation Protection Association and the World Health Organization, Geneva (1982).

17. ANSI Safe Use of Lasers, Standard Z-136.1-1993, American National Standards Institute, Laser Institute of America, Orlando, FL (1993).

18. D.H. Sliney, Physical factors in cataractogenesis--ambient ultraviolet radiation and temperature, *Invest Ophthalmol Vis Sci,* (1987).

19. S. Lerman. Radiant Energy and the Eye, MacMillan & Co., New York (1980) .

20. W.T. Ham, Jr. (1983). The photopathology and nature of the blue-light and near-UV retinal lesion produced by lasers and other optical sources, *in:* "Laser Applications in Medicine and Biology," M. L. Wolbarsht, ed., New York, Plenum Publishing Corp. (1989).

21. W.T. Ham, Jr., H.A. Mueller, and D.H. Sliney, Retinal sensitivity to damage by short-wavelength light. Nature, 260(5547): 153-155, March 11, (1976).

22. H.A. Kues, Microwave Biological Effects Program Review, Report JHU/APL, SR 90-2, Johns Hopkins University Applied Physics Laboratory, Laurel Maryland, April 1, (1990).

23. W.T. Ham, Jr., H.A. Mueller, M.L. Wolbarsht, and D.H. Sliney, Evaluation of retinal exposures from repetitively pulsed and scanning lasers, *Health Phys.,* 54(3): 337-344 (1988).

24. D.H. Sliney and J. Marshall, Tissue specific damage to the retinal pigment epithelium: mechanisms and therapeutic implications, *Lasers Light Ophthalmol.* 5(1):17-28 (1992).

25. K.R. Foster and H.P. Schwann. Dielectric properties of tissues, in (C. Polk and E. Postow, Eds.), CRC Handbook of Biological Effects of Electromagnetic Field, Boca Raton, CRC Press, PP. 27-96 (1985).

# NEW IEEE STANDARDS ON MEASUREMENT OF POTENTIALLY
# HAZARDOUS RADIOFREQUENCY/MICROWAVE
# ELECTROMAGNETIC FIELDS

R. C. Petersen

AT&T Bell Laboratories
Murray Hill, NJ 07974

## ABSTRACT

In September 1991, the Institute of Electrical and Electronics Engineers (IEEE) Standards Board approved two important standards that pertain to radiofrequency (RF)/microwave safety: IEEE Standard for Safety Levels with Respect to Human Exposure to Radio Frequency Electromagnetic Fields, 3 kHz to 300 GHz, C95.1-1991,[1] and IEEE Recommended Practice for the Measurement of Potentially Hazardous Electromagnetic Fields - RF and Microwave, IEEE C95.3-1991.[2] The first standard is primarily a safety standard specifying maximum permissible exposure (MPE) values expressed in terms of field parameters (electric field strength, magnetic field strength and power density) and, for frequencies below 100 MHz, induced current (body and contact). Exclusions expressed in terms of maximum whole-body-averaged and peak specific absorption rates (SAR) are also included. The second standard recommends measurement techniques and procedures that should be followed to ensure compliance with the safety standard.

Just as the 1991 safety standard is considerably more complex than its predecessors, so to is the measurement standard. Instruments and techniques for measuring electric and magnetic fields in both near and far field situations, and for measuring induced (body) current, contact current, and specific absorption rate (SAR) are described. Techniques for spatial averaging are discussed and minimum measurement distance from radiating and re-radiating (passive) structures and other scattering objects in the field are specified. Techniques for calibrating field measuring instruments, theoretical methods for determining exposure fields, and numerical and experimental techniques for determining SAR are also described in the 1991 measurement standard. However, since the latter are for the most part laboratory techniques, neither the determination of SAR nor instrument calibration techniques will be discussed here. Instead, only those practical techniques and instruments that are used in the field to ensure compliance with the C95.1-1991 standard will be discussed.

# INTRODUCTION

The IEEE C95.1-1991 safety standard, which was approved by the American National Standards Institute (ANSI) Board of Standards Review (BSR) in November of 1992 and is now being considered for adoption by the Federal Communications Commission, is the latest in a series of consensus safety standards that address the issue of the safe use of electromagnetic energy at RF and microwave frequencies. It is a revision of American National Standard Safety Levels with Respect to Human Exposure to Radiofrequency Electromagnetic Fields, 300 kHz to 100 GHz, C95.1-1982.[3]

The IEEE C95.3-1991 measurement standard, which also has been approved by the ANSI BSR in 1992, combines and extends two earlier measurement standards: American National Standard Techniques and Instrumentation for the Measurement of Potentially Hazardous Electromagnetic Radiation at Microwave Frequencies, ANSI C95.3-1973 (reaffirmed in 1979)[4] and American National Standard Recommended Practice for the Measurement of Potentially Hazardous Electromagnetic Fields - RF and Microwave, ANSI C95.5-1981.[5] The IEEE C95.3-1991 measurement standard is used in conjunction with the IEEE C95.1-1991 safety standard.

The purpose of the measurement standard is to recommend techniques and instrumentation for the quantification of potentially hazardous electromagnetic fields both in the far field and in the usually more complex near field.[2] Since modern RF/microwave exposure limits are primarily based on limiting the whole-body-averaged specific absorption rate (SAR) to values significantly below reported thresholds (in experimental animals) for biological effects considered adverse (at least over the frequency range where SAR is meaningful), methods for estimating SARs from field measurements are discussed in the standard and experimental and numerical methods for determining SAR are described. SAR determination is usually carried out in the laboratory and will not be discussed here. Instead, only those techniques and requirements will be discussed that are applied in the field, and are new or unique with respect to ensuring compliance with the new safety standard.

The practical problem of ascertaining compliance with the provisions of an exposure guideline is to characterize a generally complex electromagnetic environment, usually by measurement, in a manner that permits the resulting potential exposure to be meaningfully compared with the appropriate maximum permissible exposures (MPEs). In many situations the electromagnetic environment is not complex, e.g., far field and plane wave fields. In these situations, repeatable meaningful measurements can readily be made and interpreted provided the limitations of the measuring instruments are known and appropriate procedures are followed.

In many situations of interest, however, the electromagnetic environment may be poorly defined, e.g., a near-field leakage situation where the electric and magnetic field strengths are complex in space and time and are not related in any simple manner. The polarization may not be known a priori, there may be components at a number of widely different frequencies, the fields may be pulsed and/or intermittent, multipath components associated with nearby scatterers may be present, the fields may be reactive and may be affected significantly by objects in the field such as the person performing the measurements. In many of these situations the fields can be meaningfully quantified using modern near-field survey instruments.

There are cases, however, where the only meaningful way to ensure compliance with SAR derived exposure limits would be by the use of dosimetric techniques, e.g., the measurement of the SAR in a full-size or scaled (the frequency must also be scaled), or partial model human (phantom) placed at the point of interest. This also applies to cases where the fields can be measured but the interpretation of the results, in terms of potential hazard, is not clear. Included would be situations where the measured fields exceed the MPE values, but the corresponding whole-body-averaged and peak SARs are expected to be below the limiting values upon which the MPEs are based. Examples of such situations would be exposures in proximity to certain hand-held transceivers, mobile antennas, and exposures in reactive leakage or re-radiated fields.

The IEEE C95.3-1991 standard provides guidance toward correctly characterizing the exposure environment and interpreting the results. The following aspects of the standard are briefly discussed below: parameters to be measured, techniques for spatial and time averaging, measurement distance, reactive fields and measurements near reradiators, instrument types and limitations, potential sources of error, and the measurement of induced (body) current and contact current.


## MEASUREMENT PARAMETERS

The MPEs in the IEEE C95.1-1991 safety standard, which covers the frequency range of 3 kHz to 300 GHz, are expressed in terms of: the root-mean-squared (rms) electric field strength E in V/m (frequencies up to 300 MHz); rms magnetic field strength H in A/m (frequencies up to 300 MHz); rms equivalent plane-wave free-space power density S in mW/cm$^2$ (frequencies between 100 MHz and 300 GHz); induced current in the body (through the feet) and contact current (through the hand in contact with a conducting object in the field). Determination of induced and contact current I (in amperes) is required only for frequencies below 100 MHz. For whole-body exposure, the appropriate values of E, H and S are the spatially averaged values (averaged over an area equivalent to the vertical cross-section of the human body) obtained by measurement.

To ascertain compliance with the exposure guidelines, the measured spatially averaged value of S, or the mean-squared spatially averaged values of E and H ($|E|^2$ and $|H|^2$) are time-averaged over the appropriate frequency-dependent averaging time and compared with the MPEs. The averaging time for induced current and contact current is one-second; the averaging time of the measuring instrument should be "no greater than one-second."[2]

While ensuring compliance with the MPEs at frequencies below 300 MHz requires determination of both E and H (because of the reflective exposure environments commonly encountered where E and H are spatially out of phase with one another), it is usually only required to measure both at frequencies below 30 MHz. At frequencies between 30 and 300 MHz it may be possible through analysis to show that measurement of only one field parameter is sufficient. An example of where this generally would be appropriate is in the far field of a well-characterized antenna. At frequencies above 300 MHz it is rarely required to measure anything other than E or E$^2$ (which is what most survey meters measure at microwave frequencies). S can be determined (if not displayed directly by the measuring instrument) from the expression S = E$^2$/377 W/m$^2$, where the unit of E is V/m and 377 (ohms) is the free-space wave impedance.


## SPATIAL AVERAGING

The MPEs specified in the IEEE C95.1-1991 safety standard are whole-body averages. Thus, when the exposure fields are substantially nonuniform over the body, the corresponding measured fields may be spatially averaged over an area equivalent to the vertical cross-section (projected area) of the human body. This not only provides a more meaningful assessment of exposure in nonuniform fields, it also removes an ambiguity associated with earlier safety standards where neither partial-body exposure relaxation nor spatial averaging was explicitly considered. Frequently, the inexperienced surveyor would overstate a potential exposure by comparing MPEs for whole-body exposure with values measured at a small localized point in space, or inappropriately made within a few cm of a reradiating object.

Spatial averaging is especially important at frequencies below 300 MHz where the exposure fields may be relatively nonuniform over the projected area of a human. At higher frequencies, e.g., above a few GHz, it is also important to determine detailed field distributions over the body, if focusing is possible that could result in localized exposures that are considerably greater than the spatial average. In neither case is the spatial peak of the mean-squared field strengths (or power density) allowed to increase without bound as long as the spatial average is below the corresponding MPEs. Frequency dependent (above 300 MHz) limits for the localized spatial peak mean-squared field values, derived from dosimetric studies of spatial distributions of induced SAR in the body, are provided in IEEE C95.1-1991. Specifically, these are the values to which the MPEs may be relaxed under partial body or nonuniform exposure conditions.

Several methods for spatially averaging the mean-squared fields and power density are described in the IEEE C95.3-1991 standard. The most practical and repeatable method is by means of a survey meter equipped with a datalogger, i.e., a programmable device (usually connected to the recorder output of the survey meter) that samples the meter output at a predetermined rate (usually once per second), stores the data and calculates the maximum, minimum and mean values of the data over any predetermined time interval (scan time). This method, first described by Tell,[6] consists of vertically scanning the averaging area (projected area of a human) at a uniform velocity that is sufficiently slow to capture an adequate number of data points for averaging, e.g., at least ten for frequencies below 300 MHz and considerably more at higher frequencies if focusing is possible. The mean value of the sampled fields (time average) over the scan time is, thus, also the spatial average. Alternatively, a vertical scan of ten or more measurements can be made manually over the projected plane of the body, e.g., every 20 cm for frequencies below 300 MHz, and the results averaged.

In addition to ensuring compliance with the MPEs for whole-body exposures, these techniques are also applicable for ensuring that the partial body relaxation limits are met under highly nonuniform exposure conditions. The use of a datalogger also provides a repeatable method for rapidly and accurately mapping fields over any areas and paths, e.g., antenna fields.

The accuracy of the results depends in large part on the calibration accuracy of the survey instrument and the care and technique of the surveyor. After a few practice scans it is relatively easy to repeatedly scan a specific vertical path (2 m) in a specific time, e.g., 20 s to yield 20 data points. (Another advantage of the datalogger is that the stored data can be downloaded into a PC for graphical presentation.) Whether done manually or with the aid of the datalogger, the measurements should be made in absence of the person whose exposure is being assessed, when feasible, and at least 20 cm from any nearby object. Both points are extremely important for reasons described below and also for providing a standardized uniform methodology that will allow results obtained by different surveyors to be compared in a meaningful manner.

## TIME AVERAGING

While few RF/microwave exposure guidelines specify spatial averaging explicitly, most specify time averaging, e.g., the time over which exposures are averaged for comparison with the MPEs. The averaging time for the controlled environment specified in the 1991 IEEE C95.1 standard is 6 min for frequencies between 3 kHz and 15 GHz. The averaging time decreases with increasing frequency above 15 GHz to a value of 10 s at 300 GHz (which is equal to the averaging time for infrared radiation that is specified in most laser standards, e.g., ANSI Z136.1[7]). The corresponding averaging time for the uncontrolled environment is considerably more complex. At frequencies below 1.34 MHz (for $E^2$) and below 30 MHz (for $H^2$) the averaging time is 6 min; for frequencies between 3 MHz and 3 GHz (for $E^2$ and S) and between 100 and 300 MHz (for $H^2$) the averaging time is 30 min; and from 3 GHz to 300 GHz the averaging time decreases from 30 min to 10 s (for $E^2$ and S). Note: the parameters that are time averaged are S and the mean-squared field values ($E^2$ and $H^2$); not E and H (the rms values). (As indicated above, rms-induced currents and contact currents are averaged over a period of one second or measured with an instrument with an averaging time of no greater than one second.)

As with spatial averaging, in many situations time-averaging can be greatly facilitated by means of a datalogger (the purpose for which it is most often used). Again, the datalogger is connected to the recorder output of a suitable survey instrument and the averaging interval set to the averaging time specified in the standard, e.g., 6 min. The datalogger provides a record of the time average of the fields (averaged any user-specified time interval), plus a record of the maximum, minimum and time average of the fields for each contiguous averaging interval, over the entire duration of the survey. Some devices independently provide a real-time running time-average (called current six-minute average - CSMA) that is updated every few seconds. The highest value is retained and displayed as a short-term exposure level. This feature is particularly important to the surveyor if the fields being measured are close in value to the MPEs.

Care should be taken when interpreting time-averaged results for complicated exposures with large time variations that approach or momentarily exceed the MPEs. For example, depending on the levels measured and their temporal characteristics (associated with the source or with movement of the surveyor in the field), one can easily imagine situations where the time-averaged values of the fields measured at the same point in space can either exceed or not exceed the MPEs, depending on exactly where the averaging-time interval begins or ends. Sampling over a number of contiguous intervals with a datalogger and keeping track of the real-time running time-average, removes much of the ambiguity that may otherwise complicate interpretation. In these situations, the use of a datalogger is almost mandatory.

## MEASUREMENT DISTANCE

As indicated above, measurements are to be made no closer than 20 cm to an object and in absence of the person whose potential exposure is being assessed. The latter is to provide a standardized measurement methodology; the former is to minimize measurement errors associated with source-probe interactions, e.g., loading and coupling. Before the 1982 ANSI C95.1 standard, the minimum measurement distance from a radiating source, a reradiating (passive) structure or any nearby scattering objects in the field was not specified. Moreover, guidance was not given as to whether or not the person whose exposure was being assessed should or should not remain at the location of interest during the survey. As can be imagined, this led to inconsistent and often controversial survey results, particularly with respect the interpretation of measurements made with the measuring probe in close proximity to (if not in contact with) reflecting and reradiating objects.

To help alleviate what was becoming a recognized measurement problem, a minimum measurement distance of 5 cm from any object was first recommended in the 1982 ANSI

C95.1 standard. While this at least called attention to the problem, 5 cm was found not to be adequate, particularly when measuring leakage fields with high spatial gradients (where the radial component of the field varies significantly over the length of the antenna element aligned with it) and when the measurement probe couples with large reflecting surfaces in its proximity. This causes probe-loading errors resulting from changes in the antenna output impedance and detector input impedance. An example of a situation where the fields could have steep spatial gradients would be leakage from a crack or slot at frequencies below several hundred MHz; an example of when the effects of probe loading would be important is when measurements made in proximity to reflecting and reradiating metal walls, railings, cabinets, etc.

A number of analyses have been made of measurement error associated with the interaction of a measurement probe and nearby sources and scatterers.[8,9] The results of these analyses indicate that for a typical RF/microwave hazard probe with an array of 5 to 10 cm dipoles or loops, the worst-case measurement error associated with proximity effects can be kept below 3 dB ($E^2$, $H^2$ or S) for frequencies below 500 MHz if a separation distance between probe and scatterer of 3 to 5 antenna lengths is maintained.[2] The largest errors would be associated with measurements made near small active radiators such as an electrically small half-wave dipole. The maximum error (associated with field curvature) generally decreases with increasing frequency because the extent of the region of steep spatial gradients in the radial component of the field also decreases, i.e., the fields are more uniform over the volume of the probe. Based on these analyses, a minimum measurement distance of 20 cm from any scattering object and the probes used with popular survey meters was considered reasonable and sufficient. Thus, 20 cm is specified as the minimum measurement distance in the IEEE C95.3-1991 standard. The distance is measured from the scattering object to the closest part of the sensing antenna. For a typical near-field hazard meter, where the antenna elements are enclosed in a Styrofoam radome, this would generally mean the distance between the object in the field and the closest part of the radome.

## TYPES OF FIELD MEASURING INSTRUMENTS AND THEIR LIMITATIONS

The type of instrument or measurement system employed to characterize potential exposure is dictated by the characteristics of the fields being measured. For example, if the purpose of the measurements is to characterize fields that are known to be well below the MPE values but concern dictates quantification, narrowband instruments, such as field strength meters and spectrum analyzers, and the appropriate calibrated antennas could be used. Similar instruments also may be employed in conjunction with broadband survey instruments in multifrequency fields to ensure compliance with a frequency-dependent MPE. The use and limitations of narrowband instruments for hazard assessment is discussed in detail in a number reports and books, the most pertinent being the recent National Council on Radiation Protection and Measurements (NCRP) report, A Practical Guide to the Determination of Human Exposure to Radiofrequency Fields,[10] and will not be discussed further.

### Frequencies Above 1-10 MHz

In situations where a potential exposure may approach or exceed the MPEs, commercially available instruments specifically designed for characterizing this type of environment are usually employed. These include the broadband field-strength meters, commonly called survey or hazard meters. The most popular types can be used for characterizing RF electromagnetic fields at frequencies from below one MHz to 40 GHz (depending on the manufacturer). Generally more than one probe (each with a broadband isotropic response) is provided; one that responds to E (or $E^2$), the other to H (or $H^2$). A single E-field probe usually covers a wide frequency range, e.g., from a few hundred kHz to more than 40 GHz (in some cases); H-field probes usually cover much narrower frequency ranges, e.g., 0.5 to 10 MHz or 5 to 300 MHz. The upper limit of most commercial H-field probes is 300 MHz. E-field probes contain three orthogonal linear antenna elements, which may be dipoles terminated with a diode, or some other configuration such as orthogonal linear arrays of thermocouple elements; H-field probes usually contain three orthogonal loops. Sometimes more than one probe is needed to achieve a large dynamic range.

There are a number of desirable electrical and performance characteristics for devices such as these including: an isotropic response; the probe should be non-perturbing; the device should respond only to the field quantity being measured, i.e., the E-field probe should not respond to the magnetic component of the field and vice versa; the electronics should be shielded to prevent "pickup"; the instrument should indicate the rms values of the parameter being measured independent of the modulation characteristics of the field; zero-drift should be minimum to eliminate the need for re-zeroing during surveys; the response time should be known and selectable multiple response times provided; a peak-hold function should be provided; the out-of-band performance should be known (ideally the response should roll off smoothly above and below the usable frequency band of the instrument and should not exhibit a resonance-enhanced response); the display should indicate the quantity being measured, e.g., mean-squared electric field strength $E^2$ in $V^2/m^2$, mean-squared magnetic field strength $H^2$ in $A^2/m^2$ or, average plane-wave equivalent-power density S in $mW/cm^2$. Some instruments have a frequency-weighted response that is the inverse of the frequency dependence of the MPEs and display "Percent of MPE."[2,10]

The instrument manufacturers have been very responsive to the needs of the measurement community (as small as it is) and modern instruments meet most of the desired characteristics above. However, enhanced out-of-band response, particularly with some magnetic field probes, RF susceptibility (particularly in pulse-modulated fields) and "pickup" induced in the probe leads from cable flexure, etc. (particularly important when the peak-hold function is used on the most sensitive ranges), thermal drift on the most sensitive range of some instruments, the lack of a true rms response of some diode-based instruments when subjected to multiple frequency fields of approximately the same amplitude and to pulse modulated fields of high peak to average ratio, e.g., pulsed radar, are still problems that have not been resolved to the extent that would be desirable. Each of these undesirable characteristics can lead to serious measurement error if not understood and considered.

Broadband survey instruments of the types described have been in use for more than a decade and there is a wealth of information available on their proper use and, more important, on their limitations and the potential sources of measurement error. Much of this information is detailed in the IEEE C95.3-1991 standard[2] and the report of NCRP Scientific Committee SC78.[10] Only key points will be summarized here.

One frequently overlooked source of error inherent with the broadband isotropic survey type of instrument becomes important at frequencies below several MHz. Since the lower usable frequency for many of these instruments is specified by the manufacturer to be of the order of several hundred kHz, many use these devices indiscriminately at frequencies near the low end of their response. The problem here is that at frequencies below about 10 MHz, the typical high-resistance signal-carrying leads between the detector and the electronics package become a more efficient antenna than the dipole antenna elements in the probe. The currents induced in the probe leads when inadvertently aligned with the E-field can lead to large measurement errors.[11] (Alignment of the probe leads orthogonal to the E- field vector is good practice at any frequency.)

Further, at frequencies below a few MHz, there is evidence that the body of the person holding the probe of some broadband instruments becomes part of the antenna[12] that also can lead to significant measurement error. Both of these sources of error can be minimized by supporting both the probe and the electronics package with dielectric supports with the connecting lead held orthogonal to the direction of the E-field vector. A better way is to use an instrument of the active, self-contained monopole (or triaxial) antenna configuration (described below) that eliminates the lead and surveyor interaction problem altogether.

Another potential source of measurement error is encountered when certain broadband isotropic survey meters are used for making measurements of pulse amplitude modulated fields with high peak-to-average ratios. If certain diode-based instruments are used and the peak fields are significantly higher than the recommended upper amplitude limit for the probe (which may not be recognized because of the low average power indicated on the meter), the response may no longer be proportional to the rms value of the fields. This also could lead to significant errors. Under the same conditions, the thermocouple junctions of a thermocouple-based instrument (which essentially always has a true rms response) could easily be destroyed by high peak powers even though the average power is well below the burnout rating for the probe (typically three times the maximum cw rating). Generally, the duty factor of the radar should first be determined or measured, e.g., with a spectrum analyzer and an appropriate antenna or with an antenna, a diode detector and an oscilloscope, so that a probe with the appropriate dynamic range and operating point can be selected to minimize measurement error or prevent burnout. RF susceptibility of the instrument is also more of a problem in pulse-modulated fields and precautions should be taken to ensure that the meter readings are not being affected by interference.

In mixed frequency fields, a potential source of error is an enhanced out-of-band response for the probe. This is somewhat common with magnetic field probes and has also been reported for the self-contained active antenna device. With some instruments the enhancement at frequencies greater than the stated upper usable frequency for the probe could be more than 10 dB. This could lead to serious errors if sources are present that operate at the frequency where the response is enhanced. While this information is usually not found in the manufacturer's literature, excellent evaluations of the most widely used instruments have been published that include the measured out-of-band frequency response.[13,14]

When care is exercised using the types of instruments discussed above, accuracy of the order of 3 to 4 dB can be expected for most field measurements.[10] Most of the error, however, is associated with measurement technique and not with the calibration accuracy of the instruments.

## Frequencies Below 1-10 MHz

Both the electric and magnetic fields also can be measured at frequencies below a few MHz using narrowband instruments, such as field strength meters and spectrum analyzers, with the appropriate calibrated antennas. As above, narrowband instruments are generally used when the fields are well below the MPEs but perception dictates that they be quantified. They would also be used in multifrequency environments where the fields at some frequencies approach the limits of a frequency dependent MPE. Because of their complexity, size, RF susceptibility (particularly spectrum analyzers), and the experience required to make accurate measurements, narrowband instruments are not generally used for routine hazard surveys.

Because of some of the problems discussed above, e.g., pickup on the probe leads, errors associated with the operator becoming part of the antenna circuit, and the increased difficulty of providing adequate shielding to reduce the effects of RF interference at the lower frequencies, instruments such as broadband isotropic survey meters are not generally used to measure fields at frequencies below a few MHz. Instead, self-contained active monopole (or triaxial) type devices and self-contained instruments in the displacement sensor configuration are usually used to measure the electric field, and specially designed self-contained instruments containing single or triaxial loops are used to measure the magnetic field.

These self-contained instruments consist of the antenna element(s) and the electronics and battery supply all in one package. For example, the active monopole device consists of a small metal box in the shape of a cube (approximately 10-15 cm on edge) with an antenna element projecting from one surface. Some versions contain an antenna element on each of three orthogonal faces of the cube. The box contains a high impedance amplifier (between the antenna element(s) and the diode detector) to provide the necessary sensitivity, the electronics necessary to process the signal and drive the display, and a battery to power the instrument. While the antenna elements appear as monopoles, they are actually one-half of a dipole; the other half is the metal box. A meter, usually calibrated in units of field strength, is located on one face of the box. This self-contained instrument can be placed at any point in the field (or held with a dielectric support of sufficient length) and the field strength measured. A useful option is a light emitting diode (LED) transmitter that transmits the output, via an optical fiber cable, to a remote readout meter. This device all but eliminates problems of pickup and interaction with the operator (although its use taxes the patience of the operator because the sensitivity is changed in decade increments by replacing the antenna elements with longer or shorter elements). While the frequency response given by the manufacturer is 10 kHz to 220 MHz, the user should be aware of a significant resonant-enhanced response reported to occur above 220 MHz.

Self-contained instruments that measure the displacement current (which is proportional to the time derivative of the component of the electric field perpendicular to the surface of the device) are also commonly used at frequencies below a few MHz. Commercially available devices have options such as remote LED/fiber optic readouts. These instruments respond correctly only to the component of the E-field that is perpendicular to surface of the sensor and, therefore, must be oriented in the field. Some devices, generally used for measurement of the fields from visual display terminals, contain a shielded loop around the edge of the circular displacement-current sensing element and can be used for measuring both the E field and the H field.

Instruments and techniques for field measurements at frequencies between 10 kHz or so and a few MHz are not nearly as standardized as at the higher frequencies. While there is considerable experience with E-field instruments, there is considerably less with H-field measuring devices. Questions that have to be addressed, particularly at frequencies below a few tens of kHz, are: where should magnetic field measurements be made (minimum distance from the source) and what should be the diameter of the loop? These questions are important when interpreting measurements made near sources with steep spatial field gradients. For example, a small diameter probe placed near a source (where the spatial gradients of the field generally would be large) averages over a smaller area and, consequently, the indicated value of the field measured with the small diameter loop will generally be larger than that measured with a loop of larger diameter. Since the MPEs for the magnetic field at the lower frequencies are based on induced current (to some extent), the use of a larger diameter loop may provide a more meaningful measurement with respect to current induced in the body and, hence, to the MPE. These issues, and others, are now being addressed by the subcommittee developing the future revision of IEEE C95.3.

## INDUCED CURRENT AND CONTACT CURRENT

As the frequency decreases below a few MHz, whole-body-averaged SAR becomes less important and the SAR in the wrists and ankles (frequencies above 70 to 100 kHz) and surface phenomena such as shocks and burns (frequencies below 70-100 kHz) associated with induced (body) and contact current, becomes more important. For frequencies below 100 MHz, the MPEs expressed in terms of field strength in the IEEE C95.1 standard will generally limit the current induced in the body to values below those that will produce SARs in the ankles exceeding 20 W/kg. However, because of the wide and varied nature and size of conducting objects encountered in ambient RF fields, e.g., guy wires, buses, trucks, vans, and the diverse opportunities for individuals to come in contact with these objects, the MPE field values, particularly at frequencies below 70 to 100 kHz, may not preclude all possible shock and burn effects.[1] Therefore, limits for induced current (in a freestanding individual not in direct contact with any conducting object other than the ground upon which he or she is standing), and contact current (current flowing through an individual in contact with a conducting object), are also given for frequencies below 100 MHz. The determination of induced current and contact current generally is not amenable to analytic techniques and in most cases must be measured.

Induced current can be measured through one or both feet by means of commercially available foot-current meters. Most of these devices consist of a parallel plate capacitor arrangement upon which the individual stands. The capacitor is shunted by a low inductance resistor. A voltmeter measures the voltage developed across the capacitor. This voltage is proportional to the current flowing through the individual to ground and, hence, to the induced current. Shoe-insertable devices are also available. Induced current measurements using any of these devices is relatively straightforward but the results may vary considerably depending on the size, shape, orientation of the arms, etc., of the individual standing on the meter. To remove this ambiguity, instrument manufacturers have developed a human-equivalent antenna that is placed on the foot-current meter. In mixed-frequency fields where some of the fields are at frequencies below 100 kHz, a means must be provided for examining the frequency components of the current, e.g., the use of a field strength meter or spectrum analyzer connected across the low-inductance resistor. This is because the induced-current and contact-current limits are frequency dependent below 100 kHz. Similar standardized devices for measuring contact current through the hand are also becoming available.

## CONCLUSIONS

The new IEEE standards address many of the bothersome issues not addressed in the earlier standards. Questions of interpretation of partial body exposures have been addressed by providing explicit guidance on spatial averaging and criteria for partial body relaxation have been provided. The potential for unnecessary concern, based on incorrect interpretation of inappropriate measurements made with the probe of a survey meter within a few cm of (or in contact with) a radiating source or a passive radiator, has been minimized with the recommendation of a more reasonable minimum measurement distance, i.e., 20 cm. Practical means for addressing the potential for shock and burn at the lower frequencies has been addressed by limits on induced and contact current, and recommendations have been made for measurement. Instrument manufacturers have responded to the needs of the measurement community by tailoring their instruments to the requirements of the standards. While the measurement standard adequately addresses field measurements at frequencies above a few MHz and induced current measurements at frequencies between 100 MHz and 10 kHz, there remains a need for standardized instruments and measurement techniques at frequencies between a few tens of kHz and a few MHz. These areas are now being addressed by the subcommittee responsible for the IEEE C95.3 measurement standard.

## REFERENCES

1. IEEE, Standard for Safety Levels with respect to Human Exposure to Radio Frequency Electromagnetic Fields, 3 kHz to 300 GHz, Institute of Electrical and Electronics Engineers, New York, NY (1991).
2. IEEE, Recommended Practice for the Measurement of Potentially Hazardous Electromagnetic Fields - RF and Microwave, IEEE C95.3-1991, Institute of Electrical and Electronic Engineers, New York, NY (1991).
3. ANSI, American National Standard Safety Levels with Respect to Human Exposure to Radiofrequency Electromagnetic Fields - 300 kHz to 100 GHz, ANSI C95.1-1982, American National Standards Institute, New York, NY (1982).
4. ANSI, American National Standard Techniques and Instrumentation for the Measurement of Potentially Hazardous Electromagnetic Radiation at Microwave Frequencies, ANSI C95.3-1979, American National Standards Institute, New York, NY (1979).
5. ANSI, American National Standard Recommended Practice for the Measurement of Hazardous Electromagnetic Fields - RF and Microwave, ANSI C95.5-1981, American National Standards Institute, New York, NY (1981).
6. R.A. Tell, Real-Time Data Averaging for Determining Human RF Exposure, *in:* Proceedings, 40th Annual Broadcast Engineering Conference, National Association of Broadcasters, Dallas TX, April 12-16, 388-394 (1986).
7. ANSI, American National Standard for the Safe Use of Lasers, ANSI Z136.1-1993, American National Standards Institute, New York, NY (1993).
8. A. Rudge and R. Knox, Near Field Instrumentation, US Department of Health, Education, and Welfare Publication BRH/DEP 70-16 (1970).
9. G. Smith, The electric field probe near a material interface with application to probing of fields in biological bodies, *IEEE Transactions on Microwave Theory and Techniques,* MTT-27:3: 270-278, (1979).
10. NCRP, A Practical Guide to the Determination of Human Exposure to Radiofrequency Fields, Report of Scientific Committee 78, National Council on Radiation Protection and Measurements, Bethesda, MD (1993 - in press).
11. E. Aslan, Non-ionizing radiation - measurement methods and artifacts, *in:* "Proceedings of the 39th Annual Broadcast Engineering Conference, "National Association of Broadcasters, Las Vegas, NV (April 13-17, 1985).
12. H.R. Kucia, Accuracy limitation in measurements of HF field intensities for protection against radiation hazards, *IEEE Transactions on Instrumentation and Measurements,* IM-21:4 (1972).
13. B. Nesmith and P. Ruggera, Performance Evaluation of RF Electric and Magnetic Field Measuring Instruments, HHS Publication FDA-82-8185, National Technical Information Service, Springfield VA (1982).
14. E.D. Mantiply, An Automated TEM Cell Calibration System, US EPA Publication EPAS20/1-84-024, National Technical Information Service, Springfield, VA (1984).

# RISK ASSESSMENT OF HUMAN EXPOSURE
# TO LOW FREQUENCY FIELDS

J.H. Bernhardt

Institute for Radiation Hygiene
Federal Office for Radiation Protection
85762 Oberschleißheim-Neuherberg
Germany

## INTRODUCTION

Although several epidemiological studies suggest a weak association between the exposure to extremely low-frequency (ELF) fields and an increase in various kinds of cancer, a final risk assessment of long-term continuous exposure to ELF fields is so far not possible. It has not been proven definitely, that the electric and especially magnetic ELF fields occurring at working places or in every-day life are mutagenic or carcinogenic. The main critical points are those concerning statistical evaluation, insufficient determination of the field strength during exposure, dose-effect relationships, inadequate demarcation of concomitant factors, and - as one of the most important points - the absence of known interaction mechanisms. Final clarification of the question of possible late effects requires further elucidation. The non-stochastic ELF field effects, therefore, are of major importance when deriving standards are given priority.[1]

The International Commission on Non-Ionizing Radiation Protection (ICNIRP) reviewed the data about possible carcinogenicity of power-frequency magnetic fields at its first annual meeting on May 7-12, 1993 held in Neuherberg, Germany. This review considered all scientific data that have been published or publicly presented since the "Interim Guidelines on Limits of Exposure to 50/60 Hz Electric and Magnetic Fields" were published in 1990 by the predecessor International Non-Ionizing Radiation Committee (INIRC) of the International Radiation Protection Association (IRPA). The major reason for the interim nature of these guidelines was the inability to arrive at a scientifically based judgement concerning any causal relationship between 50/60 Hz magnetic-field exposures and the excess occurrence of cancer. The most recent data reflect some improvements in methodology in laboratory studies and in epidemiological studies of both occupational and general populations. After careful consideration of this evidence, the Commission concludes that the data related to cancer do not provide a basis for health risk assessment of human exposure to power-frequency field. Accordingly the Commission confirms the interim guidelines published in 1990.[1]

## ACUTE FIELD EFFECTS

Electric and magnetic ELF fields - due to different interaction mechanisms - may generate electric field strengths (in V/m) and electric current densities (in mA/m$^2$) within the body. These tissue field strengths and current densities can produce biological effects depending on their intensity and frequency. Table 1 summarizes the most important steps involved in the interaction mechanisms of ELF fields on the cellular levels presently discussed. Although the induced electric tissue field strength is the basic quantity responsible for biological effects, dose-effects relationships are expressed in terms of current density in most cases. Both of these quantities are connected by the electric conductivity.

**Table 1.** ELF fields and interaction mechanisms on the cellular level.

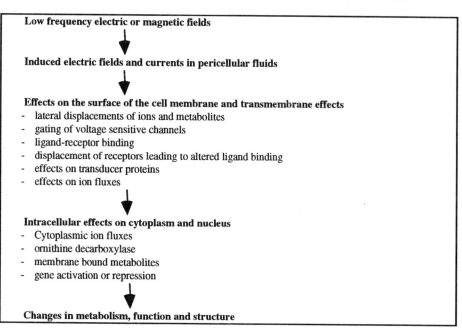

Up to now, no biological effects below about 1 mA/m$^2$ have been confirmed scientifically. Current densities between 1 and 10 mA/m$^2$ correspond to the endogenous background level of current densities in most organs and tissues of the body.[2,3] On the surface of electrically active nerve or muscle cells current densities of up to 1000 mA/m$^2$ can occur for short times.

From in-vitro laboratory studies using current densities between about 1 and 10 mA/m$^2$, some minor biological effects have been reported (see Table 2). Some of these effects were observed only at distinct frequencies and field strengths ("windows")[4,5]. Examples include changes in the calcium-efflux from preparations of brain-tissue following exposure of 16 Hz-electric or magnetic fields, modified calcium-uptake of lymphocytes following magnetic field exposure, and inhibition of melatonin synthesis by the pineal gland following exposure to weak static magnetic fields. The significance of these findings and possible adverse health effects to human beings are not clear.

**Table 2.** Examples of biological effects at low induced field strengths (field strengths in extracellular medium between about 3 and 50 mV/m, corresponding to current densities between 1 and 10 mA/m$^2$). *Some of these effects were observed only at distinct frequencies and field strengths ("windows"). The biological significance of these findings and possible adverse health effects to human beings is not clear.*

- Several transcription processes
- Calcium-efflux from preparations of brain-tissue following exposure to 16 Hz electric or magnetic fields
- Modified Calcium-uptake of lymphocytes following magnetic field exposure
- Inhibition of melatonin synthesis by the pineal gland
- Accumulation of counter ions; Ion-density fluctuations

ELF fields of relatively high intensity, producing internal body current densities exceeding about 10 mA/m$^2$, can cause some biological effects that cannot be ignored. Examples are enhancement of DNA synthesis, changes in the molecular weight distribution during protein synthesis, delay of the mitotic cell cycle, blocking of the action of parathyroid hormone at the site of its plasma membrane receptor, and inhibition of the cytotoxicity of T-lymphocytes (see Table 3).

**Table 3.** Observed biological effects at electrical field strengths exceeding 50 mV/m in the extracellular medium (induced current densities of 10 - 100 mA/m$^2$).

- Enhancement of DNA synthesis
- Change of transcription of DNA into mRNA
- Changes in the molecular weight distribution during protein synthesis
- Delay of the mitotic cell cycle
- Blocking of the action of the parathyroid hormone at the site of its plasma membrane receptor
- Inhibition of the cytotoxicity of T-lymphocytes
- Transient increase of the activity of ornithine-decarboxylase
- Visual and nervous system effects
- Facilitation of bone fracture reunion is reported

The thresholds for stimulation of excitable cells are above 100 mA/m$^2$; for frequencies above about 300 to 1000 Hz, these thresholds increase proportionally with the frequency[2,6] (see Figure 1).

A systematic evaluation of the actually induced current densities and field strengths at the tissue and cellular level of these findings is complicated by the following facts:

- large variations of exposure conditions and,
- lack of details on the geometry of the biological samples.

Furthermore, the lack of reproducible results among different laboratories limit the interpretation. Since dose-response relationships have not yet been identified, systematic assessments of threshold values for tissue field strengths and their frequency dependence, as well as clearing up of frequency and intensity "windows," are urgently needed.

Controlled laboratory studies on volunteers exposed for short periods to electric field strengths up to 20 kV/m or magnetic flux densities up to 5 mT revealed no adverse clinical or significant physiological changes[3,7]. These data do not exclude that health effects may occur by long-term exposure.

In addition to effects caused by induced tissue field strengths, there are surface effects due to electric field exposure resulting in sensory perception and, furthermore, perception of transient or steady electric currents occurring from touching charged objects in electric fields. At 50/60 Hz, a field strength of 20 kV/m is the perception threshold of 50 % of people for sensations from their head hair or of tingling between body and clothes. A small percentage of people can perceive a field strength of 2 to 3 kV/m.

Electric charges induced in a conducting object exposed to electric fields may cause current to pass through a human in contact with it (indirect effects). The effects of steady or transient ("spark discharges") contact currents depend upon many factors, e.g., the size and geometry of the object, the electric field strength, the body impedance, the size of the contact area and the strength and duration of the contact current. Typical electric field strength levels leading to spark discharges that are felt as an annoyance in our daily environment are between 2 and 7 kV/m. A small percentage of people can perceive a field strength of 0.5 kV/m via spark discharges. Some data on direct and indirect effects of 50 and 60 Hz electric fields are summarized in Table 4 (see Bernhardt,[6] and references).

A further group of indirect effects results from possible effects of ELF fields on electric or electronic implanted medical devices. A typical example is the implanted pacemaker.

## BASIC AND DERIVED EXPOSURE LIMITS

When establishing frequency dependent exposure limits and safety factors, the following factors should be considered: duration of exposure, presence of controlled or uncontrolled environments, existence of risk groups (e.g., with medical implants). The evaluation of acute effects has lead to recommendations of exposure limits that are different for occupationally exposed persons and the general public. The main reasons for adopting lower exposure limits for the general public than for the occupationally exposed population are the following[1]: the general public comprises individuals of all ages and different health status; individuals or groups of particular susceptibility may be included in the general population; in many instances, members of the general public are either not aware of being exposed, or may be unwilling to take any risks (although slight) associated with exposure; finally, the public cannot be expected to accept effects such as annoyance and pain due to transient discharges or hazards from contact currents.

The International Non-Ionizing Radiation Committee (INIRC) of IRPA recommends that the ELF-field induced current density should not exceed 10 mA/m$^2$ in the body. Since most evidence is based on short-term observations and because of the limited knowledge about the possible effects of long-term exposure, the INIR Committee recommends to limit the induced current density to 4 mA/m$^2$ for continuous occupational exposure, and to 2 mA/m$^2$ for the general public (a factor of 5 below 10 mA/m$^2$). The current densities should be averaged over a period of 1 s and a cross-section of 1 cm$^2$ perpendicular to the current direction. This averaging seems to be sufficient to include spatial and temporal peak values in view of the fact that the effects of current densities are occurring at the cellular levels, that a plurality of cellular effects are resulting in an action on the whole organization and, that the safety zone from stimulating effects is sufficiently large. In Table 5 proposals for basic restrictions for current densities and induced electric field strengths for the frequency range between 10 and 1000 Hz and frequency-dependent values for frequencies exceeding 1000 Hz are given, taking into account the basic exposure limits of IRPA/INIRC[1] at 50/60 Hz and other proposals and recommendations. For frequencies exceeding 1000 Hz, a linear frequency dependence was assumed, following the frequency dependence of the stimulation thresholds of nerve and muscle cells. The current density f/250 mA/m$^2$ (f in Hz) is lower by a factor of 100 approximately than the stimulation threshold. The intersection of f/250 mA/m$^2$ and 80 A/m$^2$ (approximately 10 W/kg local SAR, using 0.6 S/m) occurs at 20 MHz (see Figure 1).

**Table 4.** Direct and indirect effects of 50 and 60 Hz electric fields[9,10,11].

| Field strength kV/m | Effect |
|---|---|
| > 50 | Direct perception of short circuit body current of grounded man |
| 20 - 24 | Median pain perception for men, finger contact, car |
| 20 | Perception threshold for 50 % of men with sensations on their head, head hair or tingling between body and clothes |
| 16 - 20 | 0.5 % let-go threshold for men, truck contact |
| 14 - 16 | Median pain perception for women, finger contact, car |
| 11.5 - 14 | 0.5 % let-go threshold for children, bus contact |
| 11 - 13 | 0.5 % let-go threshold for women, truck contact |
| 10 - 12 | Median pain perception for children, finger contact, car |
| 8 - 10 | 0.5 % let-go threshold for children, truck contact |
| 4 - 7 | Median annoyance level for spark discharges, the person acting as charge collecting object (170 pF) |
| 4 - 5 | Median touch perception for men, finger contact, car |
| 2.5 - 6 | 90 % - Perception levels for spark discharges, the person acting as charge collecting object (170 pF) |
| 2 - 3.5 | 10 % - annoyance levels for spark discharges, the person acting as charge collecting object (170 pF) |
| 3 | Perception threshold for 5 % of men with sensations on their head, head hair, or tingling between body and clothes |
| 2 - 2.5 | Median touch perception for children, finger contact, car |
| 2.5 | Threshold for interference with extremely sensitive unipolar cardiac pacemakers (0.5 mV peak to peak voltage sensitivity) |
| 1.2 - 2.5 | Median perception levels for spark discharges, the person acting as charge collecting object (170 pF) |
| 0.6 - 1.5 | 10 % - perception levels for spark discharges, the person acting as charge collecting object (170 pF) |

In Figure 1 the proposals for the basic restrictions for the current density from Table 5 are shown together with threshold curves for the stimulation of peripheral nerves/muscles and for the cardiac muscle, which are taken from WHO[12].

From these basic restrictions for the current density the dosimetric quantities, which are necessary for practical purposes, must be derived. The derived secondary exposure limits, i.e., the external electric and magnetic field strengths, must be deduced in such a way that the protective aim is guaranteed even under worst case conditions. Numerical and measuring methods exist for the derivation of field strength exposure limits from the basic restriction. For both methods considerable simplifications have been used up to now that did not account for phenomena such as the inhomogeneous distribution and anisotropy of the electric conductivity and other factors (compare Table 6). As a result of such simplifications, possible spatial increased values of the induced-field strengths and currents remain disregarded.

**Table 5.** Proposal for basic restrictions for current densities (induced-field strengths), whole body average and local peak SAR. (Different recommendations and proposals were taken into account, i.e., Allen et al.[13], Bernhardt[6], IRPA/INIRC[1,8], WHO[12]).

| Exposure Characteristics | Frequency Range | Current density (mA/m$^2$) | (field strength) (in mV/m) | Whole body average SAR | Local peak SAR (head and trunk) |
|---|---|---|---|---|---|
| Occupationally exposed population (controlled exposure conditions) | 10 - 1000 Hz[1] | 4 | (20) | - | - |
| | | | | - | - |
| Whole working day | 1000 Hz- 100 kHz[2] | f/250 | (f/50) | | |
| | 100 kHz - 10 MHz[3] | f/250 | (f/50) | 0.4 W/kg | 0.1 W/10 g |
| General public | 10 - 1000 Hz | 2 | (10) | - | - |
| up to 24 h d$^{-1}$ | 1000 Hz - 100 kHz[2] | f/500 | (f/100) | - | - |
| | 100 kHz - 10 MHz[3] | f/500 | (f/100) | 0.08 W/kg | 0.02 W/10 g |

f = frequency in Hz

[1] Current densities of up to 10 mA/m$^2$ are allowed for exposures of up to 3 hours per working day in the frequency range 10 - 1000 Hz.

[2] Current densities must be averaged over periods shorter than 1 s and over cross sections of 1 cm$^2$, perpendicular to the current direction.

[3] In this frequency range both the induced current density and SAR limits should not be exceeded. The peak value of the current density should not exceed the tenfold value of the given current density limits.

**Table 6.** Difficulties in determining the induced field strengths ($E_i$) and current densities ($J_i$) and their distribution within the body exposed to external electric ($E_o$) or magnetic ($B_o$) fields.

- Heterogeneity of the electric conductivity of different tissues (variation up to a factor of 50)
- Anisotropy of the electric conductivitiy of muscle- and brain tissue
- Local enhancement of field strength in tissue due to inhomogeneity of conductivity
- Exact current paths unknown
- Strong dependence on geometric factors and orientation of electric or magnetic field components
  Rough estimates are possible by using considerable simplifications with the following relations

E-field: $E_i = k \cdot f \cdot E_o/\sigma$; $J_i = k \cdot f \cdot E_o$ (strong variation of k for different body parts)

B-field: $E_i = \pi \cdot R \cdot f \cdot B_o$; $J_i = \pi \cdot R \cdot \sigma \cdot f \cdot B_o$

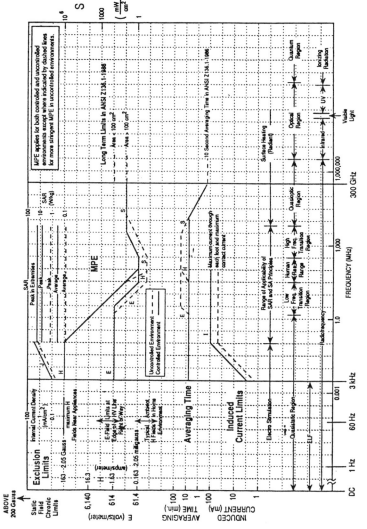

**Figure 1.** Threshold current densities for stimulation of nerve/muscle cells (curve A) and cardiac muscle (curve B), from WHO[12]. The curves C and D are proposals for the basic restriction for current densities for occupationally exposed population (C) or the general public (D), using the data of Table 5. The restriction for peak pulses should be based on the stimulation curve A including a safety factor..

Table 7 summarizes the results of measurements and calculations of derived values of the electric-field strength and magnetic-flux density at 50 or 60 Hz[8,14,15,16]. The table partially reflects the heterogeneous distribution of the current density within the human body produced by external electric or magnetic fields. For example, an electric-field strength of 5 kV/m (the IRPA/INIRC recommendation for the limit of continuous exposure of the general public to 50/60 Hz electric fields) produces current densities of up to 1.7, 0.5, 2.5 and 10 mA/m$^2$ in the trunk, head, neck and ankles, respectively. A magnetic flux density of 0.1 mT induces current densities of up to 1.7, 0.4 and 0.15 mA/m$^2$ in the trunk, head and wrists/ankles, respectively. A refinement of such model estimations, however, is urgently needed.

**Table 7.** Summary of derived values for the electric field strength and magnetic flux density approximately producing a current density of 1 mA/m$^2$ in different body parts at 50 or 60 Hz. Left: Values of the electric field strength; Right:Values of the magnetic flux density (peripheral regions), a homogeneous conductivity of 0.2 S/m is assumed. For tissues in the center of the body (with exception of the heart), the magnetic flux densities for inducing the same current density are larger

| Electric field strength in kV/m | | Magnetic flux density in µT | |
|---|---|---|---|
| Head | 8.5 - 12.5 | Head | 250 |
| Neck | 2 - 3.5 | peripheral areas of the trunk | 60 - 100 |
| Thorax, cardiac region, trunk | 2.5 - 10 | cardiac region | 90 - 200 |
| Ankles, both feet grounded | 0.25 - 0.5 | wrists/ankles | 600 |

The results of the Tables 5, 6 and 7 lead to the proposal of frequency-dependent derived-limit values given in Table 8.

Some summarizing issues for a new NATO STANAG 2345, concerning the frequency range from 10 kHz for 10 MHz are presented in Table 9.

**Table 8.** Proposal of derived values (reference levels) for the electric and magnetic field in the frequency range 10 Hz to 1 MHz. (The results of the Tables 5, 6 and 7 and different recommendations and proposals were taken into account, i.e., IRPA/INIRC[1] draft document of a NRPB-proposal 1992, German draft standard[17]).

| Exposure characteristics | Frequency range | Electric field strength[1) 2)] (V/m) | Magnet flux[2)] density (µT) |
|---|---|---|---|
| Occupational, whole working day | 10 - 1000 Hz<br>1000 Hz - 1 MHz | 500/f<br>500 | 25/f<br>25 |
| General public | 10 - 1000 Hz<br>1000 Hz - 1 MHz | 250/f<br>250 | 5/f<br>5 |

f in kHz
1) Where ungrounded conducting objects are present in an electric field, these will result in the flow of current to a person coming into contact with them. For a sufficiently large ungrounded conducting object, shock or burn may result from such a contact at levels of electric field strength below the exposure limits. Such effects will be avoided if contact currents are less than 1 mA between 10 Hz and 2 kHz, 0.5 f(kHz) between 2 and 100 kHz, and 50 mA between 100 kHz and 1 MHz.
2) Interference with the normal operation of electronic devices can arise at levels below those given in the table. Advice on acceptable electric and magnetic field levels should be given for people with medical electronic devices such as pacemakers.

# ASPECTS OF PREVENTIVE HEALTH CARE

Generally, the simultaneous occurrence of other physical agents, noxious chemicals or biological factors is not considered when standards are established. For example, the exposure limit of 5 kV/m at 50 or 60 Hz provides substantial protection for the public from annoyance caused by contact currents or transient discharges, which is considered acceptable for occupationally exposed persons. An electric field strength of 5 kV/m, however, cannot completely eliminate perception of electric field effects. Additionally, there is a small probability that a malfunction of some sensitive unipolar cardiac pacemakers will occur under worst-case conditions at the electric field-strength values and

**Table 9.** Issues for a new NATO STANAG 2345, concerning the frequency range from 10 kHz to 10 MHz.

| |
|---|
| - Define basic restrictions; derive action levels, |
| - 10 to 100 kHz: restrictions should be based on J or E, |
| - 100 kHz to 10 MHz: restrictions should be based on J or E and additionally on SAR (whole body and local SAR), |
| - peak pulses: should be based on the stimulation curve ($J_{peak} = f/2.5$, f in kHz, J in A/m$^2$) with an additional safety factor, |
| - Define averaging times for induced currents and for partial body exposure, |
| - Define averaging masses for local SAR (i.e., 10 g instead of 1 g), |
| - Define contact currents for finger- and grasping contacts. |
| - Consider only occupational exposure limits (EL), let public EL be the responsibility of WHO/ICNIRP |
| - If both, occupational and public EL are considered, use general principle of different EL (IRPA/INIRC/ICNIRP). Preventive health care measures for the public should be taken into account, if possible (see table 10). |

magnetic-flux densities given in Table 8. For occupational exposure, hazardous body currents and contact voltages can and must be avoided by special measures. For the general public, a further limitation of the limits of Table 8 should be discussed, if special measures for avoiding indirect effects are not possible. Furthermore, in view of this and some other unknowns and uncertainties concerning the complete understanding of the interacting mechanisms and the final clarification of possible long-term effects, it may be prudent not to exhaust the exposure limits. Some aspects of preventive public health care are summarized in Table 10. Such aspects of preventive public health care should be taken into consideration especially on developing new technologies by using electric energy or building strong sources of electromagnetic fields near public dwellings.

**Table 10.** Aspects of preventive public health care.

---

There are several aspects that point to the fact that it may be prudent not to exhaust the long-term exposure limits of the general public:

- A small percentage of people can perceive electric field strengths below the derived exposure limits and feel it as an annoyance

- The function of some sensitive unipolar cardiac pacemakers (about 2 - 5 % at 50/60 Hz) may be influenced at field strengths below the exposure limits

- There are deficits in our knowledge about the induced-field strengths and their distribution within the body

- The interaction mechanisms of the induced-field strengths on the cellular level are not fully understood

- There are numerous uncertainties about the biological significance of the results of in-vitro experiments on the cellular level and especially on their impact for possible health effects for long-term exposure

- For the establishment of standards and exposure limits, the simultaneous occurrence of other physical agents, noxious chemicals or biological factors is not being considered

Due to these aspects of preventive health care, a minimizing of exposure and with that possible detrimental consequences for the human health should be taken into account especially, i.e., on developing new technologies by using electric energy or building strong sources of electromagnetic fields near public dwellings.

---

## REFERENCES

1.  International Radiation Protection Association/International Non-Ionizing Radiation Committee. Interim guidelines on limits of exposure to 50/60 Hz electric and magnetic fields, *Health Phys* 58: 113-122 (1990).
2.  J.H. Bernhardt. The direct influence of electromagnetic fields on nerve and muscle cells of man within the frequency range of 1 Hz to 30 MHz, *Radiat. Environ. Biophys* 16: 309-23 (1979).
3.  World Health Organization. Environmental Health Criteria 69: Magnetic fields. WHO, Geneva (1987).
4.  E. Postow and M.L. Swicord. Modulated fields and "window" effects, *in*: CRC Handbook of Biological Effects of Electromagnetic Fields, C. Polk and E. Postow, eds., Boca Raton, Florida, CRC Press, pp. 425 (1986).
5.  Z.J. Sienkiewicz, R.D. Saunders, C.I. Kowalczuk. "Biological Effects of Exposure to Non-ionizing Electromagnetic Fields and Radiation: II Extremely Low Frequency Electric and Magnetic fields. Chilton, NRPB-R239, London HMS0 (1991).
6.  J.H. Bernhardt. The establishment of frequency dependent limits for electric and magnetic fields and evaluation of indirect effects, *Radiat. Environ. Biophys.* 27:1-27 (1988).
7.  R. Sander, J. Brinkmann, B. Kuehne. Laboratory studies on animals and human beings exposed to 50 Hz electric and magnetic fields. International Conference on Large High Voltage Electric Systems. Paris, September 1982, Cigre No. 36 (1982).
8.  International Radiation Protection Association/International Non-Ionizing Radiation Commitee, Guidelines on limits of exposure to radiofrequency electromagnetic fields in the frequency range from 100 kHz to 300 GHz, *Health Phys.* 54: 1154-23 (1988).
9.  Institute of Electrical and Electronics Engineers: Working Group on Electrostatic and Electromagnetic Effects. Electric and magnetic field coupling from high voltage AC power transmission lines - classification of short-term effects on people, *IEEE Trans. Power Appar. Syst.* 97: 2243-2252 (1978).
10. L.E. Zaffanella and D.W. Deno. Electrostatic and Electromagnetic Effects of Ultra-high-voltage Transmission Lines. Palo Alto, CA: Electric Power Research Institute, Final report, EPRI EL-802 (1978).
11. World Health Organization. Envrionmental health criteria 35: Extremely Low Frequency (ELF) fields. WHO, Geneva (1984).

12. World Health Organization. Environmental health criteria 137: Electromagnetic Fields (300 Hz - 300 GHz). WHO, Geneva (1993).

13. S.G. Allen, J.H. Bernhardt, C.M.H. Driscoll, M. Grandolfo, G.F. Mariutti, R. Matthes, A.F. McKinlay, M. Steinmetz, P. Vecchia, M. Whillock. Proposals for basic restrictions for protection against occupational exposure to electromagnetic non-ionizing radiations. Recommendations of an International Working Group set up under the auspices of the Commission of the European Communities, *Physica Medica* VII, No. 2: 77-89 (1971).

14. P.J. Dimbylow, Finite difference calculations of current densities in a homogeneous model of a man exposed to extremely low frequency electric fields, *Bioelectromagnetics* 8: 355-73 (1987).

15. W.T. Kaune, W.C. Forsythe, Current densities measured in human models exposed to 60 Hz electric fields *Bioelectromagnetics* 6:13-22 (1985).

16. T.S. Tenforde, W.T. Kaune, Interaction of extremely low frequency electric and magnetic fields with humans, *Health Phys.* 53:595-606 (1987).

17. German Draft Standard, DIN VDE 0848, Part 2: Safety in electromagnetic fields; Protection of persons in the frequency range from 30 kHz to 300 GHz. Berlin, Beuth Verlag (in German) October (1991).

**Session C: Military Operations and RFR Standards**

**Chair: L. Court**

Navy Issues Surrounding Department of Defense
Electromagnetic Radiation Safety Standards
*John de Lorge*

An Overview of the Proposed Industrial Hygiene Technical Standard for
Non-Ionizing Radiation and Fields for the U.S. Department of Energy
*John A. Leonowich*

Radiofrequency Safety Practice in the UK Ministry of Defence
*Bob Gardner*

Practical Control of Non-Ionizing Radiation Hazards
*Rick Woolnough*

DOD Implementation of New ANSI/IEEE Personnel
Radiofrequency Standard
*David N. Erwin and John Brewer*

# NAVY ISSUES SURROUNDING DOD-EMR SAFETY STANDARDS

John de Lorge[*]

Naval Aerospace Medical Research Laboratory
51 Hovey Road
Pensacola, FL 32508-1046

## INTRODUCTION

New standards for human exposure to safe levels of nonionizing radiation in the frequency range of 3 kHz to 300 GHz were recently adopted by the Institute of Electrical and Electronics Engineers (IEEE) (C95.1-1991,[1] revision of ANSI C95.1-1982). These standards will likely be incorporated into the Department of Defense Instruction (DODINST) 6055.11 and thereby into the Naval Operations Instruction (OPNAVINST) 5100.19B and 23B. Should this be the case, the new standards pose several issues of concern to the U.S. Navy. The new standards may impact the military in several ways. The standards are much more complex and will require new interpretations and more understanding of relationships between electromagnetic radiation (EMR) and humans. A two-tiered exposure standard will be imposed, and induced current will be used for the first time. In addition, guidelines will be provided for exposure to pulsed EMR and relaxations for magnetic fields will be allowed. In other words, a number of changes from the previous instruction will be imposed. These changes will provoke both known and unknown operational issues.

Among the known issues are (1) limits on radiofrequency- (RF) induced currents, (2) time duration for averaging measurements, (3) site uniqueness and measurement problems with RF heat sealers, and (4) appropriate measurement devices aboard ship. For the purposes of the issues mentioned in this paper, it is assumed that all exposures will occur in a controlled environment as defined in IEEE, C95.1-1991, and that aboard ship and at RF shore stations are controlled environments. The DODINST 6055.11 will cover both controlled and uncontrolled environments. It is essential that radiating devices be turned off at those times of public access. Otherwise, RF hazard signs will have to be posted during visits of nonmilitary personnel or personnel who have no knowledge of electromagnetic radiation issues. Moreover, exposure of the public to various physical agents aboard Navy ships is not a new concept. Ionizing radiation protection programs aboard nuclear-powered ships is a good example of the recognition of a two-tiered protection system. On the other

---

[*] These views and opinions are those of the author and do not necessarily state or reflect those of the U.S. Government.

hand, the two-tiered system might encourage some commands to impose the stricter public exposure limit in the mistaken belief that this will demonstrate safer operations or that their personnel are exposed to less risky conditions.

A number of other questions exist regarding the two-tiered system. Does the fence line at shore stations define controlled versus uncontrolled environments? Will there be a requirement for reporting overexposure of the public in the uncontrolled environment? In the uncontrolled environment, does one inform exposed individuals (staff or public) of overexposures?

## LIMITS ON RF-INDUCED CURRENTS

The C95.1-1991[1] standards established safety limits based on RF-induced current. This is the first time such limits have been imposed. If current is measured through 1 foot, a limit of 100 mA is applicable, and 200 mA if the current flows through both feet (in controlled environments). It is felt that these limits are overly conservative aboard Navy ships. It is not unusual during surveys aboard ship to frequently find these limits easily exceeded severalfold in the high frequency (hf) band (3-30 MHz) without even a sensation of ankle warming. As long as the specific absorption rate (SAR) is not exceeded, limits on electric and magnetic field strengths or power density can be exceeded. This is not allowed for induced body currents. Recent research shows that even at induced currents as great as 300 mA averaged spatial peak SARs may not exceed 20 W/kg at some HF frequencies above 20 MHz. Although the philosophy behind the induced-current limits is to protect against RF shocks and burns, there are many determinants of shock and burn other than induced or contact current. Induced current alone is probably not the best determinant of the intensity of shock or burn. A voltage and/or stored energy measurement should also be required.

Current measurements are only rough approximations to safety problems, because it is recognized that induced currents are determined not only by frequency and field strength, but also by a person's height and the types of shoes worn (sole material and thickness). Will ship personnel be assigned to radiation hazard (RADHAZ) areas because they are shorter and meet the standards? Will special shoes be issued for weather deck use?

There seems to be some confusion between induced current, contact current, and grasping current. Whereas all three involve induced current, the differentiation seems to be made by designating the body extremity through which the current flows, e.g., feet, hand or fingers. Additionally, if a very small area of the extremity is involved, a grasping current becomes a contact current. To determine the maximum contact current on a coated or insulated surface, one must penetrate the surface, i.e., break the paint; an utterly unreal situation aboard ship. Nevertheless, typical survey methodology does just that to simulate "worst case" situations, and ship certification can be affected.

In summary, the induced current limits will frequently be exceeded in many antenna locations aboard ship. Permissible exposure limit (PEL) lines would be overlapping and ignored. A more realistic method of determination of RADHAZ based on voltage and current is required.

## TIME DURATION FOR AVERAGING MEASUREMENTS

The most appropriate time period for averaging measurements for partial body exposure, especially the eye, is of concern. Averaging time is frequency dependent; minutes at resonant frequencies and seconds at near IR frequencies. The maximum permissible exposure (MPE) or PEL may be exceeded if the SAR averaged over the whole body does not exceed 0.4 W/kg in the controlled environment. The MPE may also be exceeded if the spatial peak value of the SAR averaged over any 1 g of tissue does not exceed 8 W/kg in a controlled environment. The appropriate averaging time for these measurements is 6 min up to 15 GHz, and it decreases to 10 s at 300 GHz. However, this is an assumption since the relaxation for partial body exposure does not address the issue of averaging time. Are the power densities provided applicable regardless of exposure time? If that is the interpretation, many problems will arise on the weather decks of ships as rotating radar beams briefly illuminate a variety of personnel.

## SITE AND MEASUREMENT PROBLEMS

### Heat Sealers

This issue involves RF heat sealers aboard ship and the measurement device typically used. This device is normally an isotropic probe that reads and sums three orthogonal fields. The reading that results is higher than the one SAR-producing component. The work of Dr. R. G. Olsen conducted with heat sealers located in shipyards and on subtenders has revealed a major problem associated with MPE determination. Current and SAR measurements vary widely depending primarily on the local building material. In wooden structures with a minimum of steel supports, low values are obtained with no MPE problem. On the other hand, with identical heat sealers located shipboard, up to five times as much current in the ankles can be observed. The shipboard room almost serves as a cavity-type oven with the heat sealer as a source. Depending on how the heat sealer is grounded, less of a problem occurs, but invariably the 200 mA limit is surpassed. This finding indicates that hazard determinations for heat sealers will require a site-by-site assessment.

### Practical Measures

Current Navy procedures for measurement of RF exposure require three measurements at 3, 4.5, and 6 feet above the deck, and the greatest value is used. The new standard states that 10 measurements 20 cm apart up to 2 m will suffice. Additional measurements can be included. In practice, a determination of whole-body average by this method will be time-consuming and will probably not be done consistently.

## MEASUREMENT DEVICES

The type of instrumentation to be used to obtain measures of E and H fields, SAR-induced current, and power density are not specified in the new standards although the companion volume, ANSI C95.3-1991: Recommended Practice for the Measurement of Potentially Hazardous Electromagnetic Fields - RF and Microwave,[2] does identify such instrumentation. In many cases, instruments will be used that measure only one component of the field when all components were used for setting the standard. In other cases, shipboard personnel will have no idea of what instrument to use. Specific absorption rate is a very unfamiliar term to many electronic technicians. Stand-on meters for induced current measurements are not standard issue. Without special assistance, fleet personnel will not be able to verify an MPE as new instrumentation comes onboard.

## CONCLUSIONS

The proposed adoption of IEEE C95.1-1991 safety level standards for DODINST 6055.11 will pose numerous problems for the U.S. Navy if modifications are not included. The modifications involve utilization of induced-current limits, the appropriateness of time averaging, and measurement issues. All of these problems are solvable, but the solutions are not all easy. For example, the induced-current measurements will be complicated by variations in human physiognomy and a reluctance to use the body as part of the measurement instrument. Only recently, have commercial instruments been made available for measuring induced current. Inspection and Survey (INSURV) teams and Naval Occupational Safety and Health Inspection Program (NOSHIP) personnel are typically unfamiliar with such instruments and will probably encounter many surprising situations. In particular, space constraints aboard ship will likely call for unique interpretations of a vertical E-field component. The resolution of these developing problems will probably depend entirely on a history of experience with the situations produced by the new standards.

## DISCLAIMER

The views expressed in this article are those of the author and do not reflect the official policy or position of the Department of the Navy, Department of Defense, nor the U.S. Government.

## REFERENCES

1. Institute of Electrical and Electronics Engineers. "IEEE Standard for Safety Level with Respect to Human Exposure to Radio Frequency Electromagnetics Fields, 3 kHz to 300 GHz," New York, NY, IEEE C95.1-1991 (1991).

2. Institute of Electrical and Electronics Engineers. "IEEE Recommended Practice for the Measurement of Potentially Hazardous Electromagnetic Fields RF and Microwave," New York, NY, IEEE C95.3-1991 (1991).

# AN OVERVIEW OF THE PROPOSED INDUSTRIAL HYGIENE
# TECHNICAL STANDARD FOR NON-IONIZING RADIATION
# AND FIELDS FOR THE U.S. DEPARTMENT OF ENERGY

John A. Leonowich and T. Edmond Hui

Applied Industrial Hygiene
Pacific Northwest Laboratory
Richland, WA 99352

## INTRODUCTION

The U.S. Department of Energy (DOE) is in the process of developing an industrial hygiene technical standard for non-ionizing radiation (NIR) and fields. The proposed standard aims to establish requirements for the control of occupational exposure to man-made sources of NIR and fields in DOE-owned or -leased facilities and operations. It is by far the most comprehensive standard to date proposed by any federal agency. Covered in the proposed standard is the whole spectrum of NIR and fields, including static electric and magnetic fields, sub-radiofrequency fields, radiofrequency, microwave, infrared, visible and ultraviolet radiation. Both non-coherent and coherent (laser) radiation is considered. Only exposures to NIR and fields from man-made sources, instead of those originating from natural phenomena such as sunlight, are considered.

Exposure limits specified in the DOE NIR standard are maximum permissible exposure (MPE) limits which are defined as the level of NIR or fields to which a person may be exposed to without hazardous effect or adverse biological change. The MPE limits are based on the American National Standard for Safe Use of Lasers (ANSI Z136.1-1993),[1] the IEEE/ANSI Standard for Safety Levels with Response to Human Exposure to Radio Frequency Electromagnetic Fields, 3 kHz to 300 GHz (IEEE/ANSI C95.1-1992),[2] and the American Conference of Governmental Industrial Hygienists (ACGIH) Threshold Limits Values (TLVs)© for physical agents.[3] Guidelines for high-powered microwave pulses, which are the subject of significant research in DOE, are currently under development and will be published as a supplement to the standard.

The DOE NIR standard has specified a list of requirements for an NIR protection program. These requirements include: sources of NIR and fields that need to be evaluated; MPE limits for different NIR and fields; the definition of a "competent person"; the requirements of a written safety program; the need for hazard assessments; the requirements for personal protection equipment (PPE), and warning signs; and employee medical surveillance, training, and recordkeeping. The purpose of this paper is to provide highlights of the requirements contained in the standard.

*Radiofrequency Standards*, Edited by B.J.
Klauenberg *et al.*, Plenum Press, New York, 1994

## SOURCES OF NIR AND FIELDS

The DOE has a vast number of equipment which can produce NIR and fields. Therefore, only those made-made NIR and field sources that can potentially exposure workers to the specified exposure limits are included for consideration. Sources of NIR and fields that require evaluation under the standard include:

1. Fixed radiofrequency/microwave emitters exceeding 10 mW radiated output (even if contained);
2. Lasers of class 2 or higher;
3. High current electrical equipment exceeding 1 gauss at 60 Hz;
4. High voltage electrical equipment exceeding 1 kV/m at 60 Hz;
5. DC magnets generating fields greater than or equal to 5 gauss at accessible places;
6. Walkie-talkie-type portable communications set capable of radiating over 7 W at frequencies between 100 kHz and 450 MHz or [7 X (450/f)] W at frequencies between 450 MHz and 1.5 GHz, where f is the frequency in MHz;
7. Induction heaters;
8. Satellite and permanent communications transmitters;
9. Infrared sources exceeding 10 W;
10. Ultraviolet sources exceeding 1 W, including mercury vapor lamps not used for lighting; and
11. Arc lamps.

## MAXIMUM PERMISSIBLE EXPOSURE LIMITS

The DOE MPE limits for radiofrequency and microwave radiation and sub-radiofrequency fields are established by combining the exposure limits specified in IEEE/ANSI C95.1-1992[2] for radiofrequency/microwave radiation and the ACGIH TLVs[3] for sub-radiofrequency fields. However, the IEEE/ANSI C95.1-1992[2] covers exposure limits for radiofrequency and microwave radiation from 3 kHz to 300 GHz, and the ACGIH TLVs[3] covers the exposure limits for sub-radiofrequency electric and magnetic fields from 0 to 30 kHz. For continuity between IEEE/ANSI and ACGIH, the standard considers radiofrequency and microwave radiation to be electromagnetic radiation with frequency between 4 kHz and 300 GHz. The MPE limits for radiofrequency and microwave electric and magnetic fields, and equivalent power density for controlled environments are shown in Figures 1, 2, and 3, respectively. The MPE limits for radiofrequency and microwave electric and magnetic fields, and equivalent power density for uncontrolled environments are shown in Figures 4, 5, and 6, respectively. The MPE limits for both controlled and uncontrolled environments for sub-radiofrequency electric and magnetic fields are shown in Figures 7 and 8, respectively.

The MPE limits for laser radiation are based directly on ANSI Z136.1-1993.[1] For non-coherent optical radiation which include ultraviolet, visible and infrared radiation, the MPE limits are adopted directly from the ACGIH TLVs.[3]

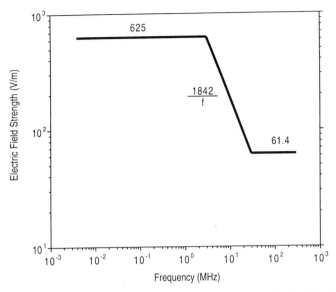

**Figure 1.** Maximum permissible exposure for radiofrequency/microwave electric fields for controlled environments.

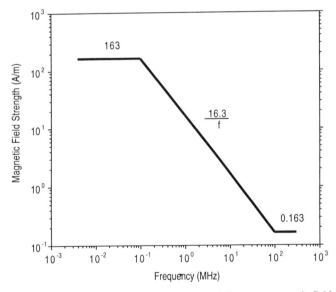

**Figure 2.** Maximum permissible exposure for radiofrequency/microwave magnetic fields for controlled environments.

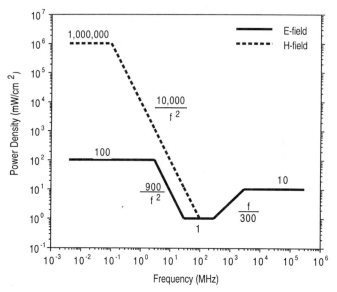

**Figure 3.** Maximum permissible exposure for radiofrequency/microwave power densities for controlled environments.

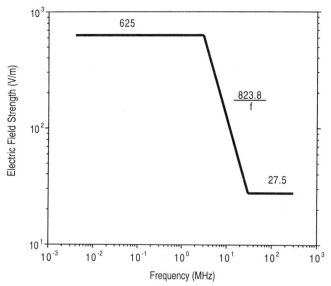

**Figure 4.** Maximum permissible exposure for radiofrequency/microwave electric fields for uncontrolled environments.

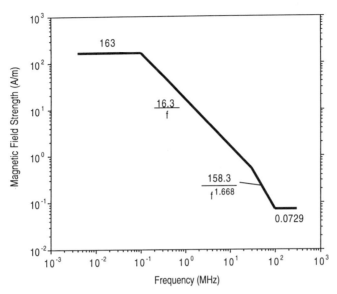

**Figure 5.** Maximum permissible exposure for radiofrequency/microwave magnetic fields for uncontrolled environments.

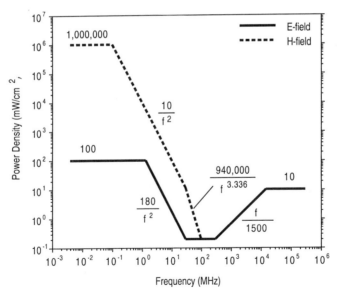

**Figure 6.** Maximum permissible exposure for radiofrequency/microwave power densities for uncontrolled environments.

123

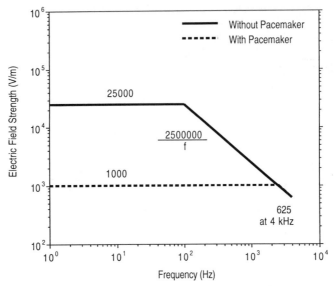

**Figure 7.** Maximum permissible exposure for sub-radiofrequency electric fields for both controlled and uncontrolled environments.

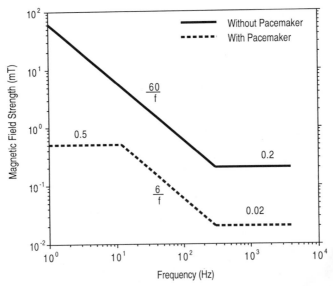

**Figure 8.** Maximum permissible exposure for sub-radiofrequency magnetic fields for both controlled and uncontrolled environments.

## COMPETENT PERSON

For facilities with operations which potentially expose workers to NIR and fields exceeding the MPE limits, a written NIR protection program as well as a person responsible to implement the program is required. The standard requires that a "competent person" be designated to be responsible for the protection of employees from the adverse health effects of NIR and fields. This "competent person" is a health and safety professional determined by the employer to be competent by reason of training and experience in one or more of the disciplines of NIR, its hazards, and means of abating them. The minimum requirements in the professional and academic experience of the "competent person" are specified in the proposed standard.

## WRITTEN SAFETY PROGRAM

The standard requires a written NIR protection program when workers are potentially exposed to levels exceeding the MPE limits. The written program includes descriptions of operations where sources of NIR are used an inventory of such emitters; engineering plans and any studies or data used to determine control effectiveness of engineering or other controls; procedures for controlled areas; procedures for emergency situations; the training program; and the experience and qualifications of the "competent person." The program is required to be reviewed and revised as necessary, and reviewed at least annually, to reflect any changes in the NIR exposure situation having the potential to pose a hazard to employees.

## HAZARD ASSESSMENT

A hazard assessment is required to be performed for each operation covered by the standard. A hazard assessment consists of two parts: a hazard evaluation and a hazard survey. Hazard evaluation is a technical evaluation conducted by a "competent person" on the basis of manufacturers' data and other information which identifies the occupational hazards potentially associated with the use of each NIR source at the facility and the controls necessary to protect employees from those hazards. Hazard evaluations are conducted prior to hazard surveys. The hazard survey is an on-site technical inspection of NIR systems or devices to determine employee exposures to NIR using appropriate monitoring equipment. Any NIR measurements taken during the hazard survey is expected to be representative of the NIR exposures of the workers.

For each workplace or work operation where employees are potentially exposed to NIR at or above the applicable MPE, the standard requires that an initial hazard assessment be performed of the workplace prior to the routine operation of a NIR source; periodic surveys to be conducted in a manner and at a frequency (in no case greater than once a year) that represents employee exposures to NIR; and additional hazard assessments to be conducted whenever there are changes in the mode of operation of the NIR source or the competent person requests such an assessment.

If the initial assessment reveals a potential for employee exposures to NIR to be at or above the MPE, controls are to be evaluated and implemented to reduce employee exposure to levels below the applicable MPE. Hazard surveys are to be repeated to confirm that employee exposures have been reduced to levels below the MPE.

The standard also requires that whenever monitoring results indicate that employee exposure exceeds the applicable MPE, each employee affected are to be notified in writing within 10 working days of receipt of the results of any monitoring performed.

## METHODS OF COMPLIANCE

For operation or workplaces where exposure may exceed the applicable MPE limits, the standard requires that engineering and work practice controls be implemented to reduce NIR to levels at or below the maximum permissible exposure limits. The "competent person," on a semi-annual basis, and more often if necessary, is required to determine that engineering and other controls are operating according to design and shall take such measures as necessary to ensure that these controls are maintaining employee exposures at or below the applicable MPE limit.

## CONTROLLED AREAS

In areas where the level of NIR or fields potentially exceed the MPE limits, the standard requires the area to be established as a controlled area. Access to, occupancy of, and activities within a controlled area are all controlled for the purposes of NIR protection. Only personnel required by work duties, after having been trained; and wearing the required personal protective equipment are allowed in controlled areas. Controlled areas are demarcated from the rest of the workplace in a manner that adequately establishes such an area and alerts employees to the boundaries of the controlled area. The names of all persons who are permitted to enter a controlled area are posted at the entrance to this area.

## EMERGENCY SITUATIONS

Any occurrence, such as the failure of engineering controls, that has the potential to immediately overexpose employees to NIR is considered to be an NIR emergency situations. Facilities engaged in NIR operations are required to develop and implement procedures for potential emergency situations and for alerting workers in the event of a NIR emergency.

## PERSONAL PROTECTIVE EQUIPMENT

Appropriate PPE, as well as training in its proper use, are to be provided to workers who may potentially be exposed at or above the applicable MPE limits. For lasers and non-coherent optical radiation, PPE includes protective eyewear for ocular exposure; and gloves and protective clothing for skin exposure. For workers working near electric fields exceeding 15 kV/m, protective clothing made of insulating material is provided to reduce the risk of electric shock. The standard does not recommend the use of protective clothing for entry into radiofrequency/microwave fields, since the technical information regarding the efficacy of such equipment is still incomplete. In all cases, the standard stresses the importance of the proper storage and care of PPE.

## MEDICAL SURVEILLANCE

The standard requires the establishment of a medical surveillance program for all employees potentially exposed to NIR at or above the MPE. Prior to initial assignment, workers will be provided with a preplacement medical examination. This examination will include an ophthalmological and dermatological assessment if the employee is working with class 3b or higher lasers, or ultraviolet sources. Medical assessments will also be performed periodically and immediately after a suspected NIR overexposure.

The examining physician is provided with descriptions of the employee's duties as they relate to the employee's exposure to NIR; any PPE to be used; and information from previous medical examinations of the affected employee that is not otherwise available to the physician. The content of the medical examinations will include a detailed medical history with particular attention to past NIR exposures, the use of photosensitizers, the presence of medical devices (e.g., cardiac pacemakers, aneurysm clips) that increase the employee's risk of experiencing NIR-induced adverse health effects, and the existence of medical conditions (e.g., aphakia, cataracts) that may be contraindications to NIR exposure; and a thorough physical examination with particular attention to the eyes and the skin.

## WARNING SIGNS

Appropriate and legible warning signs are required to be posted and maintained to demarcate controlled areas and entrances and access ways to controlled areas. Warning signs for the standard are as follows:

1. Lasers. The signs found in ANSI Z136.1[1] are adopted directly by the standard.
2. Radiofrequency/microwave radiation. The sign found in ANSI C95.2 [4] is adopted directly.
3. Ultraviolet/visible/near-infrared radiation. The sign shown in Figure 9 is for situations where the accessible levels of ultraviolet, visible, or near-infrared radiation have the potential to approach or exceed the applicable MPE.
4. Electric fields. The signs shown in Figures 10 and 11 are for alternating current (AC) electric fields where the field exceeds 5 kV/m and 15 kV/m, respectively. The sign shown in Figure 12 is for direct current (DC) electric fields where the field exceeds 25 kV/m.
5. Magnetic fields. The sign shown in Figure 13 is for alternating current (AC) magnetic fields at or approaching the MPE. The sign shown in Figure 14 is for direct current (DC) magnetic fields at or approaching the MPE.

The signs developed for non-coherent optical radiation, as well as sub-radiofrequency electromagnetic fields, are optional. Pre-existing signs which communicate the same information are allowed under the standard. The laser and microwave signs, which have been developed by consensus committees, are mandatory.

## EMPLOYEE INFORMATION AND TRAINING

A training program is required to be developed and implemented for all employees potentially exposed to NIR. The training program addresses the specific nature of operations that could result in exposure to NIR; the biological effects of NIR; the engineering controls and work practices associated with the employee's job assignment; the personal protective equipment associated with the employee's job assignment; and the specific procedures implemented to protect the employee. Training is provided to workers prior to assignment to a job involving potential exposure to NIR and at least annually thereafter.

In addition to specific job training, hazards communication awareness training is to be given to all employees working with lasers of class 3a or less, any ultraviolet source, and any source which could potentially expose an individual above the uncontrolled electromagnetic field limits stated in the standard.

## RECORDKEEPING

Exposure measurement and medical surveillance records are generated by the requirements of the standard. Exposure measurement records include the name, job classification, location, unique identifier, and exposure results of the employee monitored and of all other employees which the measurement is intended to represent; a description of the procedures used to determine representative employee exposure and the date(s) exposure was monitored; and the type of personal protective equipment (if any) worn by the monitored employee. Medical surveillance record for each employee enrolled in the medical surveillance program include the name and unique identifier of the employee; the physician's written opinion; copies of the physician's written reports relating to all evaluations and tests specified in the standard; any employee medical complaints related to NIR exposure; and a copy of the medical history and the results of any physical evaluation and all test results which are required to be provided by this standard or which have been obtained to further evaluate any condition occurring as a result of NIR exposure.

Records are required to be maintained for the duration of the worker's employment and for a period of at least 75 years beyond. These records are maintained in a manner to ensure employee confidentiality as well as all requirements of the Federal Privacy Act.

## SUMMARY

The DOE NIR standard marks an ambitious attempt to provide users with a single document which describes all the necessary requirements for an NIR protection program. This standard will be made available to other federal agencies, including OSHA, as a "model program" document for potential adoption to their particular needs.

## ACKNOWLEDGMENT

This work was prepared for the U.S. Department of Energy under Contract No. DE-AC06-76RLO 1830. Pacific Northwest Laboratory in Richland, Washington, is operated for the U.S. Department of Energy by Battelle Memorial Institute.

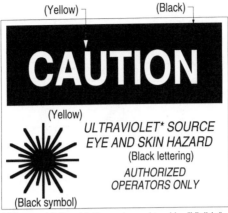

(Yellow) ⌐          (Black) ⌐

# CAUTION

(Yellow)

ULTRAVIOLET* SOURCE
EYE AND SKIN HAZARD
(Black lettering)

AUTHORIZED
OPERATORS ONLY

(Black symbol)

\* The word "Ultraviolet" may be replaced by "Visible"
or "Infrared" as needed

**Figure 9.** Warning sign for accessible ultraviolet radiation.

(White) ⌐          (Black) ⌐

# CAUTION

(White)                    (Yellow)

XX*
Hz

(Black
symbol)

Strong ac electric field

Pacemaker users keep out!
Irritating sparks possible.
Use insulating garments/equipment.
Authorized personnel only.
(Black lettering)

\* Place appropriate subradiofrequency

**Figure 10.** Warning sign for AC electric fields that exceed 5 kV/m.

* Place appropriate subradiofrequency

**Figure 11.** Warning sign for AC electric fields that exceed 15 kV/m.

**Figure 12.** Warning sign for DC electric fields that exceed 25 kV/m.

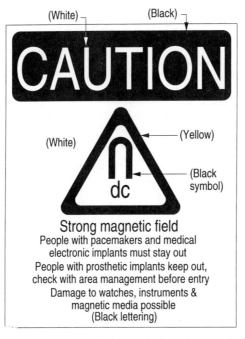

**Figure 13.** Warning sign for DC magnetic fields that exceed 0.5 mT.

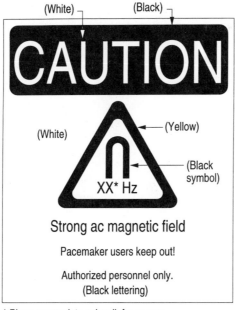

\* Place appropriate subradiofrequency

**Figure 14.** Warning sign for AC magnetic fields that exceed pacemaker MPE at the given frequency.

# REFERENCES

1. American National Standards Institute (ANSI), 1993. *American National Standard for Safe Use of Lasers*, ANSI Z136.1-1993, The Laser Institute of America, Orlando, FL.

2. Institution of Electrical and Electronic Engineers (IEEE), 1992. *IEEE Standard for Safety Levels with Response to Human Exposure to Radio Frequency Electromagnetic Fields, 3 kHz to 300 GHz*, IEEE C95.1-1992, IEEE, Piscataway, NJ.

3. American Conference of Governmental Industrial Hygienists (ACGIH), 1993. *The 1993-1994 Threshold Limit Values for Chemical Substances and Physical Agents and Biological Exposure Indices*, ACGIH, Cincinnati, OH.

4. American National Standards Institute (ANSI), 1982. *American National Standard for Radiofrequency Radiation Hazard Warning Symbol*, ANSI C95.2-1982, ANSI, New York, NY.

# RADIOFREQUENCY SAFETY PRACTICE
# IN THE UK MINISTRY OF DEFENCE©

Bob Gardner


Directorate of Defence Health and Safety, MOD
Aquila, Golf Road, BROMLEY, BR 2JB, UK

Currently there is no UK legislation specific to personal exposures to electromagnetic fields and waves below 300 GHz. However, in the past there has been and still is guidance on this matter.

In 1960 the Post Office published Safety Precautions Relating to Intense Radiofrequency Radiation. The guidance it gave came from the Home Office and was based upon recommendations made by the Medical Research Council and set an acceptable level of 100 $W.m^{-2}$ for continuous exposure. Further practical guidance on implementation was given for:

- members of the general public
- operating and maintenance personnel, and
- personnel at research and experimental establishments and factories.

This was further refined by Ministry of Defence (MOD) in conjunction with the Medical Research Council as a level of 1000 $V.m^{-1}$ in the near field for frequencies of 1 to 30 MHz. This together with a restriction on exposure of 10 $W.h.m^{-2}$ in any 0·1 h period in the frequency band 30 MHz to 30 GHz and a contact current of 200 mA was adopted by the MOD and promulgated as a standard. They are still MOD's basic restrictions for existing equipment.

In 1989 new national guidance was published by the National Radiological Protection Board (NRPB) as Guidance on Standards Report 11 (GS11) - *Guidance as to Restrictions on Exposure Time Varying Electromagnetic Fields and the 1988 Recommendations of the International Non-Ionising Radiation Committee.* It is general policy that new equipment should be designed and installed to conform to GS11. However, much equipment exists designed to conform to the previous standard. It is our policy that these should be modified or operated to conform to GS11--"wherever this is reasonably practicable."

---

I should explain the position of MOD with regard to legislation. In 1974 the UK government passed the Health and Safety at Work, etc., Act (HSWA). (N.B. The timing of this Act is between 1960 and 1989 and the "etc." implies duties in regard to persons other than employees affected by your work activities, e.g., visitors and the public.) MOD is not exempt from this Act or any Regulations that are made under it, examples of which are:

- The Ionising Radiation Regulations
- The Control of Substances Hazardous to Health Regulations
- The Noise at Work Regulations
- The Management of Health and Safety at Work Regulations

Therefore the MOD has a "duty of care" to its employees (sailors, soldiers, and airmen included) and is required to provide "a safe place of work" under HSWA. It also has to abide by specific Regulations, but may seek exemptions on grounds of national security. It is policy to use this latter provision sparingly and only where truly essential and not for administrative convenience.

Where specific regulations do not exist, MOD will conform to national or international standards and guidance wherever practicable. Where this is not possible, it must make an assessment of the risks to its employees and record this if they are "significant." Such assessments are open to inspection by the enforcing authority for such legislation, the Health and Safety Executive (HSE). After inspection, HSE may instruct the MOD to make improvements or prohibit the practice.

MOD is content, therefore, while there is only guidance and standards in the RF field. At present it is carrying out surveys to establish where assessments are necessary to determine what problems it may have if revised guidance and standards are produced, or indeed legislation.

As I have said, the guidance, at present, is NRPB GS11, Table 4 of which is reproduced here:

| Item | Restrictions |
|------|--------------|
| 1 | The continuous induced current in any arm, hand, leg, ankle or foot should not exceed: $$\left[ 1 + \frac{f\,(Hz)}{1500} \right] mA \text{ or } 100\, mA$$ whichever is smaller for frequency f less than 30 MHz |
| 2 | The average specific energy absorption rate in the body over any 6 min should not exceed $0 \cdot 4$ W.kg$^{-1}$ |
| 3 | When taken in conjunction with 2 above, the maximum value of the specific energy absorption rate in any $0 \cdot 1$ kg of an internal organ or tissue in the head or trunk over any 6 min should not exceed 10 W.kg$^{-1}$ |
| 4 | When taken in conjunction with 2 above, the maximum value of the specific absorption rate in any $0 \cdot 1$ kg of an arm, hand, leg, ankle or foot in any 6 min should not exceed 20 W.kg$^{-1}$ |
| 5 | Exposures to time integrated power densities in any pulse of duration less than 50 μs exceeding $0 \cdot 4$ J.m$^{-2}$ should be neither prolonged nor frequent |
| 6 | Radiofrequency burns from objects in the field should be avoided |
| 7 | Any uncomfortable sensation of heat in the superficial layers of the body should be avoided at frequencies above 1 GHz |

Table 5 of GS11 NRPB, gives Derived Reference Levels (DRLs) and, for the sake of completeness, is reproduced here:

(a)    For frequencies below 30 MHz: electric and magnetic fields to be separately considered

| Frequency | Root mean square values | | |
|---|---|---|---|
| | Electric field strength $(V.m^{-1})$ | Magnetic field strength $(A.m^{-1})$ | Magnetic flux density * (mT) |
| <100 | 614,000/f  (Hz) | 1630 | 2 |
| 0·1 - 1 kHz | 614/f       (kHz) | 163/f     (kHz) | 0·2/f          (kHz) |
| 1 -30     kHz+ | 614 | 163 | 0.2 |
| 0·03 - 1 MHz+ | 614 | 4·89/f   (MHz) | 6 x 10-3/f  (MHz) |
| 1 -10    MHz+ | 614/f | 4·89/f   (MHz) | 6 x 10-3/f  (MHz) |
| 10 - 30  MHz+ | 61·4 | 4·89/f   (MHz) | 6 x 10-3/f  (MHz) |

Note: f = frequency in the units shown within the brackets

\* Magnetic flux density in tissue is given as an alternative to the equivalent magnetic field strength.

+ At these frequencies the reference electric field strength in uncontrolled areas of public access should be one-third of the values in the table because of the possibility of electric shock or radiofrequency burns from exceptionally large ungrounded objects in the field.

(b)    For frequencies above 30 MHz: electric field strength, magnetic field strength and power density to be taken as equivalent alternatives under far-field conditions

| Frequency | Root mean square values | | |
|---|---|---|---|
| | Electric field strength $(V.m^{-1})$ | Magnetic field strength $(A.m^{-1})$ | Power density $(W.m^{-2})$ |
| 30 - 400 MHz | 61·4 | 0·163 | 10 |
| 0·4 - 2   GHz | 97·1√f  (GHz) | 0·258√f  (GHz) | 25f  (GHz) |
| 2 - 300  GHz | 137 | 0·364 | 50 |

Note: f = frequency in the units shown within the brackets

NRPB is currently revising this guidance. It is important to remember that these latter are DRLs with pessimistic assumptions built into them. Exposure above these levels is acceptable if properly managed, by which I mean that the risks are assessed and, if need be, balanced against social (or operational) and economic factors. A quote from GS11 will serve to illustrate this:

"*These reference levels should not be regarded as limits* and should be used to indicate a requirement for guidance as to the appropriate administrative and technical measures to limit exposure. The levels are **extremely** (my emphasis) conservative, particularly for partial and non-uniform exposures in the near field,...."

Most importantly, NRPB did not, and do not intend to, make a basic restriction for contact current, but gave only a DRL.

GS11 shows these graphically in a manner similar to Figures 1 and 2:

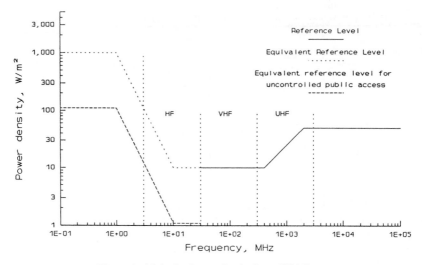

**Figure 1.** Advised reference levels above 100 kHz.

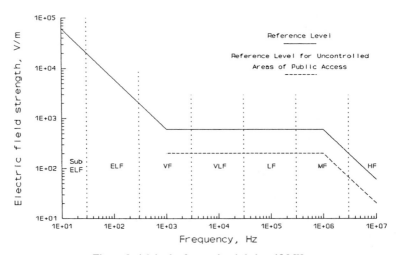

**Figure 2.** Advised reference levels below 10 MHz.

The MOD is divided into 4 main areas: Navy, Army, Air Force and the Procurement Executive. Within the latter there used to be many, mainly civilian, research and development establishments, but most of these now form the Defence Research Agency, an autonomous organization accountable to the Secretary of State for Defence. Whatever policy MOD adopts will have to be applied in all of these areas.

Outside the MF to UHF resonance region, it does not appear that MOD will have difficulty meeting NRPB GS11 or their revised reference levels. However, within that region two major problem areas have been identified where it may not be possible to move to NRPB GS11 or subsequent guidance. The first are several land-based HF stations, which when installed used the criterion of $1000V.m^{-1}$ to establish their perimeter. The land beyond the perimeter is privately owned and in some cases has been used for residential developments.

The second is on naval vessels where the confines of space are such that much of the upper deck may reach levels in excess of the guidance. In some cases this is the only available deck space that is not out of bounds for operational reasons. It is undesirable to exclude personnel unless for overriding safety reasons, i.e., the detriment caused by such a reduction of recreational space must be a factor in the decision. That the areas affected are large has been demonstrated by a program of surveys currently being carried out on ships of the Royal Navy. This may be a major problem.

The surveys consist of measurements of electric field strength and magnetic field strength around aerials, etc., to identify the reference-level boundaries for the current equipment and GS11 restrictions. Measurements of ankle current are made at those boundaries. Measurements of contact current and contact voltage are also made on metallic structures nearby. The surveys have been carried out using novel equipment specially designed for the purpose. These are robust., easy to use, easy to operate, and have a good common mode rejection.

Both these problems occur in the near field where the guidance needs careful interpretation and serve to highlight a potential problem with the use of such figures in legislation.

# PRACTICAL CONTROL OF NON-IONIZING RADIATION HAZARDS

Rick Woolnough

RADHAZ/Laser Safety Specialist
British Aerospace Defence Ltd.,
Military Aircraft Division,
Warton Aerodrome, PRESTON PR4 1AX
Lancashire, England, United Kingdom

## ABSTRACT

The levels of exposure of individuals to non-ionizing radiation (NIR) within the frequency range of 3 kHz to 300 GHz have been detailed in a number of national standards. Most of these standards within the North Atlantic Treaty Organization (NATO) countries are based on the principles of induced body currents for the lower range of frequencies, and the thermal or specific absorption for the higher frequencies.

This paper describes methods of practical control of Non-ionizing Radiation Hazards (RADHAZ) with a view to ensuring an adequate yet flexible approach to RADHAZ control measures throughout NATO. It describes the relationship between RADHAZ and other safety factors associated with typical Radio Frequency Emitting Systems (RFE).

Due to the diverse nature of RFE that may be encountered, identification of the hazards is often difficult. This paper describes an ordered approach to the control of both RADHAZ and secondary hazards which, if ratified, will permit a consistent approach to RADHAZ throughout NATO.

Reference is made in this paper to other NATO Standardization Agreements which may be found to be applicable or of assistance.

## INTRODUCTION

The levels of exposure of individuals to non-ionizing radiation (NIR) within the frequency range of 3 kHz to 300 GHz have been detailed in a number of national standards. Most of these standards within the North Atlantic Treaty Organization (NATO) countries are based on the principles of induced body currents for the lower range of frequencies, and the thermal or specific absorption for the higher frequencies.

*Radiofrequency Standards*, Edited by B.J.
Klauenberg *et al.*, Plenum Press, New York, 1994

The purpose of this presentation is to help appraise a level of practical guidance that can be included within this NATO Standardization Agreement (STANAG). In the United Kingdom (U.K.) the Ministry of Defence has already prepared guidelines in Defence Standard 05-74[1]. This Defence Standard appears to offer an excellent starting point for a practical approach to any guidance that this forum proposes.

## PRACTICAL CONSIDERATIONS

### General Considerations

Due to the flexibility of military organizations under various regimes of risk, any measures offered by this forum should be only guidance. It appears reasonable to assume that, given the risks associated with personnel under combat conditions, this forum would find it impossible to write precise operational methods that would fully evaluate all of the occurrences and circumstances within which this guidance may be applied or disregarded.

The relativity of all of the hazards associated with Radio Frequency Emitters (RFE) must be reviewed in this guidance. By this it is not proposed to assess formally the relative health risks from different physical agents, but to evaluate what the hazards and effects may be, under different circumstances. The effects of RF Radiation on Electro-explosive Devices (Draft STANAG 4238)[2], safety critical electronic systems and flammable vapors may be more catastrophic than the effects on biological systems. Due reference to standards for the reduction of risk to such systems should therefore be included.

### Hazards Associated with Radio Frequency Emitters (RFE)

As a review aid, Figure 1 illustrates in a block diagram a typical mechanically scanned radar system and some typical hazards associated with the system.

The Antenna system, as well as producing microwave radiation, is a rapid and variable scanning system. Therefore it is capable of causing severe injury, including broken limbs and death, by mechanical shock. It is a flexible, rapid reacting system and is powered by a very high pressure hydraulic system that could cause death, by injecting hydraulic fluid into the blood-stream, or the amputation of extremities. The hydraulic fluid itself is an irritant to both eyes and skin. Overexposure to the fluid may cause dermatitis and skin cancer; overexposure to the hydraulic mist can cause lung cancer.

The Transmitter/receiver, as well as producing microwave radiation, also contains extra high voltage power supplies that, as well as presenting a lethal shock hazard, may be capable of generating X-rays. There may be exposed thermionic valves that are capable of explosion and that may cause skin burns if touched. Components within the transmitter may be manufactured of hazardous materials such as beryllium.

The Display Unit has all of the above mentioned secondary hazards associated with the transmitter. However the priority of these risks must be re-arranged. The explosion hazard becomes important during normal operation in which the operator is looking at the display. An additional hazard that has become more publicized recently, is that of repetitive strain injury associated with using the controls or a keyboard at the operator console.

Looking at the overall system, all the units are heavy and, to avoid serious injury, may require lifting apparatus. They may contain potentially hazardous components, such as screws that are cadmium plated for corrosion resistance and other operational needs.

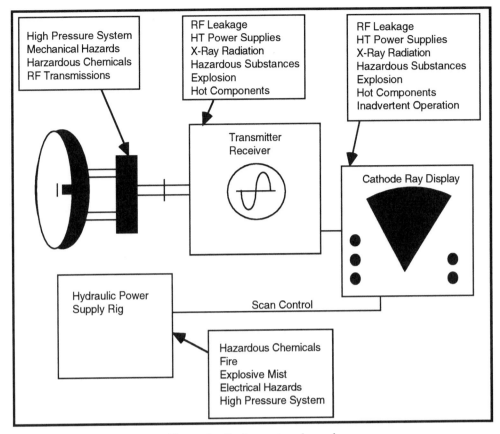

Figure 1. Radar system hazards.

As can be seen in Table 1 below, the majority of the hazards of a typical radar system may kill or severely maim. It is very likely that of all the accidents that have occurred, within this area of employment, the majority of those that have resulted in death or severe injury will have been due to the secondary hazards and not RF Radiation.

It is therefore important that a brief description of the types of secondary hazards that may occur is included within this proposed standard. Additional reference to those STANAGs that control these types of hazards should be included under Applicable Documents. Where STANAGs have not been published, the relevant International or National standards should be referenced.

## Confounding Factors & Errors

A realistic evaluation of measurement errors is readily available by reference to a large number of books, articles and papers. Theoretical assessment errors can be both difficult and time consuming to reveal. The complexity of carrying out multiple emitter assessments is further complicated by height errors when plotting zones on maps. Multiple emitter assessments may be treated in a quantitative manner where the probability of encountering a maximum in excess of the prescribed limits is evaluated. This technique may be useful under difficult operating conditions.

**Table 1.** Secondary Hazards of a Radar System. (All hazards are dependent on REAL system setup.)

| System Component | Hazard from System | Worst Possible Injury | Most Likely Injury |
|---|---|---|---|
| Radar Scanner | Mechanical Shock | Brain Damage/Death | Crushed Limbs |
| | High Pressure Oil | Lethal Oil Injection | Lethal Oil Injection |
| | Hazardous Substances | Skin/Lung Cancer | Dermatitis/Skin Disorders |
| | RF Transmissions | Heat Stroke/Death | RF Burn/Cataracts |
| | Fire | Death/Suffocation | Suffocation/Burns |
| Transmitter/Receiver | RF Leakage | Burn/Cataracts | Burn/Cataracts |
| | HT Power Supplies | Electrocution/Death | Electrocution/Death |
| | X-Ray Radiation | Cancers/Death | Cancers/Death |
| | Hazardous Substances | Death (Beryllia) | Death (Beryllia) |
| | Explosion of Valves | Blinding/Cuts | Blinding/Cuts |
| | Hot Components | Skin Burns | Skin Burns |
| Cathode Ray Display | All As Transmitter | All As Transmitter | All As Transmitter |
| | Inadvertant Operation | All Hazards at Worst | All Hazards At Worst |
| Hydraulic Power Unit | Hazardous Substances | Skin/Lung Cancer | Dermatitus/Skin Disorders |
| | High PressureOil | Lethal Oil Injection | Lethal Oil Injection |
| | Fire | Death/Suffocation | Suffocation/Burns |
| | Electrical Hazards | Electrocution/Death | Electrocution/Death |

## The Layered Approach to Control

In a military organization a consistent approach is necessary to ensure teamwork and cooperation. A reasonable approach to the needs of a military RADHAZ Control Program appears to be a layered method of control in which the following situations will occur.

1. Policy setting will take place at the highest level. This policy will determine the precise route to achieving full control of the hazards and assessing the risks under active Service conditions.
2. Control measures and hazard assessments will be collated centrally and disseminated to individual locations to form part of the overall local hazard database.
3. For particular installations and locations the local control measures should be operated on site to ensure that real-time monitoring of the local situation occurs.

However, the true realization of a safety program must lie with its acceptability to the organization implementing it. If the guidance of this proposed STANAG does not allow for such contingencies then all of the rules could be thrown away at the time when NATO forces need the most flexibility and the most effective protection of Service personnel against all risks.

## CONTROL MEASURES

### Military RADHAZ Control Policy

Military RADHAZ Control Policy should define the ground rules for the implementation of the RADHAZ control measures. It should state the exposure levels that may or may not be exceeded under the various operational conditions of training and warfare. It should also define the layers of responsibility of personnel and their training requirements when working near RADHAZ zones.

### Safety in Documentation

When preparing operational, maintenance and training manuals it is essential for them to be written with safety as a prime consideration. In today's world of litigation and liability it is beneficial to suppliers of RFE to provide as much information as possible within the manuals to ensure that customers' personnel can use the RFE both safely and efficiently.

As a general rule all documentation should be written with clarity, brevity and must always be unambiguous. The information required to carry out suitable RADHAZ assessments should always be provided to the customer.

The most prudent time to request that this information be provided is in the contract conditions. If the correct requirements are stated at the start of a project, then the cost of providing such information is minimized, which is good for both the supplier of the RFE and the procurement agency.

### Assessment of Hazards

The assessment of hazards posed by RFE should also be carried out before being operated in an environment where it may cause a hazard. In order that a supplier of RFE can carry out the assessment, approved methods and systems of assessment should be provided by the procurement agency to the supplier. The assessment itself may, upon agreement between the supplier and the procurement agency, be carried out by the procurement agency and the findings disclosed to the equipment supplier for inclusion in the safety documentation. A sample flow for a safety assessment is included in Table 2.

### Personnel Training Structure

The overall training structure will be defined in the Military Non-ionizing Radiation Safety Policy. From this policy and the assessment of the hazards will evolve the training requirements for personnel using RFE.

Training for a specific RFE should be based upon the needs of the operator and the degree of hazards posed. We at Military Aircraft Division, Warton have covered this in our Non-ionizing Radiation Hazards Control Procedures[3] have used the safety assessment system to determine the level of training requirements.

**Table 2.** Safety Assessment.

| AUTHORITY | TASK | REQUIREMENTS |
|---|---|---|
| National Military Authority | On implementation or change to RADHAZ Policy | Define Military RADHAZ Policy |
| National Defence Staff | Develop Staff Requirement or Staff Target for RFE, weapons system, new aircraft ship etc., | Ensure that the Staff Requirement or Target is forwarded to the relevant Military Specialists to carry out a full hazard evaluation which should include an assessment of risks due to Radio Frequency Radiation. |
| Military Procurement Authority | Prior to procurement or placing final contract | Ensure that supplier of RFE is provided with a copy of the RADHAZ Policy and Defence Standard. Ensure that the supplier is fully aware of the limits of their responsibility and the requirements for adequate documentation. |
| RFE Supplier | During Development | Carry out risk assessment on full system. Ensure safety risks are minimized & highlighted. Ensure documentation reflects level of risk assessment carried out. Prepare clear, precise and unambiguous Operating and Maintenance Procedures that are non-hazardous. |
| Military Procurement Authority | Prior to Release to Service | Ensure documentation is adequate to ensure safe operation |
| National Defence Staff | Prior to issue of RFE to user Commands | Ensure adequate information is passed to local RSO(NI) to update site RADHAZ Map. |
| Radiation Safety Officer (Non-ionizing) | During Use | Ensure local control measures applied and personnel are adequately prepared for use. |

## Radiation Safety Officer (Non-ionizing)

The appointment of a responsible person to oversee the safe control of Non-ionizing Radiation Hazards [RSO(NI)] is an important factor. The task of RSO(NI) is a focal one for a local situation such as an air-base or a ship. The RSO(NI) should be in charge of establishing control measures that are applicable within his area of responsibility, as well as acting as a central focus for RADHAZ queries.

The organization level at which an RSO(NI) is appointed depends upon the operational need and the perceived risks within the allocated area of responsibility. The RSO(NI) should be fully conversant with the methods of control and assessment of the hazards within his area of responsibility.

## Preparation of a RADHAZ Map

The RADHAZ Map is a very important tool for a RSO(NI). The map should show the emitters on site, susceptible device storage and movement area and prohibited areas for both emitters and susceptible devices (including personnel). Multiple emitter criteria should be included on the RADHAZ Map so as to ensure that safety margins incorporated within assessment guidelines are not eroded.

In order to assist the RSO(NI) in the assessment of hazards and preparation of a RADHAZ map, a register of all emitting RFEs and susceptible devices within the allocated area of responsibility should be kept. Prior to when a new RFE is brought onto the Unit an assessment of the impact of that RFE should be carried out, the RADHAZ Map should be amended and a letter of clearance sent to the appropriate provisioning authority and user areas.

## Responsible Person

For each RFE that presents any hazard outside of its physical enclosure, a responsible person should be appointed. The duties of the responsible person should include liaison with the RSO(NI) and ensuring that safety procedures and devices are implemented and maintained.

## Medical Surveillance

Medical surveillance within this field of NIR should be geared to the hazards associated with the RFEs in question. There is a need to assess prosthetic devices, especially active devices, for safety within an electromagnetic environment. Within the context of this STANAG the requirement to evaluate any prosthetic devices is probably the prime consideration for any medical surveillance.

Ophthalmic examination may identify the formation of cataracts within the eye. However the current state of our present knowledge indicates that cataract formation will occur only following very high levels of overexposure to NIR. Therefore ophthalmic examination appears to be unwarranted if adequate engineering and/or administrative controls limit the maximum NIR personnel exposure levels.

There may be a concern with the effectiveness of the human body to thermoregulate itself when treated with certain drugs. However, any evidence that the maximum permissible exposure levels do not fully cater for this reduction in thermoregulatory efficiency is equivocal.

Except for personnel fitted with prosthetic devices it appears reasonable to say that medical surveillance appears not to form part of a program of control measures to protect personnel exposed to NIR, and should therefore only be considered necessary for medical-legal reasons.

Following overexposure to NIR, medical treatment should be symptomatic and should be considered on a case-by-case basis. If the exposure is considered to be adequate for cataractogenesis to occur then the accident could be considered to be very serious and a full ophthalmic examination should be carried out, within a suitable timescale to ensure that the maximum effect may be observed.

## Accident and Incident Reporting

Accident and incident reporting should follow the standard national requirements for accident and incident reporting in the relevant RFE operator.

## CONCLUSIONS

This forum must provide guidance simple, practical and flexible enough to permit a realistic evaluation of the real risks.

The secondary hazards of a RFE in nearly all cases pose the greatest cumulative risk to personnel and recognition of this is vital.

A layered approach to RADHAZ control with good liaison between the procurement agency and the system supplier will permit a coherent approach to RADHAZ throughout NATO.

With the exception of the evaluation of prosthetic devices medical screening appears not to form part of a RADHAZ Control Policy

The guidance given by this forum should permit the flexibility required by NATO Forces by appreciating the relative risks and should contain enough guidance to produce a consistent approach throughout NATO and thereby avoid confusion when mixed nationality forces are operating together.

## ACKNOWLEDGMENTS

The author wishes to thank British Aerospace Defence Limited Military Aircraft Division for permission to publish this paper.

Additionally the Author would like to thank Mr. Mike Pywell, Warton Unit Radiation Safety Officer (Non-ionising), for input to and comment on this paper.

## DEDICATION

This paper is dedicated to the memory of John C. Coppola (4 Sept. 1931 - 16 June 1993), former Vice President of Marketing for Loral Microwave-NARDA. A Friend and an Inspiration.

## REFERENCES

1. Defence Standard 05-74 Guide to the Practical Safety Aspects of the use of Radio Frequency Radiation 1989. Published by and obtainable from: Ministry of Defence, Directorate of Standardisation, Kentigern House, 65 Brown Street GLASGOW G2 8EX.

2. Draft STANAG 4238 Design Principles for hard munitions against electromagnetic environments NATO Military Agency for Standardization.

3. BAe-WAD-RP-GEN-RAD-0140 ISSUE 1 The Control Of Non-Ionising Radiation Hazards Within The Warton Unit R.C.N. Woolnough 1992.

# DOD IMPLEMENTATION OF NEW ANSI/IEEE PERSONNEL RF STANDARD

David N. Erwin and John E. Brewer[*]

AL/OER, 8308 Hawks Road
Brooks AFB, TX 78235-5324

## INTRODUCTION

The United States Department of Defense (DoD) is adopting the provisions of the Institute of Electrical and Electronic Engineers (IEEE) C95.1-1991, "IEEE Standard for Safety Levels with Respect to Human Exposure to Radio Frequency Electromagnetic Fields, 3 kHz to 300 GHz,"[1] as its new Radio Frequency Radiation (RFR) protection standard under DoD Instruction 6055.11. The IEEE standard was approved by the IEEE in 1991 and by the American National Standards Institute (ANSI) in 1992, and is a revision of the ANSI C95.1-1982[2] standard. It is a significant improvement over previous standards. The Department of Defense Instruction (DoDI)[3] was drafted by the Tri-service Electromagnetic Radiation Panel (TERP), which includes representatives from each of the service's Surgeon's General offices, RF bioeffects research functions, and the RF consultation functions. Presently, the Air Force chairs the TERP.

## BACKGROUND

The DoD enforces these standards through each of the services' health and safety standards. The United States Air Force will adopt this standard after the DoDI is approved. Presently, the Air Force RF standard, Air Force Occupational Safety and Health (AFOSH) Std 161-9, 1987,[4] was based on the ANSI C95.1-1982 standard for unrestricted areas and the American Conference of Governmental Industrial Hygienists (ACGIH) 1984 standard[5] for restricted areas. The standard is summarized in Figure 1.

This dual tier standard is applied to public areas and to the workplace. The standard establishes Permissible Exposure Limits (PELs) which are expressed as far-field equivalent power densities. These PELs are based primarily on the heating potential of the incident energy in the "resonance" portion of the curve. The absorbed power is limited to a specific

---

[*] These views and opinions are those of the author and do not necessarily state or reflect those of the U.S. Government.

absorption rate (SAR) of 0.4 W/kg. Also included on the graph is a representative 0.4 W/kg SAR curve.

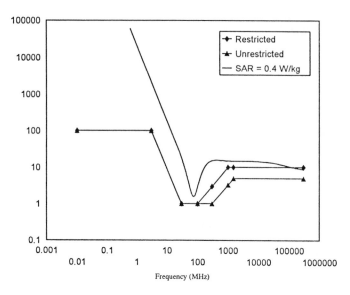

**Figure 1.** AFOSH Std 161-9, 1987.

## IEEE C95.1-1991

The IEEE C95.1-1991 standard is diagrammed in Figure 2. The new standard establishes maximum permissible exposures (MPEs). These MPEs are expressed in the standard as field strength and far-field equivalent power densities.

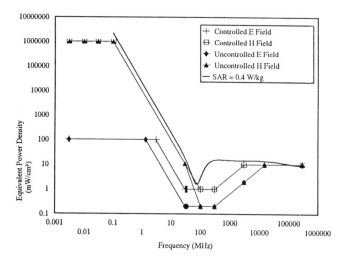

**Figure 2.** IEEE C95.1-1991.

As shown in this diagram, the standard is still based on the human body absorption curve, and has two tiers, but has been greatly relaxed for lower frequency magnetic field exposures. It should be noted that the basis of the two tier standard is not whether a person is "occupational" or not, but instead on whether the area is controlled or not. This important concept is also being adopted by other standards bodies, such as ACGIH. The lower frequency magnetic field limits were relaxed because the human body does not couple energy from magnetic fields as easily as from electric fields (at equivalent far-field power densities). This phenomenon is observed in the lack of field perturbations by personnel standing in a field during magnetic field measurements.

The standard also establishes new frequency dependent averaging times. These are shown overlaid with the MPE curves in Figure 3.

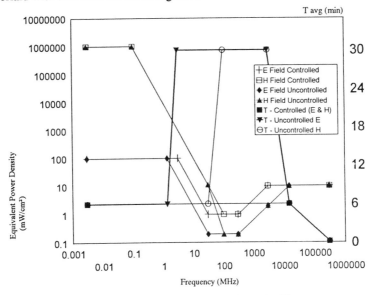

**Figure 3.** Equivalent Power Density vs Averaging Time.

IEEE C95.1-1991 allows somewhat higher exposure field values, if the exposure is only to part of the body (e.g. limbs). This relaxation does not apply to the eyes or gonads. These additional provisions are shown below.

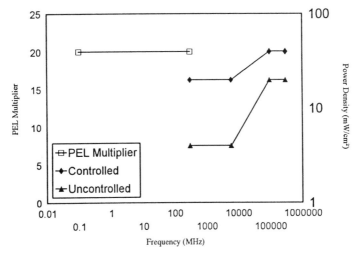

**Figure 4.** Relaxations for Partial Body Exposures.

149

Additionally, the standard establishes a new concept of induced current limits. These currents can come from induction or from contact. The new standard incorporates concerns that these currents have the potential to produce shock, burn, and heating in the narrow cross sections of the limbs (e.g. ankles). These are displayed in the following chart.

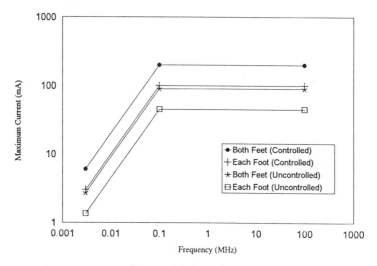

**Figure 5.** Maximum Current.

The standard also has several exclusions beyond those described already. These exclusions apply to low power devices, current densities, and several SAR issues. These are detailed in Table 1.

**Table 1.** Exclusions.

| | 3 kHz - 100 kHz | 100 kHz - 6 GHz | | | >6 GHz |
|---|---|---|---|---|---|
| | Current Density | SAR | | | |
| | | Whole Body | Spatial Peak | Hands, Wrists, Feet, & Ankles | |
| Controlled | <35f mA/cm² (1 cm², 1s) | 0.4 W/kg | 8 W/kg (1 g) | 20 W/kg (10 g) | See Partial Body Relaxation Chart |
| | | & conform with induced current MPE | | | |
| Uncontrolled | <17.5f mA/cm² (1 cm², 1s) | 0.08 W/kg | 1.6 W/kg (1 g) | 4 W/kg (10 g) | |
| | | & conform with induced current MPE | | | |

| Low-Power Devices | 100 kHz - 450 MHz | 450 MHz - 1.5 GHz |
|---|---|---|
| Controlled | <7W | <7(450/f) |
| Uncontrolled | <14W | <14(450/f) |

## SUMMARY

As shown, this standard is much more complex than previous standards. It therefore requires a better understanding of electromagnetic field measurements than ever before. However, it represents a significant improvement over previous standards

NATO's RF safety standard is quite dated and it should also be updated as most other standard setting organizations are doing. We have seen many organizations and nations moving in the same direction in personnel protection. Although they are not rallying around one standard, they are developing standards that are in remarkable agreement. To show this agreement we have overlaid three international RF safety standards for comparison. As shown, these are in close agreement to each other and the IEEE standard.

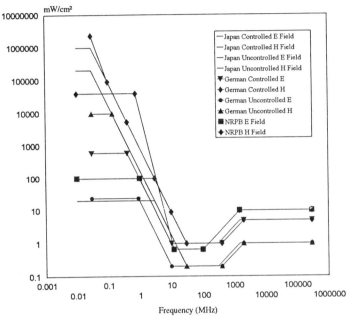

**Figure 6.** NRPB, Japanese, and German radiofrequency radiation standards.

## REFERENCES

1. IEEE C95.1-1991, "IEEE Standard for Safety Levels with Respect to Human Exposure to Radio Frequency Electromagnetic Fields, 3 kHz to 300 GHz," April 27, 1992.
2. ANSI C95.1-1982, "American National Standard Safety Levels with Respect to Human Exposure to Radiofrequency Electromagnetic Fields, 300 kHz to 100 GHz," July 30, 1982.
3. DoDI 6055.11, "Protection of DoD Personnel from Exposure to Electromagnetic Fields (EMF) at Radio Fequency (RF) from 3 Kilohertz (kHz) to 300 Gigahertz (GHz)," (Draft)
4. AFOSH Std 161-9, "Occupational Health, Exposure to Radiofrequency Radiation," 12 February 1987.
5. ACGIH, "Threshold Limit Values for Chemical Substances and Physical Agents in the Work Environment and Biological Exposure Indices with Intended Changes for 1984-1985," 1984.

**Session D: Evaluation of the Epidemiologic Database**

**Chair: P. Vecchia**

Epidemiology and What It Can Tell Us
*Leeka Kheifets*

Evaluation of Reproductive Epidemiologic
Studies
*Teresa M. Schnorr and Barbara A. Grajewski*

Epidemiology of Electromagnetic Fields and Cancer
*Charles Poole*

# EPIDEMIOLOGY AND WHAT IT CAN TELL US

Leeka Kheifets

EMF Health Studies Program
Environment Division
EPRI
Palo Alto, CA 94303

The purpose of this presentation is to describe the role of epidemiology in assessing human health risk and offer you a framework for evaluating epidemiologic research. Throughout the presentation, examples will be drawn from studies of electric and magnetic fields (EMF) and cancer.

Epidemiology is the study of the distribution of disease and its determinants in human populations. It aims to explain the causes of disease (in terms of exposures, behaviors or personal characteristics) and to predict who and how many are at risk for disease. Its ultimate goal is to promote public health. By identifying and modifying the causes of disease, epidemiology is an important factor in disease control and prevention. Because it is conducted on humans, the primary advantage of epidemiologic research is that it avoids the uncertainties associated with extrapolation from laboratory models.

A criticism often made of this discipline is that it can only demonstrate an association and cannot prove causation. Although it is not always easy to distinguish between spurious and real associations, epidemiologists attempt to accomplish this through rigorously designed studies with appropriate analyses that include powerful tools for assessing causation. Indeed, much of what we know about the etiology of human disease comes from epidemiologic research.

## STUDY DESIGNS

Most epidemiologic research falls into one of five major design categories. The choice of design depends on a number of different considerations such as time and cost limitations, frequency of disease and exposure, availability of data and specific study objectives (Figure 1). The first three designs are often referred to as descriptive: ecological, cross-sectional, and proportional mortality (PMR). They are relatively quick and inexpensive. Their value is in the generation and preliminary testing of hypotheses. Thus the data they provide is rarely considered definitive. The other two types of designs are cohort and case-control designs. They are generally more costly and time-consuming.

*Radiofrequency Standards*, Edited by B.J.
Klauenberg *et al.*, Plenum Press, New York, 1994

However, the information they yield tends to be more precise and is used to test hypotheses and evaluate causation. I will explain each of these in more detail.

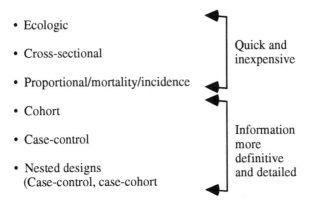

**Figure 1.** Study designs.

**Ecological Studies**

In the ecological study, both exposures (i.e., risk factors) and disease occurrence are described for groups of people rather than for individuals. Typically, some aggregate exposure information, such as per capita power consumption, is correlated with some aggregate measure of disease occurrence, such as rates of childhood leukemia. An example of an ecological argument is that: per capita electric power consumption has increased many-fold over the past fifty years while childhood leukemia rates have remained stable; thus leukemia cannot be caused or promoted by exposure to electric power. This conclusion may not be valid, however, as the link between individual exposure and per capita power consumption is unclear. For example, it is possible that per capita consumption could rise, but because of changes in technology, exposures could actually have diminished.

Ecological studies are useful in assessing broad trends and suggesting hypotheses that can be further tested in analytical studies. However, they have some characteristic limitations. One known as "ecological fallacy" occurs when a correlation between variables based on group characteristics does not necessarily hold true for individuals. Two other limitations of this study design are noteworthy: ecological studies are unable to detect weak effects and, even strong effects could be obscured by extraneous or confounding factors.

**Cross-sectional Studies**

Cross-sectional studies, also referred to as survey or prevalence studies, examine the relationship between a risk factor (i.e., exposure or characteristic) and disease occurrence at one point in time. These are distinguished from ecological studies in that the data are obtained on individuals rather than for a group. This individual data is then used to delineate groups of people, and the prevalence of the disease among different groups is compared via prevalence ratios or differences. This is a useful technique for examining the distribution of disease within a population, particularly for those diseases that occur frequently and are of long duration. Such studies are valuable for determining the need for health services. Their primary disadvantage is that, since exposure and disease are evaluated at the same time, it becomes very difficult to determine whether exposure existed prior to disease with sufficient time to allow for the latent period of disease development. Moreover, the association might be distorted because diseases of short duration are more likely to be missed with this study design. A third issue with the cross-sectional studies is that factors that influence survival cannot be distinguished from factors that actually cause disease. A cross-sectional study of high-voltage substation workers revealed they were different from others with respect to fertility, cognitive function and education. Yet, it is possible that all three conditions were pre-existent and unrelated to EMF exposure.

## Proportional Mortality Studies

The proportional mortality study (PMR) compares the proportion of deaths from a specific cause in an exposed group of individuals with the proportion of the same cause-specific deaths in a comparison population. For example, Milham has examined the proportion of deaths among electrical workers due to leukemia. A drawback of this type of study is that the resultant mortality ratio, because it is derived from proportions, is affected by changes in the proportion of deaths due to other causes. In the case of workers occupationally exposed to EMF, a lower death rate from cardiovascular disease would have the effect of increasing the proportion of deaths attributable to other causes such as cancer. The mortality ratio is also susceptible to incomplete death ascertainment. Furthermore, it can be a challenge to adequately assign exposure status in a study of this design. Still another drawback is that the population at risk is undefined. As a result, PMR studies sometimes produce false positive results.

## Cohort Studies

The first of two types of analytical epidemiological designs is the cohort study. It focuses on a group of disease-free individuals who have something in common, such as an experience or exposure, and determines the development of disease in this group. The risk of disease is then compared for different levels of exposure. A cohort study can be retrospective, that is, a cohort defined in the past can be followed to the present for disease development. Alternatively, it can be prospective, which involves a group defined in the present and followed into the future for the development of disease. For example, a cohort study of cancer among electrical utility workers is now being conducted by researchers at the University of North Carolina. The usual measures of association are the rate ratio, relative risk and the standardized mortality ratio (SMR).

Cohort investigations are useful for hypothesis testing and for estimating disease frequency as well as for estimating attributable risk, that is, the proportion of disease that can be attributed to a particular cause (Table 1). Cohort studies typically require considerable time and expense; prospective studies may require many years of follow-up, especially for diseases with long latent periods. For example, the study of the cohort of atomic bomb survivors is in its fortieth year. Additionally, cohort studies of rare diseases require very large numbers of subjects. Moreover, some proportion of subjects inevitably fail to respond to follow-up questions or, are simple untraceable (lost to follow-up), raising questions about the both validity and generalizability of the results. These constraints impose practical limitations on the implementation of cohort studies, although when done well, they often provide some of the most convincing evidence regarding causal associations.

**Table 1.** Comparison of Cohort and Case-Control Studies.

| COHORT | CASE-CONTROL |
|---|---|
| + Many outcomes | - 1 Disease |
| - 1 Exposure and confounders | + Many exposures |
| + Exposure determined prior to disease | - Exposure determined after the outcome |
| + Control group less susceptible to bias | - Difficult to choose control |
| + Bias less likely | - Bias likely |
| - Long and expensive | + Cheap and quick |
| - Large number of subjects | + Small number of subjects |
| - Relatively common dieases | + Suitable for rare disease |
| - Difficult to study past exposures | + Easier to study past exposures |

+ *Advantage*

- *Disadvantage*

### Case-Control Studies

Case-control studies are retrospective comparisons of cases (individuals with disease) and controls (those without the disease) with respect to some exposure in the past. Studies based on recently diagnosed cases (prospective case ascertainment) tend to be more precise because they avoid uncertain estimates of disease status in the remote past and reduce uncertainties of exposure extrapolation.

To assure the validity of the findings, it is important to choose controls who are representative of the population from which the cases are drawn. To assure similarity on certain characteristics, controls are often matched to cases on demographic and other characteristics. Occasionally cases and controls can be overmatched, leading to no apparent differences between the two groups. Case-control studies are useful for testing hypotheses and estimating increases in risk of disease by calculating the "odds ratio." This statistic represents the ratio of the odds of exposure among the cases to the odds of exposure among the controls and is considered to be a good estimate of risk. Case-control investigations are extremely valuable for the study of rare diseases such as childhood cancer and have frequently been used to study EMF and childhood cancer. They are quicker and less costly than cohort studies. However, they are more prone to some types of bias that will be discussed later. The most important is perhaps recall bias that is related to the difficulty of evaluating exposure after disease occurrence. This bias may have been a problem in studies that considered the use of electric appliances. Another limitation of case-control studies is that they do not permit the estimation of disease frequency.

## EXPOSURE ASSESSMENT

Exposure assessment refers to the estimation of study group members' exposure to the agent under investigation and is a crucial component of epidemiologic research. Estimates may be based on direct measurements such as those obtained with a film badge or some surrogate for exposure. In most epidemiologic studies, the correct ranking of exposure is more important than precise values and, as long as this ranking or ordered classification is maintained, the conclusions of the study will be valid. Good exposure assessment involves a workable definition of biologically meaningful exposure, regular calibration and checking of instruments, unbiased ascertainment, reproducibility and completeness. Misclassification of exposure is one of the most common pitfalls in the conduct of epidemiologic studies.

The study of EMF exposure can be used to illustrate various exposure assessment techniques. The evaluation of residential exposure might include spot and 24-hour measurements. They may be obtained inside or outside the home or by personal dosimetry such as that obtained by the use of an Emdex over a short- or long-term basis. Residential exposure has also be described by a surrogate known as "wire coding." This latter is an exposure classification scheme based on the physical characteristics of nearby power lines and proximity of the home to the lines. Homes are assigned one of five categories of exposure from very low to very high wire code configuration. This code is known to be a poor predictor of measured indoor fields explaining only 15% of the variation among homes.

Similarly, occupational EMF studies employ various exposure classification schemes ranging from quite simple to complex. Simple categorizations include: job title (e.g., electrician), industry (e.g., electronics), task (e.g., VDT operator) or a short-term measurement. The most complex efforts might involve the reconstruction of exposures based on EMF measurements for different tasks weighted by the proportion of time spent at those tasks. This strategy might further be coupled with known historical changes in work practices.

## BIAS AND CONFOUNDING

One of the biggest challenges in conducting epidemiologic research is avoidance or minimization of bias that, in this context, refers to a systematic (non-random) error. Epidemiologic bias does not have the usual connotation of a prejudice. Bias can function to enhance or diminish an observed association. Increasing sample size will produce more precise estimates of risk but will not reduce bias if it is present.

Different types of bias can affect epidemiologic studies. Selection bias refers to systematic differences in characteristics of those selected for inclusion in the study compared to those not selected. This bias can occur for several reasons. For example, individuals who volunteer for participation may differ from those who do not. Similarly, some individuals or groups may have different opportunities for diagnosis, surveillance, survival or hospitalization and therefore, different access to the study. These two forms of selection bias mostly affect the external validity or generalizability of the study results. In contrast, selection out of the study due to loss to follow-up or non-response may influence either the study's internal or external validity, depending on whether it occurs at different rates in the study subgroups.

Information bias is an error in the measurement of exposure or assessment of disease outcome. It can be differential as in recall bias (e.g., better recall of exposure information among cases than among controls) or ascertainment bias (e.g., when the diagnosis of disease is influenced by knowledge of past exposures). This bias can influence the magnitude of risk in either direction. Systematic measurement error (similar for cases and controls) and other types of exposure misclassification may result in nondifferential bias. This bias occurs frequently in EMF investigations due to difficulties both in capturing the varied exposures of the present and to an even greater extent in extrapolating these exposures into the past. Nondifferential bias usually leads to an underestimation of risk.

A third type of bias, confounding, deserves a detailed explanation as it is an issue in virtually all epidemiologic studies. A confounding bias occurs when some variable distorts the association between a study factor (exposure) and disease. For confounding to occur, the following criteria must apply: (1) the confounding variable or factor must be a risk factor for disease; (2) the factor must be associated with the exposure under study.

The factor is not considered to be a confounder when: (1) both exposure and factor cause disease but are not associated with each other; (2) the factor is associated with the exposure but not the disease of interest; (3) the factor is an intermediate step in the causal path between exposure and disease.

An example is the case of EMF and smoking as they relate to risk of cancer. Some studies have reported an elevation in the risk of certain types of cancer for those individuals employed in electrical jobs. If smoking is associated with EMF exposure, i.e., if workers in electrical jobs are more likely to smoke than persons in non-electrical jobs, and, if smoking is related to the cancers of interest, then smoking could be responsible for the elevation in cancer risk observed for individuals in electrical occupations.

Even if confounding is present, it must satisfy certain conditions in order to explain the observed association, namely, that the confounding factor must be a stronger risk factor than the study factor. For example, if the relative risk for EMF and leukemia is 2.0, a confounder must have a relative risk of 4.0 to 5.0 in order to account or be responsible for the effect attributed to EMF. Alternatively, it must be correlated very tightly with the exposure under study in order to account for its association with disease.

## STATISTICAL ISSUES

Evaluating and interpreting epidemiologic investigations requires some basic statistical concepts such as sampling, parameter estimation and hypothesis testing. Testing and estimation are two related approaches for assessing the importance of a finding. The likelihood that a test statistic is the result of chance variation rather underlying causation or association is expressed as the probability value or p-value. This p-value is used to judge whether the results are statistically significant. A confidence interval is the range of values within which the true value is thought to lie. Ninety-five percent confidence limits are commonly used. Confidence intervals are useful in showing the precision of the estimate (i.e., wide confidence limits suggest poor precision) as well as whether the estimate is significantly different from an estimate of no effect. The provision of p-values and the confidence intervals are more informative than merely reporting whether the finding is significant. Statistical significance is often confused with practical significance and, as such, is frequently misused. For example, it has been found that human heart rate can be significantly reduced by exposure to EMF, but the change is so slight that it is of no practical significance.

It is important to reduce the possibility of rejecting the hypothesis of no association when, in fact, it is true (Type I error). It is equally important to avoid concluding that there is no association when one actually does exist (Type II error). The study's probability of detecting a real association is known as its statistical power (Table 2). Statistical power is important in designing studies and, in order to insure an adequate sample size to detect associations of interest. If negative findings are observed, that is, that no statistical evidence of an association is apparent, it is important to evaluate whether the study had sufficient power to detect the association of interest if it existed. In other words, a negative study may not be very important if it was too small to exclude even a large risk.

**Table 2.** Two types of error.

|  | **Accept H** | **Reject H** |
|---|---|---|
| **H True** |  | **Type I** $\alpha$ **-Level** |
| **H False** | **Type II** $\beta$ **-Level** | **Power** |

## CRITERIA FOR EVALUATING EPIDEMIOLOGIC STUDIES

The goal of this brief description of epidemiologic and statistical methods is to facilitate the critical evaluation of a particular study or area of research. The following are important considerations:

1. Questions and Hypotheses: Was the hypothesis generated by the study or was the study designed to evaluate the hypothesis? In the latter case, findings are more compelling. Was the study question vague or well-defined? How many hypotheses were evaluated in the study? A study designed to test a specific, well-formulated hypothesis is considered more definitive.

2. Study Design: Was the study descriptive or analytical? What design was used? In general, cohort and case-control studies are considered to be more definitive with a prospective cohort design offering the best possibility for minimizing bias.

3. Definition and Selection of Study Participants: Is the definition of disease clear and precise? Is it confirmed? For example, death certificate diagnoses of brain tumors are notoriously unreliable if not confirmed by surgical biopsy. What are the criteria for inclusion and exclusion of a study subject? What is the participation rate and how does it compare for cases and controls or for exposed and non-exposed individuals? All of these questions should be considered in assessing whether the comparisons are valid and the results are generalizable.

4. Exposure Assessment: Is it complete? Are the exposure levels reasonable? Is there misclassification of exposure and, if so, is it differential?

5. Bias and Confounding: Is statistical bias present? If so, what is its direction and magnitude? Every investigation may potentially be affected by confounding factors and other types of bias. However, it is important to note that they can be present, but not of sufficient magnitude to explain a reported association of exposure and disease.

6. Analysis and Statistical Considerations: Have the assumptions on which the analyses are based been properly evaluated? Is there evidence of a dose-response relationship? Was the data grouping used in the analysis arbitrary? What is the precision of the study? What was the power of the study to detect an association?

7. Conclusions: Are the conclusions consistent with the data? Are they reasonable? Could the association be spurious? Could it be indirect? Are the conclusions reproducible and generalizable?

## WEIGHT OF EVIDENCE

As mentioned at the beginning of this talk, epidemiologists attempt to understand the influence of various factors on disease and on health in order to determine causality. To this end they weigh the available evidence based on the following criteria:

1. Validity of the existing studies.
2. Strength of the association that is represented by the magnitude of the risk estimate. The stronger the association, the more likely it is to be causal; weaker associations are more likely explained by bias.
3. Evidence of a dose-response relationship, i.e., higher risk associated with higher levels of exposure.
4. Biologic plausibility of the hypothesis. The existence of a biologic mechanism makes causality more likely.
5. Temporality such that exposure not only precedes disease, but does so with the appropriate latent period.
6. Specificity of the association. Specificity refers to the concept that a cause usually leads to a single effect. Although this criteria, when satisfied, argues in favor of causality, its absence does not refute a causal relationship. Many diseases have multiple causes and many exposures cause more than one disease.
7. Consistency of results where a given association is observed repeatedly in different populations under different circumstances. This is not a hard and fast rule as some outcomes are produced by their causes only under certain circumstances.
8. Coherence with what is known about the natural history and biology of the disease.

## SUMMARY

This presentation provides brief description of the methods and complexities of epidemiologic research. Figure 2 summarizes the epidemiologic research sequence and presents some of the considerations involved in conducting and interpreting epidemiologic studies. While epidemiologic investigations are difficult to conduct and easy to criticize, they offer unique advantages (Figure 3). In particular, epidemiologic studies permit the evaluation of the consequences of an environmental exposure or other factor in the precise manner in which it occurs in human populations, i.e., under usual rather than artificial conditions of exposure. Thus, epidemiology is the only investigative technique that does not require extrapolation between species. Moreover, since epidemiologic research is conducted in natural settings, it permits the study of the joint influence of multiple factors on disease occurrence while taking into account the defense mechanisms of the individual. Equally important is the fact that extrapolation from high doses to low doses as is usual in animal studies is not necessary since epidemiologic studies examine an exposure at its relevant dose.

Epidemiology has been successful in identifying important hazards and in characterizing susceptible groups of individuals or special circumstances in which exposures are especially harmful. These studies have provided clues for other types of research that, in turn, have led to enhanced understanding of disease processes. Finally, epidemiologic research is valuable in developing strategies for the reduction or elimination of hazardous exposures and for subsequently judging the success of these strategies.

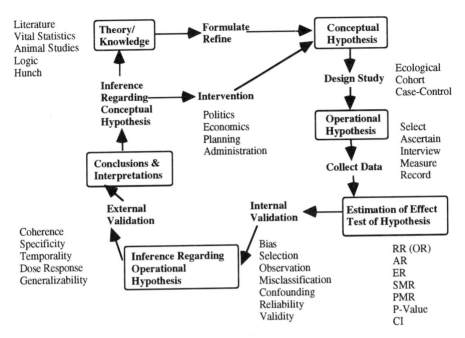

**Figure 2.** Epidemiologic Research Sequence.

| **Strengths** | **Weaknesses** |
|---|---|
| • No inter-species conversion | • Large samples required |
| • Relevant exposure | • Expensive |
| • Natural environment | • Biases and confounding |

**Figure 3.** Strengths and weaknesses of epidemiological approach.

## FURTHER READING

1. J.L.Kelsey, W.D. Thompson and A.S. Evans, "Methods in Observational Epidemiology," Oxford University Press, New York (1986).
2. D.G.Kleinbaum, L.L. Kupper and H. Morgenstern, "Epidemiologic Research Principles and Quantitative Methods," Lifetime Learning Publications, Belmont, CA (1982).

# EVALUATION OF REPRODUCTIVE EPIDEMIOLOGIC STUDIES

Teresa M. Schnorr and Barbara A. Grajewski[*]

National Institute for Occupational Safety and Health
Division of Surveillance, Hazard Evaluations and Field Studies
4676 Columbia Parkway, R-15
Cincinnati, Ohio 45226

Concern about electromagnetic fields and reproductive effects began with clusters of adverse reproductive effects among video display terminal (VDT) operators. Because of the large number of women using VDTs and the intense public concern, most of the epidemiologic studies of electromagnetic fields and reproduction have been conducted among women using VDTs. There have also been studies of men and women exposed to other electromagnetic frequencies. This review will consider only those studies that fall within the 10 kHz-300 GHz range and so will include studies of physiotherapists, VDT operators and some radar operators.

## VIDEO DISPLAY TERMINALS

In the electromagnetic frequency range of interest for the NATO Standardization Agreement, video display terminals have been studied most extensively for potential reproductive effects. Video display terminals were first associated with adverse reproductive outcomes in 1980, when a cluster of birth defects was observed among women using VDTs at the Toronto Star newspaper.[1] This report was followed by a number of other adverse pregnancy outcome clusters, primarily spontaneous abortion.[2-5] These clusters caused several investigators to conduct epidemiologic studies of the reproductive risk of VDTs.[6-18]

While physical and psychological stress have been suggested as possible explanations[1] for the observed association between VDT use and adverse pregnancy outcomes, this review will focus upon the evidence for an association of adverse reproductive effects with the electromagnetic fields emitted by the VDT.[19] Two types of electromagnetic fields are produced by the VDT: extremely low frequency (ELF) at approximately 60 Hz and very low frequency (VLF) at 15 kHz fields. Human studies of VLF electromagnetic fields and reproductive effects are limited to studies of VDTs.[6-18]

---

[*] These views and opinions are those of the author and do not necessarily state or reflect those of the U.S. Government.

Birth defects were associated with VDT use in some of the clusters. However, this association has not been shown consistently in the five epidemiologic studies that examined birth defects.[6-10] Two studies found no increased risk of major malformations among moderate or heavy VDT users.[8,10] Brandt[9] found no increased risk for major malformations as a group, but found a significant elevated risk for hydrocephalus. A fourth study observed an increasing risk of major malformations with weekly hours of VDT use[6] but saw no increased risk for specific defects. Another study found an overall excess risk of malformations as well as an increased risk of renal defects.[7] None of these studies measured the electromagnetic fields associated with the VDT.

Some investigators have also examined the relationship of VDT use with low birth weight and preterm delivery[7,11] and observed no increased risk associated with VDT use. One found a slight non-significant elevation in intrauterine growth retardation among women with greater VDT use.[12]

Spontaneous abortion was the most common outcome reported in the clusters[1-5] and has also been studied most extensively by large epidemiologic studies. To date, ten large epidemiologic studies of spontaneous abortion have been conducted.[6-8,12-18] Most of these studies examined potential risk in relation to weekly hours of VDT use and did not attempt to distinguish between specific risk factors (physical stress, psychological stress and electromagnetic fields) that might be associated with the VDT. One study that did collect data on job-stress and ergonomic work load, found that neither of these factors were correlated significantly with spontaneous abortion.[15] Only two studies have measured the electromagnetic fields associated with the VDT.[16,17]

First, I will discuss the eight studies that did not measure the electromagnetic fields of the VDTs. No increased risk of spontaneous abortion was found in six of these eight studies that examined risk in relation only to the amount of time the VDT was used.[6,12-18] One study[7] found conflicting results depending upon how exposure was defined. A 1.19-fold increased risk of spontaneous abortion was observed among women using VDTs for 15 or more hours per week when VDT use was defined according to the woman's recall, but a risk of 1.06 was found when VDT use was estimated from occupational title. A second study[8] found an 1.8-fold increased risk of spontaneous abortion for women who reported using a VDT for 20 or more hours during pregnancy. This increased risk occurred primarily among women in clerical positions (odds ratio (OR) =2.4). Other job categories had too few women using VDTs for 20 or more hours per week to draw conclusions.

Two studies have attempted to measure the electromagnetic fields associated with the VDT.[16,17] These studies give us some additional insight into the potential reproductive risks of VDT-related electromagnetic fields. In addition to measuring the fields, Schnorr[16] essentially controlled for physical and psychological stress factors by selecting a comparison group with a similar work environment to the VDT users. In this study of U.S. telephone operators, no increased risk of spontaneous abortion was found for women who used VDTs compared to those who did not, and the risk did not increase with increasing hours of VDT usage. Hours of VDT usage were recorded from detailed payroll records for each woman. The odds ratio for women using the VDT for 25 or more hours per week was 1.0, the same as the non-VDT users. Measurements of the electromagnetic fields for the two models of VDT used by the operators indicated that VDT operators had very low frequency (VLF) magnetic field exposures in excess of the general population at 30 cm from the screen (mean VLF levels = 0.28 and 1.25 mG). However, at the normal operating distance of approximately 50 cm, the VLF levels dropped to less than 0.22 mG. Extremely Low Frequency (ELF) magnetic fields also declined with distance from the screen (mean ELF fields = 2.6 and 3.9 mG at 30 cm. ELF fields dropped to below a mean of 0.78 mG at the normal operating distance. Analyses that examined risk of spontaneous abortion for the two VDT models separately found no difference in risk. Women using the IBM model (average 30 cm exposures of 0.28 mG VLF and 2.6 mG ELF) had an odds ratio of 0.98. While the CCI model had higher magnetic field levels (mean VLF and ELF 30 cm magnetic field exposures of 1.25 mG and 3.9 mG, respectively), the odds ratio did not reflect an increased risk (OR = 0.92).

Lindbohm et al.[17] conducted a case-control study of spontaneous abortion among Finnish women bank and clerical workers over a ten-year period (1975-1985). Information on amount and type of VDT use was obtained by questionnaire. No measurements were made in the workplace; however, magnetic fields were measured in the laboratory for a single unit of most models reported to be used by the women in the study. In one analysis, the authors found no significantly increased risk for women who used VDTs for greater than 20 hours per week (OR = 1.7). However, this analysis did not consider the VLF or ELF emission levels of the VDT. In other analyses, the authors found that the risk of spontaneous abortion was significantly higher for women who used VDTs in the highest tertile of ELF emissions. Women using VDTs that measured greater than 3 mG rms at 50 cm from the screen had an increased odds ratio for miscarriage (OR=3.4). The risk associated with VLF field measurements was not calculated, but one might expect similar findings since the authors reported that ELF and VLF field intensities were highly correlated. In the analyses that considered emission levels, the authors did not provide data regarding the hours of weekly VDT use. If magnetic fields are related to spontaneous abortion, one would expect to observe the highest risk of spontaneous abortion among those women who used the highest emitting VDTs for the greatest number of hours per week. The lowest risk would be expected among those women using the low-emitting VDTs for a fewer number of hours per week. Additional analyses that simultaneously provide the individual contributions of both dose components (emission level and hours of VDT use during the critical time period in gestation) may help determine whether the observed association may be related to ELF magnetic fields.

The Schnorr[16] and Lindbohm[17] studies do not necessarily contradict each other since the ELF emissions reported in the Schnorr et al. study of U.S. operators are lower than those reported by Lindbohm et al. in Finnish clerical workers. Schnorr et al. found no VDT models with ELF magnetic fields >3 mG at 50 cm, the lower bound for the high-emission group in the Lindbohm study. Ninety percent of the VDTs used in the Schnorr study correspond to the low emission group in the Lindbohm study. Other published surveys in North America and Europe[20] have found VDTs to have ELF levels similar to those in the Schnorr study. This suggests that the exposure of current VDT operators may be too similar to other typical office ELF exposures to detect any potential effect.

The weight of the evidence thus far indicates that VDTs in themselves may not increase the risk for adverse pregnancy outcomes. At this point, studies have primarily focused on clinically recognized spontaneous abortions. As ongoing reproductive investigations are completed, the potential risk of VDTs on spontaneous abortions and other aspects of reproduction will be better defined.

## RADIOFREQUENCY FIELDS

There have been a few reproductive studies of women and men occupationally exposed to high frequency magnetic fields. Most of these studies did not include electromagnetic field measurements, and the most commonly studied group has been physiotherapists.

In 1982, Källén et al.[21] examined the pregnancy outcomes of 2,043 infants born to female physiotherapists who used shortwave (330 kHz-300 MHz), microwave and ultrasound equipment during pregnancy. In this case-control study, women who had seriously malformed (n=8) or perinatally dead infants (n=3) were more likely to report using shortwave equipment "often" or "daily" than did women who had normal live births. No associations were observed with adverse pregnancy outcomes and use of microwave equipment, which was used too infrequently to allow examination, or with ultrasound equipment. The authors stressed that actual shortwave exposure was not measured. Potential shortwave exposure was determined by reported use of such equipment. Three more recent studies of physiotherapists also have not observed significant associations or dose-response relations between high-frequency electromagnetic field exposure and birth defects or spontaneous abortion.[22-24] These studies were limited in the ability to examine potential relationships because of small sample sizes and inaccurate estimation of exposure. However, one of these studies[24] found a significantly low ratio of boys to girls born to exposed women and in the risk of low birthweight boy babies born to these mothers. Only 23% of the children born to mothers with high exposure were boys, compared to 60% among the unexposed mothers. Boys born to exposed mothers were more likely to have a low birthweight. Boys born to mothers with high exposure were 5.9 times more likely to have a low birthweight than boys born to unexposed mothers. The deficit of newborn boys was an unexpected but strong finding which appeared to be consistent with simple dose-response patterns including time per week spent in the shortwave room, type of electrode (none, vs. diplode or plate) and type of exposure (none, vs. direct, indirect or both).

In 1977, Cohen and Lilienfeld[25] studied 216 Down's syndrome cases and 216 controls to determine if there was a relationship with microwave/radar exposure of the fathers. Based upon questionnaire data and examination of military records, no relationship was observed between Down's syndrome case fathers with "definite" exposure (8.3%) and control fathers (12.6%). Inclusion of "probable" exposed fathers did not change the findings; 21.6% of case fathers and 23.5% of control fathers had "probable" exposure. Exposure levels, duration and frequencies of the fathers were not given.

Lancranjan[26] examined 31 men with potential exposure to 3,600-10,000 MHz frequency microwaves and 30 control men with no microwave exposure. Exposed men had measured exposure levels between "tens to hundreds of uW/cm2" for a mean duration of 8 years. The source of microwave exposure and the methods for microwave measurement were not described. Semen analysis showed significant decreases in sperm motility, sperm count and morphologically normal sperm among the exposed men. The study was not able to differentiate between effects observed due to heating effects and nonthermal effects of exposure.

The few studies described above that have been conducted on populations exposed to high-frequency electromagnetic fields have several limitations including absence of workplace measurements that limit our ability to draw conclusions regarding potential reproductive effects.


## SUMMARY

The current literature on electromagnetic fields and reproductive effects indicates that electromagnetic fields in the range of those near most VDTs do not increase the risk of spontaneous abortion. The reproductive risks of electromagnetic fields at higher intensities and different frequencies are less clear. To examine further whether electromagnetic field exposure constitutes a risk factor for spontaneous abortion, future studies of the potential EMF-spontaneous abortion relationship should focus upon populations with exposures that are significantly greater than those in the general population. In these studies, electromagnetic field exposure should be measured and fully characterized in the workplace to account for all sources of EMF exposure.

# REFERENCES

1. U.O. Berqvist, Video display terminals and health, *Scand J Work Environ Health* 10: (Suppl 2):62-67 (1984).
2. N. Binkin, R.W. Rochat, W. Cates, and C.W. Tyler, Cluster of spontaneous abortions, Centers for Disease Control, Atlanta, Georgia. (EPI 80-131-2). Dallas, Texas (1981).
3. National Institute for Occupational Safety and Health: Health Hazard Evaluation 84-127-1337, General Telephone Company of Michigan, Alma, Michigan. (DHHS (NIOSH) Publication No. 84-297-1609). (1984).
4. National Institute for Occupational Safety and Health: Health Hazard Evaluation 83-329-1498, Southern Bell, Atlanta, Georgia. (DHHS (NIOSH) Publication No. 83-329-1498). (1983).
5. National Institute for Occupational Safety and Health: Health Hazard Evaluation 84-191, United Airlines, San Francisco, California. (DHHS (NIOSH) Publication No. 84-191). (1984).
6. A. Ericson and B. Källén, An epidemiological study of work with video screens and pregnancy outcome, II. A case-control study, *Am J Ind Med* 9:459-475 (1986).
7. A.D. McDonald, J.C. McDonald, B. Armstrong, N. Cherry, A.D. Nolin, and D. Robert, Work with visual display units in pregnancy, *Brit J Ind Med* 45:509-515 (1988).
8. M. Goldhaber, M. Polen, and R. Hiatt, The risk of miscarriage and birth defects among women who use visual display terminals during pregnancy, *Am J Ind Med* 13:695-706 (1988).
9. L.P.A. Brandt and C.V. Nielsen, Congenital malformations among children of women working with video display terminals, *Scand J Work Environ Health* 16:329-333 (1990).
10. K. Kurppa, P.C. Holmberg, K. Rantala, T. Nurminen, and L. Saxen, Birth defects and exposure to video display terminals during pregnancy, *Scand J Work Environ Health* 11:353-6 (1985).
11. T. Nurminen and K. Kurppa, Office employment, work with video display terminals, and course of pregnancy, *Scand J Work Environ Health* 14:293-298 (1988).
12. G.C. Windham, L. Fenster, S.H. Swan, and R.R. Neutra, Use of video display terminals during pregnancy and the risk of spontaneous abortion, low birthweight, or intrauterine growth retardation, *Am J Ind Med* 18:675-688 (1990).
13. A. Ericson and B. Källén, An epidemiological study of work with video screens and pregnancy outcome: I. A registry study. *Am J Ind Med* 9:447-457 (1986).
14. H.E. Bryant and E.J. Love, Video display terminal use and spontaneous abortion risk, *Int J Epidemiol* 18(1):132-138 (1989).
15. C.V. Nielsen, L.P.A. Brandt, Spontaneous abortion among women using video display terminals, *Scand J Work Environ Health* 16:323-328 (1990).
16. T.M. Schnorr, B.A. Grajewski, R.W. Hornung, M.J. Thun, G.M Egeland, W.E. Murray, D.L. Conover, and H.E. Halperin, Video display terminals and the risk of spontaneous abortion, *N Engl J Med* 324:727-733 (1991).
17. M.L. Lindbohm, M. Hietanen, P. Kyyrönen, M. Sallmon, P. von Nandelstadh, H. Taskinan, M. Pekkarinen, M. Yikoski, and K. Hemminki, Magnetic fields of video display terminals and spontaneous abortion, *Am J Epidemiol* 188:1041-1051 (1992).
18. E. Roman, V. Beral, M. Pelerin, and C. Hermon, Spontaneous abortion and work with visual display units, *Br J Ind Med* 49:507-512 (1992).
19. R. Kavet and R.A. Tell., VDTs: Field levels, epidemiology, and laboratory studies, *Health Phys* 61:47-57 (1991).
20. T.M. Schnorr, B.A. Grajewski, and W.E Murray, Letter to the Editor. Video display terminals and spontaneous abortions, *N Engl J Med* 325(11):311-313 (1991).
21. B. Källén and U. Moritz, Delivery outcome among Physiotherapists in Sweden: Is non-ionizing radiation a fetal hazard? *Arch Environ Health* 37(2):81-84 (1982).
22. H. Taskinen, P. Kyyrönen, and K. Hemminki, Effects of ultrasound, shortwaves, and physical exertion on pregnancy outcome in physiotherapists, *J Epidemiol Community Health* 44:196-201 (1990).
23. A.I. Larsen, Congenital malformations and exposure to high-frequency electromagnetic radiation among Danish physiotherapists, *Scand J Work Environ Health* 17:318-23 (1991).
24. A.I. Larsen, J. Olsen, and O. Svane, Gender-specific reproductive outcome and exposure to high-frequency electromagnetic radiation among physiotherapists, *Scand J Work Environ Health* 17:324-9 (1991).
25. B. Cohen and A.M. Lilienfeld, "Parental factors in Down's syndrome-results of the second Baltimore case-control study," *in:* Population Cytogenetics; Studies in Humans. Hook EB, Porter IH Eds., Academic Press, N.Y., pp. 301-52. (1977).
26. I. Lancranjan, M.Maicanescu, E. Rafaila, I. Klepsch, and H.I. Popescu, Gonadic function in workmen with long-term exposure to microwaves, *Health Phys* 29:381-383 (1975).

# EPIDEMIOLOGY OF ELECTROMAGNETIC FIELDS AND CANCER

Charles Poole

Boston University School of Public Health
Brown University School of Medicine

Epidemiologic research on cancer in relation to electric and magnetic fields began with the 1979 publication of a case-control study by Wertheimer and Leeper[1] of childhood cancers in Denver, Colorado, USA. This study showed moderately strong associations between all childhood cancers and a measure of exposure known as the "wiring configuration code," an index based on the distance of homes from electric power distribution and transmission lines.

The next important study was published in 1988 by Savitz et al.[2] This study was also conducted in the Denver area with somewhat different methods. Cancer diagnoses were included, rather than deaths; controls were selected by random dialing of telephone numbers, rather than from birth records; and spot measurements of magnetic fields inside homes were made in addition to wiring configuration codes.

Figure 1 shows the results for childhood brain cancer from these two studies. Based on the distinction between high- and low-current wiring configurations, both studies suggest an approximate doubling of risk. The Savitz study was larger, as indicated by the narrower confidence interval for its estimate of relative risk. For spot measurements of magnetic fields under low-power conditions, using a cutpoint of 0.1 μT (1.0 mG), a weaker association was found. The much wider confidence interval for this estimate indicates a relative problem of missing data for measurements as opposed to wiring configuration codes.

Figure 2 shows the results for childhood leukemia from the same two studies and from the third key study, conducted by London et al.[3] in Los Angeles, California, USA, with methods similar to those used in the Savitz study. Because the London study was larger and focused specifically on leukemia, the confidence interval for its estimate of relative risk based on wiring configuration codes was the narrowest of the three. In addition to wiring configuration codes and spot measurements, this study also included 24-hour magnetic field measurements. Neither the Savitz study nor the London study, regardless of the exposure measure, reported an increase in risk as great as that reported in the Wertheimer study.

The failure of the measured fields to produce sharply elevated estimates of increased risk caused some observers to question the causal hypothesis. Others questioned the premise that spot measurements or 24-hour measurements are superior to wiring configuration codes, which might be a better indicator of long-term, average exposure.

Although several studies of childhood cancer have been conducted, the next key study was reported in 1992 by Feychting and Ahlbom[4] in Sweden. This study had the advantage of focusing on children living near transmission lines and used three measures of exposure:

spot measurements, distance of the home from the nearest transmission line (analogous to wiring configuration codes), and estimated average magnetic field strengths based on historical utility records of loading conditions, etc. The authors considered this to be their best measure of exposure. As shown in Figure 3, it produced no association with childhood brain cancer and an estimated doubling to trebling of the risk of childhood leukemia.

Although a few studies of adult cancers in relation to residential exposures have been conducted, most studies of adult cancers have focused on occupational exposures. Figure 4 summarizes the results of the follow-up and case-control studies compiled in a recent review by an Oak Ridge Associated Universities (ORAU) Panel for the Committee on Interagency Radiation Research and Policy Coordination.[5] The studies display some tendency toward slight increases in risk, with most estimates of relative risk being quite imprecise and falling below 2.0.

Studies that reported results only for total leukemia, as shown in Figure 5, are most consistent with no increase in risk. A similar pattern is seen in Figure 6, for the few studies that reported results specifically for chronic myelocytic leukemia. Studies of chronic lymphocytic leukemia are somewhat more consistent with an increase in risk, as shown in Figure 7. The most striking findings are for acute myelocytic leukemia, as shown in Figure 8, with 9 of 11 point estimates falling above the null value.

The Swedish residential study by Feychting and Ahlbom[4] included adults as well as children. As in the data for children, Figure 9 shows no increase in brain cancer risk was suggested. The results for acute myelocytic leukemias in Figure 10, are not suggestive of appreciable increases in risk.

Because of hypotheses relating magnetic-field exposures to pineal function and possible alterations in melatonin levels, researchers have become interested in hormonally mediated cancers such as breast cancer. Three of the occupational studies[6,7,8] have reported extremely imprecise estimates of increased risk of male breast cancer, obviously based on very small numbers of cases or deaths.

The most serious limitation of the existing epidemiologic studies of EMF and cancer is exposure assessment. As reflected by the classification of studies into "residential" and "occupational" categories, no study to date has attempted to combine exposures from all sources--occupational, residential, appliances, and others--into a single exposure metric. Research and theory on the aspects of electric and magnetic fields to measure, and how best to measure them, are still in an extremely primitive state of development. Virtually all studies thus far have tried to focus on power-frequency (50/60 Hz) electric and magnetic fields. A very small number of attempts has been made to study radiofrequency and other exposures separately, with results that are not yet noteworthy.

In addition to their inconsistent results, most of the residential studies are limited in size and contain possible biases of control selection and confounding. The problem on confounding is especially frustrating because so little is known about the etiologies of leukemias and brain cancers, especially in children. Until recently, the occupational literature had suffered from a strong potential for publication bias, with so many occupational cohorts already under study for other reasons. It would be a relatively easy matter to add a row or column for EMF to an existing job-exposure matrix and then for the nature of the results to influence whether or not they are published. Consistent with this hypothesis, the earlier studies in the occupational literature tend to be less precise and to include more relative risk estimates comparatively far from the null value. Publication bias may also affect the view of the entire literature on male breast cancer, since presumably in many studies no such cases or deaths were observed and expected numbers consequently were not provided. As the international EMF research effort is maturing, greater resources are being expended on exposure assessment and the potential for major publication bias is lessening.

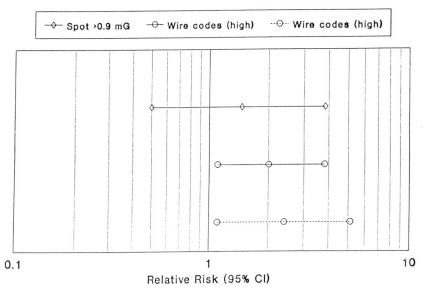

**Figure 1.** Childhood brain cancer; Savitz--solid lines, Wertheimer--broken line.

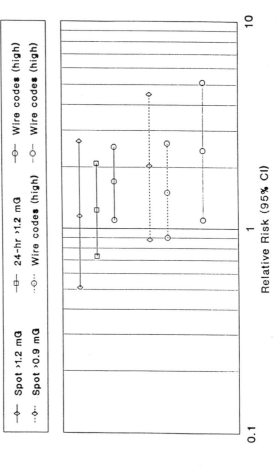

**Figure 2.** Childhood leukemia; Los Angeles--solid lines, Denver (Savitz)--broken lines; Denver (Wertheimer)--dotted line.

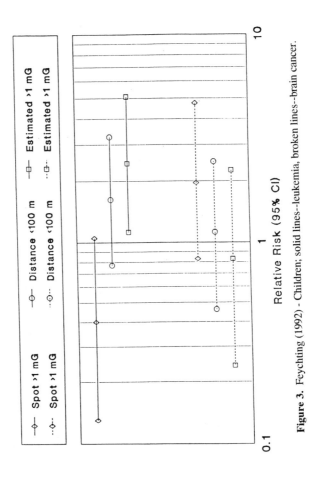

**Figure 3.** Feychting (1992) - Children; solid lines–leukemia, broken lines–brain cancer.

175

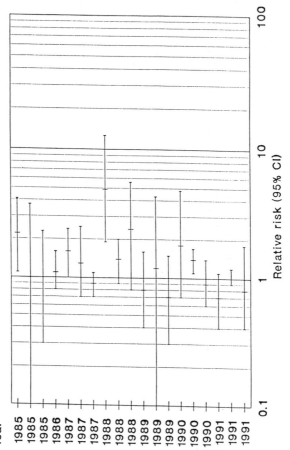

**Figure 4.** Occupational EMF studies--brain cancer.

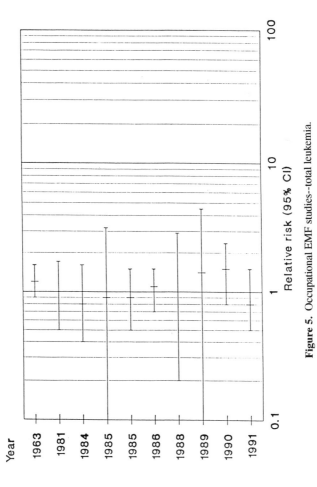

**Figure 5.** Occupational EMF studies--total leukemia.

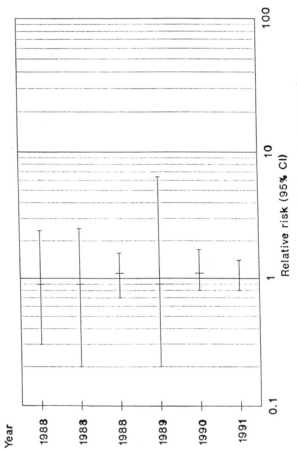

**Figure 6.** Occupational EMF studies--chronic myelocytic leukemia.

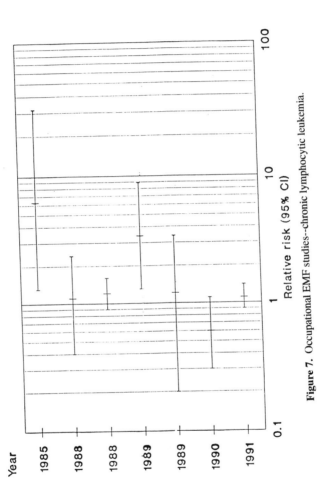

**Figure 7.** Occupational EMF studies--chronic lymphocytic leukemia.

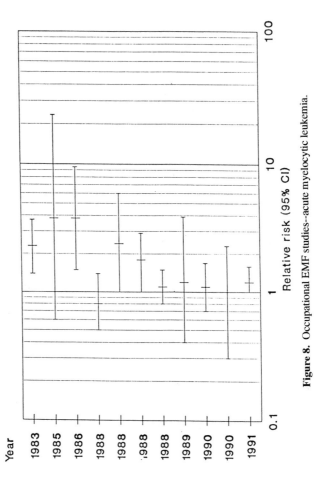

**Figure 8.** Occupational EMF studies–acute myelocytic leukemia.

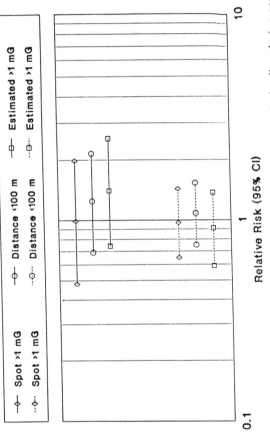

**Figure 9.** Feychting (1992)--adults; solid lines--acute myelocytic leukemia; broken lines--brain cancer.

181

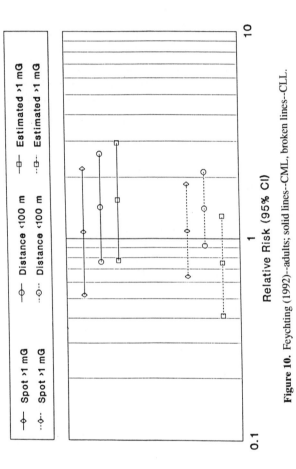

**Figure 10.** Feychting (1992)–adults; solid lines–CML, broken lines–CLL.

Childhood leukemias and breast cancers and acute myelocytic leukemia in adults seem at present to be the most potentially fruitful areas for further research, but the evidence is not sufficient to curtail research on other leukemias in adults or on brain cancers in children or adults. Although firm conclusions concerning causality would seem unwarranted at this time, the causal hypotheses have survived the first decade or so of epidemiologic tests.

## REFERENCES

1. N. Wertheimer and E. Leeper, Electrical wiring configurations and childhood cancer, *Am J Epidemiol* 109:273-284 (1979).
2. D.A. Savitz, H. Wachtel, F.A. Barnes et al., Case-control study of childhood cancer and exposure to 60-Hz magnetic field, *Am J Epidemiol* 128:21-38 (1988).
3. S.J. London, D.C. Thomas, J.D. Bowman et al., Exposure to residential electric and magnetic fields and risk of childhood leukemia, *Am J Epidemiol* 132:923-937 (1991).
4. M. Feychting and A. Ahlbom, Magnetic fields and cancer in people residing near Swedish high voltage power lines. Report prepared for the Swedish National Board for Industrial and Technical Development. Karolinska Institute, Stockholm, Sweden (1992).
5. Oak Ridge Associated Universities Panel on Health Effects of Low-Frequency Electric and Magnetic Fields. Health effects of low-frequency electric and magnetic fields, Oak Ridge, TN, Oak Ridge Associated Universities (June 1992).
6. T. Tynes and A. Anderson, Electromagnetic fields and male breast cancer, *Lancet* 1990; 336:1596.
7. G.M. Matanoski, E.A. Breysse, and P.N. Breysse, Electromagnetic field exposure and male breast cancer, (Letter), *Lancet* 337-737 (1991).
8. P.A. Demers, D.B. Thomas, K.A. Rosenblatt et al., Occupational exposure to electromagnetic fields and breast cancer in men, *Am J Epidemiol* 143:430-357 (1991).

# THE DIELECTRIC PROPERTIES OF BIOLOGICAL MATERIALS

Camelia Gabriel

Physics Department
King's College London
Strand
London, WC2R 2LS. U. K.

## INTRODUCTION

The dielectric properties of biological materials are a measure of their interaction with electromagnetic fields. When people are exposed to non-ionizing radiation, such interactions take place in the human body at various levels of organization and may initiate biological responses. While these biological effects may or may not be desirable, they should, however, be well understood for safety regulation purposes and in order to exploit beneficial effects to their full potential. Dielectric spectroscopy, or the study of the frequency dependence of the dielectric properties, helps elucidate such mechanisms of interaction at the cellular and molecular level. This subject has been widely reviewed.[1,2]

The dielectric properties of body tissues are among the factors that determine the extent of energy coupling and energy dissipation in the body from an external electromagnetic field. Knowledge of the dielectric properties is therefore of practical importance in the field of theoretical dosimetry, which aims at relating external fields to those induced in the body and to the resulting power absorption. Whereas, in electromagnetic dosimetry under conditions of continuous, monochromatic wave exposure, the dielectric properties of body tissues need only to be known at the particular frequency under consideration, for pulsed irradiation, which presents a complex frequency content, it is necessary to characterize the dielectric properties over the equivalent frequency band.[3] This creates a need for a credible database of dielectric properties of all body tissues at the relevant frequencies and temperatures. This paper will focus on the practical aspects of dielectric spectroscopy leading to the characterization of the linear responses of biological materials to electric fields.

The measurement of the dielectric properties of biological materials are increasingly being performed using vector network analyzers and open-ended coaxial probes, a technique first introduced by Burdette et al. in 1980.[4] Because of its relevance to the study of the dielectric properties of tissues, this technique will be described and evaluated and the conditions under which it can be used to perform accurate dielectric measurements will be discussed.

Examples will be drawn from recent dielectric studies on tissues to highlight the main features of the dielectric spectrum of a biological material and their interpretation in terms of interaction mechanisms and site of interaction.

## APPRAISAL OF THE MEASUREMENT TECHNIQUE

The probe is an open-ended coaxial line, measurements are made by placing it in contact with a sample and measuring its admittance or reflection coefficient using a network analyzer or equivalent instrumentation. In practice, the main advantage of the open-ended probe is that sample handling is reduced to a minimum, and that measurements of certain types of tissue can be performed *in situ* or even *in vivo*. Theoretically, the most important consideration is the derivation of the relationship between the measured reflection coefficient and the dielectric properties of the sample. The reflection coefficient $\Gamma$ is related to the admittance of the probe Y by the relationship

$$Y = Y_0\left(\frac{1-\Gamma}{1+\Gamma}\right) \tag{1}$$

where $Y_0$ is the characteristic admittance of the coaxial line.

The probe is described in Figure 1 in the cylindrical coordinate system (r, f, z) in which the e and m are the absolute permittivity and permeability, the suffixes 0, 1, and 2 refer to air, the medium inside and outside the coaxial line respectively, and r is the radial distance from the axis, f the angular displacement around the axis, and z the displacement along the axis. The dimensions of the ground plane are large compared to the dimensions of the probe. Two cases will be considered depending on the size of the sample in contact with the probe.

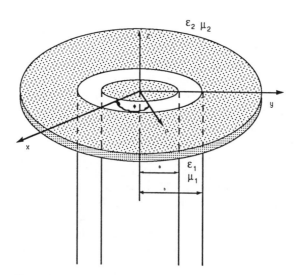

**Figure 1.** A probe of inner and outer radii a and b respectively.

## Semi-infinite Sample

The medium outside the probe is assumed to be uniform and to occupy the entire half-space beyond the ground plate.

A number of expressions have been developed for the admittance of the probe, perhaps the most rigorous derivation being achieved by Mosig et al[5] who obtained an expression for the aperture admittance Y in terms of $E_\rho^a$, the electric field at the aperture of the probe and $\Phi_n(\rho)$, the radial high-order modes potential functions for the transverse electric field in a coaxial line

$$\frac{Y}{2\pi\rho}\int_a^b E_\rho^a(\rho)d\rho = \sum_{n=1}^\infty \frac{j\omega\varepsilon_1}{\gamma_n}\frac{d\Phi_n(\rho)}{d\rho}a_n + \frac{j\omega\varepsilon_2}{2\pi}\int_a^b\int_0^{2\pi} E_\rho^a(\rho')\cos(\varphi' - \varphi)\rho'd\varphi'd\rho' \quad (2)$$

where n corresponds to the mode order, $a_n$ are amplitude terms normalized to the dimensions of the line, the primed coordinates correspond to source points at the aperture introduced to simulate the magnetic field in the region z > 0, and R is the distance from the source to a corresponding field point which at z=0 is $R^2 = \rho^2 + \rho'^2 - 2\rho\rho'\cos(\varphi' - \varphi)$. All other parameters have their usual significance.

The field component inside the line (z < 0) is expressed in terms of a superposition of the forward traveling TEM wave and its reflection from the interface at z = 0 and a series of high-order $TM_{0n}$ modes that are all evanescent along the line as a result of the line dimensions. In the dielectric material, the electromagnetic fields are radiated due to both the dominant TEM mode and the higher-order $TM_{0n}$ modes excited at the plane of the aperture.

It can be shown that for n=0, that is when TEM propagation is assumed throughout, $E_\rho^a(\rho)$ is inversely proportional to r and expression (2) for the admittance becomes

$$Y = \frac{j\omega\varepsilon_2}{[\ln(b/a)]^2}\int_a^b\int_a^b\int_0^{2\pi} \frac{1}{R}\cos(\varphi' - \varphi)e^{-jk_2R}d\varphi'd\rho'd\rho \quad (3)$$

By expanding the exponential term in (3) into an infinite series and dropping the higher-order terms in w expression (3) becomes

$$Y = j\omega C\varepsilon_2 \quad (4)$$

in which C is a frequency-independent constant whose value depends on the dimensions of the probe. By virtue of its derivation (4) is strictly valid at the limit of low frequencies. Despite its limitations it has been used, with various provisions to make up for the inadequacy of the model, with reasonable success by a number of investigators.[4,6,7] There was, however, no doubt that this model is not suitable for the measurements of high permittivity materials at high frequencies and that a better model should be sought. Expressions (2) and (3), derived in accordance with transmission line theory, should be expected to describe more adequately the aperture admittance of the probe.

A comparative study of all three admittance models given by expressions (2), (3) and (4) shows (Figure 2) that there is a systematic difference between the predictions of (2) and (3) that is attributed to the contribution of higher order modes. It also shows that the simple expression for the admittance (4) agrees with the other two models only at low frequencies and for low-loss, low-permittivity materials.

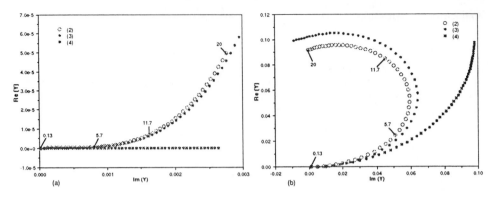

**Figure 2.** Complex admittance of probe in (a) air, (b) water at 20°C. (The numbers are frequencies in GHz).

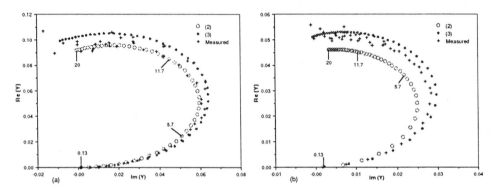

**Figure 3.** Complex admittance of probe in (a) water, (b) formamide. (The numbers are frequencies in GHz).

The predictions of models (2) and (3) are compared to corresponding experimental values of the admittance of the probe in contact with a number of standard liquids, as shown in Figure 3. The measurements were made using a network analyzer calibrated using standard components prior to connecting the probe. Under these conditions the experimental values are expected to have scatter due to unwanted reflections from uncalibrated components. Two typical sets of results are given in Figure 3 and show that the experimental values do not agree closely with either models. In fact, the probe admittance predicted by the two models appear to delineate an upper and lower bound for the measured values. It therefore appears that, under these experimental conditions, neither model is suitable for absolute measurements. However, relative measurements can be simply implemented by performing a calibration at the plane of the probe using the admittance model of choice to calculate the reflection coefficient of the standards used. This method has the added advantage of eliminating all unwanted reflections. Measurements using this technique and either of the two models produces accurate dielectric measurements in the frequency range up to 20 GHz, as evident from Figure 4 that shows the dielectric spectrum of two standard liquids measured using (2) and (3). An analysis of these data showed that the two set measurements are not significantly different from each other and from the previously published and accepted values.

The Models are therefore equally suitable for performing dielectric measurements on biological materials. The choice between them may therefore be made on the grounds of convenience since neither has been shown to be superior to the other. Model (2) requires a rather complex numerical solution for its implementation and, for this reason, is not practical for real-time measurements. By contrast, model (3) can be solved directly using integral transformation and Gaussian integration techniques.

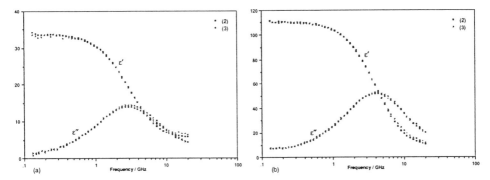

**Figure 4.**   Dielectric Properties of standard liquids (a) Methanol, (b) formamide at 20°C.

### Finite Thickness Sample

It is sometimes necessary to perform measurements, on small samples due to the nature of the material or its scarcity. To allow such measurements, expressions (2) and (3) can be adapted to the case of a sample of uniform but finite thickness backed by a conducting plane parallel to the ground plane of the probe.

A comparative study similar to that performed for the semi-infinite sample showed that for sample thicknesses h ≥ (b - a) results and conclusions similar to those obtained for the semi-infinite sample can then be reached. Both models have been used successfully to perform dielectric measurements and, provided that h ≥ (b - a), the accuracy of the measurements is not compromised by the choice of model.

### Experimental implementation

The technique described in the previous section is particularly suited to the measurements of the dielectric properties of biological materials. On the practical side, the technique can cope with the various tissues that come in a range of consistencies, from body fluids to semi-solids and hard tissues such as bone. On the technical side, the procedure is broad band as the probe dimensions can be selected for measurements over 2 to 3 frequency decades. For example, the dielectric spectrum of muscle tissue is shown in Figure 6, in the frequency range 300 kHz to 20 GHz. It was obtained using two network analyzers and different size probes and the technique describes in this paper.

It is possible to extend the measurements to higher frequencies provided the dimensions of the coaxial structure and the upper frequency of operation are selected to allow the propagation of the dominant TEM mode only. In practice, such measurements could not be made beyond 40 GHz. For measurements at higher frequencies it is usual to use equipment based on waveguides that could be built for high precision but would cover a relatively narrow frequency band. Measuring equipment, such as LCR meters and impedance analyzers, can be used to extend the coverage to lower frequencies.

### Dielectric Properties of Biological Materials

The experimental characterization of the dielectric properties of materials is obtained from measuring their complex relative permittivity $\hat{\varepsilon}$ . The relative permittivity has no units; it represents the ratio of the absolute permittivity of the material to the permittivity of space. It can be expressed as

$$\hat{\varepsilon} = \varepsilon' - j\varepsilon''$$

The real part $\varepsilon'$ is a measure of the amount of polarization produced by an electric field and the imaginary part $\varepsilon''$ is the loss factor associated with it. The loss factor $\varepsilon''$ is equivalent to a displacement conductivity $\sigma_d$ and, in the case of biological material, an ionic conductivity $\sigma_i$ due to the drift of free ions under the action of the field. The total conductivity of the material $\sigma$ is given by

$$\sigma = \sigma_d + \sigma_i$$

which is related to the measured loss factor by

$$\varepsilon'' = \sigma / \varepsilon_0 \omega$$

where $\varepsilon_0$ is the permittivity of free space and $\omega$ the angular frequency of the field. The SI unit of conductivity is siemens per meter (S/m), which presumes that in the above expression $\varepsilon_0$ is expressed in farad per meter (F/m) and w in radian per second.

Some field-induced polarizations may not occur instantaneously, rather, such effects are characterized by time constants determined by the mechanism producing them. For a polarization that approaches its final value exponentially with a time constant $\tau$ , the response to a time dependent field is

$$\hat{\varepsilon} = \varepsilon_\infty + \frac{\varepsilon_s - \varepsilon_\infty}{1 + j\omega\tau}$$

This is the well-known Debye dispersion formula where $\varepsilon_0$ is the permittivity at field frequencies such that $\omega\tau \gg 1$, and $\varepsilon_s$ the permittivity at $\omega\tau \ll 1$. Materials containing a single type of polarizable entity are likely to exhibit such dielectric behavior. For example, methanol and formamide, which have polar molecules, exhibit Debye type dispersions (Figure 4). For methanol, the dispersion parameters, corresponding to the data in Figure 4, are $\varepsilon_s = 34.6$, $\varepsilon_0 = 4.9$ and $\tau = 51.7$ps; for formamide, they are $\varepsilon_s = 111$, $\varepsilon_0 = 5.7$ and $\tau = 40$ps; and, for comparison, the parameters for water are $\varepsilon_s = 80.1$, $\varepsilon_0 = 5.0$ and $\tau = 9.17$ps. The magnitude of the dispersion is $\Delta\varepsilon = \varepsilon_s - \varepsilon_\infty$. From the above relationships, the frequency dependence of the permittivity and conductivity are given by

$$\varepsilon'(\omega) = \varepsilon_\infty + \Delta\varepsilon / (1 + \omega^2 \tau^2)$$

and

$$\sigma(\omega) = \sigma_s + \varepsilon_0 \omega^2 \tau \Delta\varepsilon / (1 + \omega^2 \tau^2)$$

Within a dispersion the conductivity increment is $\Delta\sigma = \sigma_\infty - \sigma_s$ in which $\sigma_s$ and $\sigma_0$ are the low and high frequency values of $\sigma$ and

$$\Delta\sigma = \frac{\varepsilon_0}{\tau}\Delta\varepsilon$$

The above expression shows that for a dispersion associated with a single relaxation time, the change in conductivity is proportional to the change in permittivity.

The interaction of a biological tissue with electromagnetic radiation is characterized by several polarization mechanisms involving its constituents at the cellular and molecular level. The theory underpinning these interactions is well established and each is characterized by its own dispersion. The main features of the dielectric spectrum of a high-water-content tissue (Figure 5) can be predicted over a wide frequency range.[8] By collating dielectric data published over the past decades it can be shown that the predicted characteristics of the spectrum are manifested experimentally.[9] The relative permittivity of a tissue may reach values of up to $10^6$ or $10^7$ at frequencies below 100 Hz. It decreases at high frequencies in three main dispersions designated $\alpha$, $\beta$ and $\gamma$. The $\gamma$ dispersion, in the GHz region, is due to the polarization of water molecules. The $\beta$ dispersion, in the MHz region, is due mainly to the polarization of cellular membranes that act as barriers to the flow of ions between the intra- and extra-cellular media. Other contributions to the $\beta$ dispersion come from the polarization of protein and other organic macromolecules. The low frequency $\alpha$ dispersion is a manifestation of interactions at the cellular and intracellular level, of which tangential counterion motion is a principal component.

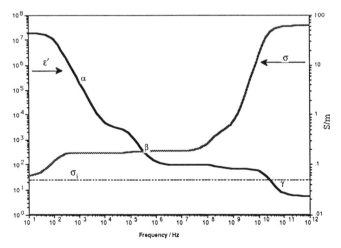

**Figure 5.** Predicted dielectric properties of a high water content tissue.

Using the experimental setup described in the previous section, it is possible to study the β and γ dispersions of tissues. Figure 6 shows the measured dielectric spectrum of animal (pig) muscle at 20 °C. Muscle is probably one of the most widely measured tissues. By contrast, there are very few published data for bone. The results in Figure 7 show well-defined β and γ dispersions. The measurements were made on the surface of a freshly excised flat bone from a sheep's skull. Measurements of the spongy center of the bone yield higher permittivity and conductivity values that is consistent with a higher water content. The dielectric method is a powerful tool for the study of water in biological systems.[10] The data in Figure 7 can thus be used to determine the nature of water in bone, which has been identified as an important but rather complex subject.[11] Tooth enamel and cartilage are two other tissues that have not been sufficiently investigated; a dielectric study would provide new data and help elucidate the nature of their water content.

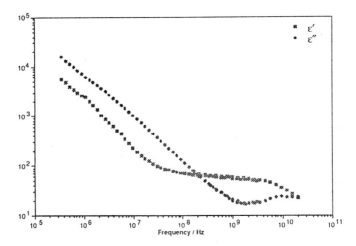

**Figure 6.** Dielectric Properties of Muscle at 20°C.

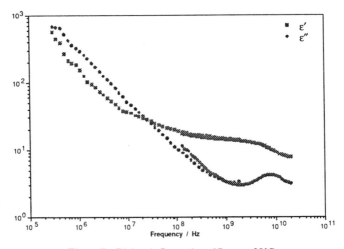

**Figure 7.** Dielectric Properties of Bone at 20°C.

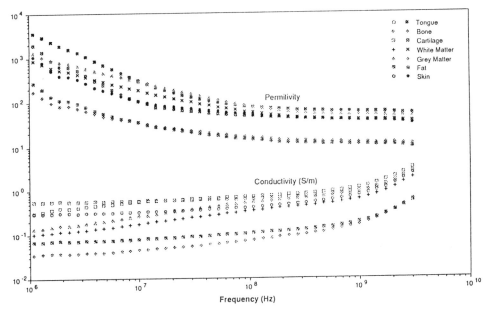

**Figure 8.** Dielectric Properties of Animal Tissue at 20°C.

Finally, Figure 8 shows the dielectric spectra of seven different animal (pig) tissues. The dielectric properties of the bone compare well with those reported in Figure 7 despite the fact that the samples originate from different species. The samples of cartilage were from the septum of the nose. Its dielectric properties suggest that it has a relatively high water content. This subject calls for further investigations and analysis.

The dielectric properties of tissues reported here consolidate our knowledge and understanding of the mechanisms of interaction giving rise to the β and γ dispersions. There is, however, room for a better experimental characterization and better understanding of the specific mechanisms responsible for the a dispersion.

## ACKNOWLEDGMENTS

This paper draws on work performed under contract AFOSR-91-0122 with the United States Air Force Office of Scientific Research.

## REFERENCES

1. E.H. Grant, R.J. Sheppard and G.P. South, "Dielectric Behaviour of Biological Molecules in Solution" Oxford, Clarendon (1978).
2. R. Pethig, "Dielectric and Electronic Properties of Biological Materials", Wiley, Chichester (1979).
3. J.Y. Chen and O.P. Gandhi, Currents induced in an anatomically based model of a human for exposure to vertically polarized electromagnetic pulses, *IEEE Trans. Microw. Theory Tech.* MTT-39:31 (1991).
4. E.C. Burdette, F.L. Cain and J. Seals, *In vivo* probe measurement technique for determining dielectric properties at VHF through microwave frequencies, *IEEE Trans. Microw. Theory Tech.* MTT-28:414 (1980).
5. J.R. Mosig, J.C.E. Besson, M. Gex-Fabry and F. E. Gardiol, Reflection of an open-ended coaxial line and application to non-destructive measurement of materials, *IEEE Trans. Instrum. Meas.* IM-30:46 (1981).
6. T.W. Athey, M.A. Stuchly and S. S. Stuchly, Measurement of radio frequency permittivity of biological tissues with an open-ended coaxial line. Part I, *IEEE Trans. Microw. Theory Tech.* MTT-30:82 (1982).

195

7. C. Gabriel, E.H. Grant and I.R. Young, Use of time domain spectroscopy for dielectric properties with coaxial probe, *J. Phys. E: Sci. Instrum.* 19:844 (1986).
8. R. Pethig and D.B. Kell, The passive electrical properties of biological systems: their significance in physiology, biophysics and biotechnology, *Phys. Med. Biol.* 32:933 (1987).
9. W.D. Hurt, Multiterm Debye dispersion relations for permittivity of muscle, *IEEE Trans. Biomed. Eng.* BME-32:60 (1985).
10. E.H. Grant, The dielectric method of investigating bound water in biological material: An appraisal of the technique, *Bioelectromagnetics* 3-1:17 (1982).
11. P.A. Timmins and J.C. Wall, Bone Water, *Calc. Tiss. Res.* 23:1 (1977).

# INTERACTION OF CALCIUM IN BIOLOGICAL SYSTEMS WITH ELECTROMAGNETIC FIELDS

James H. Merritt[*]

Radiofrequency Radiation Division
Armstrong Laboratory
Brooks Air Force Base, Texas 78235

It has been nearly 20 years now since Bawin et al.[1] described effects of amplitude-modulated radiofrequency radiation signal on calcium in the central nervous system. This study indicated that 147-MHz signals amplitude modulated in the range of 6-20 Hz, increased the release of preloaded $^{45}Ca^{2+}$ *in vitro* from the brain of neonatal chicks compared to sham-exposed tissue (Figure 1). The forebrains of 1-7 day old chicks were re-

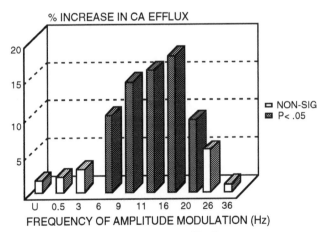

**Figure 1.** Effect of 147-MHz fields, applitude modulated on $^{45}Ca^{2+}$ efflux from isolated chick brain. Data is from Ref. 1.

moved and divided at the midline. Each brain hemisphere was then incubated in a physiological buffer containing $^{45}Ca^{2+}$ for 30 minutes. After rinsing in buffer, the

---

[*] These views and opinions are those of the author and do not necessarily state or reflect those of the U.S. Government.

hemispheres were placed in tubes containing the same buffer less tracer and exposed or sham-exposed for 20 minutes. The radioactivity in this last buffer was then assayed to determine the extent of $^{45}Ca^{2+}$ exchange during treatment. This exchange is usually referred to as "efflux" in the articles describing these studies, although it should more properly be called "net efflux." In the original study,[1] calcium efflux from nonneural tissue (gastrocnemius muscle) was not affected by exposure to the signal. However, the effect on efflux was also seen in chick brain poisoned with cyanide, indicating that the effect was independent of ongoing metabolism. Blackman et al.[2] have shown that the chick cerebral tissue becomes hypoxic during the course of the incubation and treatment, 90% of the oxygen ($O_2$) being depleted in 10 minutes. However, the brain was still capable of some level metabolism activity throughout the experimental period.

Bawin and Adey[3] next showed that extremely low frequency electric fields inhibited efflux of $^{45}Ca^{2+}$ from preloaded avian and cat cerebral issue. Again, the effective frequencies were 6-16 Hz (Figure 2). Significant differences were noted only at 10 and

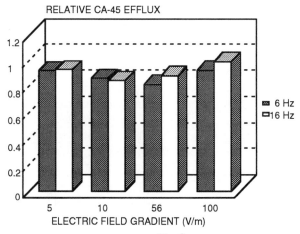

**Figure 2.** Effects of extremely low frequency fields on $^{45}Ca^{2+}$ efflux from isolated chick brain. Statistically significant inhibition of efflux is seen at 10 and 56 V/m for both 6 and 16 Hz exposures. Data from Ref. 3.

56 V/m. The electric field in the cerebral tissue exposed to the extremely low frequency fields that induced inhibition of $^{45}Ca^{2+}$ efflux was estimated to be $10^{-5}$ V/m. Thus, the effect appears to be strongly dependent upon the frequency and magnitude of the incident field. Using a 450-MHz carrier amplitude-modulated at 16 Hz, Bawin et al.[4] again showed an increase of $^{45}Ca^{2+}$ efflux from exposed chick brain tissue. This increase was further enhanced by an decrease in pH of the incubation medium but decreased in the absence of bicarbonate (Figure 3). It was suggested that transduction of low frequency, weak extracellular electric fields may occur at specific negative membrane $Ca^{2+}$ binding sites that are competitive with $H^+$.

Later, Sheppard et al.[5] showed that the effective power densities for enhancement of efflux by 450 MHz-fields amplitude modulated at 16 Hz were between 0.05 mW/cm$^2$ and 2 mW/cm$^2$, constituting an amplitude "window". The power density and modulation-frequency "windows" were subsequently confirmed by Blackman, et al.[6] as shown in Figure 4. Blackman et al.[7] then reported that the configuration of the sample holders was important in defining the width of the power density "window." In addition, this group[8] also demonstrated two separate effective power density ranges for enhancement of $^{45}Ca^{2+}$ at 50 Hz one at 1.44 mW/cm$^2$ and another including 3.14 mW/cm$^2$.

In 1982, Blackman et al.[9] attempted to corroborate the original study by Bawin and Adey[3] on extremely low frequency field effects on calcium efflux. In the replication study, emphasis was placed on identifying effective frequencies and electric fields amplitude parameters. Figure 5 shows some results of that study. In contrast to the inhibition of cal-

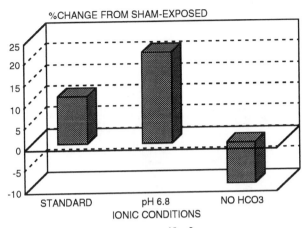

**Figure 3.** Effect of media ionic factors on efflux of $^{45}Ca^{2+}$ from chick brain. Data is from Ref. 4.

cium efflux noted by Bawin and Adey, Blackman et al. found an efflux enhancement as a result of exposure. The authors were unable to give a definitive explanation for the directly opposite results of the two studies, but have made various suggestions (vide infra).

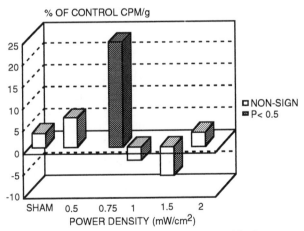

**Figure 4.** Effect of 147-MHz fields, amplitude modulated at 16 Hz, on $^{45}Ca^{2+}$ efflux. Data is from Ref. 6.

Blackman and his group later extended these studies to identify electromagnetic field conditions that induce changes in calcium efflux.[10] Exposures were made at an electric fields intensity of 42.5 V/m, identified previously as the midpoint of a range of effective intensities, and at a wide range of frequencies (1-120 Hz). Two frequency regions were shown to enhance efflux, viz. one centered on 15 Hz and the other in the range 45-105 Hz.

This research group also demonstrated that the AC magnetic component of an electromagnetic field is essential for enhancement of efflux. Thus, when an electric field alone was used in the exposure there was no enhancement of efflux.[11] In addition, the local geomagnetic field appears to be a determinant. A 15-Hz magnetic field that is normally effective in altering calcium efflux at 38 $\mu$T becomes ineffective when the local geomagnetic field was reduced by half (from 38 $\mu$T to 19 $\mu$T) with a DC magnetic field parallel to the local vector. A normally ineffective 30-Hz magnetic field is rendered effec-

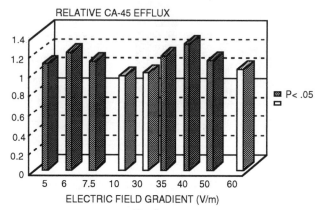

**Figure 5.** Effects of 16 Hz electric fields on $^{45}Ca^{2+}$ efflux from chick brain. Significant enhancement of efflux is seen at 5-7.5 and 35-50 V/m. Data is from Ref. 9.

tive when the net magnetic density was changed to 0.67 and 2 times ambient, but not when increased to 1.33 and 2.183 times ambient. Since the intensity and orientation of the geomagnetic field varies widely from place to place, the authors suggested that this variation may be "a probable cause for highly variable results and for inability to corroborate effects reported by different investigators."

In one study $^{45}Ca^{2+}$ efflux was studied across a frequency range of 1-510 Hz at an electric field intensity of 15 V/m.[12] The investigators believe that by examining the calculated P-values, a basis for hypothesizing the existence of three frequency-dependent patterns in the data is evident; one pattern between 15 and 315 Hz at 30-Hz intervals, with a break at 165 Hz dividing the pattern into two groups of five frequencies. The even spacing of the positive responses was said to be consistent with the magnetic resonance phenomena. The second pattern at 60, 90, and 180 Hz was said to be compatible with a Lorentz-force mechanism. The third pattern (with positive response at 405 Hz) could be a nuclear magnetic resonance (NMR) interaction.

Blackman et al.[13] have recently reported that incubation temperature during exposure or sham exposure is a critical determinant of the experimental outcome. These investigators were able to induce enhancement, inhibition, or no change in efflux from brain tissue exposed to 16 Hz fields at 14 V/m by altering incubation temperature profiles. During the 20-minute exposures, the incubation medium was allowed to rise to final temperatures of 35, 36, 37, 38 and 39°C. Enhancement of calcium efflux was observed in this ascending profile at final temperatures of 35, 36 and 37°C, but not 38 or 39°C. (Figure 6). When the temperature was not allowed to vary more than 0.3°C (i.e., stable), inhibition was noted at 36 and 37, but not at 35 or 38°C. When a descending profile was used, no change in efflux was seen at any final temperature. The authors believe that these data are consistent with the response of a system poised at phase or cooperative transition, and that a "temperature window" exists for electromagnetic fields effects on calcium efflux.

Attempts to replicate the studies that have shown electromagnetic effects on calcium efflux have not always been successful. Albert et al.[14] exposed chick brain hemispheres and brain slices to 147-MHz radiofrequency radiation, amplitude modulated at 16 Hz. These investigators were unable to show significant effects on calcium efflux. In these studies, two different experimental protocols were used. The first protocol utilized 1-2 mm slices of chick forebrain incubated in Krebs-Hensleit media that had been previously gassed with 95% $O_2$-5% $CO_2$ (Figure 7). The second protocol was essentially the same as that reported by others (Bawin and Blackman and their colleagues), in which chick brain hemi-

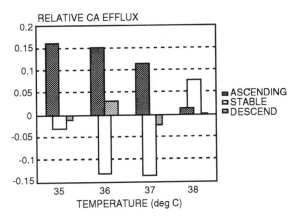

**Figure 6.** Effect of temperataure on $^{45}Ca^{2+}$ efflux. Mean relative efflux is shown as a function of final temperature reached during exposure. Data is from Ref. 13.

spheres were incubated in bicarbonate buffer. Another attempt at replication of the original Bawin et al. study[1] was likewise a failure.[15] In this study no effect was noted when the preloaded tissue was washed with media at 37°C, but an enhanced efflux was seen when the tissue was rinsed at 26.5°C.

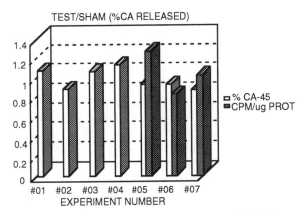

**Figure 7.** Effect of 147-MHz exposure, amplitude modulated at 16 Hz, on $^{45}Ca^{2+}$ efflux from chick brain. No significant effectss were noted in the 5 experiments. Data is from Ref. 14.

Shelton and Merritt[16] studied the effect of 1-GHz radiation, pulsed at 16 Hz and 32 Hz on calcium efflux from rat brain slices pre-loaded with $^{45}Ca^{2+}$. No effect of any combination of pulse repetition rate and power density on calcium efflux was seen. Merritt et al.[17] reported on pulsed microwave radiation on $^{45}Ca^{2+}$ preloaded in rat brain by intraventricular injection and exposed both *in vivo* and *in vitro*. No differences in calcium efflux were found in this study. There are fundamental differences between these studies and those showing effects, not the least of which is pulsed versus continuous exposure that is amplitude modulated.

Adey et al.[18] showed that amplitude-modulated (16 Hz) 450-MHz fields altered calcium efflux from the brains of awake, immobilized cats. In this study, $^{45}Ca^{2+}$ was applied directly in 1 ml of medium to the cortex via a plastic well fitted through the skull. After a 90-minute incubation, the medium was removed and replaced by tracer-free media; this was repeated at 10-minute intervals throughout the experiment. Exposures at 3 mW/cm$^2$ were initiated at intervals ranging from 80 to 120 minutes after $^{45}Ca^{2+}$ superfusion. The authors state that the efflux curves show epochs of increased $^{45}Ca^{2+}$ efflux compared with the sham-exposed condition that began shortly after the start of exposure (Figure 8). The periods of enhanced efflux were irregular in duration and ampli-

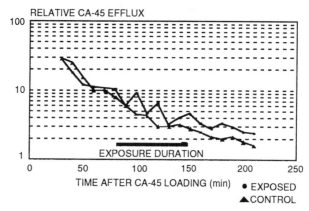

**Figure 8.** Effect of 147-MHz exposure, amplitude modulated at 16 Hz, on $^{45}Ca^{2+}$ from the cortex of awake cats. Data is from Ref. 18.

tude and continued into the post-exposure phase. The authors point out the possibility of multiple tissue pools that bind calcium and believe that their data indicate a pool with a weak linkage to cell surface molecules that can be easily disrupted by imposed electric fields. The $^{45}Ca^{2+}$ was shown to have been localized in the first 3 mm of the cortex which is about the thickness of the cortical gray matter. There appears to have been no significant leakage of the tracer to the surrounding tissue.

Lin-Liu and Adey[19] studied release of preloaded $^{45}Ca^{2+}$ from rat brain synaptosomes. Synaptosomes allow the study of the neuronal membrane since the membranes of these terminal bodies reseal after separation from the neuron. $^{45}Ca^{2+}$ efflux from these synaptosomes is biphasic with fast and slow components. This investigation utilized a continuous perfusion technique to monitor calcium efflux from the synaptosomes during exposure in a Crawford cell. Rate constants were calculated for both the fast and slow components of the efflux curves. Exposures were made at 450 MHz, either continuous wave or amplitude modulated at 16 Hz and 60 Hz. Of these exposures, only the rate constant of the slow component in the preparation exposed at 16 Hz was significantly different from the sham-exposed preparation. Further, these investigators showed that the microwave-induced enhancement in efflux was probably not due to changes in intracellular calcium; this efflux could be distinguished from CaCl$_2$-stimulated $^{45}Ca^{2+}$ efflux, which is probably derived intracellularly.

In a study of human-derived neuroblastoma cells preloaded with $^{45}Ca^{2+}$, Dutta et al.[20] exposed cultured cell monolayers to 915-MHz radiation amplitude modulated at 16 Hz. $^{45}Ca^{2+}$ efflux from these cells was enhanced by exposure at SARs of 0.05, 0.75 and 1 W/kg but not at other SARs below, above, or between these values. Likewise, this research group[21] used a 147-MHz carrier frequency amplitude modulated at various frequencies and showed narrow SAR "windows" for $^{45}Ca^{2+}$ efflux at 16 Hz (i.e., 0.005 and 0.05 W/kg). In addition to human neuroblastoma cells, they tested a hybrid hamster-mouse-derived neuroblastoma cell line. They also reported two "windows" for frequency of amplitude modulation; one at 13-16 Hz and another at 57.5-60 Hz.

Effects of magnetic fields on uptake of $^{45}Ca^{2+}$ in a variety of cells have been studied over the past few years. Liboff et al.[22] reported an enhancement of $^{45}Ca^{2+}$ uptake by unstimulated human lymphocytes as a result of exposure to 20.9 µT fields at an alternating current frequency of 14.27 Hz. Calcium uptake in these lymphocytes was 3 times greater when cells were exposed at this frequency than under control conditions. In addition, no change was observed at 1 Hz higher or lower, and uptake rose monotonically up to a field strength of 21 µT and then decreased; at 50 µT there was no effect of exposure. Prasad et al.[23] attempted to replicate this study but failed to obtain these effects conducted under the purported cyclotron resonance exposure conditions (Figure 9). This study was carried out with scrupulous attention to details and was conducted "double blindly." A positive control was included in their study that showed a highly significant increase in $^{45}Ca^{2+}$ uptake.

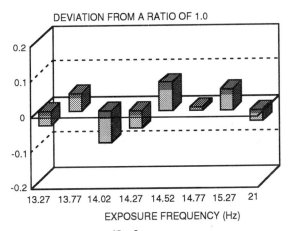

**Figure 9.** Effect of 20 µT magnetic fields on $^{45}Ca^{2+}$ uptake in cultured human lymphocytes. No significant changes were noted at any of the exposure frequencies tested. Data is from Ref. 23.

Increased calcium uptake in normal and leukemic lymphocytes has been observed as a result of exposure to 13.6 Hz, 20 µT magnetic fields.[24] These parameters constitute "near cyclotron resonance" for calcium. Apparent mistuning resulted in the field being 2.4 Hz above cyclotron resonance. All of the cell lines tested (murine, cytotoxic T-lymphocytes; spleen lymphocytes; lymphoma) showed an increase in calcium uptake during exposure to the magnetic field. Exposures were also made at 60 Hz at which frequency a weak effect was noted. On the other hand, Parkinson and Hanks[25] were unable to find evidence for changes in cytosolic calcium in a variety of adherent cell lines exposed to "cyclotron resonance" conditions (50 µT; 38.15 Hz). Carson et al.[26] investigated the effect of the radiofrequency radiation field, the static magnetic field, and the time-varying field generated by a magnetic resonance imaging unit on cytosolic free calcium in cultured HL-60 cells. Cytosolic free calcium was significantly increased in the cultures exposed to the

MRI field (combination of RF field, static magnetic field and time-varying field) and to the time-varying magnetic field alone, but not in the RF field or static magnetic field alone (Figure 16). Liburdy[27] exposed thymic lymphocytes to a 60 Hz, 22 mT magnetic field. After mitogen stimulation, field exposure enhanced calcium uptake significantly. The 22 mT magnetic field induced an electric field of 1.7 mV/cm; exposure to a pure electric field of this magnitude also enhanced calcium influx. Liburdy believes that his data are consistent with interaction of the electric field at the cell surface.

From the foregoing discussion of the data it is clear that electromagnetic effects on calcium in biological systems are not firmly established. For instance, positive effects have been reported by Blackman and his colleagues[7, 8, 10, 11] and negative effects by other research groups,[14, 15, 16, 17] as well as opposite direction of efflux changes induced by ex-

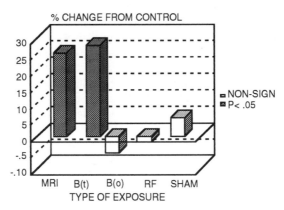

**Figure 10.** Effect of nuclear magnetic imaging exposure (MRI) on calcium uptake in HL-60 cells. Exposure to the complete MRI signal and the time-varying mangetic fields [B(t)] significantly enhanced uptake. The static magnetic field [B(o)] and radiofrequency field (RF) had no effect. Data is from Ref. 26.

tremely low frequency fields, i.e., enhancement or inhibition of efflux reported by Blackman et al.[9] and Bawin and Adey[3] respectively. Blackman et al.[13] have suggested that the difficulties in verification of results on calcium efflux could be due to differences in conditions during exposure; incubation media composition, local geomagnetic fields, prior electromagnetic field exposure history of samples, and the temperature profiles before and during exposure.

The failure of attempts to replicate studies showing electromagnetic fields effects on calcium uptake in various cells lines has further complicated interpretation of these studies. Additionally, many of the articles dealing with calcium efflux and calcium uptake in various biological systems intermix the two phenomena in discussions of mechanisms and sites of action. Surely, however, the calcium efflux phenomena is very much different from calcium influx (uptake). The calcium involved in efflux is presumably loosely bound to extracellular matrices such as glycoproteins. Efflux, in the sense used in the studies considered here, apparently does not depend upon intact cellular metabolism since the effect of electromagnetic fields is seen in tissue poisoned by cyanide.[1] In general, the metabolic state of the chick brain tissue is severely compromised during the exposure period by depletion of $O_2$ in the incubation media during experimental manipulations,[28] at least for most of the reported studies. Calcium uptake, on the other hand, is an energy-requiring metabolic process involving active transport through channels in the cell membrane.

Various mechanisms have been proposed to explain the purported effects of time-varying electromagnetic fields on calcium dynamics in biological systems. Included are cyclotron resonance, nuclear magnetic resonances and a Larmor frequency resonant response. Serious objections to cyclotron resonance as a mechanism has been raised by Sandweiss,[29] who indicated that such an explanation violated physical principles. A modification of the cyclotron resonance mechanism has been proposed to account for some of these objections,[30] but this too has been criticized.[31]

We are left then, with a set of biological data that is not entirely replicable and with no good explanation for the purported effects. As Greengard has stated;[32] "an extremely interesting set of facts and a highly speculative set of theories do not seem to impinge on each other to any reasonable extent". To evaluate the biological significance of the putative efflux or influx induced by electromagnetic fields, studies must be designed to test for effects of such alteration of calcium in biological systems. One such study[33] found no effect of microwave exposure on calcium-dependent phosphorylation of synapsin I, a neuron specific phosphoprotein, localized in the presynaptic terminal of all neurons. Another study could have tested a functional effect on frog heart rate.[34] Instead, the investigators clamped the beat rate of the heart, and only collected data on calcium efflux from the beating heart. The results do not give us any data on possible chronotropic effects of electromagnetic fields correlated with the calcium alterations induced by these fields. Studies along such functional lines are critically needed in order to assess electromagnetic field effects on calcium in biological systems and determine what weight such effects should have on the development of human exposure standards.

## REFERENCES

1. S.M. Bawin, L. Kaczmarek and W.R. Adey, Effects of modulated VHF fields on the central nervous system, *Ann N Y Acad Sci* 247:74-80 (1975).

2. C.F. Blackman, W.T. Joines and J. Elder. Calcium ion efflux induction in brain tissue by radiofrequency radiation, *in:* "Biological Effects of Radiofrequency Radiation," K. Illinger, Ed., ACS Symposium Series 157, pp. 299-314 (1981).

3. S.M. Bawin and W.R. Adey, Sensitivity of calcium binding in cerebral tissue to weak environmental electric fields oscillating at low frequency, *Proc Nat Acad Sci USA* 73:1999-2003 (1976).

4. S. Bawin, W.R. Adey and I. Sabbot, Ionic factors in release of $^{45}Ca^{2+}$ from chicken cerebral tissue by electromagnetic fields, *Proc Nat Acad Sci* 75:6314-6318 (1978).

5. A. Sheppard S.M. Bawin and W.R. Adey, Models of long-range order in cerebral macromolecules: Effects of sub-ELF and of modulated VHF and UHF fields, *Radio Science* 14(6S):141-145 (1979).

6. C.F. Blackman, J. Elder, C. Weil, S. Benane, D. Eichinger and D. House, Induction of calcium-ion efflux from brain tissue by radio-frequency radiation: Effects of modulation frequency and field strength, *Radio Science* 14(6S): 93-98 (1979).

7. C.F. Blackman, S. Benane, J. Elder, D. House, J. Lampe and J. Faulk, Induction of calcium-ion efflux from brain tissue by radiofrequency radiation: Effect of sample number and modulation frequency on the power-density window, *Bioelectromagnetics* 1:35-43 (1980).

8. C.F. Blackman, S. Benane, W. Joines, M. Hollis and D. House, Calcium-ion efflux from brain tissue: Power-density versus internal field-intensity dependencies at 50-MHz RF radiation, *Bioelectromagnetics* 1:277-283 (1980).

9. C.F. Blackman, S. Benane, L. Kinney, W. Joines and D. House, Effects of ELF fields on calcium-ion efflux from brain tissue, *Radiat Res* 92:510-520 (1982).

10. C.F. Blackman, S. Benane, D. House and W. Joines, Effects of ELF (1-120 Hz) and modulated (50 Hz) RF fields on the efflux of calcium ions from brain tissue *in vivo, Bioelectromagnetics* 6:1-11 (1985).

11. C.F. Blackman, S. Benane, J. Rabinowitz, D. House and W. Joines, A role the magnetic field in radiation-induced efflux of calcium ions from brain tissue *in vitro, Bioelectromagnetics* 6:327-337 (1985).

12. C.F. Blackman, S. Benane, D. Elliot, D. House and M. Pollock, Influence of electromagnetic fields on the efflux of calcium ions from brain tissue *in vitro:* A three-model analysis consistent with the frequency response up to 510 MHz, *Bioelectromagnetics* 9:215-227 (1988).

13. C.F. Blackman, S. Benane and D. House, The influence of temperature during electric- and magnetic-field-induced alteration of calcium-ion release from in vitro brain tissue, *Bioelectromagnetics* 12:173-182 (1991).

14. E.N. Albert, F. Slaby, J. Roche and J. Loftus, Effect of amplitude-modulated 147 MHz radiofrequency radiation on calcium ion efflux from avian brain tissue, *Radiat Res* 109:19-27 (1987).

15. Q.P. Lu, A.W. Guy, H. Lai and A. Horitz. The effects of modulated radiation on calcium efflux from chick brain *in vitro*. Abstract D-1, 9th Annual Meeting of the Bioelectromagnetics Society, Portland, OR, (June 1987).

16. W.W. Shelton and J.H. Merritt, Attempts to alter $^{45}Ca^{2+}$ binding to brain tissue with pulse-modulated microwave energy, *Bioelectromagnetics* 2:161-167 (1981).

17. J.H. Merritt, W.W. Shelton and A.F. Chamness, *In vitro* study of microwave effects on calcium efflux in rat brain tissue, *Bioelectromagnetics* 3:475-478 (1982).

18. W.R. Adey, S. Bawin and A. Lawrence, Effects of weak amplitude-modulated microwave fields on calcium efflux from awake cat cerebral cortex, *Bioelectromagnetics* 3:295-307 (1982).

19. S. Lin-Liu and W.R. Adey, Low frequency amplitude modulated microwave fields change calcium efflux rates from synaptosomes, *Bioelectromagnetics* 3:309-322 (1982).

20. S. Dutta, B. A. Subramoniam, Ghosh and R. Parshod, Microwave radiation-induced calcium ion efflux from human neuroblastoma cells in culture, *Bioelectromagnetics* 5:71-78 (1984).

21. S. Dutta, B. Ghosh and C.F. Blackman, Radiofrequency radiation-induced calcium ion enhancement from human and other neuroblastoma cells in culture, *Bioelectromagnetics* 10:197-202 (1989).

22. A. Liboff, R.Rozek, M. Sherman, B. McLeod and S. Smith, $Ca^{2+}$-45 cyclotron resonance in human lymphocytes. *J Bioelectricity* 6:13-22 (1987).

23. A. Parsad, M. Miller, E. Carstensen, C. Cox, M. Azadniv and A. Brayman, Failure to reproduce increased calcium uptake in human lymphocytes at purported cyclotron resonance exposure conditions, *Radiat Environ Biophys* 30:305-320 (1991).

24. D. Lyle, X. Wang, R. Ayotte, A. Sheppard and W.R. Adey, Calcium uptake by leukemic and normal T-lymphocytes exposed to low frequency magnetic fields, *Bioelectromagnetics* 12:145-156 (1991).

25. W.C. Parkinson and C.T. Hanks, Search for cyclotron resonance in cells *in vitro*, *Bioelectromagnetics* 10:129-145 (1989).

26. J. Carson, F. Prato, D. Frost, D. Diesbourg and S. Dixon. Time-varying magnetic fields increase cytosolic $Ca^{2+}$ in HL-60 cells, *Am J Physiol* 259:C687-C692 (1990).

27. R. Liburdy, Calcium signaling in lymphocytes and ELF fields: Evidence for an electric field metric and a site of interaction involving the calcium ion channel, *FEBS Letters* 301:53-59 (1992).

28. C.F. Blackman. ELF effects on calcium homeostasis, *in:* Extremely Low Frequency Electromagnetic Fields: The Question of Cancer, B. Wilson et al., eds., Battelle Press, Columbus, WA, pp. 184-205 (1988).

29. J. Sandweiss, On the cyclotron resonance model of ion transport. *Bioelectromagnetics* 11:203-205 (1990).

30. V. Lednev, Possible mechanisms for the influences of weak magnetic fields on biological systems. *Bioelectromagnetics* 12:71-75 (1990).

31. R. Adair, Criticism of Lednev's mechanism for influence of weak magnetic fields on biological systems. *Bioelectromagnetics* 13:231-235 (1992).

32. P. Greengard, Effects of Electromagnetic Radiation on Calcium in the Brain. Aeromedical Review 2-82, U.S. Air Force School of Aerospace Medicine, Brooks Air Force Base, TX, July 1982.

33. M. Browning and J. Haycock, Microwave radiation, in the absence of hyperthermia, has no detectable effect on synapsin I levels or phosphorylation. *Neurotoxicology and Teratology* 10:461-464 (1988).

34. J-L. Schwartz, D. House, and G. Mealing, Exposure of frog hearts to CW or amplitude-modulated VHF fields: Selective efflux of calcium ions at 16 Hz. *Bioelectromagnetics* 11:349-358 (1990).

# EFFECTS OF WEAK HIGH-FREQUENCY ELECTROMAGNETIC FIELDS ON BIOLOGICAL SYSTEMS

Robert K. Adair

Department of Physics
Yale University
New Haven, CT 06520-8121

## ABSTRACT

I examine the constraints on the biological effects of the interactions of radio-frequency and microwave radiation imposed by thermodynamic noise. An analysis of the interaction of radiation with small biological elements at the cellular level shows that at power densities of 10 mw/cm$^2$ (or 200 V/m), a level characteristic of occupational exposure limits, the interaction of electromagnetic fields with elements holding permanent charges or charge distributions will be masked by thermal noise and, hence, cannot be expected to generate biological effects.

However, I cannot exclude the possibility that energy transfers to large free cells (with radii greater than 20 $\mu m$) exceed $kT$.

Moreover, as pointed out by Schwan, the interactions of AC fields of 200 V/m with charges *induced* by the fields may generate energy transfers in cellular systems of the order of kT. I also examine the possible actions of enzymes as rectifiers, as suggested by Astumian, and show that AC fields of 200 V/m might drive molecular concentrations in cells away from equilibrium beyond that expected from noise drifts. Hence, the possibility of biological effects from such interactions is not, therefore, definitely excluded by thermodynamic considerations.

It is of practical interest that, for *any* interaction of electromagnetic fields at this power level, sharp resonances are excluded and biological effects, if they should exist, should not change radically over differences in frequency of a factor of two.

## INTERACTION OF ELECTROMAGNETIC FIELDS

Electromagnetic fields interact with matter only through forces generated on charges. The vector force $\mathbf{F}$ on a charge $q$, is described by the Lorentz force law,

$$\mathbf{F} = q(\mathbf{E} + \mathbf{v} \times \mathbf{B}) \tag{1}$$

where the electric field $\mathbf{E}$, the magnetic field $\mathbf{B}$, the velocity of the charge $\mathbf{v}$, and the

*Radiofrequency Standards*, Edited by B.J.
Klauenberg *et al.*, Plenum Press, New York, 1994

charge $q$, are all expressed in SI units. Though it is often useful to describe the effects of fields in terms of scalar and vector potentials, a description of effects in terms of forces on charges must be complete. Since at microwave frequencies, $\nu$, $h\nu \ll kT$, quantum effects cannot be expected to be important.

Upon the exposure of biological material to radio-frequency and microwave radiation, the fields exert forces on charges that induce changes in the motion of biological elements. If those changes are to affect biology in a significant manner, they must not be so small as to be masked by endogenous noise, especially thermal noise. The changes induced by the fields can be separated into two parts:

1. Radio-frequency and microwave fields may generate changes in the energy of specific elements that are greater than $kT/2$ per degree of freedom expected from random thermal fluctuations. For the purposes of this discussion, I assume that such increases may result in significant biological effects and an examination of the conditions under which radio-frequency and microwave fields might generate such increases is the primary subject of this paper, which extends an earlier paper on effects of ELF fields[1].

2. The interactions of the fields may increase the mean energy of all elements in a macroscopic sector of biological material by an amount that is a small fraction of $kT/2$ per degree of freedom, thus raising the temperature $T$ of the sample. Such *thermal effects* of radio-frequency and microwave radiation have been studied extensively and are not discussed.

While any biological effects of electromagnetic fields must be induced through changes in the mechanics of material elements and cannot follow from the fields per se, there are fluctuating noise electric fields as a consequence of thermodynamic fluctuations in charge densities. Though these Johnson-Nyquist fields are precisely defined in terms of mean-square noise voltages, $\overline{V_{kT}^2}$, across elements with a well-defined resistance, $R$, the mean electric fields extracted from such a calculation can be used as a sensible measure of local field noise. In particular,

$$\overline{V_{kT}^2} = 4R\,kT\,\Delta\nu \qquad (2)$$

where $\Delta\nu$ is the relevant frequency band to be considered.

At very low-frequencies, there is effectively no penetration of the electric field past the cell membrane into the central region of the cell. Hence, any biological effects of external fields on the cellular level must derive from the fields generated in the membrane. If those fields are to affect cells significantly, they must be greater than the Johnson-Nyquist noise fields in those membranes. Taking the specific resistance of the membrane material as $\rho \approx 10^6 \ \Omega$, and estimating the noise field as $E_{kT} \approx (\overline{V_{kT}^2})^{1/2}/d$, where $d \approx 10^{-8} \ m$, is the membrane thickness of a cell with a radius of $r$, I estimate,

$$\overline{V_{kT}} \approx 3.6 \cdot 10^{-6} \frac{\sqrt{\Delta\nu}}{r} \ V, \quad E_{kT} \approx 360 \frac{\sqrt{\Delta\nu}}{r} \ V/m \qquad (3)$$

where $r$ is measured in $\mu m$.

At low-frequencies, the bandwidth, $\Delta\nu$, can be taken sensibly as equal to the frequency, $\nu$, and the field induced in the membrane can be taken as $E_{mem} \approx E_L r/d$, where $E_L$ is the field in the tissue outside of the cell. Then the signal to noise ratio, $S/N = E_{mem}/E_{kT}$ will be greater than one for $E_L > E_{kT}d/r \approx 3.6 \ \nu^{1/2}/r^2 \ V/m$. But, since the body is a conductor, at low-frequency the field in the tissues inside of the body will be much smaller than the external field, $E_{ext}$. For a sphere, $E_L \approx E_{ext}\epsilon_0\rho\nu$, for the human body I take $E_L \approx E_{ext}\nu/6 \cdot 10^7$. Then for the canonical value of $E_{ext} = 200$

V/m, I can estimate the signal-to-noise in the cell membrane as $S/N \approx 10^{-7} \nu^{1/2} r^2$. For the canonical cell, $r \approx 10 \ \mu m$, the signal-to-noise calculated on this basis will be less than one at frequencies less than 100 MHz – though for large cells, $r = 100 \ \mu m$, the limit will be but one MHz.

At high-frequencies, $\nu \geq 100$ MHz, $E_{mem} \approx E_L \approx E_0$. Taking a bandwidth, $\Delta \nu = 10^6$ Hz, or $Q = 100$, where $Q = \nu/\Delta \nu$, $E_L > 3.6 \cdot 10^4$ V/m for a signal-to-noise ratio greater than one. The factor $Q$ is a kind of lifetime factor for the process that must persist for a time $Q/\Delta \nu$.

These limits on signal-to-noise hold for processes that do not, in themselves, define frequency limits narrower than those assigned. However, $S/N \propto (\Delta \nu)^{-1/2} \propto \sqrt{Q}$, and for a sufficiently high $Q$, a signal-to-noise value greater than one can always be achieved – though the signal must last for a time $Q/\Delta \nu$. Hence, though the Johnson-Nyquist noise places very strong constraints on the effects of radiofrequency or microwave radiation on that biology that takes place in cell membranes, without auxiliary conditions imposed by the biological process influenced by the field, that constraint is not absolute.

Although thermodynamic noise constrains the effects of high-frequency radiation in cell membranes, the Johnson-Nyquist noise is generally not as important in the interior of the cell even as the specific resistance of the cytoplasm is not as great as that for the material of the membrane. Hence, at high-frequencies, it is usually more illuminating to compare energies transmitted to elements by the action of the fields with the thermal agitation energy, kT.

Any attempt to calculate general results, applicable to all systems, is complicated by the variety of systems. Different biological elements have different internal dynamics and are subject to different interactions with their surrounding environment. Consequently, I choose to illuminate general properties of all elements through specific calculations of energy transfers in a model system where the fluctuations of the microscopic interactions are ignored and there is no thermal noise and then compare that transfer to thermal noise. One could calculate the effects of the external fields on noisy systems but such calculations are less general and more difficult; hence the results are less transparent. In the model system, the mean effects of the random Brownian interactions are subsumed by an effective viscosity and thermal conductivity of the surrounding environment.

For quantitative purposes, in the following calculations, I consider the effects of an exposure to radio-frequency and microwave radiation at a canonical power density of 10 mW/cm$^2$, which corresponds to an electric field of $E_{rms} = 200$ V/m or $E_0 = 282$ V/m peak voltage. Many of the numerical results pertain to effects on a canonical cell-size sphere of radius 10 $\mu m$ and of the density of water. Moreover, for the most part, I neglect the effect of the dielectric properties of the media and take these fields as those that act on the biological elements I consider.

## FORCES ON PERMANENT CHARGES

### Translational Motion

I first examine the effects of fields on the translational motion of elements carrying charges in a inertial system where the mean or initial velocity of the charged element is zero and the neglect of effects of the magnetic part of the field introduces no significant error. In particular, I describe the effects of an electric field of angular frequency $\omega$ and amplitude $E_0$, on an element of mass $m$ holding a charge $q$. The total force on an element consists of a binding force, $F_b = -Ky$, a resistive force, $F_r = -\gamma \, dy/dt$, and a driving force $F_d = E_0 q \cos \omega t$, where $y$ is the displacement of the element from an equilibrium position in the direction of the field $\mathbf{E}_0$. For free elements, such as ions in solution, $K = 0$. Equating the sum of these forces to the acceleration,

$$m\ddot{y} = -\gamma\dot{y} - Ky + E_0 q \cos \omega t \tag{4}$$

For small displacements, $y$, and small velocities, $\dot{y}$, this description – essentially Newton's second law – is quite general.

Writing the variation of $y$ with time as $y = y_0 \cos(\omega t + \phi)$,

$$y_0^2 = \frac{(E_0 q)^2}{m^2(\omega_0^2 - \omega^2)^2 + \gamma^2 \omega^2} \quad \text{where } \omega_0^2 = \frac{K}{m} \tag{5}$$

The mean kinetic energy, $W_T$, and potential energy, $W_V$, of the element are,

$$W_T = \frac{1}{4}m\omega^2 y_0^2 \quad \text{and} \quad W_V = \frac{1}{4}K y_0^2 = \frac{1}{4}m\omega_0^2 y_0^2 \tag{6}$$

In general,

$$W_T \le \frac{(E_0 q)^2 m}{4\gamma^2} \quad \text{and} \quad W_V \le \frac{(E_0 q)^2 m}{4\gamma^2} \frac{\omega_0^2}{\omega^2}$$

$$\text{for } \omega^2 \ll \omega_0^2, \ W_V \le \frac{(E_0 q)^2}{4m\omega_0^2} \quad \text{or} \quad \frac{(E_0 q X)^2}{8\,kT} \tag{7}$$

where $X$ is the thermal excursion amplitude such that $\frac{1}{2}KX^2 = \frac{1}{2}m\omega_0^2 X^2 = kT$.

For small elements on the cell level, the minimum resistive effects on the motion of such elements will derive from interactions with their neighborhood environment. I describe that interaction, in minimum, by a viscosity $\eta$ of the media about the element. Further dissipation effects will only reduce the energy transfer. From Stoke's Law, $\gamma = 6\pi\eta r$, where $r$ is a characteristic length of the element; for a sphere, $r$ is the radius. Outside of cells, $\eta$ will be the viscosity of the tissue electrolyte; inside of cells, $\eta$ will be the viscosity of the cytoplasm; for elements held in the membranes, $\eta$ will be the viscosity of the dilipids making up the membrane. For general estimates, I take that viscosity $\eta$, conservatively, as that of water, where $\eta = 10^{-3} \ Ns/m^2$. The viscosity of cytoplasm has been reported to be much greater than water[2], and the viscosity of the dilipid cell membrane material is not likely to be much less than that of water.

**Kinetic Energy:** For the cell-size system, $q/\gamma = \mu$ where $\mu = 3 \cdot 10^{-8} \ m^2/Vs$ is the mobility that is (Appendix B) independent of the size of the charged sphere, its composition, or the magnitude of the charge, if the charge is large enough. With this substitution in Eqs. 7, the maximum kinetic energy for a cell of mass $m$ is $W_t = (E_0\mu)^2 m/4$, and for the canonical cell with a radius, $r = 10 \ \mu m$, and $E_0 = 282$ V, $W_t = 0.018 \ kT$.

Since $W_T \propto m \propto r^3$, for very large spheres, or large cells, $W_T$ may exceed $kT$. Though the relation for $\mu$ probably overstates the effective charge and the kinetic energy of systems where $r > 10 \ \mu m$, from these arguments, we cannot categorically exclude the possibility that our canonical field strength may result in energy transfers to very large cells in excess of $kT$. Conversely, smaller elements will have energies much less than $kT$.

The maximum value of $W_t$ is largely set by the dissipative factor $\gamma = 6\pi\eta r$, where,

$$\tau_\gamma = m/\gamma \approx 2.2 \cdot 10^{-5} \quad \text{seconds} \tag{8}$$

is the characteristic viscous time constant for the canonical cell with a radius, $r = 10\ \mu m$. Time constants derived in different ways should have similar values. The thermal relaxation time can be expressed as,

$$\tau_t = \frac{C}{4\kappa}a^2 = 1.6 \cdot 10^6 \cdot a^2 \text{ seconds} \tag{9}$$

where $C$ is the volume thermal capacity and $\kappa$ the conductivity of water while $a$ is a characteristic length. Taking $a = r = 10\ \mu m$ for the cell, $\tau_t \approx 1.6 \cdot 10^{-4}\ s$. The rough equivalence of $\tau_\gamma$ and $\tau_t$ suggests that the estimate is reliable to within an order of magnitude.

**Potential Energy.** Although the kinetic energies are small, it is possible that a small bound element might be displaced by the field to an extent that would not occur through normal thermal buffeting and induce some biological effect. This displacement against the binding force would generate a potential energy proportional to the square of the displacement, and that potential energy cannot be small compared to kT or normal thermal effects would mask the field effect. I then consider the potential energy of bound systems.

For $\omega^2 > \omega_0^2$, the constraints on the potential energy are more severe and are constrained by the arguments applied to kinetic energies. Writing $\omega_0^2 = k/m$, where $k$ is the effective spring constant, for physiologically reasonable limits on $k$, $\omega_0$ will not exceed the audio range and, therefore, for radio and microwave frequencies, $W_V \ll W_T$. For elements with very small masses, such as ions, the condition that they must be bound so that thermal agitation does not move them with excessive amplitudes, $X$, constrains their potential energies. For a calcium ion, with $q = 2e$, constrained so that $X \le 0.1\ \mu m$, $W_V < 6 \cdot 10^{-7}$ kT.

Hence, radio-frequency or microwave radiation at levels below 10 mW/cm$^2$ will not induce significant translational kinetic or potential energy changes in biological elements smaller than $r \approx 20\ \mu m$.

## Rotational Motion

Effects may follow from torques produced by the interaction of electric fields **E** with elements carrying permanent electric dipole moments **p** and the interaction of magnetic fields **B** with elements carrying magnetic dipole moments $\mu$. The respective torques will be $\mathbf{N} = (\mathbf{p} \times \mathbf{E})$ or $\mathbf{N} = (\mu \times \mathbf{B})$. I consider maximum torques – and maximum energy transfers – and write $N = Ep$ or $N = B\mu$. The total torque on an element with a electric dipole moment **p** or magnetic dipole moment $\mu$ will be made up of a resistive torque, $T_r = -\beta\dot{\theta}$, a binding torque, $T_b = -N_b\theta$, and a driving torque, $T_d = N \cos \omega t$, where $\theta$ is the angle of rotation. The equation of motion takes a form like Eq. 4:

$$I\ddot{\theta} = -\beta\dot{\theta} - N_b\theta + N \cos \omega t \tag{10}$$

where $I$ is the moment of inertia of the element.

Writing $\theta(t) = \theta_0 \cos(\omega t + \phi)$, for $\theta_0^2 \ll 1$ which will hold for the small fields considered;

$$\theta_0^2 = \frac{(N)^2}{I^2(\omega_0^2 - \omega^2)^2 + \beta^2\omega^2} \quad \text{where} \quad \omega_0^2 = \frac{N_b}{I} \tag{11}$$

The mean kinetic energy $W_T$ and mean potential energy $W_V$ are:

$$W_T = \frac{I\omega^2\theta_0^2}{4}, \quad W_V = \frac{N_b\theta_0^2}{4}, \quad \text{and} \quad \frac{W_V}{W_T} = \frac{\omega_0^2}{\omega^2} \tag{12}$$

The maximum potential and kinetic energies are:

$$W_T \le \frac{N^2 I}{4\beta^2} \text{ and } W_V \le \frac{N^2 I}{4\beta^2} \frac{\omega_0^2}{\omega^2}, \text{ for } \omega^2 \ll \omega_0^2, \ W_V \le \frac{N^2}{4I\omega_0^2} \text{ or } \frac{(N\Theta)^2}{8\,kT} \quad (13)$$

where $\frac{1}{2}N_b\Theta^2 = \frac{1}{2}I\omega_0^2 = kT$ and $\Theta$ is the thermal excursion angle taken as 0.01 radians.

Again it is important to couch conclusions in terms of some particular, but simple, structure in order to provide insights into general behaviors. Hence, I state energies and torques in units of $kT$, and I use the easily visualized sphere of radius $r$ and the density of water as a surrogate model for both the moment of inertia and the viscous resistance. Using Stokes Law as a guide, I estimate the resistive torque, $\beta\dot\theta$, on a rotating system by taking $\beta = 6\eta v$ where $v$ is the volume of the system. This relation is exact for a sphere.

**Electric Field Effects.** It is helpful to make some numerical estimates of possible interactions of electric fields, such that $E_0 = 282$ V/m, with elements holding permanent electric dipole moments.

In particular, I consider the hemoglobin macromolecule, which has an effective dipole moment of $d = 1.6 \cdot 10^{-27}$ $Cm$ and a moment of inertia $I \approx 8 \cdot 10^{-40}$ $kg\,m^2$. The length of the molecule is $L \approx 7 \cdot 10^{-9}\,m$ and I take the volume of the molecule as $v \approx 10^{-25}\,m^3$, $\beta \approx 6 \cdot 10^{-28}\,Nms$. With these values, $W_T \le 2.8 \cdot 10^{-14}\,kT$. The energy transfer is small and can be significant only if the relaxation time were about 7 orders of magnitude longer. Dimensionally, $W_T \propto Lq^2$, and again smaller systems have even stronger energy constraints.

For $\omega > \omega_0$, the constraints on the potential energy are more severe than for the kinetic energy. For $\omega^2 \ll \omega_0^2$, I take a minimal value of $\omega_0 = 10^5\ s^{-1}$ and find $W_V \le 1.5 \cdot 10^{-11}$ kT.

**Magnetic Field Effects.** Various species have been shown to manufacture magnetite, a ferrimagnetic material used by some species in conjunction with the earth's magnetic field to determine direction[3,4]. Though magnetite is not found commonly in human cells, the interaction of magnetic fields with that magnetite might be considered a possible source of biological effects of radiative fields. Biological magnetite is usually found in sets of single domain grains, about 500 Å in diameter, where all grains are magnetized in the same direction. Typically, such grains are enclosed by a membrane and are called *magnetosomes*. Taken as a sphere with a typical radius of $2.5 \cdot 10^{-8}$ m, the magnetic moment of such a single magnetosome will be about $\mu_m = 6 \cdot 10^{-17}$ $A \cdot m^2$. In the earth's field, $B_e \approx 50\ \mu T$, and $N_b = B_c\mu_m \approx 0.7\ kT$. Magnetosome sets with magnetic moments $\mu$ such that the alignment energy $N_b = B_e\mu \approx 10\ kT$ are well known, and there are systems with much larger moments. Such magnetite sets will be imbedded in some matrix (e.g., a cell), with a moment of inertia $I$, which may be free to rotate under the effects of the magnetic forces and initiate biological effects.

To consider a maximum plausible effect, I assume that the unbound magnetite structure is aligned by the earth's field and is subject to an RF magnetic field $B$ of peak amplitude $B_0 = 1\ \mu_m T$ (the peak magnetic field associated with the canonical radiative field where $E_{rms} = 200\ V/m$) directed at right-angles with the earth's field. The field $B_0$ will induce an alternating torque, $N\cos\theta = B\mu_m\cos\theta$, on the magnetic structure that will cause it to oscillate. The maximum induced kinetic and potential energies can be estimated from Eq. 13.

To gain some concept of the magnitude of the interactions of magnetic fields with magnetite, I consider a plausible, if somewhat extreme, system where the cell with a radius of $10\,\mu_m$ holds sufficient magnetite in aligned grains such that $N_b = B_e\mu_m = 1000\ kT$. (This value of $\mu$ is about 100 times that held by magnetotactic bacteria.) Then, $N = B\mu_m \approx 20\ kT$ and for the canonical $r = 10\ \mu_m$ cell, $W_T \leq 1.1 \cdot 10^{-9}$ kT.

For $\omega \geq \omega_0$, the maximum potential energy, $W_V$ is more strongly constrained than the kinetic energy. For $\omega^2 \ll \omega_0^2$, I can take $\omega_0 \geq 10^5\ s^{-1}$. Then, I find $W_V \leq 2.5 \cdot 10^{-7}\ kT$.

For smaller systems, the maximum energies will be smaller yet. In summary, the interaction of RF fields with biological magnetite can have no biological consequences.

## Resonance Absorption

Previous discussions emphasized the non-resonant interactions of electromagnetic waves with systems damped by viscous impedance. Generally the damping was sufficient to preclude strong resonant effects. I consider here the absorption of electromagnetic energy in a formally different manner through a partial wave analysis which emphasizes resonant absorption[5]. Here the damping of the system is described in terms of the removal of energy from the system through thermal conductivity.

The absorption of energy by a resonant system from a plane wave is expressed in terms of an absorption cross section, $\sigma_a$ where,

$$\sigma_a = 3\frac{\lambda^2}{\pi} \frac{\Gamma_s \Gamma_a}{(\nu - \nu_r)^2 + \Gamma^2/4} \tag{14}$$

where $\Gamma_a$ is the absorption width, $\Gamma_s$ is the scattering, or emission, width, $\Gamma = \Gamma_a + \Gamma_s$ is the total width; the lifetime $\tau = 1/\Gamma$. The resonant frequency is $\nu_r$, and $\lambda$ is the wavelength of the radiation. The cross section $\sigma$ is defined as the power absorption per unit incident power flux and has the dimensions of area.

The energy absorption can be calculated using the relation,

$$\int_{\nu_i}^{\nu_j} \sigma_a\, d\nu \leq 6\lambda^2 \Gamma_s \tag{15}$$

where the equality holds if the frequency spread of the incident radiation, $\Delta\nu = \nu_j - \nu_i$ encompasses the whole resonance, $\Delta\nu > \Gamma$, and $\Gamma_a \gg \Gamma_s$. Here, $\Gamma_s = 1/\tau_s$ where $\tau_s = W/P \approx Q/\omega$ where $W$ is the energy stored in the resonance and $P$ is the power radiated through the oscillating electric dipole moment. If there were no dissipative effects and $\Gamma_a = 0$, $\tau_s$ would be the lifetime of the state.

From Eqs. 14 and 15, for $\Gamma_a \gg \Gamma_s$ and for an incident power density $I$, the power absorbed by the system will be,

$$P_a \leq 6I\lambda^2 \frac{\Gamma_s}{\alpha} \tag{16}$$

Where $\alpha$ can be taken as the larger of the quantities, $2\pi\Gamma_a$ or $\Delta\nu$.

Since, biological systems tend to be much smaller than the microwave wavelengths, which by definition are greater than 1 $mm$, the coupling of the dipole moment to the radiative field tends to be small and the maximum absorption by the system will be small. I estimate $\Gamma_s$, and then the absorption cross section integral, using classical electrodynamics. The power $P$ radiated by an oscillating electric dipole is,

$$P = \frac{1}{4\pi\epsilon_0} \frac{p_0^2 \omega^4}{3c^3} \quad \text{where} \quad p_0 = qA_0 \tag{17}$$

where $\omega = 2\pi c/\lambda$ is the radial frequency of the oscillator, $p_0$ is the maximum dipole moment which I describe in terms of an amplitude $A_0$ and charge, $q$. With this model,

213

I consider the energy of the oscillator as,

$$W = \frac{1}{2}m\omega^2 A_0^2 \text{ and } \Gamma_s = \frac{1}{\tau_s} = \frac{P}{W} = \frac{q^2\omega^2}{6\pi\epsilon_0 mc^3} \qquad (18)$$

where $m$ is a characteristic mass. Combining Eqs. 16, 17, and 18,

$$P_a \leq \frac{I}{\alpha} \cdot \frac{4\pi\,q^2}{\epsilon_0 mc} \text{ and } W_a = P_a \cdot \tau \approx \frac{P_a}{\Gamma_a} \leq \frac{I}{\alpha\Gamma_a} \cdot \frac{4\pi\,q^2}{\epsilon_0 mc} \qquad (19)$$

where $W_a$ is the maximum absorbed energy and $\Gamma_a = 1/\tau_a \approx 1/\tau$.

I use these relations to consider energy transfers to several specific systems to provide general quantitative insights. (I note the claims of Grundler and Kaiser[6] that there are biological resonances in the frequency range of $45 \cdot 10^9$ Hz, hence, $\lambda = 6.67$ mm.) I use the canonical power level of $I = 10$ mW/cm$^2$ = 100 W/m$^2$ for numerical calculations and assume conservatively that the incident electromagnetic waves are nearly monochromatic and that $\Delta\nu$ is very small.

**1.** Assume the effective oscillator is a $^{40}Ca^{++}$ ion. In Eq. 7, I have estimated the relaxation time as as $\tau_r = 1/\Gamma_a \approx 10^{-14}$ s. With this value, the maximum energy transfer to the system will be $5.6 \cdot 10^{-15}$ $kT$. Only if the relaxation time is 8 orders of magnitude longer, will the absorbed energy be greater than $kT$ and larger than thermal fluctuations.

**2.** I consider a macromolecule where the effective charge is $q = 10e = 1.6 \cdot 10^{-18}$ $C$ and the mass is $5 \cdot 10^{-22}$ kg. Taking the characteristic length as $a = 5 \cdot 10^{-9}$ m (the radius of a volume of water of mass $m$.), the thermal relaxation time from Eq. 9 is $\tau_r = 4 \cdot 10^{-11}$ seconds and the maximum energy transfer as $1.5 \cdot 10^{-10}$ $kT$. Again the energy transfer is small and can be significant only if the relaxation time were about 5 orders of magnitude longer.

## Coherence

If two systems close together resonate at the same frequency, their interaction with electromagnetic fields is proportional to the sum of their vector dipole moment *amplitudes,* and the systems are *coherent.* Otherwise, the interaction is proportional to the sum of the squares of the amplitudes; the *intensities* add and the systems are *incoherent.* Hence, the interaction of $N$ equal coherent systems results in energy absorption and reemission equal to $N^2$ times that for one system while the absorption and reemission will be equal to $N$ times one system if the systems are incoherent. Therefore, the energy absorbed from an electromagnetic wave by a $N$ resonant systems can be as much as $N$ times greater if the systems act coherently than if they are incoherent. Frölich[7] has emphasized the possibility that coherent effects may be important at frequencies near the infrared level if the incident energy is sufficiently large.

If systems are to be coherent, two conditions must apply. (i) The systems must be sufficiently close together so that the phase of the incident electromagnetic wave, of wavelength $\lambda$, acting on them will be nearly the same. This will be the case if the distance $a$ between the systems will be such that $a < \lambda/2\pi$. (ii) The frequency, $\nu$, of the systems must be nearly the same. For two systems $i$ and $j$, $|\nu_i - \nu_j| < 1/(2\pi\tau_r)$, where $\tau_r$ is the relaxation time. If that frequency equivalence is to hold over large regions of generally heterogenous structure, the elements must be mechanically coupled. That coupling will extend over a characteristic length that I can estimate as $\lambda \approx v_s/\nu$, where $\nu$ is the frequency of the electromagnetic wave and $v_s$ is the velocity of sound in the system. However, if the system is sufficiently homogenous, the dipole-dipole coupling of the individual oscillators may be sufficient and the limitation on the characteristic length may be relaxed.

I consider typical systems that may act coherently.

**3.** A sector of the cell membrane, with its known electric dipole moment density, could interact with a high frequency electromagnetic wave. I estimate the relevant area of membrane $A_m \approx \lambda^2$, where $\lambda = v_s/\nu$ over which coherence can be expected to obtain where $v_s \approx 1500$ m/s, the velocity of sound in water, is an estimate of the velocity of sound in the membrane and $A_m \approx 1.1 \cdot 10^{-15} \ m^2$.

Taking the thickness of the membrane as $d = 10^{-8} \ m$ and the potential across the membrane as $dV = 100 \ mV$, I estimate the dipole charge density as $\sigma = K \, dV \, \epsilon_0/d$ where I take the dielectric constant of the membrane, $K \approx 5$. For the mass, I assume a membrane density twice that of water. Estimating the thermal relaxation time $\tau_r = 1/\Gamma_a \approx 10^{-9}$ seconds from Eq. 9, the maximum energy transfer is $3.7 \cdot 10^{-12} \ kT$. Following Frölich, the vibrational frequency must be near $\nu = v_s/d \approx 1.5 \cdot 10^{11} \ s^{-1}$ – or a wavelength, $\lambda \approx 2 \ mm$.

**4.** Perhaps the region of coherence is larger. I modify the above model by assuming a coherent region $A = r^2$ where $r = 10 \ \mu m$ is the radius of a typical cell. For a sheet, the thermal relaxation time will depend only upon the thickness of the membrane and that time will be the same as for the smaller membrane sector considered in (3) and the maximum energy transfer is, $W_a = 1.2 \cdot 10^{-7} \ kT$, again too small to be significant.

## FORCES ON INDUCED CHARGES – INDUCED DIPOLES

At frequencies less than 1 MHz, the electric polarizability of cells leads to dielectric constants of assemblies of cells, $K = \epsilon/\epsilon_0$, of the order of $10^4$ – or even greater at sub-audio-frequencies. Writing the polarizability of the individual cell as $p = \alpha E$, from the Clausius-Mosotti relation, $\alpha = 3\epsilon_0 \mathcal{V}(K-1)/(K+3)$, where $\mathcal{V} = \frac{4}{3}\pi R^3$, the volume of the spherical cell. This will be a useful approximation for other compact shapes. The large values of $K$ suggest that $\alpha$ is nearly equal to the largest possible value of $3\epsilon_0 \mathcal{V}$, which is the value for a conducting sphere.

### Dipole-Dipole Interactions – the Pearl Chain Effect

Writing the field intensity as,

$$E^2 = E_0^2 \cos^2(\omega t) = E_{rms}^2 [1 + \cos(2\omega t)] \tag{20}$$

For effects that are proportional to $E^2$, there is a DC component of the average field intensity which leads to constant forces in some circumstances. Schwan[8], who first pointed out the equivalence of AC and DC effects on induced dipoles, noted, as a homely analogy, that the force between the plates of a charged capacitor, which is proportional to the energy density, $E^2$, is attractive for both AC and DC capacitor potentials and as large for AC as for DC fields.

I take the induced dipole moment of the cell as that of a conducting sphere of the size of the cell, which is the maximum possible moment. Then, $\mathbf{p} = 3\epsilon_0 \mathcal{V} \mathbf{E_s}$, where $\mathbf{E_s}$ is the vector field at the site of the cell; $E_s = E_L (K+2)/3$, where $E_L$ is the local field and $K = 80$ is the dielectric constant, which I take as about equal to that of water.

The electric field of the dipole is,

$$\mathbf{E_p} = \frac{1}{4\pi\epsilon_0 r^3} [3(\mathbf{p} \cdot \hat{\mathbf{r}})\hat{\mathbf{r}} - \mathbf{p}] \tag{21}$$

and the translational force $\mathbf{F}$ on a dipole in a field $\mathbf{E}$ is,

$$\mathbf{F} = (\mathbf{p} \cdot \nabla)\mathbf{E} \tag{22}$$

From these relations, I find the attractive force $F$, and the potential energy $w$, between two dipoles with moments induced by an external field $E$ aligned in the direction defined by the two elements separated by a distance $z$,

$$F = \frac{6p_1 p_2}{4\pi K \epsilon_0 z^4} \quad \text{and} \quad w = \frac{2p_1 p_2}{4\pi K \epsilon_0 z^3}, \quad \text{for } z = 2r, \quad w = \pi \epsilon_0 r^3 E_L^2 \frac{(K+2)^2}{9K} \tag{23}$$

Thus, if I consider a closest distance of approach of two cells with a radius $r = 10^{-5}\,m = 10\,\mu m$ and dipole moments $p = 3 V \epsilon_0 E_s$, as $z = 2r$, then for an energy $w = kT$, $E_L = 125\ V/m$, which is in accord with the observation that there is dipole-dipole attraction for fields, $E_L \approx 200$ V/m. Perhaps, more important, taking $w = kT$, I have $E_L r^{1.5} = 4 \cdot 10^{-6}\ V/m^{1/2}$. Schwan[8] has summarizes a large set of measurements of thresholds for the effects of induced dipoles, including pearl-chain formation, for different small spheres, biological and non-biological, to find a mean value of $E_L r^{1.5} = 8.5 \cdot 10^{-6}\ V/m^{1/2}$, in good agreement with the results of my calculation[a]. Using Schwan's experimental mean value, I find a threshold value, $E_L = 280$ V/m for cells with a radius $r = 10\ \mu m$.

The general relations for the force between two dipoles and the associated energy are valid only for $z \gg r$. Also, the change in the direction of the dipole moment will take a time of the order of $\epsilon \rho$, where $\epsilon$ is the permittivity and $\rho$ the specific resistance of the cytoplasm. Then dipole-dipole attractions cannot be expected to be important for frequencies, $\nu > (2\pi \epsilon \rho)^{-1} \approx 100$ MHz. And at low-frequencies, $\nu < 10$ kHz, the fields from the induced dipoles will generally be reduced by the diffusion of counter-ions. Hence, though the above relations properly represent the magnitudes of effects, they will be modified by considerations of detailed structure so that they cannot be taken, *per se,* to set precise limits on biological effects. However, the results do suggest that the simple physical limits applied here do not completely exclude non-thermal biological effects of radio-frequency and microwave fields in the ranges allowed by present standards.

**The AC Component of the Field.** There will also be oscillatory forces on the charged elements of the dipoles that will take the general form $\mathbf{F} = \epsilon_m E^2 A$, where $A$ is a characteristic area about equal to $4\pi r^2$ for a sphere of radius $r$. These forces act to stretch and compress the elements carrying the dipoles. The oscillatory amplitude will take the form, $y(t) = y_0 \cos 2\omega t$, where,

$$y_0^2 = \frac{(\epsilon_0 E_0^2 A/2)^2}{m^2(\omega_0^2 - 4\omega^2)^2 + 4\gamma^2 \omega^2} \tag{24}$$

and $\omega_0$ and $\gamma$ have the same meaning as in Eq. 4. To put this relation into the context of the previous calculations, I note that $\epsilon_0 E_0^2 A/2 = E_0 q$, if $q = 10\,e$, $E_0 = 282\ V/m$, and for a cell radius of $10\ \mu m$.

I write relations for the maximum kinetic and potential energies in a manner parallel to Eqs. 6 and 7;

$$W_t \leq \frac{(\epsilon_0 E_0^2 A/2)^2 m}{4\gamma^2} \quad \text{and} \quad W_V \leq W_t \frac{\omega_0^2}{\omega^2}, \quad \text{for } \omega \ll \omega_0, \quad W_V \leq \frac{(\epsilon_0 E_0^2 AX/2)^2}{8\,kT} \tag{25}$$

---

[a]Schwan has also estimated the minimum field $E_L$ for alignment of an induced dipole and finds a relation for that field such that $\mathbf{p} \cdot \mathbf{E}_L = kT$ similar to the results of Eq. 23.

where $\frac{1}{2}KX^2 = \frac{1}{2}m\omega_0^2 = kT/2$ and $X$ is the thermal excursion amplitude. For the canonical cell-size system, $W_t \leq 1.4 \cdot 10^{-9}\, kT$, which is negligible. Taking $X = 0.1\,\mu m$, for the thermal agitation amplitude of the bound cell, $W_V \leq 1.5 \cdot 10^{-5}\, kT$ for $\omega \ll \omega_0$, which is sufficiently small to suggest that the oscillatory part of field will not generate biological effects through displacements of induced charges.

## RECTIFICATION AND MODULATION

If there are non-linear elements in a system, the interaction of the system with an electromagnetic field cannot be described properly if the field is defined only in terms of the spectrum of intensities derived from a Fourier analysis. Phase information is also important. In particular the interactions of a high-frequency carrier wave modulated by a lower-frequency signal can be different than that found by considering only the intensities of the carrier and side-bands.

The cell membrane is known to interact non-linearly[9] very much as a diode rectifier. Also, enzymes, perhaps held in the membrane, which act as reaction catalysts generating product currents into (out of) the interior of the cell from a substrate outside (inside) of the cell, may be modified by the electric field in the membrane acting on the enzyme so as to affect its catalytic action. That field might induce changes in the enzyme protein configuration through a transition dipole moment modifying the effectiveness of the enzyme as a catalyst[10,11].

In this discussion of rectification and modulation, for clarity, I will consider explicitly a model where the membrane acts as a rectifier. Whether the current is that of electric charge, rectified by non-linear electrical properties of the membrane, or of a neutral (or charged) current of product molecules generated from a substrate by an enzyme catalyst that is affected by the field in the membrane, the properties of the rectifier can be described approximately in terms of an equivalent circuit of a parallel membrane resistance and membrane capacitance in series with a diode and external voltage source.

For clarity, I first describe the diode effect in terms of the equivalent circuit. Here, the diode represents the non-linear conductance of the membrane for charge crossing the membrane barrier. The resistor and capacitor have the values of the normal linear membrane resistance and capacitance and play a role largely through the dominant time constant $\tau_m$ where $\tau_m = RC \approx K\rho\epsilon_0 \approx 5 \cdot 10^{-5}\, s$, where $K \approx 5$ is taken as the dielectric constant of the dilipid membrane and $\rho \approx 10^6\,\Omega m$ is the specific resistance of the membrane. In general, for successful modulation, $\tau_s$ should be long compared with the carrier period and short compared to the modulation period.

For the enzyme rectification of the non-charged product current, the diode represents the catalytic action. Since the system is nominally at an equilibrium in the absence of the perturbing field, I can presume that there is a mechanism restoring the equilibrium if the system drifts from equilibrium as a consequence of thermal fluctuations. Assuming that this restoring effect is nearly linear for very small drifts, the restoring force can be defined by a time-constant, $\tau_m$ which plays the same role as $\tau_m = RC$ for the electrical system.

In the low-field region I am considering, the properties of the biological diode, for either the electrical or the enzyme-catalysis system, can be described by the Nernst equation which is based on the Second Law of Thermodynamics,

$$I = I_0 \left[ \frac{Vq}{kT} + \frac{1}{2}\left(\frac{Vq}{kT}\right)^2 + \right] \text{ and } \bar{I} = I_0 \frac{1}{4}\left(\frac{E_0 rq}{kT}\right)^2 \text{ for } V = E_0 r \qquad (26)$$

where $\bar{I}$ is the time-average current and the field in the membrane is expressed as

$E_{mem} = E_0(r/d) \cos \omega t$ where $r$ is the cell radius and $d$ the membrane thickness. In the consideration of charge transfer, $q = e$, the electronic charge; for the enzyme system, I take $q = 5e$, to cover the possibility of large charges and dipole moments of the biological elements. It is useful to note that $I_0$ is the instantaneous dark current flowing in each direction – and cancelling in total – in the absence of a field.

I first consider charge transfers. From the work of Montaigne and Pickard[9] who found a DC offset of about $10^{-4}$ upon the application of an AC voltage of about 0.2 V across the membrane of a large cell, I set the value of $I_0$ using a current density of $j_0 = 1 \ mA/m^2$.

Then, using the canonical field, $E_0 = 282 \ V/m$, a cell radius $R = 10^{-5} \ m$, and $q = e$, I find $\bar{I} \approx 290 \, e \ s^{-1}$ and a total charge transfer of about $Q \approx 1.5 \cdot 10^{-2} \ e$ in a time $\tau_m$. Ignoring the quantization of charge, the rectified voltage will be of the order of $V_r \approx Q/C_m \approx 3.3 \cdot 10^{-10} \ V$ where $C_m \approx 5 \cdot 10^{-12} \ F$, which is quite negligible. For carrier fields as small as 200 V/m, I cannot expect modulated fields to affect cells significantly through charge transfer across membranes.

The arguments concerning neutral currents are somewhat similar except here I find it useful to compare the rectified transfer $\bar{I}t$ with the thermal noise drift $\sqrt{2I_0 t}$. In particular, I postulate that the product current can have no biological effect unless the transfer exceeds that from statistical fluctuations. That condition can be expected only after a time, $t_c$, such that $\bar{I}t_c = \sqrt{2I_0 t_c}$. Then,

$$t_c = \frac{32}{I_0} \left( \frac{kT}{V_{mem} q} \right)^4 \quad \text{where} \quad V_{mem} = E_{mem} d \leq E_0 r \tag{27}$$

If I take $j_0 = 6.25 \cdot 10^{15}$ particles per $m^2 s$ as a plausibly large value ($1 \ mA/m^2$), and $q = 5e$, then for a canonical cell of radius $r = 10 \ \mu m$, and a maximum value of $V_{mem}$, I can estimate $I_0 = 4\pi r^2 j_0 \approx 8 \cdot 10^6$ molecules per second (or a rate-of-change in internal concentration in the cell of about $3 \ \mu M/s$). With this base current, I find $t_c = 4.4 \cdot 10^{-5}$ seconds. Since I expect that physiological homeostatic time constants $\tau_m$ will be much longer, these thermodynamic constraints would not seem to preclude biologically significant effects for effective fields.

However, $t_c \propto V^{-4}$, and for $\nu < 1$ MHz, and for $\nu > 100$ MHz, the critical time, $t_c$, will be greater than an hour as the transmembrane potential $V$ will be reduced by factors greater than 100. At the lower frequencies, the potential will be reduced by the small admittance of the conducting body to the external field; at the higher frequencies, the potential will be reduced by the increased capacitative transmittance of the membrane [9]. In the intermediate region, the effective field will still be reduced, but it is not clear that this reduction will be such as to preclude biological effects from this mechanism for external radiative field of 10 $mW/m^2$. However, since the critical time varies as the square of the power density, weaker fields can be assumed to be substantially less likely to initiate non-thermal processes through this mechanism.

## APPENDIX A: Coupling of External Fields to Cell Elements

In the previous discussions, for definiteness, I considered the interaction of canonical electric fields with a strength of $E_s = 200 \ V/m$ acting on biological elements. However, for practical purposes, I am interested in the biological effects of fields, $E_{ext}$, external to the body. Hence I must consider the relation of these external fields to the fields at the biological sites. These relations depend on the detailed electrical permittivity and conductivity of biological tissues as well as the electrical properties of the biological cells and microstructures that make up these tissues. The study of the electrical properties

218

of tissues is over a hundred years old and I know that these systems are very complex. Any analysis of general limits, which is the intent of this study, must then address a set of model systems broad enough to include all real systems. The following comments are meant then to describe some of the boundaries that apply to the set of real systems and to avoid the necessity of addressing the complexity of that whole reality. That complexity has been described in some detail in an admirable tutorial and review by Foster and Schwan[12] and by other reviews by Schwan[13,14] that largely describe work by his group. My following comments are largely derived from information drawn from those reviews.

At frequencies, $\nu = \omega/2\pi$, less than 1 MHz, the conductivity of the body severely reduces the admittance of the external field $E_{ext}$;

$$E_L \approx \rho\omega\epsilon_0\,E_{ext} \tag{28}$$

where $\rho \approx 1\ \Omega m$ is the effective resistivity of the body tissues. The critical frequency, $\nu = 1/(2\pi\rho\epsilon_0)$, marks the region where the relation breaks down; $\nu_c \approx 10^{10}$ Hz. The internal field $E_L$ in the body, is then much smaller than the external field. The admittance is proportional to the frequency and, typically, at 1 MHz, $E_L/E_{ext} \approx 10^{-3}$.

At low-frequencies, $\nu < 100$ kHz, where the impedance of the cell membrane is largely resistive, the electric field in the membrane of thickness $d$ will be greater than the internal electric field outside of the cell of radius $r$ by a factor as large as $r/d \approx 10^3$. However the combination of these effects act generally so as to hold the field in the membrane to a value smaller than the external field, $E_{mem} < E_{ext}$. At such low-frequencies, fields in the cytoplasm and nucleus, internal to the cell, are much smaller than the local field in the tissues and very much smaller than the external field.

At frequencies in excess of 100 MHz, the fields at all biological elements can be taken as about equal to the external field in the absence of absorption or refraction of the radiation. Refractive effects, such as whole-body resonances, can increase field strengths by factors[13] as large as 5; absorption will reduce the internal intensity. In general, however, the internal fields will not be larger than the external fields.

At intermediate frequencies, from 1 MHz to 100 MHz, detailed calculations show that the the fields at biological elements are not greater – and usually smaller – than the external fields.

As a consequence of the polarization of the nearby media, the field $E_s$ at the site of a small biological element will be somewhat larger than the macroscopic local field $E_L$, which will, in turn, be somewhat smaller than the external field, $E_{ext}$.

As an illustrative example, I consider the relations between the external field $E_{ext}$ about a non-conducting sphere of dielectric constant $K_L$, the local field $E_L$ inside of the sphere, and the effective field $E_s$ in a small spherical cavity inside the large sphere.

$$E_L = \frac{3}{K_L/K_{ext} + 2}E_{ext} \quad \text{and} \quad E_s = \frac{3}{K_s/K_L + 2}E_L, \quad \text{hence} \quad E_s < \frac{9}{4}E_{ext} \tag{29}$$

and for the ordinary situation where $K_{ext} \approx 1$ and $K_L \approx 80$, $E_s \ll E_{ext}$.

When the conductivity of the body is considered, the local field will generally be reduced further. Therefore, the effective electric field at the site of the cell will usually be smaller than the applied, external field.

When all effects are considered, the electric fields at any biological site will usually be smaller than the fields external to the body and can be taken as no greater than those fields. Hence, my use of the external field strength, $E_{ext}$, as the field at the biological site can be considered as adequately conservative.

## APPENDIX B: Cell Charge Structure

According to the Singer-Nicholson model of the cell membrane, the membrane consists of a dilipid layer in which proteins are imbedded. I know that there is a potential difference $\Delta V \approx 0.1$ V across the membrane where the exterior is positive and the interior negative. Taking the membrane thickness as $d = 10^{-8} m$ (100 Å), the electric field in the membrane is about $\Delta V/d = 10^7$ V/m. Ascribing the potential difference to an accumulation of negative ions in the interior of the cell, the potential is consistent with a $P(ion) \approx 5.4$ for a cell of radius $r = 10 \ \mu m$ (taking the dielectric constant of the membrane as about 3). This corresponds to an excess of about $10^7$ electronic charges (or singly charged ions) in the interior cytoplasm.

This internal negative charge is enhanced by a very large external negative charge held in an extra-membrane glycocalix cell coating, typically more than 300 Å thick scialic acid residues attached to the carbohydrate side-chains generate a negative charge density of the order of 0.01 C/m². Though the intrinsic charge might be quite large[16], in conducting biological fluids that charge must be neutralized by the formation of a $\zeta$ layer of positive counter ions to produce a net charge, $Q_t \approx 0$. Since those ions held by potentials such that $Vq < kT$, where $q$ is the ionic charge, will not be firmly held, the cell, in vivo, can be expected to retain a small negative charge. The characteristic Debye screening length in typical biological fluids is estimated[17] to be about 10 Å, small compared with the thickness of the glycocalix layer.

The magnitude of the effective charge held by the cell is best addressed experimentally through electrophoresis measurements of the mobility of the cell, $\mu = v/E$, where $v$ is the velocity of the cell under the influence of a field $E$. Nominally, $\mu = q/6\pi\eta r$, for a sphere of radius $r$ holding an effective charge $q$ in a liquid of viscosity $\eta$.

If the effective charge represents the excess charge not neutralized by the $\zeta$ layer, and that excess is the charge bound by a potential energy less than $kT$, we would expect that $q \propto r$ independent of the intrinsic charge held by the sphere (if that charge be greater than $q$)[b]. Moreover, the mobility will then be independent of the radius of the sphere, and the composition – or other characteristics. These properties are known to hold reasonably well experimentally, hence, for all biological elements, we can take the measured value[18] of $\mu = 3 \cdot 10^{-8} \ m^2/(V s)$.

## ACKNOWLEDGEMENTS

I thank Prof. Herman Schwan for his patient tutoring in the basic electrical properties of the cell and for his elucidation of his work establishing the importance of induced cell dipole-dipole interactions at radio-frequencies. And I learned much about the effects of electric fields on the chemical kinetics of enzyme catalysis from R. Dean Astumian.

## REFERENCES

1. R. K. Adair, Constraints on biological effects of weak extremely-low-frequency electromagnetic fields, *Phys. Rev.* **A43**:1039 (1991).
2. A. D. Keith and W. Snipes, Viscosity of cellular protoplasm, *Science* **183**:666 (1974).
3. R. P. Blakemore, Magnetotactic bacteria, *Science* **190**:377 (1975)

---

[b]I would expect the effective charge $q$ to be defined by the relation, $kT = qe/4\pi\epsilon r$ where $r$ was the radius of the cell and $e$ the charge of the ions making up the $\zeta$ layer. Hence, the mobility, $\mu \approx (2 \, kT \, \epsilon)/(3\eta e)$, is independent of $r$ – or any other property of the sphere excepting the charge be greater than $q$. The mobility so calculated, $\mu_{calc} \approx 10^{-8} \ m^2/V s$, for $e$ taken as a unit charge, is in good agreement with the measured value.

4. M. M. Walker and M. E. Bitterman, Conditioned responding to magnetic fields by honeybees, *J. Comp. Physiol.* **A157**:157 (1985).

5. These standard techniques are described in, J. B. Marion and M. A. Heald. "Classical Electromagnetic Radiation," Harcourt Brace Jovanovich, New York (1980).

6. W. Grundler and F. Kaiser, *Nanobiology* 1:162 (1992).

7. H. Frölich, Long range coherence and energy storage in biological systems, *International Journal of Quantum Chemistry* 2:641 (1968).

8. H. P. Schwan, EM-field induced force effects, in "Interactions between Electromagnetic fields and cells," Editors, A. Chiabrera, C. Nicolini, and H. P. Schwan, Plenum Publishing Corp. (1985).

9. K. Montaigne and W. F. Pickard, Offset of the vacuolar potential of charocean cells in response to electromagnetic radiation over the range of 250 Hz to 250 kHz, *Bioelectromagnetics* 5:31 (1981).

10. J. C. Weaver and R. D. Astumian, The response of living cells to very weak electric fields: the thermal noise limit *Science*, **247**:459 (1990).

11. R. D. Astumian and B. Robertson, Quadratic response of a chemical reaction to external oscillations, *J. Chem. Phys.* **96**:6536 (1992).

12. K. Foster and H. Schwan, Dielectric properties of tissues, "CMC Handbook of Biological Effects of Electromagnetic Systems," Editors, C. Polk and E. Postow, CMC Press, Boca Raton, Florida (1985).

13. H. P. Schwan, Interactions of ELF fields with excitable tissues, and Biophysical principles of the interaction of ELF fields with living matter, II coupling considerations and forces, "Biological Effects and Dosimetry of Static and, EMF Electromagnetic Fields," Editors, M. Grandolfo, S. M. Michaelson and A. Rindl, Plenum Publishing Corp. (1985).

14. H. P. Schwan, Dielectric spectroscopy and electrorotation of biological cells, *Ferroelectrics,* **86**:205(1988).

15. J. C. Lin, Computer methods for field intensity predictions, "CMC Handbook of Biological Effects of Electromagnetic Fields," Editors, C. Polk and E. Postow, CMC Press, Boca Raton (1986).

16. C. L. Mieglaff and J. Mazur, Electrophoretic mobility and electrochemistry of latex systems, *J. Colloidal Sci.,* **15**:437 (1960).

17. F. S. Barnes, Extremely low-frequency (ELF) and very low-frequency electric fields; rectification, frequency sensitivity, noise, and related phenomena, "CMC Handbook Biological Effects of of Electromagnetic Fields," Editors, C. Polk and E. Postow, CMC Press, Boca Raton (1986).

18. L. D. Sher and H. P. Schwan, Microphoresis with alternating electric fields, *Science* **148**:229 (1965).

# EFFECTS OF RADIOFREQUENCY RADIATION (RFR) ON NEUROPHYSIOLOGICAL STRESS INDICATORS

G. Andrew Mickley[*]

Department of Psychology
Baldwin-Wallace College
275 Eastland Road
Berea, OH 44017-2088

## INTRODUCTION TO THE CONCEPT OF STRESS

Although the scientific community has been attempting to define the concept of stress for several hundred years, we have yet to devise a scientifically rigorous definition. The origins of the concept flow directly from Darwin's formulation of natural selection. In his view, the environment is in constant change (seasonal, chemical, geological, etc.) or it is continually being altered by its inhabitants. In the organism's struggle for existence, the environment is potentially threatening or dangerous due to the withdrawal of resources, the disruption of health by infection, starvation, heat or cold, and the threat of predators or competitors. Thus, stressors are selective pressures that derive from the physical and social environment. The environmental challenge, called stress, elicits physiological, as well as behavioral and psychological responses, which are specific and appropriate to the stressful situation.[1]

This definition differs from the one provided by Hans Selye in 1970,[2] who defined stress as the "non-specific response of the body to any demand made upon it." Several important aspects are missing from Selye's definition. Perhaps most prominent is the fact that specific stressors do indeed produce specific patterns of hormonal and neurochemical responses. The response evoked depends on several factors, including the animal's ability to cope with the stress (e.g., whether the stressor is escapable or inescapable[3,4]), the physical and psychological consequences of the stress,[5] the social conditions,[6] age,[7] and the genetic makeup[8] of the animal. These factors modulate stress responses that must be appropriate if they are to succeed. Were they random or indiscriminate (nonspecific), survival would not be assured. When the threat to survival is infection, the appropriate response is immunological. The appropriate response to physical attack must involve the motor system. It is also now clear that the interpretation of the threatening signal, and its value as a "predictor or correlate" are factors that tune the stress response.[9] Thus we may expect that, through an analysis of the neurophysiological correlates of RFR exposure, we might discern a typical pattern of responses to this particular stressor.

[*] These views and opinions are those of the author and do not necessarily state or reflect those of the U.S. Government.

It is recognized by many that the stress response is not a unitary phenomenon and may, in fact, be protective of the organism. This is true especially when the stress is short lived - the so-called "eustress" state. There is increasing evidence that early responses to stress enhance functioning, while prolonged stress leads to suppression of beneficial functional responses (for review see Morley et al.[10]). This may also be true of microwave exposure.

Thus, any paper with the present title must make some assumptions about what is, and is not, a stress response. I have limited the current discussion to microwave-induced alterations of classic neurophysiological stress indicators (e.g., corticosteroids, catecholamines, endogenous opioids, etc.). I have also included what are presumed to be indisputable indicators of a stressful event (i.e., changes in nervous system histology), while leaving disputed putative indicators of stress (e.g., calcium efflux) for another review. The current chapter focuses on the nervous system and therefore leaves untouched stress-induced alterations in the immune and cardiovascular systems. Finally, this paper focuses on the stress responses following RFR exposure and ignores some very interesting data that have been reported following exposure to magnetic fields and/or frequencies other than microwaves.[11,12]

## BASIC ASPECTS OF THE CLASSIC STRESS RESPONSE

The classical response to stress is the activation of the hypothalamic-pituitary-adrenal axis including the release of corticosteroids. The rise in corticosteroids follows stressor-induced activation of corticotropin releasing factor (CRF). CRF is the major regulator of adrenocorticotropin (ACTH) and beta-endorphin release from the pituitary. However, catecholamines also act directly on the pituitary to increase ACTH secretion through both alpha-1 and beta adrenergic mechanisms (see review by Axelrod and Reisine[13]). While ACTH is carrying the message to the adrenal to release cortisol into the bloodstream, beta-endorphin circulates in the blood and carries the stress message to a variety of target organs.[10]

## CATECHOLAMINES

Merritt et al.[14] reported a decrease in norepinephrine and dopamine in discrete areas of the rat brain after a 10-minute whole-body exposure to 1600 MHz microwaves at 80 mW/cm$^2$ (specific absorption rate [SAR] estimated to be 24 W/kg) that raised rectal temperature 4.1$^{\circ}$C. Both hyperthermic control rats (3.7$^{\circ}$C temperature rise for 10 minutes) and rats exposed to microwaves showed a decrease in hypothalamic catecholamines. Merritt concluded that, in general, hyperthermia was responsible for these changes in neurotransmitters. In another study, Merritt et al.[15] observed a decrease in rat hypothalamic norepinephrine and dopamine after a 10-minute whole-body exposure to 1600 MHz microwaves at 20 mW/cm$^2$ (SAR estimated at 6.0 W/kg, causing a temperature increase) but not at 10 mW/cm$^2$ (which did not change body temperature). More recently, Inaba et al.[16] also reported a significant decrease in hypothalamic norepinepherine following a 1-hour exposure to 2450 MHz at 10 mW/cm$^2$. Although dopamine levels did not differ between irradiated and control animals, there was a significant increase in the main dopamine metabolite (DOPAC) in the pontine area of the brain. Again, this effect was only observed in the high dose group (exposed to 10 mW/cm$^2$) but not in a group of rats exposed to 5 mW/cm$^2$. Thus, within the context of the studies available, it would seem that changes in body temperature, rather than microwave exposure itself, are the best predictor of depleted catecholamines in selected brain areas.

## NERVOUS SYSTEM HISTOLOGY

Investigators have observed changes in brain morphology following frankly thermal (>40 mW/cm$^2$) exposures to 3000 or 10,000 MHz for 40 minutes. More marked changes (edema, hemorrhages, vacuolation) were seen following 3000 MHz than 10,000 MHz.[17,18] Exposure of cats for 1 hour to 10,000 MHz, 400 mW/cm$^2$ resulted in injury to cerebral and spinal cord neurons.[19] However, Austin and Horvath[20] did not observe similar changes in brains of convulsive rats (rectal temperature increase of 2.3$^o$C; brain temperature of 43.7$^o$C) during a single short (7-minute maximum) exposure to 2450 MHz. These authors observed only mild pyknosis and hyperemia mostly in the pyramidal cells of the hippocampus. Albert and DeSantis[21] observed swollen neurons with frothy cytoplasm in the hypothalamic and subthalamic regions of the brain. However, these histological changes were not seen in the more posterior cerebellum, pons or spinal cord.

Histologic changes in the rat brain have been reported after multiple (35 or more) 30-minute exposures to 3000 MHz at power densities of <10 mW/cm$^2$ (SAR estimated to be 2 W/kg). The histological alterations included cytoplasmic vacuolation of neurons, axonal swelling and beading, and decreased numbers of dendritic spines.[17,18]

Baranski[22] reported that exposure of groups of guinea pigs to 3000 MHz at 3.5 mW/cm$^2$ (SAR estimated at 0.53 W/kg) and exposure of 20 rabbits to 3000 MHz at 5 mW/cm$^2$ (SAR estimated to be 0.75 W/kg) for 3 hours daily for 30 days resulted in myelin degeneration and increased proliferation of glia in the cerebellum and cerebrum. According to this investigator, temperatures in unspecified parts of the body were reported never to increase more than 0.5$^o$C.

The morphological effects on the CNS reported above are qualitatively similar in the range of 10 to 50 mW/cm$^2$ (SARs above 2 W/kg), but effects were quantitatively greater at the higher power densities. Most would recognize that irradiation at these power densities can cause a significant rise in body temperature. However, scientists from the former Soviet Union have reported similar morphological changes at power densities less than 10 mW/cm$^2$ and do not consider these alterations of thermic origin. We can conclude that histological changes in neuronal morphology is most reliable following doses > 2 W/kg. In most of the lower-dose studies, effects were rarely observed, or were less severe, in animals that were allowed to recover for 3 or more weeks following microwave exposure.

## CORTICOSTEROIDS

There seems to be some consensus that sufficiently intense microwave exposures can cause the release of corticosteroids in the bloodstream. Lu et al.[23] exposed rats to 2450 MHz for 1 hour (1-70 mW/cm$^2$) or 4 hours (0.1-40 mW/cm$^2$; measured SAR = 0.21 W/kg per mW/cm$^2$). The results for the 1-hour exposure indicated a dose-dependent increase in corticosteroid release, with evidence of a threshold between 20 and 40 mW/cm$^2$. Following the 4-hour exposure, the corticosteroid levels were significantly enhanced after 40 mW/cm$^2$, but not after $\leq$25 mW/cm$^2$. Consistent with this result is the finding by Lovely et al.[24] that rats exposed to 918 MHz microwaves at 2.5 mW/cm$^2$ (SAR approximately 1.04 W/kg) showed no change in corticosteroid levels.

In a study by Lotz and Michaelson[25] rats were irradiated with 2450 MHz microwaves at power densities of 13-40 mW/cm$^2$ for 30, 60 or 120 minutes and at 50 and 60 mW/cm$^2$ for 30 or 60 minutes (SAR estimated to be 0.21 W/kg per mW/cm$^2$). Plasma corticosterone levels were increased at power densities at or above 50 mW/cm$^2$, but not at 40 mW/cm$^2$ or less (for the 30- and 60-minute exposures). At the longest exposure time (120 minutes), increased levels were seen at or above 20 mW/cm$^2$ but not at 13 mW/cm$^2$. There was a significant correlation (r = 0.90) between colonic temperature (measured after exposure) and plasma corticosterone levels.

A subsequent study by Lotz and Michaelson[26] suggests that the change in corticosterone levels is due to the stimulation of the pituitary gland-probably due to hyperthermia. They exposed hypophysectomized rats to 2.45-GHz microwaves at 60 mW/cm$^2$ (9.6W/kg) and found significantly lower levels of corticosterone compared to normal and sham-hypophysectomized rats. In another study, dexamethasone effectively suppressed the corticosterone response in rats exposed to 50 mW/cm$^2$. In summary, these results provide evidence that a sufficiently high dose of microwaves can stimulate pituitary activity and cause an increase in plasma corticosteroids through the normal ACTH mechanism. Further, changes in corticosterone levels have been reported after exposure to microwaves at SARs as low as 10 W/kg, but not at 6.25 W/kg.

## CORTICOTROPHIN-RELEASING FACTOR(CRF)

Novtskii et al.[27] studied the CRF of the median eminence and ACTH of the hypophysis in Wistar rats exposed to whole-body dose of 0, 0.01, 0.1, 10 and 75 mW/cm$^2$ of 2.6 GHz microwaves for 30 minutes. Results indicated that the threshold intensity was 0.1 mW/cm$^2$ for increases in CRF and ACTH. In a more-recent study, Lai et al.[28] reported a decrease in high-affinity choline uptake in the rat brain following a 45-min exposure to pulsed low-level microwaves (2450 MHz, 1 mW/cm$^2$, SAR = 0.6 W/kg). This effect was blocked by the specific CRF receptor antagonist alpha-helical-CRF$_{9-41}$. The authors note that, although the energy absorption would be low under these circumstances, focal heating of a specific brain region may be sufficient to trigger the stress response.

## BENZODIAZEPINES AND ANXIETY

Quock et al.,[29] have shown that behavioral indicators of anxiety (i.e., number of rears on the stair-step test) are significantly reduced by a high dose (32 mg/kg) of the benzodiazepine chlordiazepoxide. This anxiolytic effect is antagonized by a sufficiently intense dose (36 W/kg) of 4.7 GHz microwaves. However, this effect was not observed at lower doses of 4.7 GHz nor following 4-36 W/kg of 1.8 GHz microwaves. An acute, low-level dose (average of 1mW/cm$^2$, estimated SAR of 0.2 W/kg) of 2450 MHz may also cause different phenomena. Thomas et al.[30] have reported that pulsed microwaves can actually potentiate the tranquilizing effects of chlordiazepoxide. Consequently, Lai et al.[31] have demonstrated that exposure to 0.6 W/kg of 2450 MHz can cause a selective increase in benzodiazepine receptors in some selected brain areas (e.g., cortex) but not others (e.g., hippocampus or cerebellum).

## ENDORPHINS

Endorphins are released during a variety of stressful events including exposure to ionizing radiation.[32] Evidence now suggests that low-level microwave radiation may also evoke an activation of endogenous opioids. Lai et al.[33] studied rats exposed for 45 minutes in a circularly polarized, pulsed microwave field (2450 MHz; SAR 0.6 W/kg) and found that this treatment potentiated morphine-induced catalepsy and lethality. The hyperthermia evoked by this microwave irradiation was attenuated by treatment with the narcotic antagonist naltrexone[34]. Further, low-level microwave irradiation attenuated the naloxone-induced withdrawal syndrome (e.g., rats showed significantly less wet-dog-shakes and higher body temperature than sham-exposed animals during withdrawal) suggesting RFR-induced activation of the endorphinergic systems.[35] There is some evidence to suggest that microwave-induced activation of endorphins may take place in particular neural loci (e.g., hippocampus), but that other brain areas (e.g., frontal cortex) may not be involved.[36]

## CONVULSIONS

Microwaves have been shown to influence convulsive behaviors. Edelwejn[37] demonstrated that chronic exposure of rabbits to 3 GHz pulsed microwaves (7 mW/cm$^2$; estimated SAR of 1.0 W/kg) for 3 hours a day (70-80 hr total) resulted in convulsions after injection of a dose of Cardiasol that did not normally cause convulsions in non-exposed control animals. Servantie et al.[38] investigated the effect of microwave exposure on Pentrazol (a convulsant drug) that was administered to CD1 mice exposed to 3 GHz pulsed microwaves a few minutes a day for 8-36 days. At the end of the exposure 50 mg/kg Pentrazol was administered i.p. Microwave exposure affected the time for convulsion onset and mortality rate after 15 days of irradiation (but not $\leq$ 8 days). Microwaves delayed the appearance of convulsions and rendered animals less susceptible to the epileptic action of the drug. However, after 27-36 days of exposure, this effect was reversed. These results suggest a biphasic response of mice over 36 days.

## RECENT FINDINGS

New data from studies being conducted at the USAF Armstrong Laboratory will contribute significantly to the established data base concerning the effects of RFR on neurophysiological stress indicators. For example, recent experiments by Drs. Ryan, Frei, Berger and Jauchem[39] have focused on the role of nitric oxide (NO) in RFR-induced stress. It has become clear that NO is an extremely potent vasodilator released by endothelial cells, vascular smooth muscle cells and macrophages.[40] An increased synthesis and release of NO appears to contribute to a variety of pathogenic forms of shock, including endotoxic and septic shock. Frei, Ryan and Jauchem are currently testing the hypothesis that an increase in NO synthesis mediates the visceral vasodilation and shock induction produced by sustained exposure to millimeter wave irradiation. Ketamine-anesthetized rats are exposed to 35-GHz RFR until shock is induced, at which point a bolus of an inhibitor of NO synthesis is administered. Preliminary results suggest that inhibition of NO synthesis provides no beneficial effects on the survival time of the animal. However, studies providing more chronic NO synthesis inhibition are underway and are necessary to completely explore this question.

In a different set of studies, Mickley, Cobb, Mason and Weigel[41] have begun to evaluate the RFR-induced distribution of brain c-fos and heat shock proteins (HSPs). C-fos is an early proto-oncogene that seems to initiate a cascade of compensatory physiological responses to stress (for review see Marx[42]). Twenty-minute, 600 MHz exposures at a variety of power densities produced increases in brain temperature from 0-2.5$^\circ$C. Preliminary results indicate that microwave-exposed rats (but not sham-irradiated subjects) exhibit selective expression of c-fos in several peri-ventricular brain structures (e.g., medial preoptic region of the hypothalamus, central gray, paraventricular thalamic nucleus) as well as portions of the pyramidal tract.[41] The dose-response characteristics of this phenomenon are currently being explored. Mason and his collaborators[43] have also begun to determine the localization of HSP production in microwave-exposed rats. HSPs are well established indicators of thermal stress and may be a reflection of compensatory responses to heating.[44] Early results suggest that there are two main differences between the expression of HSPs (72/73 kd) and c-fos in the brain. First, HSPs are found throughout the brains of microwave-exposed rats (2.06 GHz) and these HSPs do not appear to be region specific as is being found for the c-fos proteins. Second, there is virtually no expression of HSPs in sham-exposed rats. This is in sharp contrast to the high levels of c-fos expression found in some sham-exposed rats. From these early results, it appears that the expression of HSPs is a more specific indicator of heat stress than is the appearance of c-fos proteins. However, the regional specificity of c-fos expression makes it an essential marker of loci of heating effects. Thus, these preliminary results may suggest that the expression of HSP and c-fos proteins are in response to different components of microwave-induced heat stress.

## SUMMARY

For the most part, the available data suggest that a variety of neurophysiological stress indicators (e.g., catecholamines, corticosterone, CRF, histologic measures of brain damage, convulsions, etc.) are altered only following sufficiently sustained RFR-induced heating (see Table 1 for summary). For example, when dose-response relationships have been defined (e.g., Lotz and Michaelson;[25] Lu et al.[23]), corticosterone levels were reported to increase only following temperature rises of $>0.5^{\circ}C$. Several investigators[33,34] have reported data suggesting that microwave exposed rats release endorphins following a 45-minute exposure to 2450 MHz that caused a $1.0^{\circ}C$ increase in rectal temperature. Some scientists from the former Soviet Union have reported histologic changes in the CNS following minimal ($\leq0.5^{\circ}C$) temperature increases. The application of modern dosimetric techniques may help reconcile why these data differ from others suggesting that much higher microwave doses are required to alter CNS morphology. It should be noted, however, that significant compensatory changes in neurophysiology may be made to sustain homeostasis and resist an increase in body temperature.[45] If the thermal load can be dissipated by the host's physiological and behavioral responses, an increase in deep body temperature may not be detectable. Thus a small rise in body temperature may actually reflect a significant stress response by an animal. As our candidate neurophysiological stress indicators (e.g., c-fos or HSP expression) become more sensitive we may be able to detect quite subtle responses of the nervous system to RFR exposure. If it is possible to document specific neurophysiological indicators of the stress response to RFR, we may be able to use this information in accidental over-exposure follow-ups and RFR dose estimation.

## ACKNOWLEDGMENT

The author gratefully acknowledges the assistance of Ms Brenda Cobb, Dr. Patrick Mason and Dr. James Jauchem in the preparation of this manuscript. The animals involved in the studies conducted at the Armstrong Laboratory were procured, maintained and used in accordance with the Animal Welfare Act and the "Guide for the Care and Use of Laboratory Animals" prepared by the Institute of Laboratory Resources - National Research Council.

**Table 1.** Summary.

| Stress Indicator | Experimental Result | Microwave Frequency | Microwave Dose | Duration of Exposure (Min) | Temperature Measure (°C) | Reference |
|---|---|---|---|---|---|---|
| Catecholamines | Decreased hypothalamic norepinephrine and dopamine | 1.6 GHz | 24 W/Kg 80 mW/cm$^2$ | 10 | 3.7 | Merritt et al., 1976 |
| Catecholamines | Decreased hypothalamic norepinephrine and dopamine | 1.6 GHz | 6 W/Kg 20 mW/cm$^2$ (Neg. 10mW/cm$^2$) | 10 | 1.3 (brain) (Neg=0 change) | Merritt et al., 1977 |
| Corticosterone | Increase in corticosterone levels | 2450 MHz | 8.4-14 W/Kg (Neg. 0.21-4.2); 40-70 mW/cm$^2$ (Neg. 1-20) | 60 | 1.3-3.0 (Neg. 0-0.6) | Lu et al., 1981 |
| Corticosterone | Increase in corticosterone levels | 2450 MHz | 2.1-8.4 W/Kg (Neg. 0.2-1.0); 10-40 mW/cm$^2$ (Neg. 1-5) | 240 | 0.3-2.1 | Lu et al., 1981 |
| Corticosterone | Increase in plasma corticosterone levels | 2450 MHz | 11.5-13.8 W/Kg (Neg. 3-9.2); 50, 60 mW/cm$^2$ (Neg. 13-40) | 30-60 | 1.6-2.4 (Neg. 0.5-1.3) | Lotz & Michaelson, 1978 |
| Corticosterone | Increase in plasma corticosterone levels | 2450 MHz | 20-40 W/Kg (Neg. 13); 4.6-9.2 mW/cm$^2$ (Neg. 13) | 120 | 0.7-1.4 (Neg. 0.5) | Lotz & Michaelson, 1978 |
| Corticosterone | No change in plasma corticosterone levels | 918 MHz | Multiple (91 days) exposures 1 W/Kg; 2.5 mW/cm$^2$ | 600 | 0 | Lovely et al., 1977 |

**Table 1.** Summary (continued).

| Stress Indicator | Experimental Result | Microwave Frequency | Microwave Dose | Duration of Exposure (Min) | Temperature Measure (°C) | Reference |
|---|---|---|---|---|---|---|
| Corticosterone | Increase in plasma corticosterone levels | 2450 MHz | 8.3, 9.6 W/Kg; 50,60 mW/cm² | 60 | Not reported | Lotz & Michaelson, 1979 |
| Convulsions | Lowered convulsion threshold in Cardiosol-exposed rabbits | 3GHz | 7mW/cm²; 1 W/Kg (est.) | 180/day (70-80 hours total) | Not reported | Edelwejn, 1968 |
| Convulsions | Delayed penterazol-induced convulsions | 3000MHz | 5W/Kg (600kW peak, 350 W ave. 1µs pulse | Unspecified daily duration for 15 days | Not reported | Servantie et al., 1974 |
| Convulsions | Shorter latency to penterazol-induced convulsions | 3000MHz | 5W/Kg (600 kW peak, 350 W ave., 1µs pulse | Unspecified daily duration for 20, 27, or 36 days | Not reported | Servantie et al., 1974 |
| Histologic changes in the brain | Negative | 2450 MHz | 60-90 W power output | 0.5-6.0 | 5.5 (Est.) | Austin & Horvath, 1954 |
| Histologic changes in the brain | Myelin degeneration | 3000MHz | Multiple (90) exposures to 0.53 W/Kg (3.5 mW/cm²) | 90/day | 0.5 | Baranski, 1972 |
| Histologic changes in the brain | Vacuolation of neurons, swelling and beading of axons and dendrites | 3000MHz | Multiple (>30) exposures to 2 W/Kg (<10 W/cm²) | 30 | Not reported | Gordon, 1970 |

**Table 1.** Summary (continued).

| Stress Indicator | Experimental Result | Microwave Frequency | Microwave Dose | Duration of Exposure (Min) | Temperature Measure (°C) | Reference |
|---|---|---|---|---|---|---|
| Histologic changes in the brain | Vacuolation of neurons in hypothalamus | 2450 MHz | Multiple (1/day for 22 days) exposures to 7.5 W/Kg (25 mW/cm²) | 840 | Not reported | Albert & DeSantis, 1975 |
| Histologic changes in brain and spinal cord | Changes in nissel bodies | 10 GHz | 40 mW/cm² | 60 | ? | Bilokrinitsky, 1966 |
| Brain Benzodiazepine receptors | Increase in receptors in cortex but not hippocampus or cerebellum | 2450MHz | 0.6 W/Kg (3-6 mW/cm² in waveguide) | 45 | Not measured | Lai et al., 1992 |
| Change in benzodiazepine effects | Dose-dependent and frequency-dependent antagonism of chlordiazepoxide's (CD) behavioral effects | 1.8 GHz 4.7 GHz (Positive effects only with high dose of 4.7 GHz and 32 mg/kg CD) | 4-36 W/Kg | 5 | Not reported | Quock et al., 1994 |
| Corticotropin-Releasing Factor (CRF) | CRF antagonist blocks microwave-induced decrease in high-affinity choline uptake in brain | 2450MHz | 0.6 W/Kg (3-6 mW/cm² in waveguide) | 45 | Not measured | Lai et al., 1990 |
| Corticotropin-Releasing Factor (CRF) | Increase in CRF of the median eminence | 2.6 GHz | 0-75 mW/cm² (0.1 mW/cm² threshold) | 30 | ? | Novitskii et al., 1977 |

**Table 1.** Summary (continued).

| Stress Indicator | Experimental Result | Microwave Frequency | Microwave Dose | Duration of Exposure (Min) | Temperature Measure (°C) | Reference |
|---|---|---|---|---|---|---|
| Endorphins | Potentiation of morphine-induced catalepsy and lethality | 2450 MHz | 0.6 W/Kg (1 mW/cm$^2$ in waveguide) | 45 | Not reported | Lai et al., 1983 |
| Endorphins | Microwave-induced hyperthermia reversed by naltrexone | 2450MHz | 0.6 W/Kg (1 mW/cm$^2$ in waveguide) | 45 | 1° C increase in rectal temp, post-exposure | Lai et al., 1984 |
| Endorphins | Microwaves attenuate naloxone-induced opioid withdrawal syndrome | 2450 MHz | 0.6 W/Kg (1 mW/cm$^2$ in waveguide) | 45 | 0.5° C increase in rectal temp, post-exposure | Lai et al., 1986 |
| Endorphins | α, δ and κ opioid receptor antagonists block microwave-induced cholinergic activity | 2450MHz | 0.6 W/Kg (1 mW/cm$^2$ in waveguide) | 45 | Not reported, but presumed to be like above | Lai et al., 1992 |

# REFERENCES

1. R. Yirmiya, Y. Shavit, S. Ben-Eliyahu, R.P. Gale, J.C. Lebeskind, A.N. Taylor and H. Wagner, Modulation of immunity and neoplasia by neuropeptides released by stressors, *in:* "Stress: Neuropeptides, and Systemic Disease," J.A. McCubbin, P.G. Kaufmann and C.B. Nemeroff, eds., Academic Press, San Diego, 261-286 (1991).

2. H. Selye, The evolution of the stress concept, *Am. Sci.* 61:692-699 (1970).

3. H. Anisman, A. Pizzino and L.S. Sklar, Coping with stress, norepinephrine depletion and escape performance, *Brain Res.* 191:583-588 (1980).

4. J.M. Weiss, Effect of coping behavior in different warning signal conditions on stress pathology in rats, *J. Comp. Physiol. Psychol.* 77:23-30 (1971).

5. D.M. Gibbs, Vasopressin and Oxytocin: Hypothalamic modulators of the stress response: A review, *Psychoneuroendocrinology*, 11:131-140 (1986).

6. N.B. Thoa, Y. Tizabi and D.M. Yacobowitz, The effect of isolation on catecholamine concentration and turnover in discrete areas of the rat brain, *Brain Res.* 131:259-269 (1977).

7. S. Ritter and N.L. Pelzer, Magnitude of stress-induced norepinephrine depletion varies with age, *Brain Res.*, 152:170-175 (1978).

8. R.E. Wimer, R. Norman and B.E. Eleftheriou, Serotonin levels in the hippocampus: Striking variations associated with mouse strain and treatment, *Brain Res.* 63:397-401 (1974).

9. R. Levins and R. Lewontin, "The Dialectical Biologist." Harvard University Press, Cambridge, MA, (1985).

10. J.E. Morley, D. Benton and G.F. Solomon, The role of stress and opioids as regulators of the immune response, *in:* "Stress Neuropeptides and Systemic Disease," J.A. McCubbin, P.G. Kaufmann, C.B. Nimeroff, eds., Academic Press, San Diego, CA, 221-231 (1991).

11. M. Kavaliers and K.-P. Ossenkopp, Stress-induced opioid analgesia and activity in mice: Inhibitory influences of exposure to magnetic fields, *Psychopharmacology* 89:440-443 (1986).

12. G.C. Teskey, F.S. Prato, K.-P. Ossenkopp and M. Kavaliers, Exposure to time varying magnetic fields associated with magnetic resonance imaging reduces fentanyl-induced analgesia in mice, *Bioelectromagnetics* 9:167-174 (1988).

13. J. Axelrod and T.D. Reisine, Stress hormones: Their interaction and regulation. *Science* 224:452-459 (1984).

14. J.H. Merritt, R.H. Hartzell and J.W. Frazer, The effect of 1.6 GHz on neurotransmitters in discrete areas of the brain, SAM-TR-76-3, USAF School of Aerospace Medicine, Aerospace Medical Division, 1-11, (1976).

15. J.H. Merritt, A.F. Chamness, R.H. Hartzell and S.J. Allen, Orientation effects on microwave-induced hyperthermia and neurochemical correlates, *J. Microwave Power* 12:167 (1977).

16. R. Inaba, K. Shishido, A. Okada and T. Moroji, Effects of whole body microwave exposure on the rat brain contents of biogenic amines. *Eur. J. Appl. Physiol.* 65:124-128 (1992).

17. Z.V. Gordon. "Biological Effects of Microwaves in Occupational Hygiene," Israel Program for Scientific Translations, Jerusalem, Israel. NASA TT-F-633, TT70-50087; NTIS N71-14632, 101 (1970).

18. M.S. Tolgskaya and Z.V. Gordon, Pathological effects of radio waves, (Trans. from Russian by B. Haigh). LC Cat. Card 72-94825. Consultant's bureau, New York, NY, 63-106 (1973).

19. V.S. Biliokrinitsky, Changes in the tigroid substance of neurons under the effect of radio waves, *Fiziol. Zh. (Kiev)* 12:70 (1966).

20. G.N. Austin and S.M. Horvath, Production of convulsions in rats by high frequency electrical currents, *Am. J. Phys. Med.* 33:141-149, (1954).

21. E.N. Albert and M. DeSantis, Do microwaves alter nervous system structure? *Ann. N.Y. Acad. Sci.* 247:87-108 (1975).

22. S. Baranski, Histological and histochemical effects of microwave irradiation on the central nervous system of rabbits and guinea pigs, *Am. J. Phys. Med.* 51(4):182-191 (1972).

23. S. Lu, N. Lebda, S. Pettit, S.M. Michaelson, Microwave-induced temperature, corticosterone and thyrotropin interrelationships, *J. Appl. Physiol.:Respirat. Environ. Exercise Physiol* 50:399-405 (1981).

24. R.H., Lovely, D.E. Myers and A.W. Guy, Irradiation of rats by 918-MHz microwaves at 2.5 mW/cm$^2$: Delineating the dose-response relationship, *Radio Sci.* 12 (6S):139-146 (1977).

25. W.G. Lotz and S.M. Michaelson, Temperature and corticosterone relationships in microwave exposed rats, *J. Appl. Physiol.* 44:438-445 (1978).

26. W.G. Lotz and S.M. Michaelson, Effects of hypophysectomy and dexamethasone on rat adrenal response to microwaves, *J. Appl. Physiol.: Respirat. Environ. Exercise Physiol.* 47:1284-1288 (1979).

27. A.A. Novitskii, B.F. Murashov, P.E. Krasnobaev and N.F. Markozova, The functional condition of the system hypothalamus-hypophysis-adrenal cortex as a criterion in establishing the permissible levels of superhigh frequency electromagnetic emissions, *Voen. Med. Zh.* 8:53 (1977).

28. H. Lai, M.A. Carino, A. Horita and A.W. Guy, Corticotropin-releasing factor antagonist blocks microwave-induced decreases in high-affinity choline uptake in the rat brain, *Brain Res. Bul.* 25:609-612 (1990).

29. R.M. Quock, B.J. Klauenberg, W.D. Hurt and J.H. Merritt, Influence of microwave exposure on chlordiazepoxide effects in the mouse staircase test, *Pharmacol. Biochem. Behav.* 47:845-849 (1994).

30. J. Thomas, L. Burch and S. Yeandle, Microwave radiation and chlordiazepoxide: synergistic effects on fixed-interval behavior, *Science.* 203:1357 (1979).

31. H. Lai, M.A. Carino, A. Horita and A.W. Guy, Single vs. repeated microwave exposure: Effects on benzodiazepine receptors in the brain of the rat, *Bioelectromagnetics.* 13:57-66 (1992).

32. G.A. Mickley, K.E. Stevens, G.A. White and G.L. Gibbs, Endogenous opiates mediate radiogenic behavioral change, *Science.* 220:1185-1187 (1983).

33. H. Lai, A. Horita, C.K. Chou and A.W. Guy, Psychoactive-drug response is affected by acute low-level microwave irradiation, *Bioelectromagnetics* 4:205-214 (1983).

34. H. Lai, A. Horita, C.K. Chou and A.W. Guy, Microwave-induced post exposure hyperthermia: Involvement of endogenous opioids and serotonin, *IEEE Trans. Microwave Theory Techniques* MTT-32:882-887 (1984).

35. H. Lai, A. Horita, C.K. Chou and A.W. Guy, Low-level microwave irradiation attenuates naloxone-induced withdrawal syndrome in morphine-dependent rats, *Pharmacol. Biochem. Behav.* 24:151-153 (1986).

36. H. Lai, M.A. Carino, A. Horita and A.W. Guy, Opioid receptor subtypes that mediate a microwave-induced decrease in central cholinergic activity in the rat, *Bioelectromagnetics* 13:237-246 (1992).

37. Z. Edelwejn, Attempted evaluation of the functional state of brain synapses in rabbits exposed chronically to the action of microwaves, *Acta Physiol. Pol.* 19:791, (1968).

38. B. Servantie, G. Bertharion, R. Joly, A. Servantie, J. Etienne, P. Dreyfus and P. Escoubet, Pharmacologic effects of a pulsed microwave field, *in:* "Biologic effects and health hazards of microwave radiation," P. Czerski, K. Ostrowski, M.L. Shore, C. Silverman, M.J. Suess and B. Waldeskog, eds., Polish Medical Publishers, Warsaw, Poland, 36-45 (1974).

39. K. Ryan, M.R. Frei, R.E. Berger and J.R. Jauchem, Inhibition of nitric oxide synthesis during hypotension induced by 35-GHz radiofrequency irradiation of anesthetized rats, "Proceedings of the Annual Scientific Session of the Bioelectromagnetics Society," in press (1993).

40. J.R. Lancaster, Nitric oxide in cells, *Am. Scientist.,* 80:248-259 (1992).

41. G.A. Mickley, B.L. Cobb, P. Mason and L.K. Weigel, Microwave-induced hyperthermia disrupts working memory and evokes selective expression of brain c-fos, *in:* "Electricity and Magnetism in Biology and Medicine," M. Blank, ed., San Francisco Press, San Francisco, CA (1993).

42. J.L. Marx, The fos gene as "master switch," *Science* 237:854-856 (1987).

43. P. Mason, Armstrong Laboratory, Brooks AFB, TX, Personal communication, (1993).

44. S. Lindquist, The heat shock response, *Ann. Rev. Biochem.* 55:1151-1191 (1986).

45. S.M. Michaelson, Thermoregulation in Intense Microwave Fields, *in:* Microwaves and Thermoregulation, E.R. Adair, ed., Academic Press, New York, 283-295 (1983).

# BIOLOGICAL EFFECTS VERSUS HEALTH EFFECTS:
# AN INVESTIGATION OF THE GENOTOXICITY
# OF MICROWAVE RADIATION

Martin L. Meltz

Dept. of Radiology
University of Texas Health Science Center
San Antonio, Texas

At the present time, two distinct questions demand our attention. First, are microwave radiation exposures harmful to human health? Second, are extremely low frequency electric or magnetic field exposures harmful to human health? As we explore each of these questions, it is important that we consider the available information at three levels:

A. Is there a biological effect under the defined exposure conditions?
B. If there is a biological effect, is it hazardous to animals or to humans?
C. If there is an adverse effect on the health of humans, what is the (human) risk factor?

## IS THERE A BIOLOGICAL EFFECT
## UNDER THE DEFINED EXPOSURE CONDITIONS?

If the answer to this first question is yes, additional questions that should be answered include:

1. Is it a permanent change, or a transient one in which the system either adjusts to the exposure conditions during the exposure, or returns to its pre-exposure condition after the exposure is terminated?
2. Is it a percentage change (less than a 100% increase or decrease), or is it at least a two-fold increase (above the background level of that measurement)?
3. Is there a dose dependence?
4. Does the response depend on a specific alignment relative to the field(s) of exposure, or some other specific exposure condition?
5. Is the effect dependent on an increase in temperature?
6. Was the temperature measured during the exposure?

## IF THERE IS A BIOLOGICAL EFFECT,
## IS IT HAZARDOUS TO ANIMALS OR TO HUMANS?

Clearly, a biological effect is not an automatic indication that there is a health hazard to either animals or humans. In fact, a question that should also be addressed is: If there is a biological effect, is it beneficial to humans? Presuming that our most important concern is to evaluate any possible hazard, we can undertake an examination of those studies that have already been performed. These can generally be classified into three categories:
1. In Vitro Studies (e.g., effects on bacteria, yeast, fungi, rodent and/or human cells);
2. In Vivo Studies (e.g., effects on rodents or other animals, including non-human primates); and
3. Human studies (e.g., evidence of clinical changes; epidemiological studies).

## IF THERE IS AN ADVERSE EFFECT ON THE HEALTH OF HUMANS,
## WHAT IS THE (HUMAN) RISK FACTOR?

In the event that an adverse human health effect is established, we still must deal with some very important practical and societal questions:
1. How does the risk compare to that resulting from purposeful or allowable human medical, occupational or environmental exposures to other physical agents known to be hazardous?
2. If we try to avoid the exposure, will our actions place us at a greater risk from some factor associated with the change?

### Criteria for Examination of the Available Literature

My subsequent comments at this meeting are addressed predominantly to the question, "If there is a biological effect, is it hazardous to animals or humans?" I will give particular attention to reports in the literature on the use of certain *in vitro* cell culture assays that are accepted worldwide in the area of chemical genotoxicity testing as an indication of potential carcinogenicity. These studies in effect by-pass the first above question; they go directly to the issue of whether a health hazard is indicated by measuring whether or not the genetic material of the cell is damaged or altered.

In reviewing the existing literature, I have applied criteria based on those printed in the Newsletter of the Bioelectromagnetics Society.[1] These in part include:
1. Was the biological organism identified?
2. Were the experimental methods reported in enough detail to allow the study to be reproduced in another laboratory, and also to allow an investigator competent in the field to determine if it was performed properly?
3. Was the assay performed in accord with accepted (standard) protocols?
4. Was the microwave exposure system described?
5. Were the physical parameters of the exposure reported, including frequency, mode (continuous wave [CW] or pulsed wave [PW]), power, power density, location in the near or far field?
6. Was the dose reported by describing the specific absorption rate (SAR)?
7. Was the temperature measured continuously during the exposure (in contrast to measurements being made before and after the exposure), and was the temperature measurement technique described?
8. Was the temperature and the time at that temperature stated in the report?
9. Were independent treatment flasks exposed as replicates (to the same condition)?

10. Was the experiment repeated?
11. Were appropriate positive and negative controls performed?
12. Was the data statistically analyzed, and was the analysis appropriate?
13. Did the authors accept their own statistical result, or go on to make comments about specific changes that were not statistically significant?

# EVIDENCE ABOUT THE GENOTOXIC ACTIVITY OF MICROWAVE RADIATION

Applying these criteria, I wish to address the question of whether or not microwave radiation is genotoxic. An agent would be considered genotoxic if it can be shown to damage the DNA in cells, or cause chromosome damage (such as aberrations) in cells, or cause mutations in cells. While we would be most interested in answering these questions after exposure of human cells, in the area of *in vitro* genetic toxicology, there is widespread acceptance of information generated by testing the effects of chemicals in a variety of cellular organisms, including bacteria, fungi, yeast, and rodent cells (e.g., mouse cells, chinese hamster cells). A positive result in one of these assays by itself would not serve as a definitive indicator of a health hazard; it would, however, serve as a warning signal that additional information is needed.

Before I begin my summary, I must emphasize the importance of knowledge of the temperature of the biological sample during the microwave exposure. When experiments are performed *in vitro* to study the effects of hyperthermia (elevated temperature) on cancer cell killing, using water bath (convection) heating, the temperature achieved needs to be known within 0.1°C. The duration at that temperature also must be accurately known. This is because over the duration of an exposure, small temperature differences can lead to measurable differences in cell survival. Since it is possible for microwave radiation exposure to increase the temperature, it is important that the temperature be known during the exposure; any effect that is observed could be the result of incubating the cells at an elevated temperature, and have nothing to do with microwaves as a form of radiation.

## Direct Measurement of DNA Damage

I am not aware of any report in the literature showing that DNA is directly damaged by microwave radiation. This is in contrast to two known physical genotoxic agents, ionizing radiation and ultraviolet light, both of which damage DNA, cause chromosome aberrations and phenotypic mutations in cells, and also cause cancer in humans.

## Indirect Measurement of DNA Damage: DNA Repair Induction

If repair of DNA could be induced by microwave radiation exposure, and measured using a radio-labeling technique, it would mean that the DNA had been damaged by the exposure. This was studied *in vitro* in a normal human diploid fibroblast cell line by Meltz et al.,[2] at 350 MHz, 850 MHz, and 1200 MHz. No DNA repair was observed. In addition, if cell DNA was first damaged by ultraviolet light, and then exposed to microwave radiation at these same three frequencies, no inhibition of the UV-induced DNA repair was observed. This means that the microwave exposures did not interfere with the ability of the enzymes in the cells to recognize the UV-induced DNA damage, or nick the DNA near the damage, or repair synthesize the DNA (synthesize a new piece of DNA to replace the piece that contained the damage).

## Measurement of Sister Chromatid Exchange (SCE) Induction

Sister chromatid exchange induction may not ultimately be shown to be a direct measure of mutagenic activity by the agents that cause it, but the induction of SCEs does indicate an undesirable interaction between such agents and at least one of the processes involved in maintaining chromosome integrity. When an attempt was made to induce SCEs by exposing chinese hamster ovary (CHO) cells to 2.45 GHz microwave radiation, pulsed wave (PW), at power densities and SARs that increased the temperature in the medium by 3°C over a 2-hour exposure period, no SCE induction was observed. In addition, when SCEs were induced by simultaneous treatment of the cells during the microwave exposure with mitomycin C, there was no change from the number of SCEs induced by the chemical treatment without the microwave exposure.[3] The same result was seen with simultaneous adriamycin treatment.[4]

## Chromosome Aberration Induction

The one genotoxicity endpoint for which positive results have often been reported is in the area of chromosome aberration induction. Unfortunately, it is here where the criteria I listed earlier have most frequently been violated, and where individuals summarizing the literature have been less than critical. As early as 1970, Yao and Jiles[5] reported chromosome aberrations in bone marrow cells irradiated in flasks. Unfortunately, they removed the medium from the flasks, and exposed the flasks (standing upright) to conditions of microwave exposure that would have (likely) boiled the medium had it been present in the flask. In 1974, Chen et al.[6] first stated that their exposures of chinese hamster or human amnion cells to 20-85 mW/cm$^2$ of 2450 MHz radiation (which increased the temperature of the medium from 22° to temperatures ranging from 37° - 42°C) did not increase the number of chromosome aberrations with statistical significance. However, they then proceeded to talk about increases in certain types of chromosome aberrations (a misleading and scientifically inappropriate approach). When these authors exposed whole human blood at moderate-to-high SARs, with no increase in sample temperature, they reported that no increase in chromosome aberrations above control was observed. In 1978, Alam et al.[7] reported that exposure of a CHO cell line to 2450 MHz radiation at incident powers of 25-200 W for 30 min resulted in increased frequencies of chromosome breakage; however at the lowest power, the conditions of exposure allowed the temperature to increase to 49°C. The increased breakage was not observed at incident powers of 75-200 W (power density exceeding 200 mW/cm$^2$), when the temperature was kept at hypothermic conditions (29°C). As in the paper by Chen,[6] the SAR was not reported. In 1986, Lloyd et al[8] exposed whole human blood to 2450 MHz microwave radiation at 4 - 200 W/kg during a 20 min period; with no increase in sample temperature, there was no increase above control in chromosome aberrations. In 1990, Kerbacher et al.[9] exposed CHO cells to 2450 MHz PW radiation at an SAR of 33.8 W/kg. The conditions were such as to allow an increase in medium temperature from 37° to 40°C. These carefully performed experiments, with replicate treatment flasks and examination for 10 different types of chromosome aberrations, showed no increase in chromosome aberrations above control in the microwave exposed cells. In an expansion of this type of study, when chromosome aberrations were induced in the CHO cells by treatment with the genotoxic cancer chemotherapeutic agent mitomycin C at two different concentrations, and the cells were simultaneously exposed to the microwave radiation, no effect of the microwave exposure on the number of aberrations induced by the chemical treatment was observed.[9] In one instance, when a second genotoxic chemotherapeutic agent, adriamycin, was used for the simultaneous exposure, a very slight (but statistically significant) increase was observed by the simultaneous microwave and chemical treatment. However, exactly the same increase (with statistical significance) was observed when the cells were heated to the same temperature for the same time in a water bath.[9]

Within the past three years, three articles have appeared from one laboratory that report chromosome aberrations after exposure of chinese hamster V79 cells[10,11] and whole human blood[12] to 7.7 GHz MW for 10 - 60 min at power densities of 30 mW/cm$^2$,[10] or 0.5, 10 and 30 mW/cm$^2$.[11,12] The authors state that the temperature was "controlled" at 22°C. Unfortunately, this may be the room temperature. The authors did not describe the size of the flask (if any) holding the cells during the exposure, the volume of medium present and over the cells (if any) during the exposure, or what the temperature of the medium was during or after the exposure. They did not report an SAR in any of the papers. They do state that the cells were "irradiated in a semi-permeable membrane," and that using a surface probe, "the cell temperature measured on the surface of the semi-permeable membrane rose by 1°C after 60 min of exposure."[11] Obviously, the temperature had to be higher during the exposure; the time after exposure when the surface temperature was measured was not reported. Of importance, but not obvious to many concerned with these studies, is the statement made by the authors themselves in only one of the three papers;[11] "However, it was not possible to determine what the actual changes of temperature were within the individual membrane systems of live cells by the method applied." Clearly they did not know the temperature of the cells during the exposure. Based on examination of the clonigenic survival data in one of these papers in which V79 cells were exposed,[11] it appears that the decreasing survival levels seen with increasing time at the different power densities are similar to what one would expect from exposure of cells to water-bath hyperthermia at different elevated temperatures with increasing time. This suggests that the chromosome aberrations reported are most likely due to incubation of the cells at elevated temperatures. Unfortunately, no one will ever know! These papers should be referred to only with a simultaneous statement of their deficiencies; they should not have been published as written.

In conclusion, it appears that if and when microwave radiation does cause chromosome aberrations, it is because the biological system has been allowed to heat up to temperatures that can be lethal to the cells; this would not be likely or even possible in the human body under the existing exposure guidelines.

## Mutation Induction

The possibility of mutation induction by microwave radiation exposure has been examined at different gene loci in different cellular organisms, including bacteria, yeast, fungi, and mammalian cells. The results have been summarized.[13]

Bacterial studies after microwave radiation exposure at 1.7 and 2.45 GHz (3 - 4 hrs at 35°C) of *E. coli* strain WWU were negative, as were studies with *Salmonella typhimurium* tester strains exposed to 2.45 GHz microwave radiation at extremely high power densities for up to 6 sec (temperature increasing up to 56°C). In the latter study, for longer exposure times with still higher temperature increases, mutations were observed, with a differential between microwave radiation-induced mutants and non-microwave but temperature -induced mutants. In contrast, pulsed 1000 Hz microwave exposures at 8.6, 8.8, and 9.0 GHz of *E. coli* (Pol A⁻) did show repair-deficient cells to be more sensitive then wild type cells (indicating DNA damage), but the same result occurred with exposure to non-microwave-induced increases in temperature. For a second repair deficient strain, *E.coli* K-12, after 30-min microwave radiation exposures at frequencies of 9.4, 17, or 70 - 75 GHz (and a maximum temperature increase of 5°C at any frequency), no differential survival was observed. Similar exposures to tryptophan-dependent *E. coli* B/r strain did not result in any mutation to tryptophan independence.

Studies with different strains of fungi were negative after 2.45 GHz microwave radiation exposure (CW or PW; 28°C) of *Aspergillus nidulans*, and in a temperature-sensitive mutant of the same strain (heated to 40° - 45°C). Similarly, a negative result was obtained for *A. amstelodami* exposed to 8.7175 GHz (CW) microwave radiation.

Studies with different strains of yeast were also negative. After exposure of *S. cerevisiae* (diploid) to 2.45 GHz (CW) and 9.0 GHz (PW) microwave radiation, with temperatures increasing by up to 12°C, no mutation was observed. This was also the case in *S.cerevisiae* (haploid) exposed to 9.4 and 17 GHz microwave radiation.

Mammalian cell studies were also performed. An internationally recognized genotoxicity assay (the thymidine kinase locus mutation assay system in L5178Y mouse leukemic cells) was used to examine the effect of exposure of the cells to 2.45 GHz microwave radiation (PW) at power densities of 49 or 40 mW/cm$^2$ and SARs of 30 or 40 W/kg. These exposure conditions resulted in an increase in medium temperature from 37° to 40°C. The result was that no mutations were observed. Extending this type of study, the cells were exposed simultaneously to the same microwave radiation exposure conditions and to two different known chemical mutagens (mitomycin C or proflavin). No increases in mutation frequencies above those induced by the chemicals alone were observed.[14,15]

## SUMMARY

In summary, the *in vitro* evidence demonstrates that microwave radiation (at several different frequencies) is not genotoxic, unless extremely high temperatures are achieved. If such is the case, it is the incubation at the elevated temperature that is mutagenic, and not the radiation as such. If genotoxicity is required as an initiator step in carcinogenesis, then microwaves cannot be an initiator in that process.

## REFERENCES

1. M.L. Meltz and D.N. Erwin, Lab notes: essential RFR study information, *Bioelectromagnetics Society Newsletter,* Issue No. 78 (Sept./Oct. 1987).
2. M.L. Meltz, K.A. Walker, and D.N. Erwin, Radiofrequency (microwave) radiation exposure of mammalian cells during UV-induced DNA repair synthesis, *Radiation Research* 110:255-266 (1987).
3. V. Ciaravino, M.L. Meltz, and D.N. Erwin, Effects of radiofrequency radiation and simultaneous exposure with mitomycin C on the frequency of sister chromatid exchanges in Chinese hamster ovary cells, *Environmental Mutagenesis* 9:393-399 (1987).
4. V. Ciaravino, M.L. Meltz and D.N. Erwin, Absence of a synergistic effect between moderate-power radiofrequency electromagnetic radiation and adriamycin on cell cycle progression and sister-chromatid exchange, *Bioelectromagnetics* 12:289-298 (1991).
5. K.T.S.Yao and M.M. Jiles, Effects of 2450 MHz microwave radiation on cultivated rat kangaroo cells, *in*: "Biological Effects and Health Implications of Microwave Radiation," S.F. Cleary, ed., U.S. Dept. of Health, Education and Welfare, 123-133 (1970).
6. K.M. Chen, A. Samuel, and R. Hoopinger, Chromosomal aberrations of living cells induced by microwave radiation, *Environment. Lett.* 6: 37-46 (1974).
7. M.T. Alam, N.C. Barthakur, S.S. Lambert, and S.S. Kasatiya, Cytological effects of microwave radiation in Chinese hamster cells *in vitro*, *Can. J. Genet. Cytol* 20:23-30 (1978).
8. D.C. Lloyd, R.D. Saunder, J.E. Mocquet and C.I. Kowalczuk, Absence of chromosomal damage in human lymphocytes exposed to microwave radiation with hyperthermia, *Bioelectromagnetics* 7:235-237 (1986).
9. J.J. Kerbacher, M.L. Meltz, and D.N. Erwin, Influence of radiofrequency radiation on chromosome aberrations in CHO cells and its interaction with DNA-damaging agents, *Radiation Res.* 123: 311-319 (1990).

10. V. Garaj-Vrhovac, D. Horvat, and Z. Koren, The effect of microwave radiation on the cell genome, *Mutation Res* 243: 87-93 (1990).

11. V. Garaj-Vrhovac, D. Horvat, and Z. Koren, The relationship between colony-forming ability, chromosome aberrations and incidence of micronuclei in V79 Chinese hamster cells exposed to microwave radiation, *Mutation Res* 263: 143-149 (1991).

12. V. Garaj-Vrhovac, A. Fucic, and D. Horvat., The correlation between the frequency of micronulei and specific chromosome aberrations in human lymphocytes exposed to microwave radiation *in vitro*, *Mutation Res* 281: 181-186 (1992).

13. M.L. Meltz, Physical mutagens, *in*: "Genetic Toxicology: A Treatise," A. Li and R. Heflich, eds., The Telford Press, Caldwell, NJ (1991).

14. M.L. Meltz, P. Eagan and D.L. Erwin, Absence of mutagenic interaction between microwaves and mitomycin C in mammalian cells, *Environ. and Molec. Mutagenesis*, 13:294-303 (1989).

15. M.L. Meltz, P. Eagan and D.L. Erwin, Proflavin and microwave radiation: absence of a mutagenic interaction, *Bioelectromagnetics* 11:149-157 (1990).

**Session F:  Evaluation of the Bioeffects Database II:
Systems Physiology**

**Chair: B. Veyret**


Thermal Physiology of Radiofrequency Radiation
Interactions  in Animals and Humans
*Eleanor R. Adair*


Effects of Microwave Radiation Exposure on Behavioral
Performance in Nonhuman Primates
*John A. D'Andrea*


Effects of Radiofrequency Radiation Electromagnetic Field
Exposure on Pineal/Melatonin
*Peter Semm and T. Schneider*


Cardiovascular Responses to Radiofrequency Radiation
*James R. Jauchem*


Frequency and Orientation Effects on Sites of Energy
Deposition
*Melvin R. Frei*


Extrapolation of Animal Radiofrequency Radiation
Bioeffects to Humans
*Michael R. Murphy*

# THERMAL PHYSIOLOGY OF RADIOFREQUENCY
# RADIATION (RFR) INTERACTIONS
# IN ANIMALS AND HUMANS

Eleanor R. Adair

John B. Pierce Laboratory and Yale University
New Haven, CT 06519

## INTRODUCTION

In his 1988 review of the biological effects of radiofrequency (RF) fields, Erwin[1] counseled his readers that ". . . Protecting humans from the real hazards and allaying groundless fears requires a self-consistent body of scientific data concerning the effects of the fields, levels of exposures which cause those effects, and which effects are deleterious (or beneficial or neutral). With that knowledge, appropriate guidelines for safety can be devised, while preserving the beneficial uses of radiofrequency radiation (RFR) energy for military or civilian purposes." Erwin's good counsel forms the framework for the present critical review of recent data on the thermal physiology of RFR interactions in animals and humans.

While the goal of standards-setting activities is the prevention of harmful effects in human beings exposed to RFR, it is important to recognize at the outset that there is a paucity of relevant data collected on human subjects. Most of the published thermophysiological research has been conducted on laboratory animals, with a heavy emphasis on small rodents (e.g., mice, rats and hamsters). The large surface-area-to-volume ratio of such small mammals requires a high metabolic heat production to maintain thermal balance in the cold; however, such small mammals are at great disadvantage in warm environments because they lack efficient mechanisms for heat dissipation. Basic information about the thermoregulatory capabilities of animal models relative to human capability is essential to the appropriate evaluation and extrapolation of animal data to man. In general, reliance on data collected on humans and nonhuman primates, however fragmentary, will yield a more accurate understanding of how RF fields interact with biological systems, knowledge that will best serve the needs of standards-setting.

In this review, we first consider the fundamentals of human thermophysiology and thermoregulatory behavior as context for a synopsis of the consequences of RFR exposure on these systems. Summarized data from animal experiments will demonstrate effects of many variables including, among others, field strength, frequency, polarization, whole *vs* partial body exposure, and ambient conditions. These data will be supplemented with assessments of human responses to RF exposure collected both in the laboratory and in the clinic. The similar disposition of heat generated in the human body by exposure to RFR and heat generated by muscular exercise will be emphasized throughout. The proposed use of RFR for incubation or rewarming from hypothermia will be briefly mentioned. Finally, the value of computer simulation models as predictors of human thermoregulatory response to RFR exposure will be evaluated, with emphasis on the unique thermoregulatory capabilities of human beings.

## HUMAN THERMOREGULATION IN THE PRESENCE OF RF ENERGY

Human beings are typical endotherms, organisms that regulate their body temperature through a controlled rate of metabolic heat production in the face of rather wide fluctuations in the thermal environment. Thermoregulation in humans is accomplished by two types of response, physiological (or autonomic) and behavioral. Assorted behavioral strategies provide a hospitable microclimate within which fine adjustments in appropriate autonomic responses control the rate at which heat is gained or lost from the body. When the behaviorally generated microclimate is thermally neutral, autonomic mechanisms are minimally involved and usually restricted to adjustments in the caliber of peripheral blood vessels (vasomotor activity). If thermoregulatory behavior is restricted or unavailable, autonomic thermoregulation will predominate; the particular autonomic response that may be ongoing at any given time is dictated by the prevailing ambient conditions (temperature, humidity and air movement). Thus, people shiver in the cold and sweat in the heat, and will avoid doing either if an efficient behavioral maneuver is available to them.

Humans function most efficiently when the vital internal organs are maintained at a relatively constant temperature near 37°C (98.6°F). While the temperature of individual body parts may vary characteristically from this norm, significant departures are usually associated with disease states or possibly lethal conditions. The usual range of "normal" body temperature (35.5 to 39°C) encompasses circadian variation, vigorous exercise, variations in ambient temperature, sequelae of food intake, age factors, cyclical variation in women, and emotional factors. Temperatures outside this range must be related to the presence of disease, pharmacological intervention, unusual activity, or extraordinary environmental conditions.

RFR in the environment may be regarded conveniently as part of the thermal environment to which humans may be exposed. Figure 1, adapted from Berglund,[2] represents schematically the sources of heat in the human body and the avenues by which thermal energy may be exchanged between the body and the environment. Heat is produced in the body through metabolic processes and may also be passively generated in body tissues through the absorption of RFR ($A_{rfr}$). In order for the body core to remain at a stable temperature, this thermal energy must be continually transferred to the environment.

The figure shows that energy may be lost through the evaporation of water from the respiratory tract ($E_{res}$) or from the skin ($E_{sw}$) and by dry heat transfer from the skin surface via radiation ($Q_r$), conduction ($Q_k$), or convection ($Q_c$). When the environment is thermally neutral, dry heat loss predominates in the form of convective transfer to the air and radiant transfer to the surrounding surfaces. The small amount of heat lost by the diffusion of water through the skin is not shown in the figure. When the temperature of the environment rises above neutrality, or during vigorous exercise or defervescence, the evaporation of sweat from the skin ($E_{sw}$) is mobilized to dissipate large amounts of body heat. Sweating would be anticipated if excessive thermalizing energy is deposited in the body during RF exposure. The rate of heat loss is governed by the characteristics of the environment; these include not only the air temperature ($T_a$) but also air movement (v) and humidity (RH). Other important environmental variables (not shown here) are the mean radiant temperature of surrounding surfaces and sources, especially those close to the body, and the amount of insulation (fat, clothing, etc.).

When the thermal energy produced in the body, including that derived from absorbed RFR, is equal to that exchanged with the environment, heat storage (S) is zero and the body is in thermal balance. Under these conditions, the body temperature remains stable. This is represented by the energy (or heat) balance equation shown in Figure 1. When heat production exceeds heat dissipation, thermal energy is stored in the body (+S) and the body temperature rises. Conversely, when more heat is transferred to the environment than can be produced or absorbed (-S), the body temperature falls.

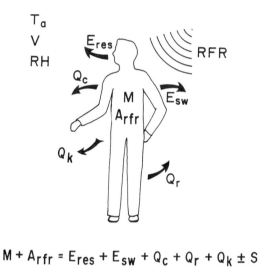

$$M + A_{rfr} = E_{res} + E_{sw} + Q_c + Q_r + Q_k \pm S$$

**Figure 1.** A schematic diagram of the sources of body heat in a human being, including absorbed RFR energy (Arfr) and the important energy flows between the human and the environment. The body is in thermal equilibrium if the equation is balanced. See text for details.

The thermoregulatory system functions as a negative-feedback control system with a reference or "set" temperature. Thermosensors are distributed around the body to provide information about the local temperature of body tissues. The sensors located in the outermost layers of the skin are most important; other sites include the medial preoptic/anterior hypothalamic area of the brainstem (locus of the "central thermostat"), posterior hypothalamus, midbrain, medulla, spinal cord, cortex, and deep abdominal structures. Neural signals from the sensors are integrated by a central controller, the integrated signal is compared with the internal reference, and an output command is generated to energize appropriate responses whenever a load error occurs. A negative load error (body temperature lower than set point) will increase heat production and conservation; a positive load error will increase heat dissipation.

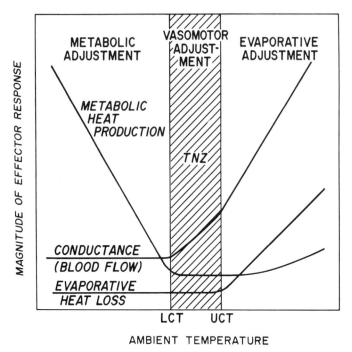

**Figure 2.** Thermoregulatory profile of a typical endotherm to illustrate how the principal types of autonomic response depend on the prevailing ambient temperature. LCT = lower critical temperature; UCT = upper critical temperature; TNZ = thermoneutral zone.

The particular effector response that is mobilized, as well as its strength, depends on the prevailing environmental conditions. A schematic "thermo-regulatory profile" of a typical endotherm (Figure 2) illustrates how the principal autonomic responses of heat production and heat loss depend on the $T_a$. We consider the responses to be steady-state, rather than transient, and the ambient air to have minimal movement and water content. Three distinct zones are defined in terms of the prevailing autonomic adjustment. Below the lower critical temperature (LCT), thermoregulation is accomplished by changes in metabolic heat production (M), other responses (conductance and evaporative heat loss) remaining at minimal strength. As the $T_a$ falls further and further below the LCT, heat production increases proportionately. Many researchers have demonstrated experimentally that RF energy absorbed by an endotherm in a cool environment will substitute for M in direct proportion to the strength of the RF field.[3,4,5,6,7,8,9,10,11] Details of these and other relevant studies will be presented later in this review.

At ambient temperatures above the LCT, M is generally at a low, resting level that is characteristic of the species, evaporative heat loss is minimal, and thermoregulation is accomplished by changes in thermal conductance. Conductance is a measure of heat flow from the body core to the skin and reflects the vasomotor tone of the peripheral vasculature. As the constricted peripheral vessels begin to dilate, warm blood from the body core is brought to the surface so that the heat may be lost to the environment by radiation, convection, and conduction. These vasomotor adjustments take place within a range of $T_a$ called the thermoneutral zone (TNZ) that is unique to each species. The TNZ for humans is extremely narrow, encompassing only a few degrees around 30°C.[12] On the other hand, the

TNZ for the rhesus monkey extends from 24.5 to 31°C,[13] that for the squirrel monkey from 26 to 35°C,[14] and that for the mouse from 30 to 33°C.[15] (Comparable data for other laboratory animals are given by Durney et al.[16]) When an endotherm at thermoneutrality is exposed to RFR, augmented vasodilation occurs so that the heat generated in deep tissues may be quickly brought to the skin surface for dissipation to the environment.[3,5,6,9,10,17]

The upper limit of the TNZ is called the upper critical temperature (UCT). At this $T_a$ the endotherm is fully vasodilated and dry heat loss is maximal. Further increases in $T_a$ stimulate the mobilization of heat loss by evaporation either from the skin (sweating) or the respiratory tract (panting) at a rate that is proportional to the deviation of $T_a$ from neutrality. In humans, whole-body sweating can attain rates of 2-3 liters/hr and 10 - 15 liters/day.[18] Assuming normal hydration, it is difficult for a human to increase M (by exercise) to levels that cannot be dissipated by sweating. Since human evaporative heat loss is controlled by both internal and peripheral thermal signals,[19] only an extraordinarily hostile thermal environment that may include a source of RF energy can be expected to pose a serious threat to a human's thermoregulatory system. Stolwijk[20] and others[21,22] have predicted minimal increases in brain and other body temperatures during local absorption of significant amounts of RF energy; this occurs because of the rapid mobilization of sweating and a substantial increase in tissue blood flow. A few reports indicate that nonhuman primates sweat efficiently during RF exposure in thermoneutral and warm environments,[3,5,23,24] but several studies purporting to measure evaporative heat loss in RFR-exposed rodents[25,26,27,28] are characterized by imprecise methodology and highly variable data.

It is important to remember that many small furred mammals (e.g., mouse, rat, hamster, guinea pig), commonly used as subjects in laboratory experiments, neither sweat nor pant when exposed to $T_a$ above the UCT. Indeed, measured increases in respiration rate of these animals reflect the general speeding up of all bodily processes as body temperature rises (often called the "Q-10 effect"). If these species are heat stressed, they must depend on behavioral maneuvers, such as spreading saliva or urine on the fur, or exiting from the environment, to achieve some degree of thermoregulation. Opportunities for behavioral thermoregulation are vital when these species undergo exposure to RFR, especially at ambient temperatures above the UCT.

Any organism may adopt thermoregulatory behavior as an alternate strategy to counter the thermalizing effects of RF exposure. Changes in certain behaviors can alter the thermal characteristics of the air/skin interface and maximize the efficiency of heat transfer to the environment. Examples are the selection of a more favorable thermal environment, the resetting of a thermostat, and the putting on or taking off of clothing (insulation). These behaviors also minimize the involvement of autonomic mechanisms of heat production and heat loss, conserve bodily stores of energy and fluid, and generate a state of maximal thermal comfort (see, for example, Ref. 29). Because behavioral responses may be mobilized quickly and are of high gain, they must always be considered in any discussion of the thermoregulatory consequences of exposure to RFR.

## MOBILIZATION OF AUTONOMIC RESPONSES DURING RFR EXPOSURE

For any given species, under any given environmental conditions, an intensity of imposed RFR can be determined that will reliably initiate or alter whatever thermoregulatory response is appropriate to those environmental conditions. Such determination requires, for comparison, adequate baseline or control data collected under identical conditions but with RFR absent. The thermoregulatory profile for the species in question (cf. Durney et al.[16]), exemplified by Figure 2, should be invoked as a guide to correct response selection. Differences in humidity and air movement, as well as dry bulb temperature, should also be considered. The RF intensity so determined can be designated a threshold for response mobilization. However, others (e.g., Refs. 30 and 31) prefer a statistical definition of "threshold" derived from a "hockey-stick" analysis of extensive experimental data collected across a wide range of RF intensities at a specific $T_a$. For either method, coherent data are essential for accurate determination of a threshold. By definition, sub-threshold intensities will not produce response alteration or mobilization.

As a general rule, RF intensities above an experimentally determined threshold level will also alter the response in question, usually to a degree that is intensity dependent.[4,6,9,10,17,31] If the strength of the RF field is great enough, the response under observation will be altered maximally and the next response in the thermoregulatory hierarchy will be mobilized. This is comparable to moving the endotherm past one of the critical temperatures (i.e., LCT or UCT in Figure 2) in its thermoregulatory profile. Adair[32] provides a detailed discussion of these concepts and their implications for human thermoregulation in the presence of RF fields.

## Adjustments in Metabolic Heat Production During RFR Exposure

Figure 2 shows that the M of an endotherm equilibrated to a $T_a$ well below the LCT will be elevated above the resting level by an amount that is directly proportional to $T_a$. During acute (a few minutes to a few hours), far-field exposure of the whole body to RFR, the elevated M of nonhuman primates (Macaca mulatta or Saimiri sciureus) will be reduced by an amount proportional to the field strength or the specific absorption rate (SAR).[3,4,5,6,9,10,24,33] As a result of this response adjustment, the internal body temperature is usually regulated with precision within the limits normal for the species. Similar results have been demonstrated in rodents after a period of whole-body irradiation in either a multimodal cavity[11] or a waveguide.[7,8] Further, chronic low-level RFR produces no measureable alteration in the normal metabolism of infant rats,[34] of rats irradiated throughout their lifetimes, or of squirrel monkeys irradiated for 15 weeks.[36]

A threshold SAR must be surpassed before a reliable reduction in M occurs; this threshold is between 0.5 and 1.5 W/kg in nonhuman primates,[5,9,10] but has not been systematically explored in other species. Clearly, it will vary with $T_a$ in accordance with the characteristics of the thermoregulatory profile for each species. Abundant evidence indicates that both the threshold and the magnitude of the M reduction depend, in regular fashion, on the magnitude of the cold stress.[3,4,8,9,10,24]

During whole-body exposure, the maximal absorption of RF energy occurs when the long axis of the body is parallel to the electric field vector (E-polarization) and the long dimension of the body is about 0.4 of the free-space wavelength (resonant frequency).[16] RF exposure of nonhuman primates to their resonant frequency yields somewhat less efficient thermoregulation than does exposure to sub-resonant or supra-resonant frequencies.[37,38] Although the threshold for M reduction may be lower at resonance,[10] the magnitude of the response change may be less for a given SAR than at non-resonance and the body temperature may rise. The hyperthermia is modest, however, and well regulated even at SARs similar (in W/Kg) to the level of resting metabolic heat production. The situation is identical to that occurring in humans during exercise.[5] Some have expressed concern that human exposure at resonance may pose a greater hazard than exposure at other frequencies. The above-cited studies on nonhuman primates should be reassuring because, even though thermosensors in the skin (necessary for thermal perception and behavioral avoidance) may be inefficiently stimulated, autonomic mechanisms are rapidly mobilized to dissipate heat generated deep in the body.

If only part of the body is exposed to RFR, the magnitude of the change in M reflects the total absorbed energy, as though it were integrated over the whole body.[39] If an endotherm is exposed to RFR at SARs greater than that which reduces M to the resting level, thermoregulation will be accomplished by mobilization of the next response in the hierarchy, changes in vasomotor state or conductance.[3,6,9,24]

## Vasomotor Adjustments During RF Exposure

When an endotherm is briefly exposed to RFR at a $T_a$ just below the LCT (Figure 2), the stage is set to initiate the peripheral vasomotor response as soon as M has been reduced to the resting level. In laboratory animals the vessels of the tail and ears usually vasodilate before those of the extremities. Generally, once the field strength is sufficient to induce vasodilation (threshold), the response occurs rapidly at field onset and the degree of vasodilation is a direct function of SAR[9,17,31] or the total heat load.[27] As $T_a$ increases above the LCT, the SAR at threshold is reduced.[9,31,24,32] Extinction of the RF field induces rapid vasoconstriction. In nonhuman primates, both the threshold and the degree of vasodilation depend on the imposed frequency, although in a species-dependent manner. For rhesus monkeys, the closer the exposure frequency to whole-body resonance, the less energy is required to induce vasodilation at a given $T_a$ and the greater the response magnitude at a given SAR.[10] However, for squirrel monkeys exposed to RFR in cold $T_a$, higher SARs are required to induce tail vasodilation during exposure at resonance than at a frequency above resonance.[5]

Peripheral vasodilation can also occur spontaneously during the course of a prolonged RF exposure that is carried out at a $T_a$ below the LCT.[6,9,24,32,40] In this case, vasodilation is mobilized because the internal body temperature slowly rises during the exposure, eventually surpassing a threshold for initiation of the response. All of the data cited above support the thesis that vasomotor control is exerted by a combination of central and peripheral thermal inputs. In addition, changes in the caliber of blood vessels deep in the body, as well as in the periphery, accompany RF exposure and the rate of local blood flow increases dramatically whenever the temperature of the heated tissue exceeds 41 to 42°C. This phenomenon forms the basis of the treatment of localized malignancies by microwave hyperthermia.[41]

On the other hand, attempts to demonstrate "thresholds" for vasodilation in laboratory animals held at $T_a$ above the UCT or at the upper end of the TNZ are doomed to failure. Figure 2 shows that partial increases in conductance occur as $T_a$ increases within the TNZ; at the UCT, by definition, vasodilation is complete. Rapid and significant changes in the local temperature of the skin in highly vasoactive areas provide a common laboratory index of peripheral vasodilation in animal subjects. Very low SAR thresholds for an increase in the temperature of the ear skin have been reported for rabbits held at $T_a$s of 20 and 30°C inside a waveguide. Concomitant measures of the SAR threshold for an increase in body temperature ($T_{co}$) in the same animals were low at $T_a = 20°C$ and virtually indeterminate at $T_a = 30°C$.[31] An evaluation of these results in terms of the thermoregulatory profile for this species reveals certain methodological difficulties. In rabbits, the TNZ ranges from 15 to 25°C and the upper survival limit is 32°C.[42] Thus, at $T_a = 20°C$ the animal subjects were already partially vasodilated and at $T_a = 30°C$ they were fully vasodilated and close to the upper limit of survival. Since the rabbit pants when heat stressed, a measure of respiratory rate developed for use in microwave-exposed rodents[28,43] could have been used to advantage in the cited study, but was not.

## Evaporative Adjustments During RF Exposure

When the peripheral vasculature of an endotherm is fully vasodilated, dry heat loss from the body nears its maximum. To prevent significant heat storage and a rise in body temperature, the organism must initiate heat loss by evaporation. Figure 2 shows that this occurs when $T_a = UCT$; it also occurs at $T_a$ within the upper reaches of the TNZ, during RFR exposure at SARs sufficiently high to produce full vasodilation.[32] Because cellular processes in the body speed up as tissue temperature rises, a small increase in heat production, concomitant with the initiation of evaporation, also occurs at high $T_a$. Any attempt to predict the evaporative capability of a particular endotherm during RF exposure must consider the thermoregulatory profile, the avenue of evaporative heat loss available (panting or sweating), or whether such capability may not exist (as is the case with rodents). Fortunately, human beings have an extraordinary capacity to lose body heat via sweating when $T_a$ rises above 30 to 31°C or the deep body temperature exceeds 37°C.[18]

Sweating from the foot of squirrel monkeys equilibrated to a $T_a$ just below the UCT can be reliably initiated by RF exposure at a threshold SAR equivalent to 20% of the animals' resting M.[23,24] Below this $T_a$, the threshold SAR is linearly related to the exposure $T_a$. Like heat production, the magnitude of the sweating response depends on the integration of absorbed RF energy over the whole body, not energy deposited in some locus such as the "central thermostat" of the hypothalamus. Indeed, thermoregulatory sweating occurs when a RF field is present, even when the hypothalamus is artificially cooled to prevent its temperature from rising.[39] During partitional calorimetric studies at cool (20°C), neutral (26°C), or warm (32°C) $T_a$, steady-state tolerance limits were determined for squirrel monkeys, exposed to 2.45 GHz CW microwaves, in terms of both power density and SAR.[3] These tolerance limits are shown in Figure 3. The limiting autonomic response in all cases is sweating, which is very limited in this species.[14] Certain other nonhuman primates (e.g., Erythrocebus patas) and, of course, humans, are far better equipped to dissipate large body heat loads through sweating.[18] For humans in particular, the tolerance limits should be correspondingly enhanced.

Sweating from the calf of rhesus monkeys, during RFR exposure at the resonant frequency in thermoneutral $T_a$ of 26 and 32°C, occurs at somewhat higher SARs, equivalent to 80% of the animals' resting M.[44] In $T_a = 26$°C, peripheral vasodilation preceded the onset of sweating, as predicted by the thermoregatory profile. However, sweating in thermoneutral $T_a$, like reduced M in cooler environments,[9] failed to prevent a substantial rise in deep body temperature of the rhesus monkey during RF exposure at the resonant frequency. As Lotz has remarked, careful control over ambient conditions ($T_a$, RH, and v) is essential during such exposures to ensure that heat loss from the skin by convection, radiation, and evaporation is not impeded.

Gordon[25,45] has reported a much higher threshold (SAR = 29 W/kg) for initiation of evaporative heat loss in microwave-exposed mice. In these studies mice were irradiated in a waveguide and the increase in RH of the air flowing through the waveguide was taken as a measure of heat lost by evaporation of body water. As noted above, mice neither pant nor sweat but are said to increase respiratory frequency somewhat when heat stressed[27] in addition to spreading saliva over the fur. None of these responses could be observed, nor could the body temperature be recorded, to provide evidence that the animals were indeed thermoregulating normally. Since the $T_a$ inside the waveguide was 22°C, well below the TNZ for the mouse, changes in M and conductance, not changes in evaporative heat loss, would first be anticipated during RF exposure. The latter should occur only after M was reduced to resting level and the animal's peripheral vasculature was fully vasodilated. With this perspective the high evaporative threshold for the mouse may be more easily understood.[46,47]

## Extrapolation of RFR Data from Animals to Humans

Similar waveguide exposure systems have been used to provide data in support of the hypothesis that a lower SAR is required to initiate heat loss responses, or produce a given increase in body temperature of a large animal, than is required to produce the same outcome in a small animal.[48] Studies of mice, rats, and hamsters,[26,28,49,31] conducted at $T_a$ ranging from 10 to 35°C and SARs as high as 140 W/kg, rely principally on $T_{co}$ measurements made <u>after</u> the test animal was removed from the waveguide. The unique thermoregulatory profile of each species is consistently ignored in a manner similar to that described above.

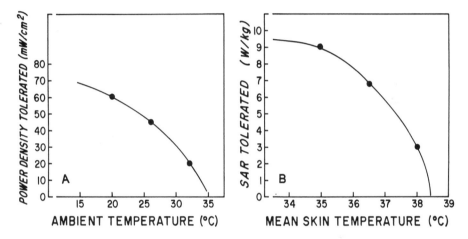

**Figure 3.** Tolerance functions for the exposure of the squirrel monkey to 2.45 GHz CW microwaves as a function of the prevailing ambient temperature (Panel A) and mean skin temperature (Panel B). Panel A shows the power density tolerated and Panel B shows the SAR tolerated.

This hypothesis, originally couched in terms of SAR vs body mass,[25] more recently in terms of SAR vs body surface area,[50] was proposed as a convenient means to extrapolate animal data to humans. Gordon and Ferguson[50] argue that large mammals have a reduced capacity to dissipate internal heat loads, such as absorbed RFR energy, because of a reduced ratio of surface area to body mass. They conclude from an analysis of extant data that an SAR of only 0.2 to 0.4 W/kg is required to initiate a thermoregulatory response in a 70 kg man. This conclusion results from their equation: threshold SAR = $aM^b$, where a is a constant, M is body mass (kg), and b is approximately -0.5. Further extrapolation of this function leads to the doubtful conclusion that turning on a 160-Watt light bulb (representing an SAR of approximately 0.02 W/kg) will initiate a thermoregulatory response in an 8000 kg elephant.

It should not be necessary to point out that a human being is neither a 70-kg mouse, a 70-kg dog, or even a 70-kg rhesus monkey. Paying lip service to the dependence of thermophysiological responses on ambient conditions, experimental variables, species differences, etc. is not sufficient. These and many other factors must play a central role in any attempt to extrapolate animal data to humans. It may turn out that a simple power function, of the type generated by Gordon and Ferguson,[50] applies to the order Rodentia, small fur-bearing animals that neither sweat nor pant. However, given the enormous differences in the capacity for heat dissipation among the members of other mammalian orders (e.g., ungulates, primates), it would be surprising if a single SAR/body-mass function could be given so much precision that extrapolation over several orders of magnitude would become the method of choice for predicting human thermoregulatory responses to RFR.

Other data on animal subjects indicate either the opposite or no functional relationship between SAR and body mass in the initiation of specific thermoregulatory effectors, such as the elevation of $T_{co}$ by 1°C. A comparison of three species (rat, squirrel monkey, rhesus monkey) by de Lorge[51] revealed an increase in SAR to elevate $T_{co}$ by 1°C when ongoing behavior was disrupted by several frequencies of RF exposure.[52,53,54,55] On the other hand, no body-mass dependence was detected when the SAR necessary to elevate $T_{co}$ was determined for a single species (rat) at specific ages from weanling to adult (Merritt, personal communication). Other data collected on the laboratory rat point to the specific effects of carrier frequency,[56] orientation in the RF field,[57,58] pulsed vs CW exposure,[59] sensitivity[60] and treatment with psychotropic and other drugs.[61,62,63] Much more animal data, carefully collected under identical experimental conditions and in full cognizance of the unique thermoregulatory capabilities of each species, is needed to resolve the extrapolation controversy.

A recent analysis of the SAR/body mass predictive relationship by Gordon,[64] predicated on the false premise that current exposure guidelines[65,66,67,68] were developed primarily on bioeffects data from the rat, concludes that body surface area is the more appropriate variable for normalizing the thermal effects of RFR. The author claims that "... the ANSI '82 guideline of 0.4 W/kg SAR in man is about a factor of 7 too high" (op. cit., p. 117) and concludes, primarily on the basis of data collected in his laboratory on mice, rats, hamsters and rabbits, that an SAR related to surface area (i.e., heat flux density in $W/m^2$) would provide a more reliable metric between species, including man. Apart from serious problems associated with the application of this proposed metric to nonuniform RFR exposure of humans (e.g., near field, partial body exposure), the unique thermoregulatory profiles of individual species, as well as many other relevant variables, are still being ignored in this new formulation. It would appear that, at the present state of our knowledge, data on the thermoregulatory capabilities of human beings and their responses to exercise and/or specific environmental heat loads would be far more reliable predictors of human thermophysiological responses to RFR exposure.

## SIMILARITIES BETWEEN MUSCULAR EXERCISE AND RFR EXPOSURE

During muscular exercise, the internal body temperature of humans rises because heat generated in the working muscles is distributed throughout the body by increased blood flow. This increased blood flow, combined with peripheral vasodilation, also brings excess body heat to the skin surface for dissipation to the environment. In the steady state, the heat produced by moderate exercise is efficiently lost to the environment so that the internal body temperature stabilizes at an elevated level that depends principally on the workload.[69] The rise in body temperature is essentially independent of the ambient temperature.[70]

During passive exposure to RFR, thermalizing energy may be selectively deposited in specific tissue beds; the particular pattern of energy deposition varies with many physical factors of both the radiation and the target. Similarly, during exercise the source of heat lies in specific groups of muscle fibers; the particular pattern of heating varies with the activity. However, some have expressed the view that these two situations may generate dissimilar thermoregulatory responses because the absorption of RF energy is somehow "unique."[48] Such a view prohibits the application of voluminous data on exercise physiology to the prediction of human thermophysiological responses to RFR.

The equivalence of thermophysiological responses during exercise and during diathermic heating was demonstrated unequivocally in the classical study by Nielsen and Nielsen.[71] In their experiments, short-wave diathermy was used to deposit heat directly into the deep tissues of the trunk of human subjects. In other test sessions, the same subjects exercised on a bicycle ergometer at a work rate adjusted so that the heat load during cycling and diathermic heating was the same (approximately 5 times the resting M). Four $T_a$, ranging from cool to warm, were studied. In the steady state, at all $T_a$, the rectal temperature increased by the same amount during the two procedures. Thermal conductance, assessed by changes in skin blood flow, and sweating rates were also comparable. Thus, passive heating by diathermy and the heat generated by active exercise produce the same kind of thermal disturbance in the body as a whole, although the distribution of heat in individual tissue compartments of the body may have been very different in the two cases. These important findings demonstrate that the thermoregulatory consequences of whole-body RFR energy deposition may be predicted by the consequences of equivalent heat loads produced by exercise.

It was stated earlier that the operating characteristics of the thermoregulatory system have much in common with those of an automatic control system involving negative feedback. The body temperature appears to be regulated at a "set" or reference level. In the presence of an internal disturbance, this system generally operates with an offset in its regulated variable. The control system for thermoregulation operates identically under passive heating by RFR and active heating during exercise because there is no change in the "set point" for the body temperature. Figure 4, adapted from Stitt,[72] diagrams the events leading to an elevation in body temperature during both exercise and exposure to RFR. At the initiation of exercise, metabolic heat production (M) rises immediately and produces temporarily a high rate of heat storage (S) in the body. Similarly, at the initiation of RF exposure at a thermoneutral $T_a$, energy deposition generates heat in body tissues and produces temporarily a high rate of S in the body. In both cases the rising body temperature generates an error signal which drives heat loss mechanisms at an ever-increasing rate until heat loss equals heat production/generation. When a balance is attained, the body temperature stabilizes at a level appropriate to the level of exercise or energy deposition, both of which may be quantified in W/kg. The elevated body temperature remains until the cessation of exercise or RF exposure.

In contrast, the events that occur during the generation and maintenance of a fever have little in common with those diagrammed in Figure 4. Instead, the thermoregulatory responses during fever are indicative of an elevation in the "set point" for body temperature. The febrile temperature, once attained, is defended against disturbances just as the normal body temperature is defended in the afebrile condition. Once the febrile temperature is attained, by whatever mechanisms are appropriate to the ambient conditions, effector activity returns to the level that existed before fever onset. This normal regulatory activity serves to maintain the new, higher level of body temperature until the fever "breaks."[72] Experiments on laboratory animals have demonstrated that during RF exposure mobilized thermoregulatory responses are maintained without significant alteration for the duration of the exposure and only return to normal, or pre-exposure, levels after extinction of the RF field.[3,5,6,9,10,24] Thus, the exercise model, rather than the fever model, adequately describes the disposition of heat generated in the body by RFR exposure.

Despite the fact that few data have been collected on the thermophysiological responses of human beings exposed to RFR in the laboratory setting, voluminous data on human responses during controlled exercise may be brought to bear on the prediction of safe levels of absorbed RF energy. Since both SAR and M may be quantified in the same units (W/kg), any level of M can provide a good estimate of the thermophysiological response patterns and resulting body temperatures to be anticipated during specific RFR exposures. For example, the average resting M of humans (sitting or standing quietly) is 1.4 W/kg. Double this value is the equivalent of walking on the level. Three times this value (over 4 W/kg, the basis for many exposure guidelines) is the equivalent to light or moderate

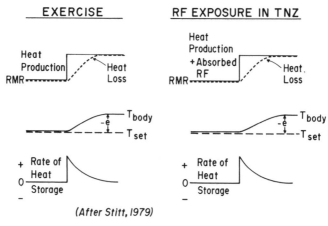

*(After Stitt, 1979)*

**Figure 4.** The events that lead to a rise in body temperature during exercise and exposure to RFR. Hyperthermia results from the sustained increase in metabolic heat production during exercise and heat generated from energy absorption during RFR exposure. The set temperature appears to remain unchanged during both conditions. TNZ = thermoneutral zone. (After Stitt[72]).

physical effort such as social dancing, house cleaning, or driving a truck.[73] The exercise literature provides accurate additional information on age and sex differences, effects of a wide range of ambient conditions, the benefits of training and acclimatization to heat, the complications of hypo- and hyper-hydration, drug effects, etc. (cf. 74, 75, 76, 77, 78, 79). While ambient RH and v influence heat losses from the body, their impact is considerably less than changes in $T_a$. The ASHRAE Handbook[73] provides details of the appropriate modifications to the heat transfer coefficients for each mode of heat loss under a wide assortment of metabolic loads and ambient conditions. The criteria for a recommended standard for exposure to hot environments[80] best summarize our knowledge of the tolerance of human beings to environmental thermal stress. The NIOSH limits are based on a criterion elevation in body temperature of 1° C. The recommended environments are characterized in terms of the Wet Bulb Globe Temperature (WBGT), which combines the effects of RH, v, $T_a$, and radiation on the body. The recommended alert and exposure limits are set as a function of both M and the percentage of total time exposed to particular ambient conditions.

Figure 5 reproduces two figures from the NIOSH guidelines, modified to include data relevant to the setting of standards for human exposure to RFR. The NIOSH limits, calculated for a "standard worker" of 70 kg body mass, can be converted to specific metabolic rate (M) in W/kg. The dashed curves added to each panel represent M minus the SAR criterion of 0.4 W/kg that has been adopted for RF exposure guidelines since ANSI 1982.[66] This maneuver assumes that absorbed RFR energy may be substituted for metabolic energy.[4,24,32,71] It is evident that the maximal shift in the NIOSH guideline for continuous exposure of the heat-unacclimatized person to RFR at 0.4 W/kg is 0.8° (WBGT); this occurs at very low workloads close to resting M. The average shift in the NIOSH guideline across workloads is only 0.3° - 0.4° (WBGT), an insignificant amount. The reader is referred to the excellent review by Tell and Harlan[81] for additional analyses of the NIOSH recommendations, together with other pertinent biological effects data that are useful for development of RFR exposure guidelines.

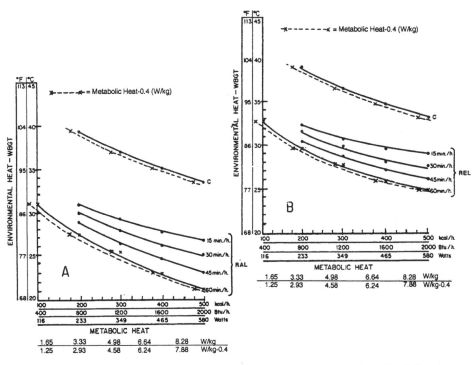

**Figure 5.** NIOSH recommended heat-stress alert limits for heat-unacclimatized workers (Panel A) and exposure limits for heat-acclimatized workers (Panel B) as modified to include the SAR criterion of 0.4 W/kg adopted for many current RFR exposure guidelines for humans (dashed curves). C = Ceiling limit. RAL = Recommended Alert Limit. REL = Recommended Exposure Limit. Recommendations for "standard worker" of 70 kg (154 lbs) body weight and 1.8 m² (19.4 ft²) body surface area. Figures reproduced from NIOSH.[80]

257

# THERMOPHYSIOLOGICAL RESPONSES IN HUMANS EXPOSED TO RFR

No mammalian species is as well equipped, both physiologically and behaviorally, as the human to withstand heat generated in the body by both exogenous and endogenous sources. Thermalizing energy deposited in the body during exposure to RFR should be no exception. A comprehensive study of 486 locations within 15 metropolitan areas of the United States estimates that more than 99% of the population is exposed to background RFR at less than 1 $\mu W/cm^2$. [82] At the resonant frequency for humans, this represents a whole-body SAR of 0.0004 W/kg or about 0.03% of the normal resting M, a completely insignificant level for thermoregulation. Even the whole-body SAR of 0.4 W/kg adopted as the basis for the new IEEE C95.1-1991[67] guideline represents only 35% of the resting M (equivalent to donning a light sweater) and is small enough to provide little more than noise during most daily activity (see Figure 5).

Thermoregulatory processes are ongoing in each of us all of the time. Small perturbations in endogenous heat production or in the thermal characteristics of the environment result in finely tuned adjustments in one or more of the thermoregulatory mechanisms discussed above. The remarkably stable internal body temperature, with its circadian tides, is the result. In addition, most humans have the capacity to cope with exercise or work loads that are the equivalent of 15 times the resting M and that are often taken in thermally stressful environments. The exceptional rates of human sweat production, coupled with efficient behavioral thermoregulation, ensure a minimal rise in deep body temperature. It is small wonder, then, that the few reports of humans exposed to modest levels of RFR in laboratory or clinic show no surprising "adverse health effects."

In addition to the extensive use of microwave energy as an adjunct to cancer treatment,[41] techniques are also under development that utilize RF energy for rewarming from hypothermia, both whole-body and extremities only. A 13.56 MHz trunk coil that provided RF warming to mildly hypothermic male subjects was found to restore normal body temperature more rapidly than the conventional methods of warm water immersion or use of a sleeping bag.[83] Special RF coils developed to resonate at 27.12 MHz have been used to warm divers' cold hands and feet. When tested in either 10°C air or 24°C seawater, the RF coils were much more efficient and tolerable than similar nichrome wire coils energized by DC.[84,85] Not only were skin areas remote from the coils warmed by heated blood, but also no adverse thermal effects were observed during the tests.

A single study, reporting an attempt to use RFR to induce temporary sterilization in adult males, also gives surprising data related to the thermal effects of long-term, partial-body exposure at high intensities.[86] Thirteen male volunteers underwent localized RF exposures of the testes at 915 or 2450 MHz in weekly sessions that lasted 30 min each. Power levels were 20 to 30 watts, sufficient to raise the scrotal skin temperature 10° C above its normal level. Seven subjects received over 100 such sessions, the remainder somewhat fewer. Six months after termination of the exposures, biopsies of testicular tissue were examined microscopically. While considerable cellular damage was evident, no significant morphological abnormalities were found. And despite evidence of reduced spermatogenesis during the treatment, sterilization was not reliably achieved by this method; indeed, a couple of the men fathered children during the course of treatment. A major conclusion: human testicular tissue appeared much more resistant to thermal injury than similar tissue in experimental animals such as rats or rabbits.

A considerable literature is now available that describes thermal responses in human subjects exposed to RF fields during magnetic resonance imaging (MRI) procedures. Although the data reported were collected primarily in the clinic, rather than in the laboratory, considerable attention has been paid in recent studies to equilibration of the patients prior to MRI scans, control of ambient conditions, specification of SAR, and assessment of a variety of thermophysiological variables. Shellock and Crues[87,88] measured skin, sublingual and corneal temperatures of 35 patients during MRI with a head coil (1.5 T at 64 MHz). Estimated peak SARs ranged from 2.54 to 3.05 W/kg. A corneal temperature rise of 0.5°C and slight elevations in the skin temperature of head regions were judged to be trivial. No change was measured in sublingual temperature after scans of at least 8 minutes duration. In another study,[89] sublingual and skin temperatures were measured with Luxtron probes in 6 subjects before (20 min), during (30 min), and after (20 min) MRI in a 1.5 T body coil at SARs from 2.7 to 4.0 W/kg. Skin blood flow (SkBF) was also measured with a laser-doppler device. Although the 30-min scans were insufficient to establish a steady state, the measured temperature changes (e.g., a maximal rise of 0.1°C in sub-lingual temperature) were judged to be physiologically inconsequential. The scans produced primarily surface heating, with modest increases in SkBF. Scrotal MRI of 8 men at 1.5 T (range of whole-body SAR from 0.56 to 0.84 W/kg) and an average duration of 23 min produced a maximum rise of 3.0°C in scrotal skin temperature.[90] This effect was well below the threshold for a reduction in spermatogenesis.

That measured tissue temperature changes during MRI are attributable to RF exposure during the procedure has been demonstrated by experiments in which 6 male subjects were exposed to 1.5 T static magnetic fields only.[91,92] Sublingual temperature and an assortment of skin temperatures were measured during 20-min scans that followed a 20-min equilibration to $T_a = 21$°C. No change from the equilibrated level occurred in any measured body temperature during the scans. The authors stress the importance of careful control over environmental and circadian variables in such experiments. The general conclusion to be drawn from data collected on subjects undergoing MRI at 1.5 T, under ambient conditions typically found in the clinic, is that tissue temperature changes are modest and far below hazardous levels.[91] It is of interest that Gordon[93] has concluded that if the ANSI 1982[66] exposure limits are applied to MRI procedures, the resulting elevations in body temperature will be inconsequential.

As an adjunct to the human data, a study was undertaken to measure the thermal effects of MRI on 12 anesthetized, fleeced sheep.[94] Exposures occurred in a 1.5 T (64-MHz) MRI system using quadrature, circularly polarized field (vector) excitation for whole-body scans and a head coil for scans of the head only. Subcutaneous temperatures (abdomen, chest), vena caval and rectal temperatures were monitored during body scans (SAR = 1.5 to 4.0 W/kg) of 20 to 104 minutes duration; other temperatures measured during head scans (4.0 W/kg) included cornea, vitreous humor, head skin and jugular vein. Five animals exposed at 4.0 W/kg (head or whole-body) were allowed to recover from the procedure and exhibited neither incipient cataracts nor ill health 10 weeks later. No controls, other than pre-scan "control periods" of variable duration were undertaken. During whole-body scans at 4 W/kg, rectal temperature rose at a rate of about 0.023°C/min while vena caval temperature rose at a slightly higher rate. Lower SARs produced proportionately lower rates of temperature increase. The temperature increase in eye or cornea during a 60-min head scan at 4 W/kg did not exceed 1.5°C. In this study behavioral thermoregulation was disabled by anesthesia, panting was prevented by an endotracheal tube, and heat loss by convection and radiation was compromised by the intact fleece. All of these constraints prevented a steady state to be reached during the scans (i.e., body temperatures continued to rise). Although the exposure conditions were generally in excess of those used in the clinic, the measured elevations in body temperature were insufficient to cause adverse thermal effects.

# SIMULATION MODELS OF THERMOPHYSIOLOGICAL RESPONSES TO RFR EXPOSURE

As we have seen, all animals, including humans, have a widely varying internal heat production and live in a thermal environment whose physical characteristics are constantly changing. In the absence of effective behavioral thermoregulation (see below), the body temperature of endotherms will be regulated by the physiological mechanisms that control heat production, the distribution of heat in the body, and the avenues and rate of heat loss from the body to the environment. These mechanisms are well understood and quantified for the human body; thus, it has been possible to develop mathematical simulation models of physiological thermoregulation in considerable detail and validate them against experimental data. Thermophysiological modeling becomes especially important when its purpose is to extrapolate variables that are not experimentally attainable and to simulate experiments that cannot be performed (e.g., those involving RFR exposure of human beings).

The basis of thermophysiological modeling is the energy (or heat) balance equation shown in Figure 1. Simulation models incorporate the physical characteristics of the body, heat production and heat loss responses, and all relevant characteristics of the environment as exemplified by the thermoregulatory profile (cf, Berglund[2] for additional details). As the heat balance equation shows, absorbed RF energy is added to metabolic energy and must be balanced by appropriate heat loss from the body to avoid changes in the body temperature.

As in all modeling efforts, thermophysiological models are based on simplifications of the actual system. As described by Stolwijk[20] and Wissler,[95] it is useful to separate the total physiological system into two major components: the regulated or passive system and the regulatory or controller system. Both systems are simplified to a degree necessitated by knowledge or required by the application. The controlling or regulatory part of the model consists of the structures that sense body temperatures and their response characteristics, the central neural integration, the neural effector pathways and the effector mechanisms themselves, including shivering or exercising muscles, secreting sweat glands, and the tone of peripheral blood vessels in the skin and other blood vessels elsewhere that control the convective heat transfer between different organs and structures of the body.

The second component, the controlled system, consists of the simplified representation of the thermal characteristics of all the body tissues, including metabolic heat production, heat capacitance, local temperature, heat transfer via conduction, convection, and radiation between the skin and the external environment. In the case of RF exposure of the whole body, the absorbed energy will be added to the metabolic energy generated in the body's core. Partial body exposure or allowance for distributive dosimetry requires the absorbed energy to be proportioned appropriately in various body compartments.

A model by Stolwijk and Hardy,[96] first formulated in 1966, has been used, by Stolwijk and others, as the basis for prediction of the effects of depositing RF energy into selected parts of the human body. The original model is fully described in Stolwijk[97] and Stolwijk and Hardy.[98] Stolwijk[20] simulated the deposition of 100 watts of RF energy into the core of the head for 30 minutes in a thermoneutral $T_a$ of 30°C. This exposure caused only a modest elevation in brain temperature, because of the high rate of brain blood flow, and the mobilization of strong heat loss through sweating. Because the rate of heat loss exceeded the rate of energy input, all body temperatures were predicted to fall. A second prediction simulated 500 watts of RF energy deposited into the trunk core at thermoneutrality, but with the sweating response disabled. In this example, the temperatures in all body compartments were predicted to rise and the thermoregulatory system was overwhelmed by the RF exposure.

Way et al.[21] introduced an additional, hypothalamic compartment to Stolwijk's model, so as to consider partial heating due to hotspots that might be introduced by RF energy absorbed in the head. They found that because of thermoregulatory action, the hypothalamic temperature would not increase drastically until the rate of energy deposition exceeded a threshold level of 50 W/kg. The principal controlling physiological mechanisms in this case were found to be sweating and blood flow.

Another adaptation of Stolwijk's model (Spiegel, et al.[99]) was not as successful. These investigators added more compartments in order to introduce RF energy distributed in accordance with a block model of man.[100] The local temperature at 61 discrete bodily loci, as well as the thermoregulatory responses of vasodilation and sweating, were computed for a number of RF field intensities at 2 frequencies, 80 and 200 MHz. The incident power density vs duration necessary to obtain a hot spot of 41.6°C was calculated for several parts of the body. Spiegel et al. claimed to have demonstrated that hot spots were produced at much lower field strengths at 80 MHz than at 200 MHz and also that a 1-hr exposure to 80 MHz at 10 mW/cm$^2$ would generate a hot spot in the core of the lower thigh. However, their version of the Stolwijk model did not incorporate the great increase in blood flow known to occur at tissue temperatures of 40°C and above,[41,101,102] a response that would have prevented elevations in tissue temperature of the predicted magnitude.

A simpler model of physiological thermoregulation has recently been adapted to predict the thermophysiological consequences of exposure to RF fields in the MRI environment.[22] The model has only two nodes (core and skin) but in most other respects is similar to the Stolwijk and Hardy formulation. Based on the fundamental heat balance equation, the model predicts physiological heat loss responses in real time as a function of selected $T_a$, v, and rate of whole-body RF energy deposition (SAR). Assuming a criterion elevation in deep body temperature ($\Delta T_{co}$) of 0.6°C, $T_a = 20°$ C, and v = 0.8 m/sec, the model predicts that a 70 kg patient could undergo an MRI scan of infinite duration at SAR $\leq 5$ W/kg. Lowering $T_a$ or increasing v permits a rise in permissible SAR for a given $\Delta T_{co}$. More restrictive $\Delta T_{co}$ criteria result in lower permissible SARs and shorter exposure durations. The limiting response is usually the rate of peripheral blood flow, although sweating can play a role in limiting $\Delta T_{co}$.

Restrictions on the rate of skin blood flow (SkBF), ranging from 0 to 89% of normal, have also been studied with this model.[103] Under conditions that are desirable in the clinic ($T_a = 20°C$, 50% RH, still air), moderate restrictions (up to 67%) of SkBF yield tolerable $\Delta T_{co}$ ($\leq 1°C$) during MRI scans (SAR $\leq 4$ W/kg) of 40 min or less. Increased $T_a$ and RH exacerbate the thermal stress imposed by absorbed RF energy. Severely impaired SkBF encourages short MRI exposures (e.g., 20 min or less) at SARs $\leq 3$ W/kg.

At a given $T_a$, a person can absorb some level (SAR) of RF energy indefinitely; i.e., thermal equilibrium with the prevailing conditions can be reached. At low SARs, $T_{co}$ will rise initially, then stabilize at some elevated level. If the cardiovascular system is impaired, the maximum SAR at which thermal equilibrium can be attained will be lower. Figure 6 (Panel A) illustrates the maximal SAR for equilibrium in two $T_a$ as a function of the impairment in peripheral blood flow. The solid line defines, for $T_a = 20°C$, the limit of the zone of stable body temperature, i.e., conditions to the left of the line yield a stable body temperature, those to the right of the line yield a rising body temperature. The dashed line similarly defines the zone of stable body temperature for $T_a = 27°C$. The figure shows that raising the $T_a$ by 7°C decreases the maximal SAR for equilibrium by about 1 W/kg. It is of considerable interest that the shape of the function closely resembles that shown in Figure 3 (which delineates experimentally determined thermal tolerance levels for a RF-exposed nonhuman primate).

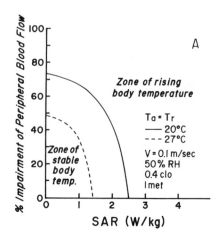

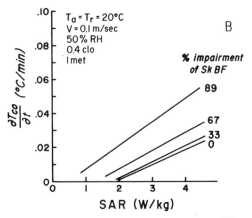

**Figure 6.** Predictions of the two node model adapted by Adair and Berglund.[103] Zones of stable and rising body temperatures in two environments during 40-min MRI scans at different SARs when peripheral blood flow is impaired (Panel A). Rate of change of core temperature as a function of SAR (W/kg) when skin blood flow (SkBF) is impaired (Panel B). See text for additional details.

If the SAR under consideration exceeds the maximal value for thermoequilibrium shown in Figure 6 (Panel A), a continuous increase in body temperature can be expected during RF exposure. The rate of this increase, based on the predictions of the model, will depend on the cardiovascular impairment, the $T_a$, the RH, and the clothing insulation (clo). Figure 6 (Panel B) shows the predicted rate of increase in $T_{co}$ that may be due to impaired cardiac output (modeled as restricted SkBF) as a function of SAR. The reader is referred to Adair and Berglund[103] for other figures which provide corrections to these predictions that depend on departures of $T_a$, RH and clo from the levels designated in Figure 6 (Panel B).

Schaefer[104] has used the predictions of Adair and Berglunds's model to calculate the effect of $T_a$ on the average rate of increase in $T_{co}$ under conditions of 40% impairment in SkBF, clo = 0.2, and a 60-min exposure duration. He predicts that a 1-hr exposure to SAR = 4 W/kg results in a $\Delta T_{co}$ of only 1°C when $T_a$ = 19°C. Further, if $T_a$ = 25°C, an SAR of 3 W/kg is appropriate. A 1-hr RF exposure to 1 W/kg, even at $T_a$=27°C, should result in no body temperature rise. These predictions do not conflict with data collected in the clinical setting and, indeed, are most reassuring for the generality of whole-body exposure to RFR.

In a recent paper, Adair and Berglund[105] report that precooling of patients to the prevailing $T_a$ prior to MRI exposure has little value in terms of preventing a rise in body temperature and use of added insulation (i.e., a blanket) should be discouraged in the normal clinical setting. Clinical tests of two normal male subjects provided limited confirmation of the model's predictions. During 20-min MRI scans at a whole-body SAR of 1.2 W/kg, core and skin temperatures, sweat rate, and judgments of thermal sensation and discomfort were very similar to the predicted values. However, the authors hasten to point out the limitations of the two-node model because the simulations are based on RF exposure of the whole body, an atypical situation for the MRI clinic. A model having many more nodes (e.g., Stolwijk[97]) would be much more useful for general predictions of the thermophysiological responses of humans to RF exposure (partial- and whole-body), as well as specific predictions for clinical use.

## BEHAVIORAL THERMOREGULATION AND THERMAL COMFORT

Humans and other endotherms continually seek a comfortable thermal environment, one that provides maximal satisfaction and minimal thermophysiological strain. When we set the thermostat, open the window, or put on more clothing, we are responding to a need for thermal comfort. The maintenance of thermal homeostasis underlies this need; as we have seen, a stable body temperature maximizes conservation of the body's energy and fluid stores.

The comfortable environment for resting endotherms is often thermally neutral -- that is, it falls within the TNZ (Figure 2). Sometimes, as after heavy exercise, a cool environment feels comfortable; at other times, as after prolonged cold exposure, a warm environment may feel comfortable. Thus the deep body temperature, as well as the skin temperature, exerts a strong bias on sensations of thermal comfort. Extensive human research[106] indicates that the thermal environment selected or preferred is that which maximizes comfort and facilitates thermoregulation. Therefore, the study of thermal comfort in any environment, including those in which a source of RFR is present, embodies study of behaviors that accomplish effective thermoregulation.

In cold $T_a$, metabolic heat must ordinarily be supplemented by exogenous energy sources to maximize thermal comfort. Similarly, in warm $T_a$, sources of convective cooling eliminate the necessity for excessive heat loss by evaporation. Most organisms readily learn to manipulate radiant or convective sources, or position themselves strategically near such sources, to thermoregulate efficiently.[107] In any $T_a$ the more intense the source of heating or cooling, the less it will be energized. If the body temperature is perturbed (as by exercise), the preferred $T_a$ (or skin) temperature will be oppositely altered.[108] In laboratory animals, heating or cooling a thermode implanted in the preoptic area of the hypothalamus (site of the "internal thermostat") will cause the animal to select a cooler or warmer environment, respectively.[109] Local temperature changes in other parts of the central nervous system (such as the medulla and spinal cord) can often stimulate equivalent behavioral changes.

Changes in behavior ensure the maintenance of a stable body temperature or the rapid restabilization of a body temperature that has been perturbed. An appropriate shift in the thermal gradient between skin and environment, or between body core and skin, accomplishes the regulation.

Research during the last 15 years has confirmed that laboratory animals energize or orient around a source of RF energy in the environment as if it were a source of radiant or convective heat. Lizards bask in the radiation from a microwave antenna and thereby regulate their body temperatures effectively.[110] Rats, trained to press a lever for infrared heat in cold $T_a$, suppress lever pressing in inverse proportion to the strength of an imposed RF field.[111] Mice select a cooler part of a thermal gradient as the imposed RF field inside a waveguide becomes more intense.[45] A discrete power density threshold governs such behavioral changes,[112] and the behavior, once initiated, persists until the RFR is extinguished.[113] Rhesus monkeys learn to press a lever for intermittent pulses of 6.4 GHz microwaves when they are cold.[114] Similarly, squirrel monkeys can be trained to turn an RF source on and off; the duration of voluntary RF exposure is inversely proportional to the field strength.[115] Even at SARs that are close to two times the resting M, the change in thermoregulatory behavior regulates the body temperature at a level close to normal.[3] All evidence indicates that a state of thermal comfort is simultaneously achieved.

If only part of the body (such as head or trunk) is exposed to RFR, the change in selected $T_a$ is governed by an integrated energy deposition over the whole body, not by energy deposited in some specific locus such as the brain. Indeed, the hypothalamic "thermostat" appears to play no more vital a role in thermoregulation during RF exposure than do other thermosensitive regions of the body,[39,116] which reinforces current views of the neurophysiological control of thermoregulation.

When squirrel monkeys undergo whole-body exposure at the resonant frequency (450 MHz, CW, E-polarization), behavioral thermoregulatory responses are mobilized in an orderly fashion.[117] The threshold SAR for the alteration of thermoregulatory behavior is nearly 2 W/kg, 40% higher than the threshold SAR determined in identical experiments during RF exposure to a supra-resonant frequency of 2450 MHz.[112] Under resonant RF exposure, behavioral responses regulate the skin temperature at the normally preferred level. Because of the deep penetration of the RF energy at resonance, a stable hyperthermic offset or bias occurs in the deep body temperature; this situation is identical to that occurring during exercise. Thermophysiological responses of peripheral vasodilation and sweating, manifested on the skin surface, are stimulated at SARs similar to the behavioral threshold, indicating the possibility that such responses could serve as auxiliary sensory cues to behavior.

In the steady state the $T_a$ selected by an RF-exposed animal is a linear function of the imposed field strength. Figure 7 (Panel A) shows this relationship for trained squirrel monkeys during whole-body exposure to a homogeneous 2450 MHz CW microwave field inside an anechoic chamber.[3] As demonstrated by Berglund,[2] this function represents the conditions that provide a constant level of thermal comfort or operative temperature ($T_o$). Because the relationships between $T_o$, comfort, and thermoregulation are well understood for humans, Berglund has been able to predict the effect of RF energy on human thermal comfort over a wide range of ambient conditions, as shown in Figure 7 (Panel B). Two families of curves are shown, one for nude persons (0 clo) and the other for persons wearing a warm vested suit or equivalent (1.2 clo). The functions are loci of constant $T_o$ (19.5 and 27.5°C) and thus assume constant thermal comfort. For each clothing level, lines are drawn for both beam (incident from one side) and diffuse RFR. The $T_o$ lines are further divided into two absorptance classes (alpha = 0.5 and 0.9) because absorptance varies with microwave frequency. The figure shows that the differences between beam and diffuse RFR are great, i.e., much larger reductions in $T_a$ for a given level of incident flux density are possible with diffuse than with beam radiation. Further, comfort for sedentary humans wearing typical winter clothing should be possible in a 10.5° C (51°F) environment with diffuse RFR at 10 mW/cm$^2$, the current safety guideline for frequencies above 3 GHz.[67] Thus the concept of microwave heating for comfort as proposed by Pound[29] appears to have great potential for improving the thermal environment in many situations and perhaps for saving energy as well.[118]

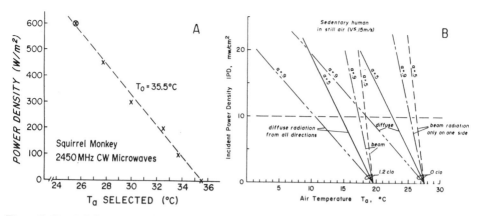

**Figure 7.** (Panel A) Steady-state air temperature chosen by sedentary squirrel monkeys exposed to RFR at different power densities. $T_o$ = operative temperature; $T_a$ = air temperature. (Panel B) Predicted incident power density required for comfort of sedentary humans in various air temperatures at two levels of clothing insulation. The effect of radiation received from one side and simultaneously from all directions is shown, as is the effect of 0.5 and 0.9 radiation absorptance. [Reproduced from Berglund[2] with permission.]

## REFERENCES

1. D.N. Erwin, Mechanisms of biological effects of radiofrequency electromagnetic fields: an overview, *Aviat. Space Environ. Med.* 59 (11, Suppl.):A21-31 (1988).

2. L.G. Berglund, Characterizing the thermal environment, *in:* "Microwaves and Thermoregulation," E.R. Adair, ed., Academic Press, New York, pp 15-31 (1983).

3. E.R. Adair, Microwave challenges to the thermoregulatory system, Technical Report No. USAFSAM-TR-87-7, USAF School of Aerospace Medicine, Brooks AFB, TX, August 1987 (1987).

4. E.R. Adair and B.W. Adams, Adjustments in metabolic heat production by squirrel monkeys exposed to microwaves, *J. Appl. Physiol: Respirat. Environ. Exercise Physiol* 52:1049-1058 (1982).

5. E.R. Adair, B.W. Adams, and S.K. Hartman, Physiological interaction processes and radio-frequency energy absorption., *Bioelectromagnetics* 13:497-512 (1992).

6. V. Candas, E.R. Adair, and B.W. Adams, Thermoregulatory adjustments in squirrel monkeys exposed to microwaves at high power densities, *Bioelectromagnetics* 6:221-234 (1985).

7. H.S. Ho and W.P. Edwards, Oxygen-consumption rate of mice under differing dose rates of microwave radiation, *Radio Sci* 12:131- 138 (1977).

8. H.S. Ho and W.P. Edwards, The effect of environmental temperature and average dose rate of microwave radiation on the oxygen-consumption rate of mice, *Radiat. Environ. Biophys* 16:325-338 (1979).

9. W.G. Lotz and J.L. Saxton, Metabolic and vasomotor responses of rhesus monkeys exposed to 225 MHz radiofrequency energy, *Bioelectromagnetics* 8:73-89 (1987).

10. W.G. Lotz and J.L. Saxton, Thermoregulatory responses in the rhesus monkey during exposure at a frequency (225-MHz) near whole-body resonance, *in:* "Electromagnetic Fields and Neurobehavioral Function," M.E. O'Connor and R.H. Lovely, eds., Alan R. Liss: New York, pp 203-206 (1988).

11. R.D. Phillips, E.L. Hunt, R.D. Castro and N.W. King, Thermoregulatory, metabolic, and cardiovascular responses of rats to microwaves, *J. Appl. Physiol* 38:630-635 (1975).

12. J.D. Hardy, Physiology of temperature regulation, *Physiol. Rev* 41:521-606 (1961).

13. G.S. Johnson and R.S. Elizondo, Thermoregulation in Macaca mulatta: A thermal balance study, *J. Appl. Physiol.: Respirat. Environ. Exercise Physiol* 46:268-277 (1979).

14. J.T. Stitt and J.D. Hardy, Thermoregulation in the squirrel monkey (Saimiri sciureus), *J. Appl. Physiol* 31:48-54 (1971).

15. J.S. Hart, Rodents, *in:* "Comparative Physiology of Thermoregulation," "Vol. II. G.C. Whittow, ed., Academic Press, New York, pp 1-149 (1971).

16. C.H., Durney, H. Massoudi, and M.F. Iskander, "Radiofrequency Radiation Dosimetry Handbook," Fourth Edition, Report USAFSAM-TR-85-73, USAF School of Aerospace Medicine, Brooks AFB, TX (1986).

17. E.R. Adair and B.W. Adams, Microwaves induce peripheral vasodilation in squirrel monkey, *Science* 207:1381-1383 (1980).

18. C.B. Wenger, Circulatory and sweating responses during exercise and heat stress, *in:* "Microwaves and Thermoregulation," E.R. Adair, ed., Academic Press: New York, pp 251-276 (1983).

19. E.R. Nadel, R.W. Bullard, and J.A.J. Stolwijk, Importance of skin temperature in the regulation of sweating, *J. Appl. Physiol* 31:80-87 (1971).

20. J.A.J. Stolwijk, Mathematical models of thermal regulation, *Ann. N.Y. Acad. Sci.* 335:98-106 (1980).

21. W.I. Way, H. Kritikos, and H. Schwan, Thermoregulatory physiologic responses in the human body exposed to microwave radiation, *Bioelectromagnetics* 2:341-356 (1981).

22. E.R. Adair, and L.G. Berglund, On the thermoregulatory consequences of NMR imaging., *Magn. Res. Imaging* 4:321-333 (1986).

23. E.R. Adair, Microwaves and thermoregulation, *in:* "Aeromedical Review USAF Radiofrequency Radiation Bioeffects Research Program-A Review," J.C. Mitchell, ed., USAFSAM Report No. SAM-TR-81-30, pp 145-158 (1981).

24. E.R. Adair, Microwave Radiation and Thermoregulation, Technical Report No. USAFSAM-TR-85-3, USAF School of Aerospace Medicine, Brooks AFB, TX, (May 1985).

25. C.J. Gordon, Effects of ambient temperature and exposure to 2450-MHz microwave radiation on evaporative heat loss in the mouse, *J. Microwave Power* 17(2):145-150 (1982).

26. C.J. Gordon, Effect of heating rate on evaporative heat loss in the microwave-exposed mouse, *J. Appl. Physiol: Respirat. Environ. Exercise Physiol* 53(2):316-323 (1982).

27. C.J. Gordon, Influence of heating rate on control of heat loss from the tail in mice, *Am. J. Physiol.* 244:R778-R784 (1983a).

28. C.J. Gordon and M.D. Long, Ventilatory frequency of mouse and hamster during microwave-induced heat exposure, *Resp. Physiol* 56:81-90 (1984).

29. R.V. Pound, Radiant heat for energy conservation, *Science* 208:494-495 (1980).

30. C.J. Gordon, M.D. Long, and K.S. Fehlner, Temperature regulation in the unrestrained rabbit during exposure to 600 MHz radiofrequency radiation, *Int. J. Radiat. Biol* 49(6):987-997 (1986).

31. C.J. Gordon, M.D. Long, K.S. Fehlner, and A.G. Stead, Temperature regulation in the mouse and hamster exposed to microwaves in hot environments, *Health Physics* 50(6):781-787 (1986).

32. E.R. Adair, Thermophysiological effects of electromagnetic radiation, *IEEE Eng. Med. Biol.* 6:37-41 (1987).

33. W.G. Lotz, Hyperthermia in radiofrequency exposed rhesus monkeys: a comparison of frequency and orientation effects, *Rad. Research* 102:59-70 (1985).

34. D.E. Spiers and E.R. Adair, Thermoregulatory responses of the immature rat following repeated postnatal exposures to 2,450 MHz microwaves, *Bioelectromagnetics* 8:283-294 (1987).

35. C.-K. Chou, A.W. Guy, L.L. Kunz, R.B. Johnson, J.J. Crowley, and J.H. Krupp, Long-term, low-level microwave irradiation of rats, *Bioelectromagnetics* 13:469-496 (1992).

36. E.R. Adair, D.E. Spiers, R.D. Rawson, B.W. Adams, D.K. Sheldon, P.J. Pivirotto, and G.M. Akel, Thermoregulatory consequences of long-term microwave exposure at controlled ambient temperatures, *Bioelectromagnetics* 6:339-363 (1985).

37. J.H. Krupp, In vivo measurement of radiofrequency radiation absorption, *in:* "Proceedings of a Workshop on the Protection of Personnel Against Radiofrequency Electromagnetic Radiation," J.C. Mitchell, ed., USAF Aeromedical Review SAM-3-81, School of Aerospace Medicine, Brooks AFB, TX (1981).

38. J.H. Krupp, *In vivo* temperature measurements during the whole body exposure of Macaca mulatta to resonant and non-resonant frequencies, *in:* "Microwaves and Thermoregulation," E.R. Adair, ed., Academic Press:New York pp 95-107 (1983).

39. E.R. Adair, Microwave challenges to the thermoregulatory system, *in:* "Electromagnetic Waves and Neurobehavioral Function," M.E. O'Connor and R.H. Lovely, eds., Alan R. Liss: New York, pp 179-201 (1988).

40. T.R. Ward and R.J. Spiegel, Thresholds of microwave-induced vasodilation in the squirrel monkey, BEMS Sixth Annual Meeting Abstracts, BEMS: Gaithersburg, MD, p 37 (1984).

41. A.W. Guy and C.-K. Chou, Electromagnetic heating for therapy, *in:* "Microwaves and Thermoregulation," E.R. Adair, ed., Academic Press, New York, pp 57-93 (1983).

42. R.R.Gonzalez, M.J. Kluger, and J.D. Hardy, Partitional calorimetry of the New Zealand white rabbit at temperatures of 5-35°C, *J. Appl. Physiol* 31:728-734 (1971).

43. C.J. Gordon and J.S. Ali, Measurement of ventilatory frequency in unrestrained rodents using microwave radiation, *Resp. Physiol* 56:73-79 (1984).

44. W.G. Lotz and J.L. Saxton, Influence of radiofrequency exposure on sweating and body temperature in rhesus monkeys, BEMS Eighth Annual Meeting Abstracts, BEMS:Gaithersburg, MD p 79 (1986).

45. C.J. Gordon, Note: Further evidence of an inverse relation between mammalian body mass and sensitivity to radiofrequency electromagnetic radiation, *J. Microwave Power* 18(4):377-383 (1983).

46. E.R. Adair, D.E. Spiers, J.A.J. Stolwijk, and C.B. Wenger, Technical Note: On changes in evaporative heat loss that result from exposure to nonionizing electromagnetic radiation. *J. Microwave Power* 18:209-211 (1983).

47. E.R.Adair, C.B. Wenger, and D.E. Spiers, Technical Note: Beyond allometry, *J. Microwave Power* 19:145-147 (1984).

48. J.A. Elder and D.F. Cahill, "Biological Effects of Radio-frequency Radiation," Report EPA-600/8-83-026F, EPA Health Effects Research Laboratory, Research Triangle Park, NC (1984).

49. C.J. Gordon and J.S. Ali, Comparative thermoregulatory response to passive heat loading by exposure to radiofrequency radiation, *Comp. Biochem. Physiol* 88(1):107-112 (1987).

50. C.J. Gordon and J.H. Ferguson, Scaling the physiological effects of exposure to radio-frequency electromagnetic radiation: consequences of body size, *Int. J. Radiat. Biol.* 46(4):387-397 (1984).

51. J.O. de Lorge, The thermal basis for disruption of operant behavior by microwaves in three animal species, *in:* "Microwaves and Thermoregulation," E.R. Adair, Ed., Academic Press, New York, pp 379-400 (1983).

52. J.O. de Lorge, The effects of microwave radiation on behavior and temperature in rhesus monkeys, *in:* "Biological Effects of Electromagnetic Waves," HEW Publication (FDA) 77-8010, C.C. Johnson and M. L. Shore, eds., pp 158-174 (1976).

53. J.O. de Lorge, Operant behavior and rectal temperature of squirrel monkeys during 2.45 GHz microwave radiation, *Radio Sci* 14:217-225 (1979).

54. J.O. de Lorge, Operant behavior and colonic temperature of Macaca mulatta exposed to radiofrequency fields at and above resonant frequencies, *Bioelectromagnetics* 1:183-198 (1984).

55. J.O. de Lorge and C.S. Ezell, Observing-responses of rats exposed to 1.28- and 5.62-GHz microwaves, *Bioelectromagnetics* 1:183-198 (1980).

56. J.A. D'Andrea, O.P. Gandhi, and J.L. Lords, Behavioral and thermal effects of microwave radiation at resonant and nonresonant wave lengths, *Radio Sci.* 12:251-256 (1977).

57. M.R. Frei, J.R. Jauchem, J.M. Padilla, and J.H. Merritt, Thermoregulatory responses of rats exposed to 2.45 GHz radiofrequency radiation: A comparison of E and H orientation, *Radiat. Environ. Biophys.* 28:235-246 (1989).

58. J.R. Jauchem, M.R. Frei, and J.M. Padilla, Thermal and physiologic responses to 1200-MHz radiofrequency radiation: Differences between exposure in E and H orientation, *Proc. Soc. Exp. Biol. Med.* 194:358-363 (1990).

59. M.R. Frei, J.R. Jauchem and F. Heinmets, Thermoregulatory responses of rats exposed to 9.3 GHz radiofrequency radiation, *Radiat. Environ. Biophys.* 28:67-77 (1989).

60. S-T, Lu, N. Lebda, S-J Lu, S. Pettit and S.M. Michaelson, Effects of microwaves on three different strains of rats, *Rad. Research* 110:173-191 (1987).

61. J.R. Jauchem, Effects of drugs on thermal responses to microwaves, *Gen. Pharmacol.* 16:307-310 (1985).

62. J.R. Jauchem, M.R. Frei, and F. Heinmets, Effects of psychotropic drugs on thermal responses to radiofrequency radiation, *Aviat. Space Environ. Med.* 56:1183-1188 (1985).

63. J.R. Jauchem, M.R. Frei, and F. Heinmets, Thermal responses to 5.6-GHz radiofrequency radiation in anesthetized rats: effect of chlorpromazine, *Physiol. Chem. Phys. Med. NMR* 20:135-143 (1988).

64. C.J. Gordon, Normalizing the thermal effects of radiofrequency radiation: body mass versus total body surface area, *Bioelectromagnetics* 8:111-118 (1987).

65. ACGIH, "Threshold Limits and Biological Exposure Indices for 1988-1989." American Conference of Governmental Industrial Hygienists, Cincinnati, OH, pp 100-103 (1988).

66. ANSI C95.1-1982, "American National Standard Safety Levels with Respect to Human Exposure to Radio Frequency Electromagnetic Fields, 300 kHz to 100 GHz," Institute of Electrical and Electronics Engineers, New York (1982).

67. IEEE C95.1-1991, "IEEE Standard for Safety Levels with Respect to Human Exposure to Radio Frequency Electromagnetic Fields, 3 kHz to 300 GHz," Institute of Electrical and Electronics Engineers, New York (1992).

68. NCRP, "Biological Effects and Exposure Criteria for Radio-Frequency Electromagnetic Fields," Publ. No. 86, National Council on Radiation Protection and Measurements, Washington, D.C (1986).

69. M. Nielsen, Die regulation der Korper temperatur bei Muskelarbeit, *Skand. Arch. Physiol* 79:193-230 (1938).

70. B. Saltin and L. Hermanson, Esophageal, rectal and muscle temperature during exercise, *J. Appl. Physiol.* 21:1757-1762 (1966).

71. B. Nielsen and M. Nielsen, Influence of passive and active heating on the temperature regulation of man, *Acta Physiol. Scand* 64:323-331 (1965).

72. J.T. Stitt, Fever versus hyperthermia, *Fed. Proc.* 38:39-43 (1979).

73. ASHRAE Handbook, Chapter 8: Physiological principles for comfort and health, *in:* "1985 Fundamentals," American Society of Heating, Refrigerating and Air-Conditioning Engineers, Atlanta (1985).

74. E.R. Nadel, A brief overview... *in:* "Problems with Temperature Regulation During Exercise," E.R. Nadel, ed., Academic Press: New York, pp 1-10 (1977).

75. L.H. Newburgh, "The Physiology of Heat Regulation and the Science of Clothing," W.B. Saunders: Philadelphia (1949).

76. L.B. Rowell, Human cardiovascular adjustments to exercise and thermal stress, *Physiol. Rev* 54:75-159 (1974).

77. R.F. Goldman, Acclimation to heat and suggestions by inference for microwave radiation, *in:* "Microwaves and Thermoregulation," E. R. Adair, ed., Academic Press, New York, pp 275-282 (1983).

78. C.H. Wyndham, N.B. Strydom, A.J. van Rensburg, A.J.S. Benade and A.J. Heyns, Relation between VO2 max and body temperature in hot humid air conditions, *J. Appl. Physiol* 29:45-50 (1970).

79. J.W. Mitchell, Energy exchanges during exercise, *in:* "Problems with Temperature Regulation During Exercise," E.R. Nadel, ed., Academic Press: New York, pp 11-26 (1977).

80. NIOSH, "Criteria for a Recommended Standard...Occupational Exposure to Hot Environments. Revised Criteria 1986," DHHS (NIOSH) Publication No. 86-113 (1986).

81. R.A. Tell and F. Harlen, A review of selected biological effects and dosimetric data useful for development of radiofrequency safety standards for human exposure, *J. Microwave Power* 14:405-424 (1979).

82. R.A. Tell and E.D. Mantiply, Population exposure to VHF and UHF broadcast radiation in the United States, *Proc. IEEE* 68:6-12 (1980).

83. R.L. Hesslink, S. Pepper, R.G. Olsen, S.B. Lewis and L.D. Homer, Radiofrequency (13.56 MHz) energy enhances recovery from mild hypothermia, *J. Appl. Physiol* 67:1208-1212 (1989).

84. R.G. Olsen, RF energy for warming divers' hands and feet, *in:* "Emerging Electromagnetic Medicine," M.E. O'Connor, R.H.C. Bentall, and J.C. Monahan, eds., Springer-Verlag: New York, pp 135-144 (1990).

85. J.R. Lloyd and R.G. Olsen, Radiofrequency energy for rewarming of cold extremities, *Undersea Biomed. Res* 19(3):199-207 (1992).

86. Y.-H. Liu, X.-M. Li, R.-P. Zou and F.-B. Li, Biopsies of human testes receiving multiple microwave irradiation: an histological and ultramicroscopical study, *J. Bioelectricity* 10:213-230 (1991).

87. F.G. Shellock and J.V. Crues, Corneal temperature changes induced by high field-strength MR imaging with a head coil, *Radiology* 167:809-811 (1988).

88. F.G. Shellock and J.V. Crues, Temperature changes caused by MR imaging of the brain with a head coil, *Am. J. Neuroradiol.* 9:287-291 (1988).

89. F.G. Shellock, D.J. Schaefer and J.V. Crues, Alterations in body and skin temperatures caused by magnetic resonance imaging: is the recommended exposure for radiofrequency radiation too conservative? *Brit. J. Radiol* 62:904-909 (1989).

90. F.G. Shellock, B. Rothman and D. Sarti, Heating of the scrotum by high-field-strength MR imaging, *Am. J. Roentgenology* 154:1229-1232 (1990).

91. F.G. Shellock, Thermal responses in human subjects exposed to magnetic resonance imaging, *Ann. N.Y. Acad. Sci.* 649:260-272 (1992).

92. F.G. Shellock, D.J. Schaefe, and J.V. Crues, Exposure to a 1.5 T static magnetic field does not alter body and skin temperatures in man, *Mag. Res. Med.* 11:371-375 (1989).

93. C.J. Gordon, Local and global thermoregulatory responses to MRI fields, *Ann. N.Y. Acad. Sci.* 649:273-284 (1992).

94. B.J. Barber, D.J. Schaefer, C.J. Gordon, D.C. Zawieja and J. Hecker, Thermal effects of MR imaging: worst-case studies on sheep, *Am. J. Roentgenology* 155:1105-1110 (1990).

95. E.H. Wissler, Mathematical simulation of human thermal behavior using whole-body models, *in:* "Environmental Ergonomics: Sustaining Human Performance in Harsh Environments," I.B. Mekjavic, E.B. Banister, J.B. Morison, eds., Taylor and Frances: Philadelphia, pp 325-373 (1988).

96. J.A.J. Stolwijk and J.D. Hardy, Temperature regulation in man-a theoretical study, *Pflugers Archiv.* 291:129-162 (1966).

97. J.A.J. Stolwijk, A Mathematical Model of Ptemperature Regulation in Man, Report No. NASA CR-1855, Washington, D.C (1971).

98. J.A.J. Stolwijk and J.D. Hardy, Control of body temperature, *in:* "Handbook of Physiology. Section 9. Reactions to Environmental Agents," D.H.K. Lee, ed., Am. Physiol. Soc: Bethesda, MD pp 45-68 (1977).

99. R.J. Spiegel, D.M. Deffenbaugh and J.E. Mann, A thermal model of the human body exposed to an electromagnetic field, *Bioelectromagnetics* 1:253-270 (1980).

100. M.J. Hagmann, O.P. Gandhi and C.H. Durney, Numerical calculation of electromagnetic energy deposition for a realistic model of man, *IEEE Trans. Microwave Theory Tech* 27:804-809 (1979).

101. D.J. Cunningham, An evaluation of heat transfer through the skin in extremity, *in:* "Physiological and Behavioral Temperature Regulation," J. D. Hardy, A. P. Gagge, and J. A. J. Stolwijk, eds., Charles C. Thomas, Springfield, IL., pp 302-315 (1970).

102. K.M. Sekins, "Microwave hyperthermia in human muscle: An experimental investigation of the temperature and blood flow fields occurring during 915 MHz diathermy," Ph.D. Dissertation, University of Washington, Seattle, WA (1981).

103. E.R. Adair and L.G. Berglund, Thermoregulatory consequences of cardiovascular impairment during NMR imaging in warm/humid environments, *Magn. Res. Imaging* 7:25-37 (1989).

104. D.J. Schaefer, Dosimetry and effects of MR exposure to RF and switched magnetic fields, *Ann. N.Y. Acad. Sci.* 649:225-236 (1992).

105. E.R. Adair and L.G. Berglund, Predicted thermophysiological responses of humans to MRI fields, *Ann. N.Y. Acad. Sci.* 649:188-200 (1992).

106. P.O. Fanger, "Thermal comfort," McGraw-Hill Book Co, New York (1973).

107. M. Cabanac, Thermoregulatory behavioral responses, *in:* "Microwaves and Thermo-regulation," E. R. Adair, ed., Academic Press, New York, pp 307-357 (1983).

108. M. Cabanac, D.J. Cunningham and J.A.J. Stolwijk, Thermoregulatory set point during exercise: a behavioral approach, *J. Comp. Physiol. Psychol* 76:94-102 (1971).

109. E.R. Adair, J.U. Casby and J.A.J. Stolwijk, Behavioral temperature regulation in the squirrel monkey: changes induced by shifts in hypothalamic temperature, *J. Comp. Physiol. Psychol* 72:17-27 (1970).

110. J.A. D'Andrea, O. Cuellar, O.P. Gandhi, J.L. Lords and H.C. Nielsen, Behavioral thermoregulation in the Whiptail Lizard (Cnemidorphorus tigris) under 2450 MHz CW Microwaves, *in:* "Biological Effects of electromagnetic waves," Abstracts URSI General Assembly, Helsinki, Finland, p 88 (1978).

111. S. Stern, L. Margolin, B. Weiss, S.-T. Lu and S. Michaelson, Microwaves: effect on thermoregulatory behavior in rats, *Science* 206:1198-1201 (1979).

112. E.R. Adair and B.W. Adams, Microwaves modify thermoregulatory behavior in squirrel monkey, *Bioelectromagnetics* 1:1-20 (1980).

113. E.R. Adair and B.W. Adams, Behavioral thermoregulation in the squirrel monkey: Adaptation processes during prolonged microwave exposure, *Behav. Neurosci* 97:49-61 (1983).

114. M.J. Marr, J.O. de Lorge, R.G. Olsen and M. Stanford, Microwaves as reinforcing events in a cold environment, *in:* "Electromagnetic Fields and Neurobehavioral Function," O'Connor, M.E. and Lovely, R.H., eds., Alan R. Liss: New York, pp 219-234 (1988).

115. V. Bruce-Wolfe and E.R. Adair, Operant control of convective cooling and microwave irradiation by squirrel monkeys, *Bioelectromagnetics* 6:365-380 (1985).

116. E.R. Adair, B.W. Adams and G.M. Akel, Minimal changes in hypothalamic temperature accompany microwave-induced alteration of thermoregulatory behavior, *Bioelectromagnetics* 5:13-30 (1984).

117. E.R. Adair, Thermophysiological effects of electromagnetic radiation, *in:* "Biological Effects and Medical Applications of Electromagnetic Energy," O.P. Gandhi, ed., Prentice Hall: Englewood Cliffs, NJ, pp 256-276 (1990).

118. E.R. Adair and L.G. Berglund, Comfort heating with microwaves: an idea whose time may have come, *in:* "CLIMA 2000," Vol. 4, P.O. Fanger, ed., VVS Kongres/VVS Messe, Copenhagen, pp 115-120 (1985).

# EFFECTS OF MICROWAVE RADIATION EXPOSURE ON BEHAVIORAL PERFORMANCE IN NONHUMAN PRIMATES

John A. D'Andrea*

Aviation Performance Division
Naval Aerospace Medical Research Laboratory
51 Hovey Road
Pensacola, FL 32508-1046

## INTRODUCTION

Research conducted during the past three decades has shown that exposure of laboratory animals to radiofrequency radiation can cause a variety of behavioral changes. These changes range from subtle effects such as perception of microwave-induced sound to complete cessation of behavioral performance due to severe hyperthermia. A central theme of this research has been to determine a relationship between specific absorption rate (SAR) and adverse consequences of exposure to microwave radiation. Studies evaluating microwave exposure on the performance of well-learned operant tasks have been the primary avenue for determining this relationship. This information provides a scientific data base from which safe exposure standards can be derived.

The nonhuman primate has served as an excellent model for this research and has advantages over other species phylogenetically more distant from man. One advantage, microwave radiation absorption characteristics that are similar to man, has often been overlooked. The whole-body and localized SAR for nonhuman primates closely resembles that of the human. Another advantage is the similarity of nonhuman primate sensory systems to that of man. The rhesus monkey (*Macaca mulatta*) visual system has nearly the same capabilities as that of man.[1] However, there are differences in auditory performance; rhesus monkeys can hear much higher sound frequencies than man.[2] A third advantage is the capability of nonhuman primates to acquire complex behavioral tasks that integrate sensory information with skilled motor tasks. Each of these advantages has been used to evaluate the effects of microwave exposure and perhaps predict human responses more accurately than with rodents.

---

* These views and opinions are those of the author and do not necessarily state or reflect those of the U.S. Government.

## MICROWAVE ENERGY ABSORPTION IN NONHUMAN PRIMATES

Microwaves occupy the portion of the electromagnetic spectrum between 100 MHz and 300 GHz. A monkey exposed in this frequency range will scatter and absorb energy depending on factors such as wavelength, body size, and body orientation in electric and magnetic field vectors.[3,4,5] Exposure to ionizing radiation would result in absorption that is directly related to the cross-sectional area of the organism.[6] Exposure to microwaves, however, results in absorption that is independent of cross-sectional area and results in resonant absorption. Several absorption curves based on analytical predictions for the E ‖ L orientation are shown in Figure 1.[7] These show a resonant peak for each species. Also, the smaller the animal, the higher the resonant frequency and the whole-body-averaged SAR produced for a given power density. As shown, the absorption profile for rhesus and squirrel monkeys more closely resembles that of man than does a rat or mouse. The primary difference between profiles for man and either monkey is a shifting of the resonant frequency to higher values for nonhuman primates (man 70 MHz; rhesus monkey 250 MHz; squirrel monkey 600 MHz). An additional factor that may be important, depending on level of exposure, is the distribution of energy absorption throughout the body.[8,9,10] The distribution is also controlled by the factors listed above such that local hotspots may be formed that are several times the whole-body average SAR. At high-power densities, these may not be ameliorated by bloodflow or thermoregulation.

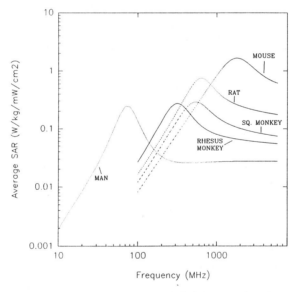

**Figure 1.** Analytical SAR profiles (E ‖ L) for several species.

Several experimental dosimetry studies have been conducted for human and nonhuman primate models to measure both whole-body as well as local SAR. Olsen and Griner[11,12,13] used a gradient-layer calorimeter for average SAR and nonperturbing temperature probes for local SAR and found good agreement with the analytical predictions[3] near resonance. A comparison of rhesus monkey and human models showed very similar absorption profiles (see Figure 2). At the higher frequency of 1.29 GHz, whole-body SAR was much higher than predicted from the analytical models. Absorption in the body showed hotspots that were twice the SAR of the whole-body average SAR. The higher absorption in the legs contributed to the overall increase in the whole-body SAR. Similar profiles have been determined for models of man (see Guy[14]).

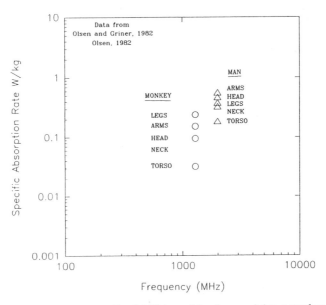

**Figure 2.** Comparison of local SAR in models of man and rhesus monkey.

## EFFECTS OF MICROWAVES ON NONHUMAN PRIMATE PERFORMANCE

Nonhuman primate behavioral studies have used the heating potential of microwaves to determine thresholds of behavioral effects. These studies have been useful in demonstrating microwave field characteristics that control the SAR, such as frequency-dependent absorption and field polarization, as well as the factors of animal shape and size. In conducting these studies, a simple test protocol has been followed to (1) establish a stable behavioral baseline of performance and then (2) determine the effects of microwaves on the baseline performance. The result of microwave exposure has generally been a body-temperature rise followed by a reduction in behavioral response rates. Stern[15] (1980) has suggested that the response reduction on a learned task during exposure may solely reflect the animals' attempts to engage in other thermoregulatory behaviors (i.e., escape) that are not compatible with behaviors such as lever pressing required by the behavioral task.

For nonhuman primate research, the studies conducted by de Lorge and coworkers are often cited in standard-setting documents.[16,17] De Lorge[18,19] summarized studies with squirrel monkeys[20] and rhesus monkeys[21] and arrived at several important conclusions. First, changes in trained behavioral performance of monkeys exposed to microwave radiation were always associated with concomitant core body temperature increases 1°C or more above the baseline body temperature. This threshold temperature increase required an SAR of near 4 W/kg. Second, the field power densities necessary to produce equal SAR in different-sized animals was confirmed. Rhesus monkeys required much higher field power densities to alter behavior than squirrel monkeys, and rats required the least. Third, de Lorge noted that with rhesus monkeys the power density required at frequencies well above resonance (5.6 GHz) was much greater than predicted from the analytical SAR absorption curves. The threshold whole-body SAR necessary to disrupt behavioral performance of an operant task is shown in Figure 3. The dotted line depicts the 4 W/kg whole-body SAR threshold often cited in safety standard documents.[16,17] As radiation frequency is increased, a somewhat higher SAR is required to disrupt behavior. Microwave frequencies where energy is absorbed primarily in the surface layers of an animal can produce different behavioral effects than deeper penetrating waves of the lower frequencies. Finally, SAR distributions in animals are not uniform, and these might interact with behavior to result in different SAR thresholds for behavioral change. For example, D'Andrea, Cobb, and Knepton[22] exposed rhesus monkeys to 1.3-GHz pulsed microwaves (see Figure 3). The exposures were conducted using an open-ended waveguide placed 7 cm behind the monkeys' head, which resulted predominately in a head-only exposure. The threshold SAR to alter a well-trained operant behavior was nearly twice that for the whole-body exposures used by de Lorge et al.[19]

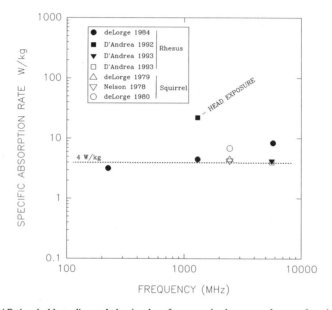

**Figure 3.** SAR thresholds to disrupt behavioral performance in rhesus monkeys and squirrel monkeys.

## HIGH-PEAK- POWER MICROWAVE PULSES

Pulsed microwaves can produce a sensation of hearing. Frey[23,24] described the effect, which required pulsed energy for its occurrence, as "clicking," "knocking," or "buzzing" behind the head. Lin[25] cites data showing that the energy density threshold for this effect in humans is 40 $\mu J$ $cm^2$ at 2450 MHz. Studies have verified the hearing sensation[26] and measured cochlear microphonics produced by pulsed microwaves.[27] Foster and Finch[28] suggested that the hearing effect was due to a thermoelastic expansion of tissue during absorption of a microwave pulse. During tissue expansion, a thermoacoustic pressure wave would be produced and propagate through soft tissues and bone, producing an acoustic sensation by mechanical stimulation of receptor cells located within the cochlea of the ear.

Does the auditory effect pose a human health hazard? Lin[25] points out that at threshold levels the auditory effect is unlikely to pose a human health hazard. However, pressure waves in the brain at significantly higher levels may be hazardous. Presently, there are insufficient data to answer this question. D'Andrea et al.[29,30] trained rhesus monkeys on a vigilance task similar to that used by de Lorge[21] and exposed them to high-peak-power microwave pulses at 2375 MHz produced by a virtual cathode oscillator. No significant changes in behavioral performance were found. Single pulses with short pulse durations (80-100 ns) resulted in very low SARs. The energy density per pulse of 640 to 800 $\mu J/cm^2$, however, was much higher than the threshold necessary for human perception. Presumably, pressure waves in the monkey brain were much higher than the human or monkey threshold, yet significant effects on monkey behavioral performance were not observed.

## CONCLUSIONS

Behavioral tests with nonhuman primates have been widely used to determine the effects of several microwave factors that control the SAR and have established a threshold for the disruption of behavioral performance to be near 4 W/kg. One tenth this value has been used as a maximum permissible safe exposure (0.4 W/kg) by several standard setting organizations.[16,17] We have a good understanding of microwave factors that control whole-body SAR, and several behavioral tests have confirmed this. However, the effects of realistic exposure situations on behavior are poorly understood. Hotspots vary dramatically with microwave frequency, animal's orientation in fields, and animal size and shape. Most of the studies cited in this paper were conducted under far-field, plane-wave exposure conditions. Human exposure situations, however, generally involve grounding factors, the presence of reflecting surfaces, and even multiple sources of irradiation.[31,32] Behavioral tests can continue to be an effective tool in corroborating analytical dosimetry predictions, and when several different animal species and sizes are studied, will provide a good data base from which to derive safe levels of exposure.

Chronic exposure to microwave radiation at low-field-power densities and whole-body SARs has not been investigated with nonhuman primates. During chronic low-level exposure, temperature rise is not measurable, as the absorbed energy should be easily dissipated by the thermoregulatory capabilities of a monkey. Behavioral changes in chronically exposed rodents have occurred at SARs of 0.40-3.6 W/kg and depended on microwave radiation frequency (see D'Andrea[33]). In all rodent studies, the effects have returned to normal on termination of microwave exposure. Mechanisms responsible for the reported behavioral effects do not seem as clear as with short-term exposures. Do rodents acclimate to microwave exposure, and does acclimatization account for the changes in behavior?[34] Chronic exposure studies of monkeys to low-level microwaves are required to answer this question.

In conclusion, thermal effects may account for all of the reported behavioral effects of microwave radiation absorption in monkeys. The reported changes in behavioral performance involve heating of tissues resulting in mild heat stress. Additional research is needed to assess the effects of more realistic exposure situations and of chronic low-level exposures.

## DISCLAIMER

The views expressed in this article are those of the author and do not reflect the official policy or position of the Department of the Navy, Department of Defense, nor the U.S. Government. Trade names of materials and/or products of commercial or nongovernment organizations are cited as needed for precision. These citations do not constitute official endorsement or approval of the use of such commercial materials or products. The animals used in this work were handled in accordance with the principles outlined in the *Guide for the Care and Use of Laboratory Animals*, prepared by the Committee on Care and Use of Laboratory Animals of the Institute of Laboratory Animal Resources, National Research

Council, DHHS, NIH Publication No. 85-23, 1985; and The Animal Welfare Act of 1966, as amended 1970 and 1976. This research was sponsored by the Naval Medical Research and Development Command under work unit 62758N MM 58524.002-0010.

## REFERENCES

1. R.S. Harwerth and E.L. Smith, Rhesus monkey as a model for normal vision of humans, *American Journal of Optometry and Physiological Optics*, 62:9:633-641 (1985).
2. G. Gourevitch, Detectability of tones in quiet and in noise by rats and monkeys, *in:* "Animal Psychophysics," W.C. Stebbins, ed, Plenum Press, New York (1970).
3. C.H. Durney, C.C. Johnson, P.W. Barber, H. Massoudi, M.F. Iskander, J.L. Lords, D.K. Ryser, S.J. Allen, and J.C. Mitchell, "Radiofrequency Radiation Dosimetry Handbook," SAM-TR-78-22, Brooks Air Force Base, San Antonio, TX (1978).
4. O.P. Gandhi, Polarization and frequency effects on whole animal energy absorption of RF energy, *Proc. IEEE* 62: 1171-1175 (1974).
5. O.P. Gandhi, Electromagnetic energy absorption in humans and animals, *in:* "Biological Effects and Medical Applications of Electromagnetic Energy," O.P. Gandhi, ed., Englewood Cliffs, New Jersey: Prentice Hall: 174-195 (1990).
6. National Council on Radiation Protection and Measurement, "Recommendations of Limits for Exposure to Ionizing Radiation," Bethesda, MD: NCRP Report No. 91 (1987).
7. D.J. Hatcher and J.A. D'Andrea, A computer program to calculate planewave average specific absorption rate in a prolate spheroidal model, NAMRL Technical Memorandum 92-3, Naval Aerospace Medical Research Laboratory, Pensacola, FL, (1992).
8. J.A.D'Andrea, R.Y. Emmerson, C.M., Bailey, R.G. Olsen, and O.P. Gandhi, Microwave radiation absorption in the rat: Frequency-dependent SAR distribution in body and tail. *Bioelectromagnetics* 6: 199-206 (1985).
9. J.A. D'Andrea, R.Y. Emmerson, J.R. DeWitt, and O.P. Gandhi, Absorption of microwave radiation by the anesthetized rat: Electromagnetic and thermal hotspots in body and tail. *Bioelectromagnetics* 8:385-396 (1987).
10. J.A. D'Andrea, J.R. DeWitt, L.M. Portuguez, and O.P. Gandhi, Reduced exposure to microwave radiation by rats: Frequency specific effects, *in* "Electromagnetic Fields and Neurobehavioral Function," M.E. O'Connor and R.H. Lovely, eds., New York: Alan R. Liss, Inc., 289-308 (1988).
11. R.G. Olsen and T.A. Griner, Partial-body absorption resonances in a sitting rhesus model at 1.29 GHz, *Rad Environm Biophys* 21:33-43 (1982).
12. R.G. Olsen and T.A. Griner, Electromagnetic dosimetry in a sitting rhesus model at 225 MHz, *Bioelectromagnetics* 3:385-389 (1982).
13. R.G. Olsen and T.A. Griner, Specific absorption rate in models of man and monkey at 225 and 2,000 MHz, *Bioelectromagnetics* 8:377-384 (1987).
14. A.W. Guy, Dosimetry associated with exposure to non-ionizing radiation: Very low frequency to microwaves, *Health Physics*, 53:569-584 (December 1987).
15. S.L. Stern, Behavioral effects of microwaves, *Neurobehavioral Toxicology* 2:49-58 (1980).
16. Institute of Electrical and Electronics Engineers: "IEEE Standard for Safety Levels with Respect to Human Exposure to Radio Frequency Electromagnetic Fields, 3 kHz to 300 GHz," New York, NY, IEEE C95.1-1991 (1991).
17. American National Standards Institute, "American National Standard safety levels with respect to human exposure to radio frequency electromagnetic fields, 300 kHz to 100 GHz," New York, NY: The Institute of Electrical and Electronic Engineers, Inc: ANSI Report C95.1-1982 (1982).
18. J.O. de Lorge, The thermal basis for disruption of operant behavior by microwaves in three animal species, *in* "Microwaves and Thermoregulation," Adair, E.R., ed., New York: Academic Press, 379-400 (1983).
19. J.O. de Lorge, Operant behavior and colonic temperature of *Macaca mulatta* exposed to radio frequency fields at and above resonant frequencies, *Bioelectromagnetics* 5:233-246 (1984).
20. J.O. de Lorge, Operant behavior and colonic temperature of squirrel monkeys during microwave irradiation, *Radio Sci.* 14:217-225 (1979).
21. J.O. de Lorge, "Behavior and temperature in rhesus monkeys exposed to low level microwave irradiation," Report NAMRL-1222; Pensacola, FL: Naval Aerospace Medical Research Laboratory: (AD A021769) (1976).

22. J.A. D'Andrea, B.L. Cobb, and J. Knepton, "Behavioral effects of high peak power microwave pulses: Head exposure at 1.3 GHz." Report NAMRL-1372; Pensacola, FL:Naval Aerospace Medical Research Laboratory, NAMRL - 1372 (1992).

23. A.H. Frey, Auditory system response to RF energy, *Aerospace Med.* 32:1140-1142 (1961).

24. A.H.. Frey, Auditory system response to modulated electromagnetic energy, *J. Appl. Physiol.* 17:689-692 (1962).

25. J.C. Lin, Auditory perception of pulsed microwave radiation, *in* "Biological Effects and Medical Applications of Electromagnetic Energy," O.P. Gandhi, ed.,Englewood Cliffs, New Jersey: Prentice Hall: 277-318 (1990).

26. J.C. Sharp, H.M. Grove and O.P. Gandhi, Generation of acoustic signals by pulsed microwave energy, *IEEE Transactions on Microwave Theory and Techniques* 22: 583-584 (1974).

27. C.K. Chou, A.W. Guy and R. Galambos, Characteristics of microwave-induced cochlear microphonics, *Radio Sci.* 12: 221-227 (1977).

28. K.R. Foster and E.D. Finch, Microwave hearing: Evidence for thermoacoustic auditory stimulation by pulsed microwaves, *Science* 185: 256-258 (1974).

29. J.A. D'Andrea, J. Knepton, B.L. Cobb, B.J. Klauenberg, J.H. Merritt, and D.N. Erwin, "High peak power microwave pulses at 2.37 GHz No effect on vigilance performance in monkeys," Joint Naval Aerospace Medical Research Laboratory Research Report, NAMRL-1348 and USAF School of Aerospace Medicine, USAFSAM-TR-89-21 (1989).

30. J.A. D'Andrea, J. Knepton, B.L. Cobb, B.J. Klauenberg, R.N. Shull, J.H. Merritt, and D.N. Erwin, "No effect of high peak power microwave pulses at 2.36 GHz on behavioral performance in monkeys," Joint Naval Aerospace Medical Research Laboratory Research Report, NAMRL-1358 and USAF School of Aerospace Medicine, USAFSAM-TR-90-14 (1990).

31. O.P. Gandhi, M.J. Hagmann, and J.A. D'Andrea, Partbody and multibody effects on absorption of radio frequency electromagnetic energy by animals and by models of man, *Radio Sci.* 14(6S):15-22 (1979).

32. O.P. Gandhi, E.L. Hunt and, and J.A. D'Andrea, Deposition of electromagnetic energy in animals and in models of man with and without grounding and reflector effects, *Radio Sci.* 12(6S):39-48 (1977).

33. J.A. D'Andrea, Microwave radiation absorption: Behavioral effects, *Health Physics*, 61:1:29-40 (July 1991).

34. R.H. Lovely, S.J.Y. Mizumori, R.B. Johnson and A.W. Guy, Subtle consequences of exposure to weak microwave fields: Are there nonthermal effects? *In* "Microwaves and Thermoregulation," E.R. Adair, ed., New York, NY: Academic Press, 401-429 (1983).

# EFFECTS OF RFR/EMF EXPOSURE ON PINEAL/MELATONIN

Peter Semm and T. Schneider

Zool. Institut
AG Magnetoneurobiologie
Universität Frankfort
Siesmayerstr. 70
6000 Frankfort 11, Germany

## ABSTRACT

It was demonstrated that the pineal organ (Epiphysis cerebri), an endocrine gland located in the epithalamus of vertebrates, is capable of responding to magnetic stimuli. After the discovery that the spontaneous electrical activity of single pinealocytes of birds and mammals could be altered by magnetic stimuli, it was shown that also the nocturnal synthesis of the pineal hormone N- acetyl-5-methoxytryptamine (Melatonin) was inhibited by a change of the ambient magnetic field. Besides the physiological responses to artificial magnetic fields, there are several indications that the orientation behavior of homing pigeons and migrating birds is influenced by the earth's magnetic field. It is believed that the photoreceptors of the retinae and the avian pineal gland may serve as magnetoreceptors. Moreover it was shown that regular daily and annual variations of the geomagnetic field, caused by solar winds acting on the magnetosphere of the earth`s sunlit side, could possibly play a role on circadian and circannual rhythms. Clinical relevant results coupled with artificial magnetic influences and possible therapeutic magnetic treatment are discussed.

## INTRODUCTION

More than a decade ago Semm and colleagues described for the first time the sensitivity of the mammalian pineal gland to earth-strength magnetic stimulation. They observed that after the inversion of the vertical magnetic field, the spontaneous discharge rate from single pinealocytes was inhibited.[1] However the reason for the electrophysiological investigations of the pineal was to study the physiological basis of the magnetic compass in migrating and homing animals, especially birds. After behavioral experiments with homing pigeons and migrating birds indicated that the pineal is not directly involved in magnetic compass functions,[2,3] the magnetic sensitivity of the gland was used to find other nerve cells that might be involved in magnetic orientation.[4,5,6] In addition to the electrophysiological results, it was observed in 1983 that the nocturnal synthesis of the pineal hormone N- acetyl-5-methoxytryptamine (Melatonin) was influenced by a change of the ambient magnetic field.[7] Moreover Wilson[8] could demonstrate that the melatonin synthesis was altered by very strong AC-electrical fields (60-Hz) in the range between 2.0 and 130 kV/m.[8] Although such fields are much too strong to fit in any "biological window" and are a sledgehammer for any sense organ, it may be a task to find out whether magnetic and electric stimuli are different or somehow related with respect to the underlying sensory mechanism.

**Characteristics of the Earth's Magnetic Field**

Before a description of the influences of earth-strength magnetic fields (MFs) on pineal physiology of mammals and birds is given, it is necessary to identify those characteristics of the geomagnetic field that may be important for animals. The earth's magnetic field is a physical field of force, which is extraordinarily reliable. Organisms that are capable of perceiving its density, polarity, direction and the temporal variations of these parameters, could determine not only their geographical location and the azimuth of any direction on the earth's surface, but they could also obtain circadian, and circannual information for synchronizing their biological rhythms. However, which of these possibilities are actually realized should be discussed later. The earth's magnetic field is a dipole whose poles are near the geographic axis and below the earth's surface. The static component of the geomagnetic field can be viewed as a giant bar magnet. The field lines go upward from the horizontal in the southern hemisphere and downward in the northern hemisphere. The equator of the earth's magnetic field runs slightly north of the geographic equator in Africa and south of it in South America. At the magnetic equator, the field lines are tangential to the earth's surface. As the geographic and magnetic poles do not coincide perfectly and because of anomalies in the earth's crust, magnetic north normally deviates from the geographic north. This deviation, called the declination, is usually less than 20% in most parts of the world, although more extreme values are found at high latitudes. The angle between the vector of the geomagnetic field and the horizon is called the inclination of dip, which is considered positive in sign when oriented downwards. The lines of equal inclination form a system of roughly parallel lines comparable to lines of latitude. The total intensity of the geomagnetic field declines from an average maximum of 60,000 nT (O.6 gauss) near the poles to minimum values of 30,000 nT (0.3 gauss) at the magnetic equator. The field is not completely regular over geographical distances, however, because of local lithographic anomalies. Temporal changes of the field occur with periods ranging from days to centuries, and therefore they are not significant in the life of an individual animal. Regular daily variations are caused by solar winds acting on the magnetosphere of the earth's sunlit side. In the northern temperate latitudes, the field intensity decreases until local noon by 30 to 100 nT, then increases again. Similar daily fluctuations in declination and inclination occur as well.

**Generation of Artificial Magnetic Fields**

Earth-strength magnetic fields could be produced by coils or by moving bar magnets. If the coils are arranged parallel with a radiuslike distance, they are called Helmholtz coils. They have the advantage of producing homogeneous magnetic fields in the center, and both intensity and direction changes can be applied. However, it is technically not easy to generate intensity and direction changes separately, because by changing for example the direction of one of the components, the intensity also changes (according to the mathemathical rules of vectors). When the coils are controlled by a computer, the intensity can be held constant and the orientation can be changed; when the earth's field is compensated by a second layer of windings, the orientation of the field can be held constant and the intensity can be changed.

## Physiological Responses to Earth Strength Magnetic Fields

The results from several experiments indicated that the geomagnetic field could influence the behavior and orientation of a variety of organisms. The behavioral responses could be categorized in two types: (1) magnetic compass orientation, or (2) some form of physiological sensitivity to weak fluctuations in natural magnetic intensity. The characterization of the avian magnetic compass, its dependence on a rather narrow intensity range centered around the actual intensity of the natural field, its plasticity, and its working as an inclination compass are well documented[9,10] (for review see Wiltschko and Wiltschko.[11] On the other side there are several time-dependent correlations of pathological phenomena and the occurence of geomagnetic activity (for review see Persinger[12]). The results from the above-mentioned experiments imply the existence of some sort of sensory receptors for sensing and transducing the magnetic field to the central nervous system. Theoretically there is no requirement for a special location of a magnetoreceptor because the magnetic field easily penetrates biological tissues. Behavioral results with homing pigeons however, indicate a magnetic-sensitive system in the head, because the initial orientation of individual birds can be altered by changes in the direction of the magnetic fields around their heads.[13]

The pineal gland might be part of the magnetic system because:
1. It is involved in the regulation of circadian rhythms and thus essential for migratory restlessness (Zugunruhe).
2. Magnetic orientation of migratory birds at Zugunruhe can be altered by an artificial magnetic field.
3. Circadian rhythms of hopping activity can be inhibited from phase shifting by compensation of the earth's magnetic field and can be influenced by an artificial magnetic field.
4. The pineal organ, especially the mammalian pineal gland, is strongly dependent on its sympathetic innervation, and the sympatho-adrenergic system as a whole is sensitive to magnetic stimuli.
5. The pineal organ is a light sensitive, time-keeping organ and could form part of a combined compass-solar-clock system, which has been postulated for maintaining orientation in birds.

Hence, the effects of a magnetic field on electrophysiological activity of the guinea pig pineal organ, which is a useful system for such studies on individual cells, was investigated. After this research the magnetic influence on the pigeon pineal was recorded.

## MATERIALS AND METHODS

The animals were narcotized and mounted in a non-magnetic stereotaxic frame and the brain exposed by boring a hole in the skull to accommodate the recording electrode. The electrode was mounted on a micromanipulator attached to the frame and stereotaxically inserted into the pineal gland. Individual spontaneously active units were located by advancing the micropipette in nm-steps using a motorized nanostepper. During recording times and magnetic stimulation the steppermotor was switched off. The recorded signals were amplified and fed to a window discriminator, an audio monitor and a storage oscilloscope, and stored on audiotape for later analysis. Three pairs of Helmholtz coils were used to alter the direction of the natural magnetic field. Two pairs of coils were arranged to exert an effect on the horizontal component of the magnetic field, one pair oriented exactly north-south and the other east-west; while the third pair affected the vertical component. The stereotaxic frame in which the animal was held was placed on a plastic tilt-table at the center of the coils. The head of the animal pointing to the magnetic northpole. In this area in the center of the globus, the artificial magnetic field is nearly homogeneous. The artificial magnetic field was controlled by the computer that was interfaced to a power supply with three outputs. The coil system was steered to invert the vertical component of the magnetic field in a gradual manner with the same intensity as in the natural field. Inversion of the vertical component moved the needle of a perpendicularly oriented inclinatorium from indicating the normal direction of inclinometer (66°), until it pointed in the opposite direction (minus 66°). The inversion was completed in 90 sec. and was followed by an immediate restoration of the natural conditions. The natural magnetic field (field strength in the experimental room = 46,000 nT) and the artificial changes were continuously monitored by a magnetometer. The intensity of the natural magnetic field within the coil system was measured just before experimentation, and the value manually entered into the computer to provide a basis for the generation of the intensity level of the artificial fields. For the biochemical studies a larger coil system was used (3.0 meter in diameter) in order to enlarge the homogeneous field part, where the animals were placed and could move freely in a wooden cage. For details of the biochemistry see Welker et al.[7]

## RESULTS

### Mammals

It could be demonstrated that there was a reproducible correlation between the onset of an artificial magnetic field stimulus and the change of the spontaneous electrical activity in individual pineal cells. Inversion of the vertical component resulted in a significant depression of single-unit activity, sometimes with a latency of minutes. Spontaneous activity returned to the outgoing level when the normal magnetic field was restored. A similar study in the rat[14] revealed that changes in the horizontal component of the earth's magnetic field had significant effects on pineal electrical activity. Some cells became activated by the magnetic field stimuli while other cells became quiescent, and most cells did not respond. For those cells that did respond the latencies in the rat were much shorter than in the guinea pig (in the range of msec.). Because the mammalian pineal message is believed to be mainly conveyed by its hormone melatonin, it was concluded that pineal melatonin synthesis also may be affected by magnetic field stimulation. A biochemical study demonstrated that magnetic field stimuli could inhibit nocturnal serotonin-N-acetyltransferase activity (NAT) and decrease melatonin content in the rat pineal.[7] An inversion of the horizontal component, resulting in a field with nearly equal intensity but different polarity, for as little as 15 minutes, had significant effects on the parameters in unrestrained animals. Small changes (approx. 5°) of the MF inclination, that is a minor change in intensity (approx. 1000 nT) and orientation, also inhibited melatonin synthesis. Additionally, it was determined that similar alterations of pineal melatonin synthesis could be obtained by MF stimuli of various orientations. Rather unexpectedly, it was also shown that hours after the onset of a continuous artificial MF pineal melatonin synthesis returned to control levels, but if the artificial MF was stopped after 24 hr, pineal function thereafter was inhibited. It would seem, therefore, that the change itself was a key aspect of the MF stimuli that the animal perceives. It is not yet totally clear, however, whether the important

change was in the MF direction or the total intensity. Confirmation of the inhibitory effect of a change in the orientation of the horizontal component was shown later (for review see Ref. 15). Another group demonstrated that the nocturnal pineal serotonin metabolism in mice and rats exposed to weak intermittent magnetic fields exhibits a clear increase.[16,17] The activity of the pineal enzyme serotonin-N-acetyltransferase (NAT) was also inhibited but not pineal and serum melatonin levels. The authors discussed that this latter result might be a consequence of the short, i.e., 1-hour exposure interval. In another study neuronal activity was measured in rats by C14 - deoxyglucose (C14 -DG) autoradiography in the presence and absence of an earth-strength magnetic field.[18] Parallel decreases were found in the C14 -DG uptake and NAT activity in the pineal glands of magnetically stimulated rats. It was suggested that low intensity magnetic fields have a selective action on the activity of the pineal gland that can indirectly alter the function of brain areas, for example the hippocampus that can bind melatonin. Indirect evidence for this hypothesis is presented by an intracellular *in-vitro* study on the action of melatonin on guinea pig hippocampal cells.[19] The following effects could be observed:
- the response to repetitive synaptic stimulation was changed drastically,
- the membrane potential was hyperpolarized,
- the duration of action potential was strongly increased,
- the threshold for triggering of action potentials was shifted to more positive levels and repetitive spiking elicited by the application of a convulsive drug, bicuculline, was reversibly abolished.

The effects of melatonin all had in common that cell excitability was lowered. It was concluded that melatonin might depress epileptic seizure activity and that magnetic influence might result in lowering of hippocampal levels of melatonin, thereby increasing seizure activity. A positive correlation between the occurrence of grand mal epileptic fits and magnetic storms has been found, indicating that there is indeed a lowering of convulsive threshold during natural magnetic field changes.[20] It may be mentioned that in case certain magnetic fields might have the capability of augmenting melatonin synthesis during daytime, the consequences for the circadian system and all relevant physiological processes will be possibly more dramatic than the nocturnal inhibition of the hormone production.

So far no melatonin responses to radiofrequencies could be observed. We tested a 150 MHz signal modulated by 8 Hz in rats and got no influence on melatonin synthesis (Reuss, Semm and Kullnick, unpublished).

**Humans**

In order to test if human melatonin synthesis can also be influenced by magnetic stimulation, 14 volunteers were placed in the center of a Helmholtz coil system, 3.0 m in diameter. Controls were sitting at the same time in a different room, far enough away not to be influenced by the magnetic fields. At 23.30 hours the first blood samples were collected. At midnight the horizontal component of the ambient MF was inverted for 30 min, without the knowledge of the experimentals. Blood samples were collected immediately afterward and melatonin content was measured according to the method of Deguchi and Axelrod.[21] 70% of the experimentals responded to MF stimulation with a significant (70%) depression of blood melatonin concentration (Semm, Welker and Bartsch, unpublished observations).

**Birds**

After the electrophysiological demonstration of magnetic sensitive pineal cells in the intact pigeon[22] (also Schneider, unpublished), it was also demonstrated that pineal cell electrical activity in blinded pigeons can be influenced by inversion of the horizontal or vertical component, even when the organ was deafferentiated with respect to the central and sympathetic innervation. This indicated that the avian pineal may be an independent magnetic sensor.[23] In order to test if other brain areas that can bind melatonin are influenced by MF via the pineal and its hormonal output, the effects of microiontophoretically applied melatonin on the electrical activity of Purkinje cells were studied in the cerebellum of pigeons. The proportion of excitatory and inhibitory responses to the pineal indolamine varied significantly depending on whether the cells were tested during the night or day. This day/night rhythm of responses was abolished if the birds were pinealectomized[24] or alternatively exposed to complete inversion of the vertical component for 1 hr at 21.00 hr.[25] This magnetic effect could be observed only in the presence of dim red light, which itself does not suppress melatonin production. These results agree with the concept that pineal involvement in endogeneous central nervous rhythms is influenced by changes in the earth`s magnetic field.

In order to test if other brain areas are also involved in the elaboration of magnetic information, a study using the C14 - DG method was conducted to reveal which parts of the pigeon brain are involved in the elaboration of magnetic information.[26] The magnetic stimulus was the inversion of the horizontal component and experiments were performed during the day and the night. The controls were treated in the same way but without magnetic stimulation. It could be shown that in all experimental animals the pineal gland was particularly intensively labeled, regardless whether stimulation occurred during the day or at night. The fact that the gland is also labeled during daytime, when its secretory activity is low, presents additional evidence for specific magnetic sensitivity. Besides the pineal, main parts of the visual and the vestibular system were labeled. In another study, NAT activity and melatonin content were measured in intact and blinded pigeons.[27] In intact animals, MF exposure resulted in a 60% reduction of NAT activity compared with controls. In blinded pigeons, which still possess intact photoreceptors in the pineal, NAT activity decreased by 19% without reaching statistical significance. It was concluded cautiously that the partly degenerated avian pineal photoreceptors may be still be sufficiently organized to respond to MF stimulation. Furthermore, it was demonstrated that the melatonin-synthesizing enzyme HIOMT (Hydroxyorthomethyltransferase) of the quail`s pineal gland and retina, in vitro, is influenced by different variations in the natural magnetic field.[28,29] Enzyme activity was inhibited by a decreased magnetic field of 50% of its original value. It can be concluded that the magnetoreceptors involved may be the photoreceptors because 1) the only receptive structures in the pineal gland under those in-vitro conditions are the photoreceptors, and 2) the retina itself is able of responding to magnetic stimulation. It can be concluded that the magnetoreceptors involved may be the photoreceptors. Microscopic studies using histological techniques revealed that only some of the pineal cells (between 20% and 30%) seemed to respond to magnetic stimulation, using a uniform 60-Hz AC field with an intensity of 20 µT for 48 hr.[30] The cells showed different ultrastructural morphological changes; the most striking effect was an apparent increase of apocrine secretion to the luminal spaces in the epiphyseal epithelium of pigeons. Regarding higher field intensities, a lack of effects of NMR-strength (Nuclear magnetic resonance) fields (0.14 T) on rat pineal melatonin synthesis has been documented.[31] On the other hand, a stable strong magnetic field (0.70 T) revealed morphodynamic reactive responses of the pineal gland.[32] Currently we are investigating the melatonin responses to a 900 MHz signal modulated by 217 Hz in pigeons. While we have to wait for these results, we already obtained data which clearly show an alteration in the spontaneous discharge rate of cortical neurons in the zebra finch in response to this stimulation (Semm and Schneider, in prep.). More than 40% of the neurones responded with either excitation or inhibition with an average latency of about 30 s. We constantly used an energy of 0.1 mW/cm$^2$. The responses were dependent on the distance between the antenna and the animal.

**Hypotheses**

The melatonin signal is believed to be transmitted to the central nervous system by membrane-bound, high-affinity melatonin binding sites, located in specific regions of the brain. These areas include the hypothalamus, pons-midbrain and retinae (for details see Rivkees et al.[33] and Weaver et al.[34]). The documented physiological effects of melatonin are possibly mediated by the modulation of central dopaminergic, noradrenergic, acetylcholinergic and serotoninergic neurotransmission. It is discussed that melatonin may exert its effects via inhibition of cAMP production by a pertussis toxin-sensitive mechanism involving a G protein of the inhibitory class.[35] Altered melatonin rhythms related to diseases like cancer, epilepsy, depression and anorexia nervosa point to an important role of the hormone in pathology.[36,37,38,39] In mammals the sleep-wake cycle, reproduction rhythms, aging, and hibernation as well as migratory unrest (Zugunruhe) in birds are influenced by melatonin.[40,41,42,43,44] Maestroni[45] demonstrated that a pharmacologically depressed melatonin level in mice has a modulating function on the primary immune response to sheep red blood cells. Moreover a supressed antibody production mediated by corticosterone was antagonized by the evening injection of melatonin.[45]

On the other hand, there are some indications that the circadian and circannual melatonin rhythm might be influenced besides photic stimuli (length and intensity of light and duration of darkness) by the geomagnetic field, i.e., natural and artificial changes of its direction and intensity[20,25,46,47,48] (also Maurer, unpublished, and Bartsch, pers. communication). It was also demonstrated that the cAMP level in the pineal was decreased after an inversion of the horizontal component for one hour.[49] Anninos et al.[50] was able to show that seizure activity in epileptic patients could be attenuated after magnetic treatment in the range of pT. Moreover evidence is accumulated that strong magnetic and electric AC-fields influence the melatonin synthesis.[8,51,52] (For a summary of the possible pineal involvement, see Figure 1).

Taken together it could be speculated that: 1) there may be some influences from the geomagnetic field and technical magnetic and electric fields by an altered melatonin level on the physiology of animals and humans, which are possibly mediated to the CNS or other peripheral targets that possess melatonin binding sites, and 2) several diseases are characterized by a change of the melatonin level. Possibly some kind of relationships exist between the electromagnetic environment, the pineal gland and pathological phenomena. Perhaps it may be possible in the future 1) to show therapeutic applications of magnetic treatment and 2) to establish reliable guidelines for protection against electromagnetic radiation.

**DISCUSSION**

Considerable evidence exists that the mammalian and avian pineal gland is capable of responding to earth-strength magnetic fields. It is believed that both the retinal and avian pineal photoreceptors may be the magnetoreceptors for the responses observed. In support of this hypothesis, recent results from photoreceptors of the fly are important.[53] The light-induced depolarization of this photoreceptor type was depressed during the presence of an artificial magnetic field (100 mT). However, the underlying mechanism for magnetic field detection is not yet known. There are two biophysical models, by Leask[54] and Schulten and Windemuth,[55] that describe mechanisms that could occur in the photoreceptors. In addition, it was believed that the biochemical alterations are not a consequence of the magnetic field itself, but of magnetically induced currents. It has been argumented that an inversion of the horizontal component of the ambient magnetic field also occurs whenever the animals are turning their heads but apparently serotonin metabolism is not influenced. These interpretations point to the insufficient knowledge of the authors with respect to the pertinent literature, basal physics and the principles of sensory physiology. The arguments are the following: (1) The authors do not refer to previous papers, which make it very likely that not the pineal itself but the retina is influenced by magnetic fields.[56] (2) The authors believe that the biochemical changes in the pineal are due to inductional currents, although

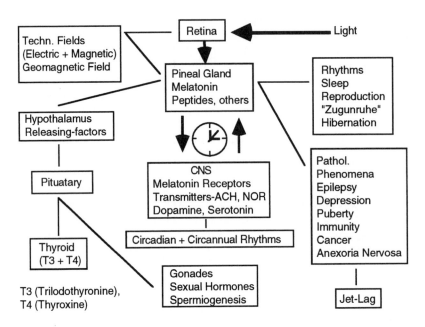

**Figure 1.** Hypothetical scheme concerning the possible pineal involvement in several physiological functions, different rhythms and some kinds of diseases. Abbreviations: CNS Central nervous system, ACH acetylcholin, NOR noradrenalin (for details see RESULTS and DISCUSSION).

they do not give any counting for that proposal. It has been shown by Welker et al.[7] that even a small (5°) change in the horizontal component results in a decrease of melatonin synthesis, which in turn would produce to less of a current to be sensed by the animal. Marine organisms such as elasmobranchs can potentially sense the ambient magnetic field by detecting the induced currents with their electroreceptors. These receptors are extremely sensitive and can detect a voltage gradient as small as 5 nV/cm.[57] This mechanism does not appear to be an option for mammals. Although mammals may move fast enough through the medium, the conductivity of the air is too poor to produce the same electrical fields as for the sharks.[58] Furthermore, mammals and most terrestrial animals lack the electroreceptors found in many marine and some other aquatic vertebrates. (3) The authors argue that changes of the natural magnetic field including its inversion also occurred whenever the animals turned their heads. However, the animals most likely interpret this differently from the artificial inversion performed by coils as they "know" that they are turning their head ("reference principle") and the relative change with respect to the magnetic field is interpreted as a directional change. Moreover, in this case the magnetic intensity remains constant, whereas the artificial change also includes an rapid change in intensity.

The important question arises as to the biological meaning of the magnetic sensitivity of the pineal gland. To our knowledge, no convincing evidence exists that magnetic fields really serve as additional "Zeitgeber" under natural conditions, as proposed by some groups. If so, animals should be able to use the MF even under constant light conditions, resulting in an new entrainment of circadian rhythmicity. However, this is apparently not the case, although under experimental conditions both the LD cycle and an imposed MF can be reflected by an animal's behavior.[47,48]

Comparing the effects of purely magnetic and electric fields on pineal function, it is clear that those electric fields never occur in nature. It is important to note that responses to those fields have nothing to do with electroreception, the existence of which has been proven in elasmobranchs.[57] Even if single neurons respond somehow to these fields, we must consider these responses as being clearly out of the range of biological sensory systems. The electric fields may affect the nerve membrane directly and thereby change electrical activity. Nevertheless, the responses can be used for monitoring low-frequency electric fields on a physiological basis.

## Considerations and Suggestions for Future Research

The innervation of the pineal gland includes the eyes, the suprachiasmatic and paraventricular nuclei, sympathetic nerve cells in the spinal cord, and the superior cervical ganglia. Because the magnetic sensitivity of the mammalian pineal gland depends on an intact retina, it is reasonable to expect that these parts of the brain and the peripheral nervous system also exhibit responses to MF stimulation. Different experimental methods could be applied to elucidate these responses. In the case of the avian pineal, many studies concerning melatonin production and the influence of different lighting conditions have already been undertaken under *in-vitro* conditions. Thus, it may be not to difficult to check the magnetic responsiveness of the pineal under those conditions, even at the single-cell level. A different valuable approach would be to measure recordings from single mammalian photoreceptors or from an "eyecup" preparation under MF stimulation. However, because the melanin of the pigment epithelium is apparently somehow involved in magnetic detection,[59] these recordings should be done in the presence of the pigment.

In future research, pineal magnetic sensitivity could be used as a tool for monitoring enviromental influences of electromagnetic nature, which might be dangerous even for humans. On the other hand, magnetic fields may be used in order to suppress melatonin secretion intentionally. For example, magnetic fields could be used to synchronize circadian rhythmicity in a person with jet lag in a shorter time than it would be possible under natural circumstances. However, we still do not know for sure whether the human pineal responds to magnetic stimulation.

Many pineal researchers are aware of the fact that on certain nights melatonin and NAT measurements are unexpectedly messy. But, having the magnetic sensitivity of the pineal in mind, there may be a partial explanation for this: during those nights the so-called "magnetic unrest" or even magnetic storms might take place, lowering melatonin secretion.[60] Data about such events can be obtained from geophysical institutions that measure magnetic fluctuations (k-factors) on a daily basis. Moreover, there is evidence that these small magnetic fluctuations are effective and can even influence behavior.[61] An important issue for the future therefore is to determine the threshold for the melatonin synthesis magnetic response. Experiments that clearly differentiate between intensity and orientation changes of the MF should be conducted. However, this requires 1) magnetically undisturbed laboratories, 2) magnetometers that are sensitive enough (flux-gate magnetometers) and 3) the technical expertise to cope with the complicated physics of electromagnetism. Thus, physicists and engineers should be involved in this research as often and far as possible and an international collaboration is desirable.

## REFERENCES

1. P. Semm, T. Schneider and L. Vollrath, *Nature* 288:607-608, (1980).
2. L. Maffei, E. Meschini and F. Papi, *Z. Tierpsychol.* 62:151-156 (1983).
3. P. Semm, H. Brettschneider, K. Dölla and W. Wiltschko, *in*: Comparative Physiology of Environmental Adaptations, ed., Pevet, P., 3:171-182 (1987).
4. P. Semm, D. Nohr, C. Demaine and W. Wiltschko, *J. Comp. Physiol.* 155:283-288 (1984).
5. P. Semm and C. Demaine, *J. Comp. Physiol.* 159:619-625 (1986).
6. R.C. Beason and P. Semm, *Neurosci. Lett.* 80:229 (1987).
7. H.A.Welker, P. Semm, R.P. Willig, J.C. Commentz, W. Wiltschko, L. Vollrath, *Exp. Brain. Res.* 50:426-432 (1983).
8. B.K. Wilson, L.E. Anderson, D.I. Hilton and R.D. Phillips, *Bioelectromagnetics* 2:371-380 (1981).
9. W. Wiltschko *in*: Animal Migration, Navigation, and Homing, K. Schmidt-Koenig & W.T. Keeton, eds. Springer-Verlag, Berlin, 50-58 (1978).
10. W. Wiltschko and R. Wiltschko, *Science* 176:62-64 (1972).
11. W. Wiltschko and R.Wiltschko, *Current Ornithology* 5:67-121 (1988).
12. M.A. Persinger, *Experientia* 43:92-104 (1987).
13. C. Walcott., *J. exp. Biol.* 70:105-123 (1977).
14. S. Reuss, P. Semm and L. Vollrath, *Neurosci. Lett.* 40:23- 26 (1982).
15. M. Villa, P. Mustarelli and M. Caprotti, *Life Science* 69:85-92 (1991).
16. A. Lerchl, K.O. Nonaka, K.A. Stokkan and R.J. Reiter, *Biochem. Biophys. Res. Comm.* 169:102-108 (1990).
17. A. Lerchl, K.O. Honaka and R.J. Reiter, *J. Pineal Res.* 10:109-106 (1991a).
18. A. Chics-DeMet, E. Chics-DeMet, H. Wu, R. Coopersmith and M. Leon, *Neuroscience* Abstract 156:17 (1988).
19. M. Zeise and P. Semm, *J. Comp. Physiol.* 157:23-29 (1985).
20. M.S. Keshavan, B.N. Gangadhar, R.U. Gautam, V.B. Ajiit and R.L. Kapur, *Neurosci. Lett.* 22:205-208 (1981).
21. T. Deguchi and J. Axelrod, *Analyt.Biochem.* 50: 174-179 (1972).
22. P. Semm, T. Schneider, L. Vollrath and W. Wiltschko, *in*: Avian Navigation, F. Papi and H.G. Wallraff eds., 329-337, Springer Verlag, Berlin-Heidelberg-New York (1982).
23. C. Demaine and P. Semm, *Neurosci.Lett.* 62:119-122 (1985).
24. P. Semm and L.Vollrath, *J. Comp. Physiol.* 154:675-681 (1984 ).
25. C. Demaine and P. Semm, *Neurosci. Lett.* 72:158-16226 (1986).
26. K. Mai and P. Semm, *J. Hirnforschung* 31:331-336 (1989).
27. S. Reuss and P. Semm, *Naturwissenschaften* 74:38-39 (1987).
28. G. Cremer-Bartels, K. Krause and H.J. Küchle, *Graefe`s Arch.Clin.Exp. Ophthalmol.* 220:248-252 (1983).
29. G. Cremer-Bartels, K. Krause, G. Mitoskas and G. Brodersen, *Naturwiss.* 71:567-574 (1984).
30. J.L. Bardasano, A.J. Meyer and L Picazo, *J. Hirnforschung* 26:471-475 (1985).
31. S. Reuss, J. Olcese, L. Vollrath, M. Skalej and M. Meves, *IRCS Med. Sci.* 13:471 (1985).
32. J. Milin, M. Bajic and V. Brakus, *Neuroscience* 26, 3:1083-1092 (1988).
33. S.A. Rivkees, V.M., Cassone, D.R. Weaver and S.M. Reppert, *Endocrinol.* 125, 1:363-368 (1989).
34. D.R. Weaver, S.A. Rivkees and S.M. Reppert, *J. Neuroscience* 9, 7:2581-2590 (1989).
35. L.L. Carlson and S.M. Reppert, *Endocrinology* 125:160a (1989).

36. H. Bartsch, L. Bartsch and D. Gupta, *in*: Neuroendocrinology: New Frontiers. Brain Research Promotion, D. Gupta, H.A. Wollmann, M.B. Ranke, eds., Tübingen, Germany, 333 (1990).

37. F. Anton-Tay, J.L. Diaz, and A. Fernandez-Guardiola, *Life Sci.* 10:841-850 (1971).

38. S.H. Kennedy, S. Tighe, G. McVey and G.M. Brown, *J. Nerv. Mental Disease* 177, 5:300-303 (1989).

39. F. Tortosa, M. Piug-Domingo, M.A. Peinado, J. Oriola, S.M. Webb and A. de Leiva, *Acta Endocrinol. Copenh.* 120:574-578 (1989).

40. S.W. Holmes and D. Sugden, *Br. J. Pharmacol.*, 76:95-101 (1982).

41. B.D. Goldman, J.M. Darrow and L. Yogev, *Endocrinology* 114, 6:2074-2083 (1984).

42. F. Waldhauser and H. Steger, *J. Neural. Transm.* 21:183-197 (1986).

43. G.C. Brainard, L.J. Petterborg, B.A. Richardson and R.J. Reiter, *Neuroendocrinology* 35:342-348 (1982).

44. E. Gwinner, *J. Comp. Physiol.* 126:123-129 (1978).

45. G.J.M. Maestroni, A. Conti and W. Pierpaoli, *J. Neuroimmunol.* 13:19-30 (1986).

46. E.D. Terracini and F.A. Brown, *Physiol. Zool.* 35:27-37 (1962).

47. V.L. Bliss and F.H. Heppner, *Nature* 261:411-412 (1976).

48. J.A. Brown Jr. and K.M. Scow, *J. interdiscipl. Cycle Res.* 9:137-145 (1978).

49. K. Rudolph, A. Wirz-Justice, K. Kräuchi and H. Feer, *Brain Research* 446:159-160 (1988).

50. P.A. Anninos, N. Tsagas, R. Sandyk and K. Derpapas, *Intern. J. Neuroscience* 60:141-171 (1991).

51. R. LaPorte, L. Kus, R.A. Wisniewski, M.M. Prechel, B. Azar- Kia and J.A. McNulty, *Brain Res.* 506:294-296 (1990).

52. A. Lerchl, R.J. Reiter, K.A. Howes, K.O. Nonake and K.A. Stokkan, *Neurosci. Lett.* 124:213-215 (1991).

53. J.B. Phillips, *Society for Neuroscience: Abstracts* 13:397 (1987).

54. M.J.M. Leask, *Nature* 267:144-145 (1977).

55. K. Schulten and A. Windemuth, *in*: Biophysical Effects of Steady Magnetic Fields. G. Maret, J. Kiepenheuer, & N. Boccara, eds. Springer-Verlag, Berlin. 167-172 (1986).

56. R.C. Beason and P. Semm, *in*: Orientation in Birds, P. Berthold, ed., Birkhäuser Verlag Basel/Switzerland, 106-127 (1991).

57. A.J. Kalmijn, *J. Exp. Biol.* 55:371-383 (1971).

58. A.J. Kalmijn, *Science* 218:916-918 (1982).

59. T. Leucht., *Naturwissenschaften* 74:441-443 (1987).

60. K.P. Ossenkopp, M. Kavaliers and M. Hirst, *Neurosci. Lett.* 40:321-325 (1983).

61. M. Lindauer and H. Martin, *Z. Vergl. Physiol.* 60:219-243 (1968).

# CARDIOVASCULAR RESPONSES TO RADIOFREQUENCY RADIATION

James R. Jauchem[*]

Radiofrequency Radiation Division
Occupational and Environmental Health Directorate
U.S. Air Force Armstrong Laboratory
Brooks Air Force Base, Texas 78235-5324

Cardiovascular changes during heating due to radiofrequency radiation (RFR) exposure have been reviewed previously.[1] That review focused on physiological responses in animals, and compared studies of conventional environmental heating. Polson and Heynick[2] also performed a recent review of bioeffects, including cardiovascular effects, due to RFR exposure. The main purpose of the present paper is to review studies of cardiovascular responses to RFR exposure in humans. In addition, cardiovascular responses to long-term, low-level RFR exposure in animals will be discussed. Cardiovascular changes during exposure to extremely low-frequency electric and magnetic fields are beyond the scope of this review.

## EFFECTS OF RFR ON THE CARDIOVASCULAR SYSTEM: EPIDEMIOLOGIC STUDIES

Epidemiologic studies of suspected RFR exposure are few in number and have generally been limited in scope. The principal groups studied have been individuals presumably exposed while assigned to the military services or working in industrial settings. Accurate estimates of dose are often difficult to obtain.[3] Michaelson[4,5] reviewed studies of workers who manufactured, operated, tested, or maintained RFR-generating equipment in the Soviet Union and other Eastern European countries. Some of the investigators[6,7,8] reported cardiovascular reactions to RFR. Resnekov[9] and Kristensen[10] noted that these studies are difficult to evaluate due to the methodology and terminology used.

One frequently described set of cardiovascular changes, supposedly attributed to RFR exposure, includes bradycardia (or occasional tachycardia), arterial hypotension (or hypertension), changes in cardiac conduction (e.g., delayed auricular and ventricular conduction), and electrocardiogram alterations. Several investigators[7,8,11,12] have reported these changes in workers in RF fields. These changes, however, do not seem to diminish the capacity to work and are reversible.[13]

---

[*] These views and opinions are those of the author and do not necessarily state or reflect those of the U.S. Government.

*Radiofrequency Standards,* Edited by B.J.
Klauenberg *et al.,* Plenum Press, New York, 1994

As Michaelson[5] has pointed out, "individuals suffering from a variety of chronic diseases may exhibit the same dysfunctions of the cardiovascular system as those reported to be a result of exposure to RFR; thus it is extremely difficult, if not impossible, to rule out other factors in attempting to relate RFR exposure to clinical conditions."

In a 12-year survey in Poland[14], no cardiovascular disturbances were found to be associated with the estimated level of RFR exposure. In another analysis,[15] no serious cardiovascular disturbances were noted in animals or humans as a result of RFR exposure.

Zaret[16] hypothesized that microwave exposure could be related to cardiovascular disease. Others[17] analyzed Zaret's work and pointed out that high levels of cardiovascular problems were reported in the study area 20 years before the microwave source in question (Russian over-the-horizon radar) was set up.

More recently, Kristensen[10] reviewed the epidemiologic literature relating cardiovascular diseases to the work environment. Relative to RFR, no studies more recent than the ones already discussed in this paper were mentioned.

## RFR POWER DEPOSITION AND CARDIAC STRESS

Observations of heart rate in humans purposely exposed to RFR have been limited to studies of magnetic resonance imaging (MRI) (which, in addition to RFR, involves static and time-varying magnetic fields). Shellock and Crues[18] found no significant changes in heart rate or blood pressure after MRI at whole-body specific absorption rates of up to 1.2 W/kg. Schaefer[19] considered whether or not RF power deposition during MRI in cardiac-compromised patients could impose potentially hazardous stresses on their cardiovascular systems. He calculated that an exposure to a specific absorption rate of 2 W/kg would elevate heart rate by less than 3%. Schaefer asserted that "such an increase is not likely to pose a safety problem" by noting that, for example, simply eating will normally increase cardiac output by 30%[20]. (One could also point out that exercise can increase cardiac output by as much as 700%[20]).

Adair and Berglund[21] predicted the effects of impaired cardiovascular function (e.g., restrictions of peripheral blood flow) on temperature changes in patients. With severely impaired skin blood flow, short MRI exposures (e.g., 20 min or less) were suggested (at specific absorption rates less than 3 W/kg). An important point to note in this discussion is that general standards for exposure to RFR (not during MRI) are set at lower levels.

## EFFECTS OF LONG-TERM LOW-LEVEL RFR EXPOSURE IN ANIMALS

Toler et al.[22] studied the effects of chronic low-level RFR exposure on cardiovascular parameters in Sprague-Dawley rats. Exposure to pulsed 435-MHz RFR 22 hours daily 7 days per week for 6 months resulted in no differences in heart rate and blood pressure between RFR- and sham-exposed animals. Estimated whole-body absorption rates ranged from 0.04 to 0.4 W/Kg. No other studies of this nature have been completed at this time.

## CONCLUDING REMARKS

In general terms, as others have mentioned (e.g., Petersen[23]), "effects" are not necessarily "hazards." The assumption that one automatically implies the other must be questioned. In considering potential hazards of RFR, one must raise the question: "Is this a transient physiological effect or is it a permanent effect?"[5]. Measurable effects may be well within the capacity of an organism to maintain a normal equilibrium or homoeostasis. If, on the other hand, an individual's ability to function properly was compromised or the recovery capability of the individual was overcome, then the effect may be considered a "hazard".

In terms of cardiovascular effects, epidemiolgic studies have not yielded any obvious cardiovascular-related hazards of long-term low-level RFR exposures. Levels of RFR that cause heating and related cardiovascular changes have been well-defined and appear to be above exposure standards.

## REFERENCES

1. J.R. Jauchem and M.R. Frei, Heart rate and blood pressure changes during radiofrequency irradiation and environmental heating, *Comp Biochem Physiol* 101A:1-9, (1992).
2. P. Polson and L.N. Heynick, Analysis of the Potential for Radiofrequency Radiation Bioeffects to Result From Operation of the Proposed ONR and Air Force High-Frequency Active Auroral Research Program Ionospheric Research Instrument (HAARP IRI): General Analysis. (Technical Report prepared for U.S. Air Force Phillips Laboratory, Contract F33615-90-D-0606) Hanscom Air Force Base, (1992).
3. C. Silverman, Epidemiologic approach to the study of microwave effects, *Bull NY Acad Med* 55:1166-1181, (1979).
4. S.M. Michaelson, Analysis of experimental and epidemiological data from exposure to microwave/radiofrequency (MW/RF) energies, *in:* "Biological Effects and Dosimetry of Nonionizing Radiation. NATO Advanced Study Institute on Advances in Biological Effects and Dosimetry of Low Energy Electromagnetic Fields. Series A, Life Sciences," M. Grandolfo, S.M. Michaelson, A. Rindi, eds. vol. 49. New York: Plenum Press. pp 589-609, (1981).
5. S.M. Michaelson, Health implications of exposure to radiofrequency/microwave energies, *Brit J Ind Med* 39:105-119, (1982).
6. S. Baranski and P. Czerski. "Biological Effects of Microwaves," Stroudsburg, PA: Dowden, Hutchinson, and Ross, 1976:234P.
7. Z.V. Gordon, Occupational health aspects of radiofrequency electromagnetic radiation, *in:* "Ergonomics and Physical Environmental Factors," Geneva: International Labour Office, 1970:159-172. (Occupational Safety and Health Series, No 21).
8. M.N. Sadchikova, Clinical manifestations of reactions to microwave irradiation in various occupational groups, *in:* "Biological effects and health hazards of microwave radiation," Czerski P, Ostrowski, Silverman C, et al., eds., Warsaw: Polish Medical Publishers, 1974:261-267.
9. L. Resnekov, Noise, radiofrequency radiation and the cardiovascular system, *Circulation* 63:264A-266A, (1981).
10. T.S. Kristensen, Cardiovascular diseases and the work environment. A critical review of the epidemiologic literature on nonchemical factors, *Scand J Work Environ Health* 15:165-179, (1989).
11. Z.V. Gordon. "Biological Effect of Microwaves in Occupational Hygiene," Leningrad: Izvestiya Meditisina Press, 1966:164. (TT 70-50087, NASA TT F-633, 1970).
12. M.N. Sadchikova and A.A. Orlova, Clinical picture of the chronic effects of electromagnetic microwaves, *Gig Tr Prof Zabol* 2: 16-22, (1958).
13. Y.A. Osipov, "Occupational Hygiene and the Effect of Radiofrequency Electromagnetic Fields on Workers," Leningrad: Izvestiya Meditisina Press, 1965:78-103.
14. M. Siekierzynski, P. Czerski, H. Milczarek, A. Gidynski, C. Czarnecki, E. Dziuk, and W. Jedrzejczak, Health surveillance of personel occupationally exposed to microwaves. II. Functional disturbances. *Aerospace Med* 45:1143-1145, (1974).
15. Z. Edelweijn, R.L. Elder., E. Klimkova-Deutschova, and B. Tengroth. Occupational exposure and public health aspects of microwave radiation. *In:* P. Czerski, K. Ostrowski, C. Silverman, et al., eds. "Biological Effects and Health Hazards of Microwave Radiation," Warsaw: Polish Medical Publishers, 1974:330-331.
16. M.M. Zaret, Electronic smog as a potentiating factor in cardiovascular disease: A hypothesis of microwaves as an etiology for sudden death from heart attack in North Karelia. *Med Res Eng* 12:13-16, (1976).

17. H.J. Healer, M.L. Shore, H. Pollack, D.S. Greensberg, L.R. Solon, E.L. Hunt, and M. Eisenbud. Public issues of nonionizing radiation. General discussion: Session V. *Bull NY Acad Med* 55:1279-1296, (1979).

18. F.G. Shellock and J.V. Crues, Temperature, heart rate, and blood pressure changes associated with clinical MR imaging at 1.5 T, *Radiology* 163: 259-262.

19. D.J. Schaefer, Dosimetry and effects of MR exposure to RF and switched magnetic fields, *Ann NY Acad Sci* 649:225-236, (1992).

20. W.F. Ganong. "Review of Medical Physiology," 6th ed. Los Altos, California: Lange Medical Publications, (1973).

21. E.R. Adair and L.G. Berglund, Thermoregulatory consequences of cardiovascular impairment during NMR imaging in warm/humid environments, *Magn Reson Imaging* 7:25-37, (1989).

22. J. Toler, V. Popovic, S. Bonasera, P.Popovic, C. Honeycutt, and D. Sgoutas. Long-term study of 435 MHz radio-frequency radiation on blood-borne end points in cannulated rats. Part II. Methods, results, and summary. *J Microwave Power Electromagn Energy* 23:105-136, (1988).

23. R.C. Petersen, Bioeffects of microwaves: A review of current knowledge. *J Occup Med* 25:103-111, (1983).

# FREQUENCY AND ORIENTATION EFFECTS ON SITES OF ENERGY DEPOSITION

Melvin R. Frei

Department of Biology
Trinity University
715 Stadium Drive
San Antonio, Texas 78212

## INTRODUCTION

The physiological and behavioral responses of animals exposed to radiofrequency radiation (RFR) are related to both the whole-body average specific absorption rate (SAR) and the distribution of absorbed energy within the irradiated subjects (the local SAR). Local and whole-body average energy absorption varies with a great number of factors-- carrier frequency, power density, physical dimensions of the organism, and orientation of the subject with respect to the electric and magnetic fields. RFR exposure at different carrier frequencies results in different SARs in prolate spheroid models,[1] and in different colonic heating rates in rats[2] and rhesus monkeys,[3] with higher SARs and faster heating occurring at or near the resonant frequency.

With regard to the effects of orientation, Chou et al.[4] and McRee and Davis[5] have shown that, in rat carcasses irradiated at 2450 MHz, E-orientation exposure (long axis of body parallel to electric field) results in higher whole-body average SARs than does H-orientation exposure (long axis parallel to magnetic field). Recently, Frei et al.[6] noted that, in anesthetized rats exposed to 2450 MHz in E and H orientations (equivalent average SARs), E-orientation irradiation resulted in greater peripheral heating, while H-orientation exposure caused greater core heating. This difference in local energy absorption patterns had dramatic effects on the animals' physiological responses to irradiation.

The objectives of this presentation are to show how exposure frequency (700 to 9300 MHz) and orientation relative to the E and H fields affect the sites of energy deposition in living animals (rats) and to show how differences in sites of energy deposition influence the animals' cardiorespiratory responses to irradiation.

*Radiofrequency Standards*, Edited by B.J.
Klauenberg *et al.*, Plenum Press, New York, 1994

## METHODS

All studies were conducted on ketamine-anesthetized male Sprague-Dawley rats weighing between 250 and 300 g. The animals were instrumented to continuously monitor and record arterial blood pressure, electrocardiogram and respiratory rate. Temperatures were monitored at five sites: left subcutaneous (lateral, mid-thoracic, side facing antenna); right subcutaneous (side away from antenna); right tympanic; colonic; and tail, using a BSD-200 thermometry system.

Irradiation was conducted under far-field conditions in environmentally controlled anechoic chambers. Prior to exposure, the normalized SARs were determined calorimetrically according to the methods of Allen and Hurt[7] and Padilla and Bixby.[8] At some frequencies, irradiation was conducted at different power densities to achieve equivalent SARs in the two orientations. Animals were exposed individually in both E and H orientations (left lateral exposure, long axis parallel to electric or magnetic field, respectively).

A colonic temperature (Tc) change of 1°C was used as the endpoint for comparing the effects of orientation and frequency variation. At a given frequency, after initial exposure in E or H orientation caused Tc to increase to 39.5°C, irradiation was discontinued. When Tc returned to 38.5°C (so that all exposures had a fixed starting point), irradiation was initiated until Tc again increased to 39.5°C. While the Tc was returning to 38.5°C, the antenna was rotated and another cycle was completed in the second orientation. The orientation in which the animals were first exposed was alternated daily.

## RESULTS

Figure 1 shows the SAR-normalized times (min/W/kg) required to accomplish a Tc increase of 1°C in rats exposed to 700-, 1200-, 2450-, 5600- and 9300-MHz RFR. A near-linear relationship between irradiation frequency and rate of core heating was noted; the higher the exposure frequency, the longer the time required for Tc increase. Interestingly, there was virtually no difference between orientations with respect to rate of core heating.

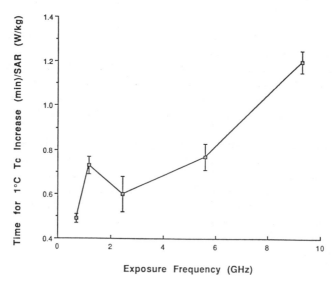

**Figure 1.** SAR-normalized times (min/W/kg) required to accomplish a Tc increase of 1°C in rats exposed to 700-, 1200-, 2450-,5600-, and 9300-MHz RFR.

The SAR-normalized left subcutaneous temperature (Ts) changes that accompanied the 1°C Tc increases during E- and H-orientation irradiation at the various frequencies are detailed in Figure 2. There was essentially a linear relationship between exposure frequency and amount of Ts change during irradiation between 700 and 5600 MHz; at 9300 MHz, a slight decrease was noted. At frequencies below 9300 MHz, there were pronounced orientation effects; E-orientation exposure resulted in greater peripheral heating than did H-orientation irradiation. During exposure at 9300 MHz, there was no significant difference between orientations.

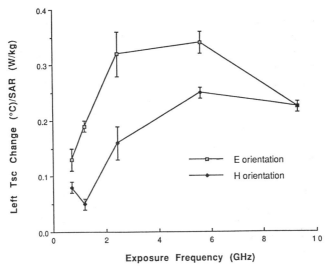

**Figure 2.** SAR-normalized Ts changes (°C/W/kg) that accompanied the 1°C Tc increases during E- and H-orientation irridiation.

The SAR-normalized mean arterial blood pressure (MAP) and heart rate (HR) changes that accompanied the 1°C Tc increases during E- and H-orientation exposure at the various frequencies are shown in Figures 3 and 4. As Tc increased from 38.5 to 39.5°C (at all frequencies and in both orientations) HR and MAP significantly increased. Results indicate

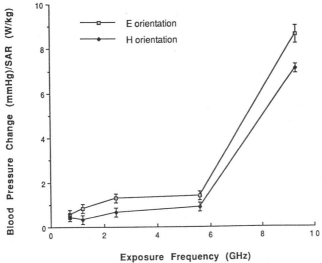

**Figure 3.** SAR-normalized MAP changes (mmHg/W/kg) that accompanied the 1°C Tc increases during E- and H-orientation irradiation.

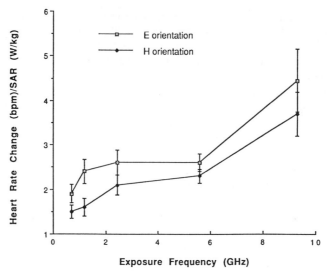

**Figure 4.** SAR-normalized HR changes (bpm/W/kg) that accompanied the 1°C Tc increases during E- and H-orientation irradiation.

a direct relationship between irradiation frequency and magnitude of HR and BP change. In addition, at frequencies above resonance, the changes during E-orientation exposure were greater than those occurring during H-orientation exposure.

Figure 5 shows the SAR-normalized respiratory rate (RR) changes that accompanied the 1°C Tc increases during E- and H-orientation irradiation at the various frequencies. An inverse relationship between irradiation frequency and RR change was seen: the higher the frequency, the less the effect on RR.

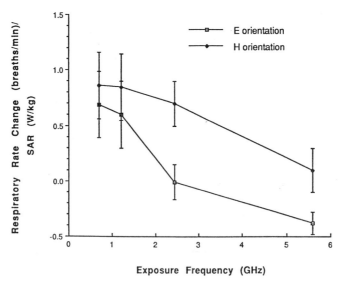

**Figure 5.** SAR-normalized RR changes (breaths /min/W/kg) that accompanied the 1°C Tc increases during E- and H-orientation irradiation.

# DISCUSSION

Earlier work by D'Andrea et al.,[2,9,10] Durney et al.,[1] and Lotz[3] showed frequency-dependent differences in sites of energy deposition and average SARs, with higher SARs and greater colonic heating occurring near resonance. Our studies, which were conducted at similar average SARs, also showed frequency-dependent differences in patterns of heat distribution within living animals (local SARs). These present and previous results are easily understandable in view of the significantly different depths of penetration: in general, the higher the frequency, the less the depth of penetration. In addition, our studies showed that at a given frequency, the local SAR is also influenced by the animal's orientation to the E and H field: in general, irradiation in H orientation resulted in greater core heating, while exposure in E orientation resulted in greater peripheral heating.

Since physiological responses are influenced by temperature change, it is no surprise that frequency- and orientation-related differences in sites of energy deposition resulted in differences in cardiorespiratory responses. In general, the magnitude of HR and MAP change was directly related to the exposure frequency, while the RR change seemed to vary inversely with frequency. The major question to be answered is: "What is causing the differences in cardiorespiratory responses?"

It has been suggested that the greater cardiovascular responses at higher frequencies might be due to the longer time required to reach the irradiation endpoint of a 1°C Tc increase. Yet, the onset of cardiovascular change at these frequencies is very rapid, often occurring before any noticeable change in Tc. At poorly penetrating supraresonant frequencies, Tc change is indirect, through circulatory transfer of heat from the periphery. Previous research[11,12,13,14] has demonstrated a direct relationship between rate of core heating and magnitude of HR and BP increase, which also counters this suggestion.

It is more likely that the frequency-dependent changes in cardiovascular function are related to the level of peripheral heating, as determined by the depth of energy penetration. Thermoreceptors, which increase firing rates in response to temperature change, are integral to thermoregulatory responses such as HR and MAP changes. These receptors are abundant peripherally (cutaneously) and also are found in the animal core (brain, spinal cord, viscera). We suggest that the degree of cutaneous thermoreceptor activation is a primary factor involved in determining levels of cardiovascular change during RF irradiation. Kenshalo et al.[15] have shown that, within limits, as cutaneous temperature increases, the average thermoreceptor discharge frequency also increases. Candas et al.[16] concluded that the rate of skin temperature change in squirrel monkeys plays a vital role in determining the magnitude of thermoregulatory responses. Recently, Adair[17] found that squirrel monkeys are less effective at maintaining core temperature during resonant frequency exposure (450 MHz) than during irradiation at a higher frequency (2450 MHz). In part, this was attributed to inadequate peripheral thermoreceptor activation at resonant frequency. In our studies, during irradiation at near-resonance frequencies in which there is little cutaneous heating, the level of HR change is small. At progressively higher frequencies, irradiation results in greater cutaneous heating and in progressively greater HR changes. This would also account for the orientation-related differences in cardiovascular response. During supraresonant irradiation, E-orientation exposure consistently resulted in greater peripheral heating than did H-orientation irradiation at an equivalent SAR. The corresponding HR and MAP increases were consistently greater during E- than during H-orientation exposure.

Earlier reports[18,19] indicated that high-intensity RF irradiation caused an increase in RR that occurred simultaneous to an increase in core temperature. Gordon and Long[20] showed in mice exposed to 2450-MHz RFR, that above a threshold SAR, RR increased as SAR increased. They also pointed out that, compared with other physiologic responses, levels of heating (core) required to cause a RR increase are relatively high. In the present study, although total changes were rather small, results indicate an inverse relationship between exposure frequency and RR change; at low frequencies, the RR significantly increased, while progressively higher-frequency exposure resulted in no change or a slight decrease. These data suggest that core energy deposition, such as occurs during lower-frequency irradiation, plays a major role in RFR-induced RR change. The importance of core thermoreceptor activation in stimulation of respiration has been previously proposed.[21,22]

## SUMMARY

1. In animals exposed to RFR, the pattern of heat distribution and the cardiorespiratory responses are both frequency- and orientation-dependent.
2. Exposure frequency and depth-of-energy penetration are inversely related; high-frequency irradiation results in greater peripheral heating, while low-frequency exposure results in greater core heating.
3. In general, at a given frequency, H-orientation exposure results in greater core heating, while E-orientation irradiation results in greater peripheral energy deposition.
4. The magnitudes of HR and MAP change during irradiation are directly related to the exposure frequency.
5. The magnitude of RR change during irradiation appears to be inversely related to the exposure frequency.
6. In general, RFR-induced HR and MAP increases are greater during E- than during H-orientation exposure.
7. Results suggest that under our experimental conditions, peripheral thermoreceptors play a more important role than do core receptors in regulation of cardiovascular responses to RF irradiation, while the reverse is true for respiratory responses.

## REFERENCES

1. C.H. Durney, C.C. Johnson, P.W. Barber, H. Massoudi, M. Iskander, J.L. Lords, D.K. Ryser, S.J. Allen and J.C. Mitchell, Radio Frequency Radiation Dosimetry Handbook, 2nd ed., Report USAFSAM-TR-78-2, USAF School of Aerospace Medicine, Brooks Air Force Base, TX (May 1978).
2. J.A. D'Andrea, O.P. Gandhi and J.L. Lords, Behavioral and thermal effects of microwave radiation at resonant and nonresonant wavelengths. *Radio Sci* 12: 251-256 (1977).
3. W.G. Lotz, Hyperthermia in radiofrequency-exposed rhesus monkeys: a comparison of frequency and orientation effects. *Radiat Res* 102:59-70 (1985).
4. C.K. Chou, A.W. Guy, J.A. McDougall, and H. Lai, Specific absorption rate in rats exposed to 2,450-MHz microwaves under seven exposure conditions, *Bioelectromagnetics* 6:73-88 (1985).
5. D.I. McRee and H.G. Davis, Effects on energy absorption of orientation and size of animals exposed to 2.45-GHz microwave radiation, *Health Phys* 52:39-43 (1987).
6. M.R. Frei, J.R. Jauchem, J.M. Padilla and J.H. Merritt, Thermoregulatory responses of rats exposed to 2.45-GHz radiofrequency radiation: a comparison of E and H orientation. *Radiat Environ Biophys* 28:235-246 (1989).
7. S.J. Allen and W.D. Hurt, Calorimetric measurements of microwave energy absorption by mice after simultaneous exposure of 18 animals, *Radio Sci* 14 (Suppl 6):1-4, (1979).
8. J.M. Padilla and R.R. Bixby, Using Dewar-flask Calorimetry and Rectal Temperature to Determine the Specific Absorption Rates of Small Rodents, Report USAF SAM-TR-86-3, USAF School of Aerospace Medicine, Brooks Air Base, TX (1986).
9. J.A. D'Andrea, R.Y. Emmerson, C.M. Bailey, R.G. Olsen and O.P. Gandhi, Microwave radiation absorption in the rat: frequency-dependent SAR distribution in body and tail, *Bioelectromagnetics* 6:199-206 (1985).

10. J.A. D'Andrea, R.Y. Emmerson, J.R. DeWitt and O.P. Gandhi, Absorption of microwave radiation by the anesthetized rat: electromagnetic and thermal hotspots in body and tail, *Bioelectromagnetics* 8:385-397 (1987).

11. M.A.B. Frey and R.A. Kenney, Cardiac response to whole-body heating, *Aviat Space Environ Med* 50:387-389 (1979).

12. J.R. Jauchem, M.R. Frei, and F Heinmets. Heart rate changes due to 5.6-GHz radiofrequency radiation: relation to average power density, *Proc Soc Exp Biol Med* 177:383-387 (1984).

13. W.P. Watkinson and C.J. Gordon, Improved technique for monitoring during exposure to radio-frequency radiation, *Am J Physiol* 250:H320-H324 (1986).

14. M.R. Frei, J.R. Jauchem and F. Heinmets, Physiological effects of 2.8-GHz radiofrequency radiation: a comparison of pulsed and continuous-wave radiation, *J. Microwave Power Electromagnetic Energy* 23:85-93 (1988).

15. D.R. Kenshalo, C.E. Holmes, and P.B. Woods, Warm and cool thresholds as a function of rate of stimulus temperature change, *Percept Psychophys* 3:81-84 (1968).

16. V. Candas, E.R. Adair and B.W. Adams, Thermoregulatory adjustments in squirrel monkeys exposed to microwaves at high power densities, *Bioelectromagnetics* 6:221-234 (1985).

17. E.R. Adair, Thermoregulatory Consequences of Resonant Microwave Exposure, Report USAF-SAM-TR-90-7, U.S. Air Force School of Aerospace medicine, Brooks Air Force Base, Texas, June 1990.

18. S.M. Michaelson, R.A.E. Thomson and J.W. Howland, Physiologic aspects of microwave irradiation of animals, *Am J Physiol* 201:351-356 (1961).

19. A.G. Subbota, Changes in functions of various systems of the organism, *in:* Petrov, R., ed., "Influence of Microwave Radiation on the Organism of Man and Animals," Washington, DC: National Aeronautics and Space Administration, pp 75-78, 1967.

20. C.J. Gordon and M.D. Long, Ventilatory frequency of mouse and hamster during microwave-induced heat exposure, *Respir Physiol* 56:81-90 (1984).

21. W.R. Beakley and J.D. Findlay, The effect of environmental temperature and humidity on the respiration rate of Ayshire calves, *J.Agric Sci* (Cambridge) 45:452-460 (1955).

22. E. Simon, Temperature regulation: The spinal cord as a site of extrahypothalamic thermoregulatory functions, *Rev Physiol Biochem Pharmacol* 71:1-76 (1974).

# EXTRAPOLATION OF ANIMAL RADIOFREQUENCY RADIATION (RFR) BIOEFFECTS TO HUMANS

Michael R. Murphy

Radiofrequency Radiation Division
Occupational and Environmental Health Directorate
U. S. Air Force Armstrong Laboratory
AL/OER, 8308 Hawks Road, Bldg 1184
Brooks Air Force Base, Texas, 78235-5324

## INTRODUCTION

This paper outlines some of the processes and issues involved in extrapolating animal radiofrequency radiation (RFR) bioeffects to humans, particularly with regard to extending the RFR database for the development of new exposure standards. A proposed framework for discussing, evaluating, and planning studies for animal-to-human extrapolation is presented in Figure 1.

## WHAT IS EXTRAPOLATION

Broadly speaking, extrapolation is the process of estimating the unknown from the known, by making inferences based on assumed continuity, correspondence, or other parallelism between the known and the unknown. For the topic at hand, we use **known** data and experience about the bioeffects of RFR in animals, i.e., the animal bioeffects database, to estimate **unknown** bioeffects of RFR on humans.

There are three primary purposes for animal-to-human extrapolation. The first is that of <u>exploration</u> of what is possible. Some hypothesis generation and testing is inherent in this process, but mostly this activity is an open-minded search for any possible effect. The extrapolation created by this process is a general one: "If RFR exposure can produce a particular bioeffect in animals, and assuming the validity of evolutionary and phylogenetic principles, then it is possible that RFR exposure can produce similar bioeffects in humans." Findings of this type lead directly to the second main purpose of animal-to-human extrapolation, the <u>analysis</u> of the mechanism of action for such effects. A supported hypothesis for a mechanism of action becomes a foundation and organizing principle for animal-to-human extrapolation, since one can then draw on many decades of research on comparative anatomy, physiology, biochemistry, pathology, behavior, etc., to assist in the extrapolation process.[1] Extrapolations based on common mechanisms take the form: "If

*Radiofrequency Standards,* Edited by B.J.
Klauenberg *et al.,* Plenum Press, New York, 1994

RFR produces a bioeffect by a certain mechanism in animals, and if this mechanism is applicable to humans, then RFR may produce a similar bioeffect in humans." The third purpose of extrapolation is <u>prediction</u>, in which we attempt to go beyond general relationships and use animal RFR bioeffects data to predict specific human bioeffects and their magnitudes.

## DEVELOPING THE RFR BIOEFFECTS
## DATABASE FOR EXTRAPOLATION

The broad procedure by which data from the animal RFR bioeffects database is extrapolated to the human condition is, in practice, accomplished by an interactive process. In this process, most of the bioeffects data are collected purposefully in formalized experiments that have, inherent in their design, the assumptions of relatedness, continuity, and parallelism employed in animal-to-human extrapolation. That is to say, these experiments are designed and the data are collected specifically for one of the purposes of extrapolation: exploration, analysis, or prediction.

The experiments generally have a common form in which (1) a stimulus (independent variable) is applied to (2) a subject (for the purposes of this paper, a whole organism) and (3) a response (dependent variable) is measured. In the laboratory, this process is highly controlled. The stimulus parameters are carefully determined and appropriately scaled. The subjects are a group of animals from the same species, the same sex, the same age, and often an inbred strain, and the response is carefully and objectively defined and readily measurable. My point here is that when the results of an experiment on animals are **extrapolated** to humans, all three aspects of the experiment must be extrapolated. How can an RFR exposure in a workplace, battlefield, or home be extrapolated from a laboratory RFR exposure? How can we extrapolate from a juvenile, male, albino rat to a NATO soldier? Does a particular behavioral, physiological, or pathological response in an animal really model some similar process in humans? Many issues are related to these three aspects of animal experimentation that are critical to the ultimate extrapolation and interpretation of bioeffects in humans.

<u>Issues Related to Stimulus Parameters:</u> It is well known to researchers of RFR bioeffects that the amount or "dose" of RFR that is received in any particular bioeffects study is an extremely complex interaction between multiple stimuli and subject parameters. The RFR literature is replete with discussions of dosimetry and scaling.[2,3] A particularly worrisome problem with respect to extrapolation is the question of hot spots. The possibility of species-specific or even experiment-specific hot spots in certain organ systems could lead to bioeffects that would have very questionable value for animal-to-human extrapolation, except in a very broad sense.[4]

Another important but often ignored stimulus issue is that the selected RFR stimulus is only one part of the subject's total environment that will influence the measured responses. All aspects of housing, handling, preparation, and response measurement are part of the stimulus context in which the RFR appears.[5] It should go without saying that experimental control, in which all groups are treated the same, except for the independent variable, is critical. But what is not so obvious is that even when "extraneous" stimuli are equated among subject groups, such stimuli can still affect the results by creating an environment that is so generally stressful that the possible effects of RFR are minor by comparison and cannot be resolved from the background. Or, conversely, if the environment is extremely carefully regulated, a preparation hypersensitive to any perturbation may be created, in which "real" but subtle bioeffects may indeed be more readily detected, but in which the effects of imperfectly controlled extraneous factors are also more likely to appear. This issue is particularly important in the study of long-term health effects, in which the brief or weak RFR exposure recedes even further into the totality of the subjects' ongoing environmental and other life experiences.

Issues Related to Subject Selection: The proper selection of an animal species for animal-to-human research depends upon the purpose of the study. If one is exploring for indications of RFR bioeffects, then a broad sampling of animal species is appropriate.[6] Species used for the analyses of mechanisms are selected from those in which significant bioeffects have been discovered or from those on which there is already an extensive research database. However, if mechanisms are to be generalized to other species, hopefully to include humans, then they need to be validated in a broader range of organisms. For the most part, primates are considered the best species for extrapolating direct predictions to man; however, other well-established animal models for specific human responses, such as those developed for cancer screening, are also especially valuable. In addition to the selection of species, there is the question of the selection of the individual subjects for the experiment. What strain? Which sex? What age? Unfortunately, final selection is often made more on the basis of convenience, cost, tradition, and what the "Animal Use Review Committee" will allow, rather than on scientific considerations.

I would like to propose that the traditional comparative approach is particularly useful for validating animal-to-human extrapolation. In this approach, several species, with proper scaling applied, are compared on the same stimulus parameters and the same responses are measured. One good example of this approach was reported by de Lorge in 1979.[7] In his study, trained rats, squirrel monkeys, and rhesus monkeys were exposed to different power densities of 2.45 GHz CW RFR and observed for behavioral disruption as well as temperature rise. A log-linear relationship was discovered for body mass versus power density for both the physiological and behavioral measures. The comparative approach can be even more powerful if human data can be included in the comparison.[8,9]

Issues Related to Response Selection: As with the selection of stimulus parameters and subjects, the selection of which responses to measure greatly affects how the results of a study can be extrapolated. I make 3 points on this topic. 1) First, it is helpful to choose a response or endpoint that has a medium rate of natural occurrence, i.e., something that occurs at a rate that can conceivably either increase or decrease; this choice immediately doubles the opportunity of finding an effect. 2) The selection of responses or endpoints that already have a large existing research database allows the RFR bioeffects to be interpreted in relation to other agent-induced bioeffects and thereby can give great insight into mechanism of action. 3) For direct prediction of human responses, validation of responses between animal subjects and humans is important.[10] How does one know that a response measure in an animal model relates in any way to a certain response in humans? At the Armstrong Laboratory, we are attempting to validate three behavioral measures, serial memory, equilibrium in space, and acoustic startle, by conducting parallel testing of both rhesus monkeys and humans performing identical tasks and then developing and validating task-specific monkey-to-man extrapolation functions.[11]


## ANIMAL-TO-HUMAN EXTRAPOLATION FOR SETTING HEALTH AND SAFETY STANDARDS

If the goal is knowledge about human responses, why bother with extrapolation from animals? Why not learn directly from humans? The most common answer is one of ethical considerations. Many of the exposures required are of unknown safety, at least, and many of the measurements (for example, neurochemical analyses) are themselves hazardous. In addition, animal data can be advantageous because of time and cost considerations and ability to have greater experimental control and more accurate measurements. However, a more satisfactory answer is that human data <u>are</u> included whenever possible.

Animal-to-human extrapolation should be viewed in context as only one of several processes and areas of knowledge that can contribute to assessing potential harmful effects of RFR and to setting standards for permissible exposure. Epidemiological research[12] and clinical data from accidental exposures are clearly relevant. Mechanistic work using cells or even simpler aspects of life provides much insight, and ever more advanced computers and mathematical models are being developed. However, I believe that human research data that can link with animal research data hold excellent promise for validating animal-to-human extrapolations. I concur with the statement made at the 1984 NATO Conference by Dr. E. Adair, who, in discussing the inclusion of human <u>experimental</u> data in the RFR bioeffects database, stated: "Further experimentation of this type will be invaluable to the evaluation of predictions based on animal data or simulation models and must ultimately be applied to the formulation of acceptable standards for RF exposure of human populations."[8] Unfortunately there has been only minimal activity in this topic in the intervening nine years.

## CONCLUSION

In this paper, I have proposed a framework for discussing issues relating to the extrapolation of animal data on the bioeffects of RFR to the human condition. Three purposes of extrapolation are described as (1) the exploration of possibilities, (2) the analysis of mechanisms, and (3) the prediction of human responses. In the experimental collection of animal data for extrapolation, different issues and concerns apply to the three major parameters of the experimental process: the stimulus, the subject, and the response. Finally, I have suggested that well-planned animal-to-human extrapolation, including validating human studies, is important to the development of valid RFR exposure standards.

**Parameters**

| Purpose | Stimulus | Subject | Response |
|---|---|---|---|
| **Exploration** | | | |
| **Analysis of Mechanism** | | | |
| **Prediction** | | | |

**Figure 1.** Framework for Discussing Animal-to-Human Extrapolation Issues. Animal data developed for extrapolation to humans is collected in experiments employing protocols that specify the parameters of the stimuli to be applied, the animal subject model to be used, and the responses to be measured. Guidelines for these selections may vary greatly according to the purpose of the extrapolation, namely to explore for possible effects, to analyze mechanisms of effects, or to make specific and quantitative prediction of human responses. Therefore, this matrix framework is suggested for use in discussing, evaluating, or planning efforts involving animal-to-human extrapolation.

# REFERENCES

1. D.N. Erwin, Mechanisms of biological effects of radiofrequency electromagnetic fields: an overview, *Aviation, Space, and Environmental Medicine* 59:21-31 (1988).
2. C.H. Durney, H. Massoudi, and M.F. Iskander. "Radiofrequency Radiation Handbook (4th Edition)," USAF School of Aerospace Medicine, USAFSAM-TR-85-73, Brooks AFB, TX (1986).
3. C.J. Gordon, Scaling of the physiological effects from exposure to radiofrequency radiation: extrapolation from experimental animals to humans, *in:* "Biological Effects of Electropollution: Brain Tumors and Experimental Models," S.K. Dutta and R.M. Millis, eds., Information Ventures, Inc., Philadelphia, PA, 131-140 (1986).
4. S.M. Michaelson, Microwave and radiofrequency radiation, *in:* "Nonionizing Radiation Protection," M.J. Suess, ed., World Health Organization, Copenhagen, 97-174 (1982).
5. M.I. Gage, Non-electromagnetic factors influence behavioral effect of microwave exposures, *in:* "Biological Effects of Electropollution: Brain Tumors and Experimental Models," S.K. Dutta and R.M. Millis, eds., Information Ventures, Inc., Philadelphia, PA, 117-130 (1986).
6. G.E. Paget, Animal selection and extrapolation - the problem defined, *in:* "Human Risk Assessment - The Role of Animal Selection and Extrapolation," M.V. Roloff, ed., Taylor & Francis, New York, 1-7 (1987).
7. J.O. de Lorge, Disruption of behavior in mammals of three different sizes exposed to microwaves: extrapolation tolarger mammals, *in:* "Symposium on Electromagnetic Fields in Biological Systems," S.S. Stuchly, ed., International Microwave Power Institute, Edmonton, Alberta Canada, 215- 228 (1979).
8. E.R. Adair, On the prediction of human responses to RFR, *in:* "Proceedings of a Workshop on Radiofrequency Radiation Bioeffects," J.C. Mitchell, ed., USAF School of Aerospace Medicine, USAFSAM-TP-85-14, Brooks AFB, TX, 227- 230 (1985).
9. E.R. Adair, Thermoregulation in the presence of microwave fields, *in:* "CRC Handbook of Biological Effects of Electromagnetic Fields," C. Plok and E. Postow, eds., CRC Press, Boston, 315-337 (1986).
10. P.B. Dews, Extrapolation by proximate interference, *Neuroscience and Biobehavioral Reviews* 15:149-151 (1991).
11. W. Alter, S.L. Hartgraves, and M. Wayner, A review of animal-to-human extrapolation: issues and opportunities. *Neuroscience and Biobehavioral Reviews* 15:1-184 (1991).
12. J.R. Jauchem and J.H. Merritt, The epidemiology of exposure to electromagnetic fields: an overview of the recent literature, *J. Clin Epidemiol* 44:895-906 (1991).

**Session G:  Evaluation of the Bioeffects Database III:**
**Overview**

**Chair: E. Adair**


Bioeffects of Long-Term Exposures of Animals
*Arthur W. Guy*


Evaluation of Electromagnetic Fields in Biology and Medicine
*Maria A. Stuchly*


Overview of the Radiofrequency Radiation
Bioeffects Database
*Peter Polson and Louis N. Heynick*

# BIOEFFECTS OF LONG-TERM EXPOSURES OF ANIMALS

Arthur W. Guy

Emeritus Professor
Center for Bioengineering RJ-30
University of Washington
Seattle, Washington 98115

## INTRODUCTION

Between 1962 and 1992 at least 10 major studies have been reported in the literature of the United States on the biological effects of long-term exposure to radiofrequency radiation (RFR) spanning a frequency range of 800 to 9270 MHz. Exposure levels in the studies ranged from 0.48 to 100 mW/cm$^2$ with specific absorption rates (SARs) ranging from 0.075 to 45 W/kg. The studies include the work of Prausnitz and Susskind[1] on the life span and pathology in mice; Spalding et al.[2] on the body weight, activity, hematopoiesis and life span of mice, Preskorn et al.[3] on tumor growth and greater longevity in mice after fetal exposure; Guy et al.[4] on ocular and physiologic effects in rabbits; McRee et al.[5] on hematological and immunological effects in rabbits; Chou et al.[6] on body weight, electroencephalogram (EEG), hematology, and morphology in rabbits; Szmigielski et al.[7] on development of spontaneous and benzopyrene-induced skin cancer in mice; Chou et al.[8] on body weight, food consumption, and hemotological, morphological, chemical, protein electrophoresis and lymphocyte blast transformation parameters in rabbits; Adair et al.[9] on metabolic rate, internal body temperature, blood indices, and thermoregulatory behavior in squirrel monkeys; and Chou et al,[10] on general health and longevity, serum chemistries, hematological values, protein electrophoretic patterns, thyroxine, plasma corticosterone levels and histopathological patterns in rats. An overview of each one of these studies is given below, followed by a table summarizing the results.

## REVIEWS

**Prausnitz and Susskind[1]** conducted the first long-term exposure of animals in this country. They exposed 200 male Swiss albino mice (in groups of 10) to 9270 MHz pulsed microwaves (2 μs, 500 pps) of 100 mW/cm$^2$ average power density for 4.5 minutes daily over a period of 59 weeks, resulting in an average body temperature rise of 3.3° C during each exposure. The animals were contained in a compartmentalized circular polystyrene cage with a plastic window-screen floor. The animals were separated from each other by wedge-shaped compartments formed by small rectangular strips of plastic with the long dimensions oriented radially every 36° in the circular cage. The cage was placed above a standard gain horn (4 x 6 cm aperture) so that the animals were ventrally exposed while being rotated at 1 rpm around the vertical axis of the cylindrical cage to randomize the effect of the multiple reflections between the adjacent animals. The exposure level corresponded to one-half of a LD$_{50}$ exposure dose. Between exposures, the colony of 300 male Swiss albino mice were housed in a cabinet where the temperature was maintained between 21° and 24° C. Histopathology was performed on all of the 200 exposed and the 100 control group mice. Based on the Radio Frequency Radiation Dosimetry Handbook (RRDH)[11] the whole-body average SAR for the animals is estimated to be 45 W/kg.

The authors reported that the rate of lethality was higher in the control group but not significantly so. Part of the colony suffered from pneumonia during the last 3 months of the exposures resulting in a number of deaths in both groups. The authors speculate that the periodically induced slight artificial fevers are of some benefit to the animal in combating disease. At the time a death of an animal was noted, it was autopsied and histological parameters of key tissues were determined. These included the liver, spleen, lymph nodes, kidneys, adrenals, gut, lungs, and testes. Other tests included random blood count, spot checks of urine for glucose, weekly weighing of all mice, and recording of body temperature. Of the total of 300 mice in the experiment, 132 were sacrificed in 3 sacrifice series to assess damage in surviving mice. Sixty eight mice that had died, because of extensive postmortem changes, were unsuitable for histological assessment. The remaining 100 animals were used for the longevity assessment. Testicular degeneration was found in 40% of the dead exposed mice and only 8% of the controls. The authors reported that cancer of the white cells was found in 35% of the exposed mice and 10% of the controls evidenced by either monocytic or lymphatic leucosis or myeloid leukemia. The authors described leucosis as a noncirculating neoplasm of the white cells and leukemia as a circulating leucosis. A three-sacrifice series was carried out at 7, 16, and 19 months from the start of the exposures. No effects in the first sacrifice series of 5% of both exposed and control groups were reported. However, in the second sacrifice involving twice as many mice, 30% of the exposed and 10% of the control animals were reported to have leucosis but no correlations were reported regarding testicular degeneration as seen in the acute death histologies. As a result of this partial contradiction, a third sacrifice series was performed on the remainder of the 67 exposed and 19 control mice. In this group 21% of the exposed and 5% of the controls were reported to show evidence of some degree of testicular degeneration, but no correlation was found with leucosis, in this sacrifice series. Elder and Cahill,[12] in examining the reported results, found that when animals in the sacrifice series 1 and 2 were ignored, the difference in survival was not significant at 14 months into the experiment but was significant at 19 months. A 2 x 2 contingency table analysis also revealed that the prevalence of leucosis in exposed animals was not significantly different from that in the control animals or any of the sacrificed groups. Elder and Cahill[12] also determined that the combined groups including all of the animals at risk in the original longevity group did not have statistically different leucosis rates.

In their analysis of the same data, Roberts and Michaelson[13] concluded that a plausible case could be made for the concept that the study of Prausnitz and Susskind[1] demonstrated that the microwaves induced beneficial rather than detrimental effects, and that the study neither confirms nor denies the correlation between the exposure to microwaves and development of neoplasia, and they asserted that the report should not be cited as suggestive of such a correlation.

**Spalding et al.**[2] reported the effects of 800-MHz RFR on body weight, activity, hematopoiesis and life span in mice. The authors divided forty-eight mature-strain RF virgin female mice into 2 groups of 24 each, one group exposed to 800 MHz RFR at an average incident power level of 43 mW/cm$^2$, 2 hours per day, 5 days a week, for 35 weeks and the other group sham-exposed for the same period of time. The mice were placed in a 12-compartment acrylic resin plastic restraining cage during sham control and RFR exposures within a standard rectangular, 9.75 x 4.875, 10 ft. long ventilated wave guide. The exposed animals were subjected to a TE$_{10}$ mode field configuration with a spatially averaged power density of 43 mW/cm$^2$ over the cross-section of the waveguide. However, with the half-sinusoidal distribution of the TE$_{10}$ mode electric field across the wave guide, the peak exposure at the center of the wave guide would be 86 mW/cm$^2$. The authors selected the mice at random for each exposure run. They stated that because of the large number of exposures, those for any one mouse should integrate out to be the average exposure level. Assuming that the mass of each mouse is approximately 20 grams and based on the estimated power absorbed by the mice of 0.5 Watts, the authors calculated a whole-body SAR as averaged over all exposure times of 2.1 W/kg. However, it should be noted that mice exposed at the center of the wave guide would have experienced a whole-body average SAR of 4.2 W/kg. The authors indicated that during the 33rd and the 34th RF exposures, four mice died from thermal effects. They found that the exposure field was much higher than estimated due to a standing wave. When the authors placed the exposure cages 1 inch closer to the source of rf power away from the maximum electric field, no further lethality was observed during the exposure. It should be noted that based on the RRDH, 43 mW/cm$^2$ and 83 mW/cm$^2$ free space exposures would produce whole-body average SARs of 1.29 to 12.9 W/kg for the former and 2.58 to 25.8 for the latter, depending on the orientation of the animals.

The authors concluded that no detrimental effects, detectable in the blood or the physical condition of the animals, were evoked from the rf exposure. Though the mean life span of the experimental group was 19 days longer than that of the control group, the difference was not statistically significant.

**Preskorn et al.**[3] reported greater longevity and retarded tumor growth in mice exposed fetally to RFR. The exposure apparatus consisted of a modified Tappan R3L multimode cavity fed by half-wave, 60 Hz sinusoidaly modulated 2450-MHz microwave energy from an overhead waveguide. Four or more maternal subjects and later, four weanling offspring were exposed for each treatment. A total of 4 in utero and 36 postnatal treatments of 20 minutes in duration were alternated across time, within days, between radiation and sham treatments. Average whole-body dose rates of 35 mW/kg were obtained calorimetrically from exposed animals. Based on the RRDH, the equivalent average whole-body plane wave exposure would be 29.2 mW/cm$^2$. The before and after treatment colonic temperatures averaged 37.4° and 39.8° respectively.

The authors' experiment was based on a 2 x 2 factorial design with 12 CFW mice each assigned to one of 4 exposure conditions including 2 exposure levels, 0 and 35 mW/g, and 2 treatment periods, in utero (11-14 day of gestation) and postnatal (19-54 day postpartum). During the 16-day postpartum each of the 48 weanlings was injected with a sterile homoginate of tumorous tissue containing the avian, fast reticuloendothelial T virus prepared from sarcomatous tumors of adult CFW donors. All radiation treatments were administered between 7 am and 10 am at an ambient temperature averaging 23° C, humidity ranging between 50% and 75% and a forced ambient air flow velocity of 0.1 m/sec. Group 1 mice were sham exposed during both exposure periods. Group 2 mice were sham exposed in utero and exposed later, Group 3 mice were exposed in utero and sham exposed later. Group 4 mice were exposed during both periods. One treatment was given daily with all days of treatment occurring consecutively during each of the two exposure periods. After implantation of the homogenate each of the 48 weanlings were examined each day both visually and by palpation until the 93rd day postpartum at which time they were euthanized by an overdose of sodium pentobarbital to permit histopathological examination. The fetally and postnatally exposed animals of group 1 developed tumors at a rate normally reported for a CFW mice of the commercially available strain. Five out of 12 of the group 2 mice developed sarcomas, indicating that the postnatal treatments had little effect on the rate of tumor induction. Two of the 12 fetally but not postnatally exposed mice of group 3 and 1 of the 12 mice of group 4 exposed during both periods developed sarcomas. The reduction of incidence of tumors in the fetally exposed animals was statistically significant. One of the mice from each of the groups 3 and 4 not only had a lower incidence of sarcomas but experienced an apparent regression of tumors. The second experiment was similar to the first except larger numbers of gravid CFW mice were used to provide larger samples of experimental and control subjects and all the treatments were administered in utero. Both exposed and control animals were observed for longevity across a 3-year span of time. Eighty-four weanlings were selected from 10 exposed dams, and 60 weanlings were selected from 8 sham-exposed dams. The 44 weanlings were implanted with the T virus homogenate on the 16th postpartum and thereafter segregated by sex into groups of 5 or fewer animals and placed in cages in a mouse vivarium. As in the first experiment, observations at 2.5 months implantation indicated that the percentage of fetally exposed mice with tumors (15%) was statistically smaller than that (37%) of controls. Later, however, the difference narrowed and by the 4th month 46% of the fetally exposed mice and 40% of the control mice had tumors, an insignificant difference. Though the tumors were delayed in the fetally exposed mice, the absolute incidence was not significantly altered. Again in this group there was evidence of regression of tumors in fetally exposed mice. The authors conclude that the data of both experiments support the notion that moderate short-term elevation of fetal body temperature by microwave radiation can delay onset of an experimental neoplasms and reduce the mortality rate of tumor-bearing animals. They further believe that the data reflects an enhanced immunocompetency that has its origin in elevation of fetal, and perhaps maternal temperature.

**Guy, et al.**[4] report the results of exposing male New Zealand white rabbits to 10 mW/cm$^2$, 2450 MHz cw microwave radiation for 23 hours per day for 180 days in a miniature anechoic chamber. A total of 8 rabbits were studied simultaneously; 4 received radiation and 4 served as controls. Each rabbit was kept in an acrylic cage within a rectangular, 16 x 16 cross section, 122 cm high chamber with 20.3 cm thick microwave absorbing fiber walls and floor. A standard microwave horn was placed 1 meter above the centerline of the body of each rabbit. The sham-exposure chambers were constructed with thinner panels of fiber packing material of the same size and similar composition as the microwave absorber. In all cases the absorber and packing material panels were covered with a thin plastic film to block outgasing that might arise from the material. With the exception of the absence of microwave horns, topography of the control chambers were similar to that of the exposure chambers. The ambient temperature of within each chamber was maintained at 24° ± 2° C by thermostatically controlled air conditioner. A watering system was designed to eliminate the problem of field intensification during drinking. As water entered the chamber through a tubing it was separated into individual, radiolucent drips by a needle valve. Each rabbit would drink by catching drops in its mouth with excess water draining immediately from the chamber, thereby breaking up any continuously conductive pathway. The floor of each acrylic box consisted of 1 cm diameter glass rods spaced at 2 cm. Urine was drained to the outside of the chamber in a collection beaker. Each animal remained in a position with its body extended along the long dimension of the cage (parallel to the electric field for exposed animals) during most of the long-term exposure. Input power to each standard gain horn was set at 32 Watts, which provided a power density of 7 mW/cm$^2$ at the position of the body axis of the animal (measured with the animal absent) and 10 mW/cm$^2$ at the usual position of the animal's head, approximately 16 cm above the long axis of the body. Thermographic measurements, based on techniques described by Guy,[14] indicated a peak SAR of 17 W/kg in the head of the exposed animal. Maximum whole-body-average SAR is estimated to be 1.5 W/kg based on the RRDH.

The animals, weighing approximately 4 kg, received periodic examinations of the eyes with a slit lamp microscope and the following parameters were monitored throughout the exposure period: body mass, urinary output, rectal temperature, hematocrit, hemoglobin, white cell count, platelet count and basic blood-coagulation studies. No significant differences other than a decrease in a percentage of eosinophils were found between the experimental and control animals. The authors noted, however, that eosinophil percentage varies widely among animals.

**McRee et al.**[5] performed additional analyses on the same animals exposed by Guy et al.[4] discussed above. Immediately after termination of exposure of the rabbits, blood samples were drawn for standard hematologic and serum-chemistry analyses. Necropsies were performed on all eight animals and histopathological analyses were made on specimens of the tongue, esophagus, trachea, lung, heart, liver (including gall bladder), stomach, small and large intestine, spleen, thymus, kidney, urinary bladder, testes, skin (ear), brain, thyroid, pancreas, adrenal, pituitary, sternum (for bone marrow) and thigh muscle. A sample of splenic tissue from exposed and control animals was taken aseptically from cadavers for immunological studies. The lymphoid cells obtained from individual spleens were cultured and stimulated with the mitogens, phytohemagglutinin-P (PHA), Concanavalin A (Con A) and pokeweed mitogen (PWM).

Of the hematological and clinical chemistry parameters analyzed from the blood samples, significantly lower levels of albumin, calcium and eosinophils were found in the samples from the exposed animals. The authors stated that since only 3 of the 41 parameters were significantly different, which is close to the number expected by chance; the validity of these changes can only be determined through independent replication. No significant changes were seen in catecholamines and creatinine content in the urine. Blood specimens taken from the animals 30 days after termination of exposures showed no significant differences. Depression in eosinophil numbers seen earlier had normalized 30 days after exposure. A reliable decrease in the albumin (albumin/total globulin ratio [A/G]) was found in the exposed animals, and was due to both the decrease in the albumin and an increase in the total levels of globulin. The authors noted, while neither of these 2 parameters has changed significantly in themselves, the combination of the trends causes a significant change in the A/G ratio. They concluded that the biological importance of this finding is difficult to interpret and may be negligible since A/G ratio of the treated and controls was essentially the same at the termination of the exposure. Examination of bone marrow revealed an abnormal myeloid/erythroid ratio (ratio of leukocytes of the granulocytic series to nucleated erythrocyte precursors) in the exposed rabbits. The authors state that the importance of the finding is questionable since the hematologic (erythrocyte and leukocyte counts) parameters did not differ between treated and control rabbits. Both absolute and relative masses of organs as a percentage of body mass showed no significant differences in the exposed and control animals. At necropsy, histopathologic examination of representative tissues showed no lesions that could be attributed to microwave irradiation, which led the authors to report that the chronic irradiation at 10 mW/cm$^2$ either did not induce pathological changes or it induced only minimal, reversible changes. In vitro response of lymphocytes to stimulation was reduced for exposed animals as compared to control levels, but the difference was not statistically significant. The same statements are true for splenic cells cultured in the presence of ConA, but at all three levels of PWM employed, there was a significant reduction in responsiveness of spleen cells to stimulation in microwave-exposed specimens as compared to controls. The authors state that due to the limited number of animals used in the study, the results are not meant to be definitive. Rather, they are aimed at providing insights into potential biological effects that should be further studied.

**Szmigielski et al.**[7] studied the effect of RFR on accelerated development of spontaneous and benzopyrene-induced skin cancer in mice. The authors exposed 2 groups of mice: C3H/HeA mice with a high incidence of spontaneous breast cancer and Balb/c mice with skin cancer, resulting from painting with 3,4-benzopyrene (BP). The animals were exposed at 5 or 15 mW/cm$^2$ for 2 hours daily, 6 sessions per week. The C3H/HeA mice were exposed from the 6th week to the 12th month of life. The BP-treated Balb/c were exposed either 1 or 3 months prior to or simultaneously over 4 months with BP treatment. The mice were exposed in an anechoic chamber with walls covered with styrofoam-graphite absorber and a 2450 MHz microwave beam directed vertically from a 30 x 30 cm horn antenna placed at a distance of 220 cm above the cages. The sham-exposed animals were exposed at different times in the same chamber without microwave radiation. The power source was a microwave generator produced in the USSR with a maximum power output of 150 Watts cw. Ten mice were contained in each of 4 polymethacrylate cages placed on the 30 x 50 cm floor of the anechoic chamber. Power densities were measured with the cages removed. The authors did not indicate whether the mice were free to move around or were restrained to a particular orientation with respect to the polarization of the horn antenna. The authors determined the SAR calorimetrically by exposing a mouse cadaver and a liquid phantom, with implanted liquid crystal probes. The cadaver and phantom were exposed to power densities ranging from 20 to 60 mW/cm$^2$ and the temperature changes after 1, 3, and 5, minutes exposure were noted. From the temperature increase and known power density, the authors calculated the SAR for 5 and 15 mW/cm$^2$ exposures as 2-3 W/kg and 6-8 W/kg respectively. No comparisons between the SARs were obtained for the cadaver and the phantoms nor was any indication given of the polarization of the field with respect to the long axis of the exposed animals. According to the RRDH, the estimated maximum SARs for the 2 exposure levels of 5 and 15 mW/cm$^2$ would be 6 W/kg and 18 W/kg respectively for E-polarization (electric field parallel to the long axis of the animal) or 2.2-6.6 W/kg respectively for H-polarization (magnetic field vector aligned parallel to the long axis of the exposed animal). The measurements appear to be consistent with exposure to H-polarized RFR. However, if the animals were free to move around so that they could be aligned with the E vector just as often they could be aligned with the H vector, the whole-body-average SAR would likely be 4.1 and 12.3 W/kg respectively for the 2 exposure power densities. The authors did not indicate whether the measurements were done on a single cadaver or phantom in the chamber or whether they were exposed in the presence of other subjects.

The authors subjected one group of male Balb/c mice (6 weeks old) to chronic stress by placing each into a 5 x 6 x 10 cm compartment with 20 contained in a 20 x 30 x 10 cm cage. Each compartment was provided with standard food and water. The animals were grown under these conditions for 8 to 10 months depending on the schedule of the experiments. These animals served as additional controls to normal sham-exposed and microwave-exposed animals. Measurement of natural antineoplastic resistance was done by injecting solid transplantable L1 sarcoma cells into recipient healthy Balb/C mice. Biologic experiments determined that development of 1-4 nodules in the lungs would result, 14 days after injection of 2 x 105 cells (in 0.1 ml of saline). In animals exposed either to microwaves or to chronic confinement stress significantly higher numbers of lung nodules were observed after 1 or 3 months of treatment. After 3 months of exposure at 15 mW/cm$^2$,

10.8 ± 2.1 lung nodules were found vs 6.1 ± 1.8 nodules found in mice exposed at 5 mW/cm$^2$; the latter was close to the 7.7 ± 2 nodules found in mice with chronic confinement stress. Nodules in sham-exposed mice (3.6 ± 2.2) was significantly less than those in mice exposed with microwaves at 5 mW/cm$^2$. Experiments on C3H/HeA mice showed similar results. Radiation with microwaves resulted in significant acceleration of breast tumors in mice exposed at 15 mW/cm$^2$. The first tumors developed in the 5th month of life with half of the animals developing tumors at 219 days. Of the animals exposed to microwaves at 5 mW/cm$^2$ or to chronic confinement stress, half of the animals had tumors at 261 days and 255 days respectively. Both were significantly different from controls or sham-exposed animals (297 days). Survival times in the mice also followed the same trends. The mean survival time of one-half of the animals (MST50) in controls was 358 days, MST50 for 15 mW/cm$^2$ exposed, 5 mW/cm$^2$ exposed, and confined mice was 231, 264 and 272 days, respectively while MST50 for control and sham exposed mice was 326 days. For control Balb/C mice the mean skin cancer development time (MCDT) was 285 days; the time where half the animals developed skin cancer (CDT50) was 296 days; and MST50 was 205 days. Microwave exposure and confinement stress prior to treatment with BP resulted in significant acceleration of the development of skin cancer and shorter survival time. For mice exposed to 15 mW/cm$^2$ microwaves for 3 months and later treated to BP, MCDT was 160 days; CDT50 was 171 days; and MST50 was 205 days. For animals sham exposed for 3 months and later treated with BP, MCDT was 256 days; CDT50 was 272 days; and MST50 was 312 days. Similar results were found in groups of mice exposed to microwaves and treated with BP at the same time. The shortened survival time was also found for these animals as a result of the earlier appearance of cancer. Also for mice exposed to 15 mW/cm$^2$ microwave radiation prior to treatment with BP and simultaneously with application of BP was early appearance of fully developed skin cancer; MCDT was 121 days; CDT50 was 131 days; and MST50 was 165 days. There was very little difference in the results for mice exposed to 5 mW/cm$^2$ and those exposed to chronic confinement stress. The authors concluded that their data demonstrated significant acceleration in the development of both spontaneous and chemically induced tumors in mice exposed with 2450 MHz microwaves for 1-6 months. The authors indicate that the results are accompanied with the lowering of the natural antineoplastic resistance both for microwaves applied before or simultaneous with benzopyrene treatment.

Chou et al.[6] reported results of an experiment on the effects of continuous and pulsed chronic 2450 MHz microwave exposure on body weight, blood, eyes, electroencephalogram, evoked potentials, and pathology of young 3-month-old adult New Zealand rabbits. Eighteen rabbits (nine males, nine females) were equally divided into 3 groups. One group was exposed to 1.5 mW/cm$^2$ CW for 2 hours daily for 3 months, another group was simultaneously exposed to the pulsed microwaves (10 μs, 100 pps) at the same power density, and the third group was simultaneously sham exposed. Exposures were made in 3 miniature anechoic chambers of the type described by Guy[15]. The RFR in each chamber was provided by an S band standard gain horn located 1 meter above the animal. The inside walls and floors of the chambers were lined with microwave absorber. The rabbits were contained in a plexiglas cage oriented so that their long-body axis was parallel to the electric field most of the time. Urine and feces were collected in a Plexiglass pan located under the cage. The chambers were ventilated by a fan mounted at the top that drew air up through the porous microwave absorbing floor. The exposure area was illuminated by 4 lights placed at the top of the chamber around the waveguide adapter above the horn so there would be no field perturbation. The animals had access to dry rabbit food chow but no water during each 2-hour exposure. The SAR patterns were obtained thermographically at the sagittal plane by exposing a sacrificed rabbit to high-power fields for a short time as described by Guy[14]. Peak SARs of 2.1 W/kg and 1.6 W/kg were measured in the back and head of the animal, respectively. The estimated whole-body average was 0.24 W/kg based on the RRDH. Three radiolucent carbon-loaded Teflon electrodes were implanted in the head of each animal so that electroencephalographs and evoked-potential recordings could be made during the experiments. Exposure began 1 week after the final surgery. Exposure for each group of 3 animals was rotated to control for possible effect of circadian rhythms.

The authors permanently implanted the EEG electrodes in the skull in contact with the dura, at the sensory motor, occipital, and nasal areas of the brain. The opposite end of each carbon-loaded Teflon fiber was terminated by a metal pin embedded in dental acrylic placed over the area of the implanted electrodes. At appropriate times between exposures, EEG signals were recorded by connecting the leads of the recording instruments to the metal pins embedded in the acrylic. Measurements were made every other day of body weight and blood samples were taken monthly for hematological, chemical and morphological studies. EEG and evoked potential recordings were made every Friday after the 2-hour exposure. The EEG from sensory motor and occipital areas was recorded after a 5-10 minute adaptation period from the time the electrodes were connected over a period of 20 minutes. Subsequently 5-minute, visual-evoked responses were recorded from the occipital cortex by stimulating the eyes with a strobe light at the rate of 1 flash per second. During the last 5 minutes the auditory-evoked responses were recorded from the sensory motor cortex by presenting 70 dB, 0.1 ms clicks at the rate of 1 per second. The EEG data were analyzed by computer, off line, to obtain the frequency spectrum and averaged amplitudes. The evoked responses were recorded by means of a computer-averaged transients.

Since there was so much variability in the frequency spectrum from animal to animal and different recording sessions, individual frequencies could not be compared so integration of the power spectrum was performed. Though there were large values of standard deviations and great variability in the results, there was a trend of decreased amplitude of the EEG signals in all of the animals in the later part of the experiment, probably due to acclimation. Statistical tests showed no difference in the data obtained from CW, pulsed and control animals. Slit lamp examinations both before and after the 3-month exposure revealed no cataract development in any of the 18 animals. Unsuccessful attempts were made to determine the effect of apomorphine induced hypothermia at the end of the 3-month exposure. Not only was there a large variability in observed temperature rise and behavioral responses on the small group of animals but five animals (1 CW exposed, 3 pulsed exposed, 1 control) died during the experiment. Thirteen survivors were sacrificed and histology was performed on lungs, heart, vessels, stomach, small and large intestines, pancreas, liver, gall bladder, adrenal gland, kidneys, bladder, testes or ovaries, bone marrow, spleen, and brain. Histopathological studies showed no consistent, significant differences among the 3 groups.

**Chou et al.**[8] studied the effect of 2450 MHz cw microwave exposure on health profile of two groups of 16 male New Zealand rabbits. The rabbits were exposed to incident power densities of 0.5 and 5 mW/cm$^2$ for 7 hours daily, 5 days a week for 13 weeks. The effects of the radiation were assessed on food consumption, body mass, blood parameters including hematology, morphology, chemical, protein electrophoresis, and lymphocyte blast transformation. The eyes were examined for cataract formation and pathological examinations of 28 specimens of organs and tissues of each rabbit were performed. Sixteen miniature anechoic chambers as described by Guy[15] and in the above paper overview were used for chronically exposing the rabbits to 2450 MHz cw microwaves. During the 7-hour exposures used for this experiment, water was supplied through a special water-dripping system. A water bag was hung outside of the chamber and water was guided by plastic tubing to drip into a small cup at the rate of 1 drop per second. Excess water flowed to a plastic bag at the bottom of the chamber. The design eliminated any field enhancement due to animal contact with a voluminous water container. A total of identical 8 exposure and 8 sham-exposure chambers were set up in a single room. In the first experiment 8 New Zealand rabbits, initially weighing approximately 2 kg were exposed to 0.5 mW/cm$^2$, 7 hours a day, 5 hours a week for 13 weeks. The other 8 animals of similar body mass were sham exposed at the same time. The SAR patterns in the sagittal plane, obtained thermographically as described by Guy,[14] indicated a peak SAR of 0.7 W/kg in the back and 0.55 Watts in the head of the exposed animal. Based on the RRDH, the whole-body-average SAR for this exposure level was 0.075 W/kg. The second group of animals, exposed to 5 mW/cm$^2$, would experience a whole-body average SAR of 0.75 W/kg and peak SARs of 7 W/kg and 5.5 W/kg in the back and head, respectively. The animals were exposed between 9 am and 4 pm at an environmental temperature of $21° \pm 1.5°$ C and relative humidity of $50\% \pm 10\%$. Body mass and food consumption during the 7-hour exposure period was measured daily. Blood samples were taken before the initial exposures and monthly thereafter from which hematological, chemical, protein electrophoresis, and lymphocyte studies were performed. Lymphocyte studies included lymphocyte blast transformation, mitotic index and stimulation index. All animals were examined for cataracts with a slit lamp before and at the end of the 13-week exposure period. Complete gross necropsy on each animal was performed, and histological examinations and evaluations were made on tissues from all organs including femur bone marrow, pituitary, adrenals thyroid, parathyroid, trachea, esophagus, brain, heart, skeletal muscle, spleen, liver, gall bladder, pancreas, lungs, salivary glands, cervical lymph node, kidneys, urinary bladder, testes, epididymis, prostate, stomach, duodenum, ileum, colon, mesenteric, and any grossly observable lesions. The tissue samples were microscopically evaluated in the blind by a pathologist. The only statistical difference found between test results on the exposed and the control animals was a significant (P < .01) decrease in food consumption in the 5 mW/cm$^2$ exposed animals. Based on this finding and similar findings by other investigators, the authors felt that the absorbed microwave energy (0.75 W/kg is 20% of metabolic rate) was used by the animals partially for metabolism causing 23% reduced food consumption during the 7 hour 5 mW/cm$^2$ exposure. It is also pointed out that the failure to find changes in blood tests was not in agreement with several other similar investigations conducted in the Soviet Union and East European countries.

**Adair et al.**[9] investigated the effect of 2450 MHz CW RFR on changes of both behavioral (stable internal temperature controlled by voluntary behavior) and physiological (stable internal temperature controlled by involuntary autonomic mechanisms) thermoregulatory responses in squirrel monkeys exposed for 40 hours/week for 15 weeks at power densities of both 1 mW and 5 mW/cm$^2$ at controlled environmental temperatures of $25°$, $30°$, or $35°$ C. The monkeys were chronically exposed, a pair at a time, in a standard 2.45 x 2.45 x 3.66 m anechoic chamber to microwave radiation from a 15-dB standard gain horn. During exposure within the anechoic chamber and sham exposure outside of the chamber, the monkeys were housed individually in exposure cages consisting of 61 cm high, 30.5 cm diameter plexiglas cylinders, each contained in environmentally controlled 5 cm thick 61 x 91 x 122 cm insulating foam chambers. A drip water system as described by Guy[15] supplied water to the animals. Although the RRDH indicates an SAR in the squirrel monkey of 0.1 (W/kg)/(mW/cm$^2$), the authors measurements on exposed phantom models indicated a whole-body-average SAR of 0.16 (W/kg)/(mW/cm$^2$). The difference is probably due to the difference in heights of the models (33 cm used by the authors versus 23 cm used by the RRDH).

The authors performed standardized tests of the physiological thermoregulation on each animal prior to and at predetermined times during and after chronic exposures to establish reliable baselines. The tests were conducted in a 92 x 102 x 66 cm insulated chamber with no microwaves present. Under different controlled environmental temperatures, four skin temperatures (abdomen, tail, leg and foot), colonic temperature, oxygen consumption, and sweating rate from the foot of a restrained monkey were measured each minute by an on-line computer. All environmental temperatures were controlled within ± 0.5° C.

The authors performed standard behavioral tests by restraining the monkey in the far field of a 15-dB standard gain horn while it was enclosed by an environmentally controlled insulated foam box within an anechoic chamber. Each monkey was trained to select between between 2 preset air (flow velocity of 0.36 m/s) temperatures of 10° and 50° by means of pulling a response cord. Each response was reinforced by a 15-second application of 50° C air followed by an application of 10° C air lasting until the animal responded again. During the behavioral tests, the monkeys were exposed to microwave power densities of 0, 4, 6, 8, 10, and 12 $mW/cm^2$ with corresponding SARs (obtained from phantom and confirmed by live-animal measurements) of $(0.15 W/kg)/(mW/cm^2)$. The test protocol required each animal to equilibrate for a minimum of 1 hour to an environmental temperature of 20° ± 0.5° C. The temperature was then successively ramped to different levels (6.6° increment for the first, maintained for 45 m, and 3.3° increments maintained for approximately 37.5 m, each, for the remaining increments) up to 36.5° C. After training, each animal underwent 3 or more 4-hour baseline test sessions of behavioral regulation to establish the normally preferred environmental temperature pattern of responding and the preferred colonic and skin temperatures. Prior to chronic microwave or sham exposure, each animal was given 3 standardized behavioral tests to determine the microwave intensity for whole-body exposure that would reliably alter thermal regulatory behavior, which was found to be 10 $mW/cm^2$.

Animals were chronically exposed to 2 power densities, 1 and 5 $mW/cm^2$, and 3 environmental temperatures, 25°, 30°, and 35° C, for a total of 6 treatment groups. All groups of animals were subjected to 3 test phases including: 1) pre-exposure phase of 8 to 12 weeks duration during which a number of physiological, baseline behavioral thermal regulation without microwaves, and behavioral with microwaves (to determine the microwave threshold to alter thermoregulatory behavior) tests were given; 2) chronic microwave or sham exposure phase of 15 weeks duration during which a number of physiological and behavioral tests were given; and 3) post exposure follow-up phase of 4 to 8 weeks where the animals remained in the home cage, except when additional physiological behavioral and tests were administered.

Blood samples were taken at 1, 5, 10, 15, and 20 weeks of the treatment. Assessments were made of cell counts, hemoglobin determination, serum-thyroxine concentration, thyroxine binding capacity, blood sodium, potassium, bicarbonate, chloride concentrations, total serum protein and albumin concentrations.

Tests showed no significant differences in metabolic rate, internal body temperature, blood indexes, or thermoregulatory behavior between sham- and microwave-exposed animals. However, the authors found that the ambient temperature prevailing during chronic exposure could produce an effect, especially in the 35° C environment where there was an increase in sweating rate. Skin temperature was found to be reliably influenced by both ambient temperature and microwaves, and the most robust effect of microwave exposure was found to be a reduction in body mass as a function of power density.

Chou et al.[10] reported the biological effects in 100 male Sprague-Dawley rats exposed over a lifetime to pulsed-microwave radiation as compared to those in an equal number of sham-exposed animals. All 200 rats were maintained under specific pathogen-free (SPF) conditions while individually housed in circular waveguides, half of which were energized to produce circularly polarized 2450 MHz pulsed (10 μs duration, 800 pps) exposure and half of which were not energized. The pulsed microwaves were square-wave amplitude modulated at 8 Hz, producing 62.5 ms wide trains of 50 pulses each separated by 62.5 ms intervals. The modulation was applied, based on evidence of modulation frequency window effects[16] which were reported to be most pronounced at dominant EEG frequencies (16 Hz) of exposed chicken and cat brains. Eight Hz is the equivalent frequency for the rat brain.

Exposed and sham-exposed animals were individually housed in single plastic cages placed in each waveguide. Exposure and sham-exposure guides were randomly placed in alcoves in the SPF rooms where air-flow rate was programmed for 22 exchanges each hour at a temperature of $21° \pm 1°$ C and a relative humidity range of 30% to 70%. The exposure waveguides were each energized by a transducer at one end that excited circularly polarized waves with power that was partially absorbed by the exposed animal, partially absorbed by the waveguide walls, partially reflected back to the feed transducer, and partially absorbed in matched terminating loads at the other end. As described by Guy et al.,[17] the whole-body-average SAR of the rats exposed to the guided circularly polarized waves remains relatively constant regardless of the orientation of the animals. Whole-body-average SAR in the exposed rats was determined by 2 methods, 1) measuring the SAR by twin-well calorimetry in sacrificed animals ranging from 200 to 800 grams in mass, exposed to high power density and 2) measuring the waveguide reflected and terminal load absorbed powers and subtracting them from the incident power to determine the total power absorbed in the waveguide walls and exposed animal. The waveguide power loss may be quantified by subtracting the the power loss in the animal, determined by the twin-well calorimetry, from the total. During the long-term exposure period, the whole-body SAR of each exposed animal was continuously monitored for one day out of each 50 days by this method. An average power density exposure level of 480 $\mu W/cm^2$ was chosen since the dosimetry measurements showed that the whole-body-average SAR ranged from 0.4 W/kg (basis of a number of exposure standards) for a 200 gm rat to 0.15 W/kg for an 800 gm rat. The exposures began at 8 weeks of age and continued daily, 21 1/2 hours per day, for 25 months. Blood samples taken at intervals of every 6 weeks up to the 60th week and every 12 weeks thereafter, were analyzed for serum chemistries, hematological values, protein electrophoretic patterns, thyroxine, and plasma and corticosterone levels. In addition, daily measures were made of body mass, food and water consumption by all animals, and $O_2$ consumption and $CO_2$ production in a sub-sample (N = 18) of each group. Behavioral activity was assessed in an open field apparatus at regular intervals throughout the study. After 18 months, ten rats from each group were euthanized to test for immunological competence and to permit whole-body analysis, as well as gross and histopathological examinations. At the end of 25 months, the survivors, consisting of 11 sham-exposed and 12 radiation-exposed rats, were euthanized for a similar analysis. The remaining 159 animals, upon either spontaneous deaths or terminated in extremis, were examined histopathologically. A total of 155 biological parameters were examined.

Microwave exposure did not produce any significant effects on activity levels of the animals nor did it cause any significant effects seen in the serum corticosterone levels, except for a significant transient elevation at the time of the first session only for the exposed animals, and at the time of the third session only for the sham-exposed animals. Exposed and sham-exposed animals had comparable levels of corticosterone on all other regular sampling sessions. A follow up study by Chou et al.[18] of two groups of 20 animals each, exposed in the same system for 6 and 12 months revealed no statistically significant differences in corticosterone levels between exposed and sham-exposed animals. Immunological competence studies of rats sacrificed at 13 months indicated a significant increase in both splenic B cells and T cells, which was not detected in the animals sacrificed at the end of the 25 months of exposure. No significant effects were seen in the percentage of complement-receptor-positive cells in the spleen either for the interim or final euthanasia. Mitogen-stimulation studies on the animals sacrificed at 13 months exposure revealed significant differences between groups in their responses to B- and T cells specific mitogens. Exposed animals had a nonsignificant increase in the response to PHA but a significant increase in response to lipopolysaccharide (LPS) and mitogen PWM as compared with sham-exposed animals. The exposed animals had a significantly increased response to ConA and a decreased response to purified protein derivative of tuberculin (PPD). Mitogen-response data were not available from the 25-month euthanized animals since lymphocyte cultures failed to grow. In the follow-up study by Chou et al.[18] no significant differences between 20 exposed and 20 sham exposed rats were observed in the proliferation of thymocytes to ConA, PHA, and PWM after 6 and 12 months of RF exposure, nor were there any differences found for splenocytes stimulated by LPS, PHA, PPD, ConA, and PWM. The follow-up study shows cytometry revealed no group alterations in the number and in the frequency of B and T cells. However, the follow-up study did show that after a 12-month exposure, there was a reduction in cell surface expression of Thy 1.1 (T-cell related) surface antigen and a reduction in the mean cell-surface density of s-IG (B-cell related) on small lymphocytes and spleen. Stimulatory effects observed in the original study were not confirmed. Evaluation of mytology, serum chemistry, protein electrophoretic patterns and fractions, and thyroxene levels revealed no significant differences between groups. Likewise, no effects of microwave exposure were observed on overall patterns of growth, food and water consumption, and body-mass loss and recovery. However, there was a highly significant elevation of adrenal mass (75% increase) observed for exposed rats as compared with sham-exposed animals, which became insignificant when the animals with benign tumors in the adrenal gland were separated from those without tumors. For the animals with tumors the adrenal mass was significantly higher in the exposed group than in the sham exposed group. Thus the analysis demonstrated that the increase in adrenal mass was related to tumors and was therefore independent of the metabolic processes in the rats. A significant decrease in $O_2$ consumption and $CO_2$ production was seen in the exposed young rats but not in the exposed mature animals. Though the survival curves indicated a lower mortality rate and longer mean survival time (668 days) for the exposed animals than for the sham-exposed (663 days), the differences were not significant. No association was found between specific cause of death and treatment conditions; however for cause of death due to urinary tract blockage (9 in the exposed group and 19 in the sham group), there was some indication that survival times were longer in the exposed animals. Histopathological studies indicated that chronic glomerulonephropathy was the most frequent cause of death and one of the most consistently encountered non-neoplastic lesions. Statistics indicated that the lesion was less frequently observed in the exposed than in the sham-exposed animals. There were no significant differences in other non-neoplastic lesions. The incidence of neoplastic lesions corresponds with that normally reported for the Sprague-Dawley rats. Due to the low incidence of neoplasia with no significant increase in specific organ or tissue, it was necessary to collapse the data and to make evaluations with respect to occurrence of neoplasms with no attention given to the site or organ of occurrence. There was no evidence of either the exposed or sham group having an excess of benign lesions, nor was there any significant difference between the total neoplastic incidence if both benign and malignant lesions were included. However, there was nearly a fourfold increase in primary malignancies in exposed animals (18 vs. 5) as compared to the sham-exposed animals. The authors state that though the overall difference in numbers of primary malignancies is statistically significant, biological significance is open to question since: 1) the detection of the difference required the collapsing of sparse data without regard for specific type of malignancy or tissue origin; 2) the incidence of the specific primary malignancies in exposed animals is comparable to specific tumor incidence reported in the literature; 3) no single type of primary malignancy was enhanced in the exposed animals; 4) the lack of any

## TABLE 1. SUMMARY OF STUDIES CONCERNING LONG-TERM RADIATION EFFECTS ON ANIMALS

| Effects | Species | EXPOSURE CONDITIONS | | | | | References |
|---|---|---|---|---|---|---|---|
| | | Frequency (MHz) | Intensity (mW/cm2) | Duration (daysxmin) | AVG. SAR (W/kg) | Peak SAR (W/kg) | |
| Increase in longevity/suggestion of testicular degeneration/ pneumonia infection during exposure. | Swiss albino mice | 9270 pulsed | 100 | 295x4.5 | 45 (HB)* | | Prausnitz and Susskind (1962) |
| Slight but nonstatistical increase in longevity | strain RF female mice | 800 | 43 avg. 86 peak | 175 x 120 | 12.9 (HB) 25.8 (HB) | | Spalding et al (1971) |
| Increase in immunocompetency/slower development of implanted tumors but final number not affected/ > mean longevity in exposed mice with and without tumors. | CFW mice | 2450 half wave 60 Hz modulation | 29.2 (HB) | 4 x 20 (in utero) | 3.5 | | Preskorn et al (1978) |
| No effects on eyes as seen by slit lamp, body mass, urinary output, rectal temperature, hematocrit, hemoglobin, white cell count, platelet count and basic blood-coagulation | New Zealand white male rabbits | 2450 | 7 - 10 | 180x1380 | 1.5 (HB) | 17 | Guy et al. (1980) |
| Decrease in albumin to globulin ratio/abnormal myeloid to erythrocyte ratio in bone marrow/no pathological changes/reduced response of spleen cells to stimulation by pokeweed but not other mitogens/ results not meant to be definitive due to low number of animals | New Zealand white male rabbits | 2450 | 7 - 10 | 180x1380 | 1.5 (HB) | 17 | McRee et al. (1980) |
| Acceleration of appearance of spontaneous mammary cancer in females, skin cancer in males treated with 3,4-benzopyrene, and lung cancer in males injected with $L_1$ sarcoma /5 mW/cm2 exposure produced same effect as cage confinement | Balb/c male mice C3H/HeA female mice | 2450 | 5, 15 | 217x120 | 2, 6 6, 18 (HB) | | Szmigielski et al.(1982) |

TABLE 1. SUMMARY OF STUDIES CONCERNING LONG-TERM RADIATION EFFECTS ON ANIMALS (continued)

| Effects | Species | Frequency (MHz) | Intensity (mW/cm²) | EXPOSURE CONDITIONS Duration (daysxmin) | AVG. SAR (W/kg) | Peak SAR (W/kg) | References |
|---|---|---|---|---|---|---|---|
| No effects on EEG, evoked potentials, eyes seen by slit lamp, body weight, blood parameters, histopathological parameters | young adult New Zealand male & female white rabbits | 2450 cw and pulsed | 1.5 | 92×120 | 0.24 (HB) | 2.1 | Chou et al. (1982) |
| No effects on body mass, blood parameters including hemotology, morpology, chemical, protein electrophoresis, lymphocyte blast transformation, mytotic index and eyes seen by slit lamp and histology/food intake was reduced by 23% during exposure which is equal to percentage of SAR to metabolic rate | New Zealand male white rabbits | 2450 | 0.5, 5 | 420×65 | 0.075, 0.75 (HB) | .7, 7 | Chou et al. (1983) |
| No effects on behavioral thermoregulatory function, metabolic rate, internal body temperature, thermoregulatory behavior and blood indices, effects on skin temperature and reduction of body weight as function of power density | squirrel monkeys | 2450 | 1, 5 | 75×480 | 0.16, 0.8 0.1, 0.5 (HB) | | Adair et al. (1985) |
| Transient changes in corticosterone levels and immunological parameters excess primary malignancies in exposed animals; but authors state, in light of other parameters in study, it is conjectural whether excess reflects a true biological influence | Sprague-Dawley male rats | 2450 pulsed | 0.5 avg 125 peak | 758 x 1290 | 0.15-0.4 | | Chou, et al (1992) |

* HB refers to estimates made from the Radiofrequency Radiation Dosimetry Handbook (Durney et al., 1986)

significant difference in benign neoplasia which, morphologically, generally precedes or is part of the process leading to malignant neoplasia; and 5) with the induction of a cancer by a carcinogen, tissue specific effects are usually induced, so that an agent is not usually considered carcinogenic unless it induces a significant response in any one tissue.

The authors conclude by indicating that the study showed no biologically significant effects on general health, serum chemistry, hematological profiles, longevity, cause of death, and lesions commonly associated with aging and benign neoplasia. Statistically significant effects in corticosterone levels and the immunological parameters 13 months exposure were not confirmed in the 25-month exposure nor the follow-up study. $O_2$ consumption and $CO_2$ production were lower in the exposed rats but the effects were not observed in the mature rats. The findings of excess primary malignancies in exposed animals is provocative but single findings considered in light of other parameters is conjectural and may not reflect a true biological influence. Positive findings need further independent experimental evaluation.

## REFERENCES

1. S. Prausnitz and C. Susskind, Effects of chronic microwave irradiation on mice, *IRE Transactions on Bio-Medical Electronics,* April:104-108 (1962).
2. J.F. Spalding, R.W. Freyman, and L.M. Holland, Effects of 800-MHz electromagnetic radiation on body weight, activity, hematopoiesis and life span in mice, *Health Physics,* Pergamon Press. 20:421-424 (1971).
3. S.H. Preskorn, W.D. Edwards, and D.R. Justesen, Retarded tumor growth and greater longevity in mice after fetal irradiation by 2450-MHz. microwaves, *Journal of Surgical Oncology,* 10:483-492 (1978).
4. A.W. Guy, P.O. Kramar, C.A. Harris, and C.K. Chou, Long-term 2450-MHz CW microwave irradiation of rabbits: methodology and evaluation of ocular and physiologic effects, *Journal of Microwave Power* 15(1):37-44 (1980).
5. D.I. McRee, R. Faith , E.E. McConnell,  and A.W. Guy, Long-term 2450-MHz CW microwave irradiation of rabbits: evaluation of hematological and immunological effects, *Journal of Microwave Power,* 15(1):45-52 (1980).
6. C.K. Chou , A.W. Guy, J.A. McDougall, and L.F. Han, Effects of continuous and pulsed chronic microwave exposure on rabbits, *Radio Science,* 17(5S):185S-183S (1982).
7. S. Szmigielski, A. Szudzinski, A. Pietraszek, M. Bielec, M. Janiak, and J.K. Wrembel, Accelerated development of spontaneous and benzopyrene-induced skin cancer in mice exposed to 2450-MHz microwave radiation, *Bioelectromagnetics,* 3(2):179-191 (1982).
8. C.K. Chou, A.W. Guy, L.E. Borneman, L.L. Kunz, and P. Kramar, Chronic exposure of rabbits to 0.5 and 5 mW/cm$^2$ 2450-MHz CW microwave radiation, *Bioelectromagnetics,* 4(1):63-77 (1983).
9. E.R. Adair, D.E. Spiers, R.O. Rawson, B.W. Adams, D.K. Sheldon, P.J. Pivirotto, and A.M. Gillian , Thermoregulatory consequences of long-term microwave exposure at controlled ambient temperatures, *Bioelectromagnetics,* 6(4):339-363 (1985).
10. C.K. Chou , A.W. Guy, L.L. Kunz, R.F. Johnson, J.J. Crowley, and J.H. Krupp, Long-term, low-level microwave irradiation of rats, *Bioelectromagnetics,* 13(6):469-496 (1992).
11. C.H. Durney, H. Massoudi, and M.F. Iskander, "Radiofrequency Radiation Dosimetry Handbook," 4th Ed. Report USAFSAM-TR-85-73, October, Brooks AFB, TX:USAF SAM (1986).
12. J. A. Elder and D.F. Cahill , "Biological Effects of Radiofrequency Radiation," Report EPA-600/8-83-O26F, September, Research Triangle Park, NC 27711, EPA (1984).
13. N.J. Roberts and S.M. Michaelson, Microwaves and neoplasia in mice: analysis of a reported risk, *Health Physics,* 44(4):430-433 (1983).
14. A.W. Guy, Analyses of electromagnetic fields induced in biological tissues by thermographic studies on equivalent phantom models, *IEEE Trans. Microwave Theory Tech.,* MTT-19, 205-215 (1971).
15. A.W. Guy, Miniature anechoic chamber for chronic exposure of small animals to plane-ave microwave fields, *Journal of Microwave Power,* 14(4):327-338 (1979).
16. W.R.Adey, Tissue interaction with non-ionizing electromagnetic fields, *Physiol Rev.,* 61:435-514 (1981).
17. A.W. Guy , J. Wallace, and J.A. McDougall, Circularly polarized 2450-MHz waveguide system for chronic exposure of small animals to microwaves, *Radio Sci.,* 14(6S):63-74 (1979).

18. C.K. Chou, J.A. Clagett, L.L. Kunz, and A.W. Guy, "Effects of Long-term Radiofrequency Radiation on Immunological Competence and Metabolism," USAFSAM-TR-85-105, May, Brooks AFB, TX 78235. NTIS publication AD-A169064 (1986).

# EVALUATION OF ELECTROMAGNETIC FIELDS
# IN BIOLOGY AND MEDICINE

Maria A. Stuchly

Department of Electrical and Computer Engineering
University of Victoria
Box 3055
Victoria, B.C.
V8W 3P6
Canada

## INTRODUCTION

While influences of electricity and electromagnetic fields on biological systems were observed as early as the 18th century (Galvani) and the 19th century (d'Arsonval), and numerous speculations have been advanced since, a rigorous inquiry started after the Second World War. By the mid-seventies a broad range of topics was addressed, including potential health hazards of human exposure to electromagnetic energy. The interest in this field has been to a large extent stimulated by the worker and public concerns and pressures regarding safety of proliferating technologies. The main effort has concentrated in two frequency ranges: power line frequencies (50-60 Hz), and radiofrequencies (RF) and microwaves (from a few kHz to hundreds of GHz).

Considerable progress has been made in understanding the interactions of RF and microwave fields with living systems. This understanding and agreement among the majority of scientists and regulators have resulted in very similar or nearly the same recommendations regarding safe exposure levels in many national standards and international guidelines.

Despite a lot of effort, particularly in the last decade, the mechanisms of interaction of extremely low frequency (ELF) electric and magnetic fields of relatively low intensities remain poorly, if at all, understood. Similarly, there is no convincing evidence if, at what level and under what conditions can these fields be harmful to human health. The following brief overview is aimed at evaluating the state of knowledge regarding electromagnetic fields in biology and medicine. A comprehensive review of health effects and medical application is not attempted; instead reference to current reviews are given. Critical unresolved issues are emphasized.

*Radiofrequency Standards*, Edited by B.J.
Klauenberg *et al.*, Plenum Press, New York, 1994

# RF AND MICROWAVE (MW) FIELDS

## Far-field Exposures - Bioeffects

Biological effects of RF/MW fields have been extensively investigated as documented in several reviews.[1-4] With the help of dosimetric modeling, thresholds of the average specific absorption rate (SAR) have been established for various effect, in animals. Some examples follow. Radiofrequency radiation is teratogenic at SARs that approach lethal levels, and a threshold for induction of birth defects is associated with the maternal core temperature of 41-42ºC. Chronic exposure of rats during gestation at 2.5 W/kg was reported to result in lowered fetal body weight at weaning. Temporary sterility in male rats occurred at an SAR of 5.6 W/kg, which produced a core temperature of 41ºC. Decreases in operant- and learned-behavioral responses occur at SAR = 2.5 W/kg in the rat and at 5.0 W/kg in the rhesus monkey. Some types of behavior are affected at SARs approximately 25 - 50% of the resting metabolic rate. These behavioral changes are reversible, with time, after exposure ceases. Changes in the endocrine system and blood chemistry at SARs greater than 1 W/kg and in hematologic and immunologic systems at SARs equal and greater than 0.5 W/kg, for prolonged exposures, appear to be associated with thermal stress. One group of researchers reported that chronic exposure at SARs of 2-3 W/kg resulted in cancer promotion or co-carcinogenesis in mice. The effect was similar to that caused by chronic stress. Neurons in the central nervous system were altered by chronic exposure at 2 W/kg. All these effects appear to be associated with thermal load due to RF exposure and are characterized by thresholds.

In summary, for whole-body exposures to RF/MW radiation the experimental data currently available strongly suggest that biological effects in mammals occur at average SARs of about 1 W/kg. This database is well established and consistent. This fact has been reflected in more recent exposure standards such as the U.S. ANSI/IEEE C95.1-1991, Canadian Safety Code, 1991 Revision, the U.K., NRPB Standard. Each of these documents contains reference either in the text or in the accompanying publication to the same database.[1-4]

## Portable Transmitters

An evaluation of spatial distributions of SAR from RF/MW transmitters presents challenging problems both theoretically and experimentally. The main difficulties are due to geometrical complexities and biological tissue heterogeneities. Several theoretical and experimental investigations of the SAR distribution in various models of the human body have been conducted.

Apart from theoretical solutions for overly simplified geometries and electrically small antennas, the first numerical evaluation was performed by the tensor integral method.[5] The human body was modelled by 180 cubical cells of varying sizes to best fit the contour of a 70-kg man. A thin resonant dipole was located with the feed point at the neck level, 7.3 cm away from the body. Calculations were performed at 350 MHz. The results were later compared with the experimental data.[6] This model did not accurately predict local and even sub-regional SARs. Essential features of the energy deposition pattern, such as an exponential decay with the distance away from the surface of the local SARs at the plane of the antenna feed, and a "hot spot" in the neck, were not identified by this model.

Because of the serious limitations of the existing numerical techniques in the early 1980s, an experimental approach was explored.[7] A full-scale heterogeneous human model, was constructed from a 1 mm thick fiberglass shell. The following organs and tissues were placed inside the shell: a skeleton comprised of a skull, spinal cord, rib cage, and all major bones except those in the feet and hands; brain; lungs; and muscle. The dielectric properties of all materials stimulating the tissues closely corresponded to those of the respective *in vivo* tissues at test frequencies. Resonant dipole antennas were placed in front of the model at distance of $0.06\lambda$, $0.07\lambda$, and $0.2\lambda$ at the three test frequencies: 160, 350 and 915 MHz, respectively. The electric field inside the model was measured using a miniature implantable electric field probe with a computer-controlled scanning and data acquisition system.[8] The main findings were that: (1) significant (an order of magnitude) differences exist in SARs between heterogeneous and homogeneous models having the same shape, (2) at all three test frequencies the maximum SAR is produced at the surface of the model (across from the feedpoint), and the SAR close to the surface of the model decreases exponentially with distance in the direction perpendicular to the surface, (3) at 160 and 350, but not at 915, MHz there is a "hot spot" in the neck, and (4) in all cases most of the energy is deposited in about 20% of the total body volume closest to the antenna feed point. It should be noted that because of the antenna position modeled, the SARs in the eye are lower than if the antenna was located closer to the eye, as is common in a normal use of portable transmitters.

Other experimental evaluations were also performed; notably, two portable transceivers were assessed, one operating in the range of 810-820 MHz with an output power of 1.8 W, and the other operating at 850-860 MHz with 1 W of output power.[9] A heterogeneous model of the head was used with simulated skull, brain, eye, and muscle tissues. The electric field in tissue was measured with an implantable probe. Similarly, as in Stuchly et al.,[7] a maximum of the electric field on the surface and decay with distance were observed. The estimated maximum SARs on the surface of the eye were 3.2 and 1.1 W/kg per 1 W of the input power for the 810-820 MHz transmitter and 850-860 MHz transmitter when the speaker was placed flush with the operator's mouth. These values decreased to 1.1 and 0.7 W/kg per 1 W when the speaker was 2.5 cm away from the mouth. Comparable SAR values were measured in the temple and forehead areas.

More recently, numerical modeling using 3-dimensional multiple multipole (MMP) method was applied to resonant-dipole antennas in the proximity of biological bodies[10,11]. The computations were verified experimentally for a planar tissue model.[12] An approximate formula for the SAR induced at the surface of a lossy semi-infinite planar tissue was derived. The peak SAR is proportional to the square of the surface incident magnetic field, and depends on the tissue electric properties and the antenna-tissue distance expressed in wavelengths. For resonant dipoles the incident magnetic field is proportional to the antenna feed-point current, and inversely proportional to the distance. These properties apply at frequencies higher than 300 MHz. It was also estimated that for a 7 W, 1.5 GHz transceiver 2.5 cm from the eye, the peak SAR would be 40 W/kg (averaged over 1 g of tissue). A conclusion reached in this work was that very close to an antenna (a small fraction of the wavelength), the SAR is not directly related to the power radiated but to the antenna current.[12]

Deposition of energy from a dipole in a heterogeneous model of human head was evaluated by finite-difference time domain method.[13] This is so far the most accurate SAR determination in the head, and especially the eye. Computations were done at 900 MHz and 1.9 GHz. These computations confirmed that higher SARs close to the surface of the eye are obtained at the higher frequency. Data are given for SARs in the eye and 1g tissue for various separations of the antenna from the head. SARs of about 3 W/kg per 1W of the power to antenna are predicted for the antenna-head separation of 3 cm at 900 MHz and at 1.9 GHz (for the higher frequency the SAR is somewhat higher).

In the case of portable transmitters, it is apparent the average SAR is not the critical parameter in terms of protection against health effects, since the power deposition (SAR) is localized in the vicinity of the antenna. In most cases that means energy deposition in the eyes and the head.

Effects of electromagnetic radiation on the three major eye components essential for vision, cornea, lens and retina, have been investigated.[1,14] The largest number of studies has been concerned with cataracts. It was established that lens opacities can form after exposure to microwaves above 800 MHz but below about 10 GHz cataract induction requires sufficiently long exposures at an incident power density exceeding 100 mW/cm$^2$. SARs in the lens large enough to cause temperatures in the lens greater than 41$^\circ$C are required. Effects on retina have been associated with rather high levels of microwave radiation, above 50 mW/cm$^2$. More recently, effects of microwaves on the cornea have been investigated.[15,16] One of the essential parts of the cornea is the endothelium.[15] Its damage can lead to corneal edema and visual loss. Studies of the effects on the corneas of monkeys have indicated that 2.45 GHz radiation causes damage to the corneal endothelium.[13] Lesions were found sixteen to twenty-four hours past exposure, for exposures to 30 mW/cm$^2$ for CW and 10 mW/cm$^2$ for pulsed radiation (10ms, 100pps). The estimated SAR in the cornea was 0.26 W/kg per 1 mW/cm$^2$. A subsequent study of the pulsed radiation and two ophthalmic drugs revealed a lowering of the thresholds for damage.[16] The drugs used were those clinically used for treatment of glaucoma. Treatment with these drugs lowered the thresholds for microwave damage to below 1 mW/cm$^2$ (SAR of 0.26 W/kg).

Dosimetric data, as well as the observed biological effects on the retina, demonstrate a need for specific recommendations with respect to exposure from portable transmitters. Such recommendations have been made in the U.S.[17] and Canada[18] However, the scientific basis behind them was relatively limited at the time they were developed.

## Amplitude Modulated Fields

Biological effects have been observed at RF and MW fields amplitude modulated at ELF at SAR levels below thresholds for effects for continuous waves.[1,19,20] Many of these effects are the same or similar to effects observed for ELF electric and magnetic fields. The observed effects are usually field frequency and intensity specific, tend to occur within relatively narrow ranges of both field parameters, and are dependent on other physical and physiological characteristics of the exposed biological system. Many of these parameters have not been fully identified and characterized. The interaction mechanisms remain unknown. The scientific database is relatively limited in this area. However, the potential importance of these effects should not be overlooked for two reasons. First, the scientific evidence with respect to health effects of ELF fields while still inconclusive, is suggestive of possible detrimental effects. Second, until the recent developments in digital communication, hardly any situations of human exposure to RF/MW fields deeply amplitude modulated at ELF occurred. This situation is going to change rather rapidly with expansion of wireless digital communication.

## Medical Applications

RF/MW fields have found a few important medical applications.[19] Hyperthermia, available for some time, is becoming a recognized form of cancer therapy, particularly when used as an adjunct to other modalities such as radiotherapy and chemotherapy. Heat has been used in medical therapy since antiquity, while in this century scientific investigations have provided a better understanding of the interaction and more effective clinical application.[20] There are numerous comprehensive reviews of electromagnetically induced hyperthermia.[20-23] Frequencies used range from hundred kilohertz to a few gigahertz.

Many devices operate at the frequencies allocated for industrial, scientific, and medical (ISM) applications to avoid potential interference with communication and other electronic systems. The selection of the operating frequency as well as the type of applicator (the device that determines the spatial distribution of the electric field in the treated tissue) depend on the location and size of the malignancy. The applicator design depends on whether localized or regional hyperthermia is required and whether the tumor is superficial or deeply located. Among a variety of applicators and methods of electromagnetic tissue heating, magnetic induction has been successfully used in many experimental and clinical applications. Various applicator designs are used to produce desired patterns of induced electric fields in the treated tissue. Particularly successful have been applications of microwave energy for tissue heating or destruction through interstitial applicators.[24] In addition to treatment of various tumors, this technique has been successful in treatment of non-malignant disorders of prostate.[25,26] Other applications include angioplasty and ablation.

In Magnetic Resonance Imaging (MRI) and Magnetic Resonance Spectroscopy (MRS), the nuclear magnetic resonance is produced by a strong static magnetic field, a time-varying magnetic field, and RF pulses and is used to image body tissue and to monitor body chemistry. Both techniques have gained wide acceptance, and there are increasing ranges and volumes of applications. Since 1982, following the first publication of the image of the human body, MRI has been available clinically.[19] MRI provides comparable, and in some cases superior, detection capabilities of abnormal tissue to those of X-ray computer tomography (CT). For abnormalities in the brain, spinal cord, various abdominal regions, breast, and the cardiovascular and musculoskeletal systems, MRI appears superior to other modalities. MRS is used to diagnose metabolic disorders in selected regions of tissue. The metabolism of high-energy phosphate compounds, C-labelled metabolites, and many other endogenous or injected compounds can be evaluated[19]

In several applications, electrical energy must be transferred through intact skin. In the first group of applications, implanted medical devices require more power than can be supplied by presently available batteries over a sufficiently long period of time. The second category includes implanted devices that are passive receivers. The third category covers various biotelemetry systems, where information about physiological parameters within the body is transmitted to an external system. Transcutaneous energy and signal transmission (TET) has a great advantage since a tethering cable poses a serious medical problem as a source of infection and complications such as air leaks, tissues tearing and bleeding. In all TET applications, a pair of coils is used. The coil outside the body couples electromagnetic energy to the coil inside the body. Telemetry is probably the oldest application of TET. In this case, the function of the coils is reversed, with the internal coil acting as a transmitter. There are many applications of telemetry, e.g., monitoring intracranial pressure, pH of the digestive tract, bioelectric activity, and many others. Recently, systems have been developed to extend the lifetime of implanted biotelemeters by recharging implanted batteries using magnetic fields.[19] In all aforementioned TET devices, only a relatively small amount of power is transferred. Since the 1960s, there has been interest and considerable progress in developing TET systems to power cardiac assist devices and a total implantable heart. Some systems have been evaluated in animals and found to be able to transfer up to 100 W over a long period of time without any noticeable detrimental effects.

Medical applications of RF and MW fields have provided information relevant to both the evaluation of energy deposition, particularly from local sources, and to understanding biological interactions and thermal tolerance. As such, they are of some use in development of human exposure standards.

# EXTREMELY LOW FREQUENCY (ELF) FIELDS

## Biological Effects

The last decade witnessed considerable scientific effort as well as public concern regarding potentially harmful effects of human exposure to ELF and more specifically the power line frequency fields. The main concern is due to a number of epidemiological studies that have shown associations between exposure to low magnetic flux densities (of the order of 0.2 - 0.3 µT) and rates of childhood leukemia and, to a lesser extent, brain cancer. Similarly suggestive evidence between the rates of some cancers and exposures associated with so called "electrical" occupations have been provided by epidemiological studies of various cohorts of workers. Several recent reviews outline the results and evaluate the limitations of the studies.[26-29] The latter range from surrogate measures of exposure to very small numbers of cases in high-exposure groups, limiting the statistical power of the findings. While most of these studies suffer from at least some limitations, the weight of the evidence cannot be dismissed. The main problem is a limited amount of experimental data that could support the findings of the epidemiology. Some experimental results are outlined below that may at least partly support the proposition that ELF fields affect the development of cancer.

Magnetic fields were found not to cause mutagenic effects, as investigated by the Ames test and studies of chromosome aberrations and sister-chromatid exchanges in human peripheral blood. There was no significant difference in DNA single-strand breaks in mammalian cells exposed to magnetic fields. DNA repair mechanism was also not affected by pulsed magnetic fields of 2.5 mT (peak) and 1 T/s.[19,27]

A few studies of carcinogenicity have been done on animals. Large scale studies are now underway. Cancer development is considered in experimental models as a multistage process involving initiation (a change in the cell genetic material - DNA), promotion and progression. Promotion is associated with repeated exposures to an agent that may also act as an initiator (e.g., X-rays) or only as a promoter (many chemicals). Promotion involves interactions at the cell membranes. Progression is the last stage involving rapid growth of tumor and metastasis. Since electromagnetic fields have not been found to cause genetic changes, most investigations have been concentrating on promotion and co-promotion. As co-promotion one considers modification either through an accelerated rate of development or a greater incidence of cancer produced by two chemical agents, both an initiator and a promoter. Three studies, two in Sweden and one in Canada, have shown that magnetic fields do not act as promoters. The Canadian study has shown that a relatively strong magnetic field accelerated the rate of development of tumors in mice, but not the final yield in terms of the number of tumors and the number of animals affected. One Swedish study did not show any co-promoting effect, and the other showed a slight inhibition of the rate of tumor growth. So while here appears to be some biological interaction, at least in some cases, the actual effect magnetic fields have, if any, in cancer development remains undetermined.

There has been a plethora of laboratory investigations of various isolated cells in culture. These studies have clearly shown that electromagnetic fields of low frequencies can interact with biological systems at moderately low and some times very low intensities. Typical responses reported were altered cell growth rate; decreased rate of cellular respiration; altered metabolism of carbohydates; proteins and nucleic acids; changes in gene expression and genetic regulation of cell function; and altered hormonal responses. Recently, further observations were reported regarding effects on cellular transcription, charge on the cell surface, ATP and oxygen levels in the slime mold, growth, proliferation, and functional differentiation of cells.[19,27] Several studies have reported changes in calcium efflux.[19,27]

Another important effect of ELF fields is suppression of melatonin production and alterations in circadian rhythms. Effects on melatonin have been shown both in-vitro and in animals.

The main problem in evaluating of biological effects of ELF fields is the complexity of observed interactions. Effects observed are not always proportional to the field strength or the magnitude of induced currents. Sometimes, they show "window" responses in the field frequency and amplitude, and cell physiological state. The critical parameters responsible for the effect are often not well defined nor are they understood. This makes the reproduction of these experiments in different laboratories difficult, and sorting out what is a real effect and what may well be an experimental artifact is quite challenging. Difficult as these experiments are, they are extremely important in elucidating how the interactions occur, and what are the physical, biophysical and biological mechanisms of action.

**Medical Applications**

Since the mid-1960s, when effects of electric currents on bone growth were described,[30] there have been many successful clinical applications of ELF currents to reverse therapeutically pathologic processes in the musculoskeletal system, among them healing of non-unions. Non-union is the failure of a bone to heal normally following a fracture. In non-united fractures, the induced currents trigger calcification of the gap tissue and result in boney union. Successful applications are in the treatment of non-united fractures, infantile non-unions, and arthrodesis. To develop further clinical applications of pulsed magnetic fields for growth and repair of various tissues, several experimental studies in animals have been conducted including ligament and tendon repair, soft-tissue wound healing, osteoporosis, nerve regeneration, and liver regeneration. Pulsed magnetic fields were found promising in lowering the serum glucose levels in diabetic rats. Unfortunately, a large number relatively poorly conducted studies here contributed little to the understanding of important therapeutic applications.

Investigations of medical applications of ELF fields and their mechanisms of action are closely related to the study of these fields from the point of view of prevention of harm.

**CONCLUSIONS**

Electromagnetic fields can interact with biological systems and under some conditions affect human health. Some of these interactions can be harmful, while others can be beneficially employed in medical diagnosis and therapy. A lot has been learned about the interaction mechanisms and biological effects of RF and MW radiation. However, mechanisms of interaction and the range of biological effects of ELF fields remain enigmatic despite an increased research effort in the last decade.

For RF/MW exposures in the far-field, interactions are well quantifiable in terms of average SARs with some additional allowances for local SARs. The existing database provides sufficient information to establish protection standards. In view of the relative coherence of recent exposure standards in various countries, it appears reasonable for STANAG to derive its recommendations directly from one of the recent standards, e.g., the U.S. IEEE/ANSI (1991) recommendations.

The situation is more complex for portable and perhaps mobile transmitters. For analog devices (no ELF amplitude modulation), the IEEE/ANSI, 1991 recommendations provide a good starting point. However, the more recent dosimetry data and biological data on ocular effects should be evaluated and incorporated. Because of a much greater variety of portable/mobile devices and their higher power in military applications than in civilian applications there is a need for careful evaluation. Since modeling techniques and computer capabilities have significantly progressed in the last few years and the existing database is limited, more dosimetric evaluations should be undertaken. Also, the ocular effects have to be confirmed in another laboratory and further evaluated.

The resolution of potentially harmful effects of ELF fields and conditions under which they occur requires still a lot of scientific effort; only some of it is underway. It would be premature to set up an exposure standard at this time. If the standard were based on well established and understood tissue stimulation thresholds, as e.g.,[31] it would not reflect the current scientific database indicating biological interactions, some of which may prove harmful at much lower field levels. On the other hand, the weak field effects are still so poorly understood and not quantifiable that the existing scientific evidence is not suitable for standard setting. There is an urgency to understand these interactions and find out whether and at what levels effects of ELF fields also occur at RF/MW field amplitude modulated at ELF. Little research is being conducted in this area at present, but in view of the important role played by wireless digital communication, this area deserves attention.

## REFERENCES

1. "Electromagnetic Fields, 300 Hz - 300 GHz," Environmental Health Criteria, World Health Organization, Geneva (1993).
2. R.D. Saunders, C.I. Kowalczuk, and Z.J. Sienkiewicz, The biological effects of exposure to non-ionizing electromagnetic fields and radiation: III Radiofrequency and microwave radiation, National Radiological Protection Board, Chilton Didcot (1991).
3. J.H. Bernhardt, Non-ionizing radiation safety: radiofrequency radiation, electric and magnetic fields, *Phys. Med. Biol* 37:807-844 (1992).
4. M.A. Stuchly, Proposed revision of the Canadian recommendations on radiofrequency-exposure protection, *Health Phys* 53:649-665 (1987).
5. R.J. Spiegel, The thermal response of a human in the near-zone of a resonant thin-wire antenna, *IEEE Trans. Microwave Theory Techn* 30:177-185 (1982).
6. M.A. Stuchly, R.J. Spiegel, S.S. Stuchly, and A. Kraszewski, Exposure of man in the near-field of a resonant dipole: comparison between theory and measurements, *IEEE Trans. Microwave Theory Techn* 34:944-950 (1987).
7. M.A. Stuchly, A. Kraszewski, and S.S. Stuchly, RF Energy deposition in a heterogeneous model of man: near-field exposures, *IEEE Trans. Biomed. Eng* 34:12:944-950 (1987).
8. S.S. Stuchly, M. Barski, B. Tam, G. Hartsgrove, and S. Symons, A computer-based scanning system for electromagnetic dosimetry, *Rev. Sci. Instrum* 54:1547-2550 (1983).
9. R.F. Cleveland and T.W. Athey, Specific absorption rate (SAR) in models of the human head exposed to hand-held UHF portable radios, *Bioelectromagnetics* 10:173-186 (1989).
10. N. Kuster and R. Ballisti, MMP method simulation of antenna with scattering objects in the close near-field, *IEEE Trans. Magn* 25:2881-2883 (1989).
11. N. Kuster and L. Bomhott, Computations of EM fields inside sensitive subsections of inhomogeneous bodies with GMT, Proc. IEE AP-S Intern. Symp., Dallas, TX, May (1990).
12. N. Kuster and Q. Bolzano, Energy absorption mechanism by biological bodies in the near field of dipole antennas above 300 MHz, *IEEE Trans. Vehic. Tech* 41:17-23, (1992).
13. P.J. Dimbylow, FDTD calculations of SAR for a dipole closely coupled to the head at 900 MHz and 1.9 GHz, *Phys. Med. Biol* 38 (1993, in press).
14. J.A. Elder and D.F. Cahil, eds., Biological effects of radiofrequency radiation, EPA report 600/8-83-026F, NTIS accession number PB 85-120-843 (1984).
15. H.A. Kues, L. Hirst, G.A. Lutty, Effects of 2.45 GHz microwaves on primate corneal endothelium, *Bioelectromagn* 6:177-188 (1985).
16. H.A. Kues, J.C. Monahan, S.A. D'Anna et al, Increased sensitivity of the non-human primate eye to microwave radiation following ophthalmic drug treatment, *Bioelectromagn* 13:379-393, (1992).
17. ANSI C95.1-1991, American National Standard Safety Levels with respect to human exposure to radio frequency electromagnetic fields, 300 KHz to 100 GHz, The Institute of Electrical and Electronics Engineers, Inc., New York, N.Y. (1992).

18. Safety Code 6, "Limits of exposure to radiofrequency fields at frequencies from 10 KHz - 300 GHz," National Health and Welfare (Canada) EHD-TR-160, Catalogue No. H46-2/90-160E (1991).

19. M.A. Stuchly, Applications of time-varying magnetic fields in medicine, *Crit. Rev. Biomed. Eng* 18:89-124 (1990).

20. J.W. Hand, Heat delivery and thermometry in clinical hyperthermia, *Recent Results Cancer Res* 104:1-23 (1987).

21. R.A. Steeves, Hyperthermia in cancer therapy: where are we today and where are we going? *Bull. N.Y. Acad. Med* 68:341-350 (1991).

22. C. Franconi, Hyperthermia heating technology and devices *in:* Physics and Technology of Hyperthermia, S.B. Fields and C. Franconi, eds., NATO ASI Series, Martinus Nijhoff Publ., 80-121 (1987).

23. R.L. Magin and A.F. Peter, Noninvasive microwave phased arrays for local hyperthermia: a review, *Int. J. Hyperthermia* 5:429-450 (1989).

24. C.F. Gottlieb et al., Interstitial microwave hyperthermia applicators having submillimetre diameters, *Int. J. Hyperthermia* 6: 707-714 (1990).

25. M. Astrahan et al., Heating characteristics of a helical microwave applicator for transurethral hyperthermia of benign prostatic hyperplasia, *Int. J. Hyperthermia* 7:141-155 (1991).

26. D. Savitz, N.E. Pearce, C. Poole, Methodogical issues in the epidemiology of electromagnetic fields and cancer. *Epidemiol. Rev* 11:59-71 (1989).

27. "Electromagnetic Fields and the Risk of Cancer," Report of an Advisory Group on Non-ionizing Radiation, National Radiological Protection Board, UK., vol. 3, no. 1 (1992).

28. G. Theriault, Electromagnetic fields and cancer risks, *Rev. Epidem Santé Publ* 40:555-562 (1992).

29. M.N. Bates, Extremely low frequency electromagnetic fields and cancer: the epidemiologic evidence, *Environ. Health Perspectives* 95:147-156 (1991).

30. C.A.L. Bassett, B.J. Pawluk, and R.P. Becker, Effects of electric currents on bone *in vivo, Nature,* 294:252-254 (1964).

31. Interim Guidelines on Limits of Exposure to 50/60 Hz Electric and Magnetic Fields, IRPA/INIRC Guidelines, *Health Phys* 58:113-122 (1990).

# OVERVIEW OF THE RADIOFREQUENCY RADIATION (RFR) BIOEFFECTS DATABASE

Peter Polson and Louis N. Heynick

Consultants,
Cupertino, California and Palo Alto, California

## INTRODUCTION

For simplicity, we have used the acronym "RFR" in our database to span the nominal frequency range from 3 kHz to 300 GHz even though that range covers both the radiofrequency and microwave regions. So far, we have not included analyses of possible bioeffects of electric and magnetic fields at powerline frequencies (50-60 Hz).

Our database contains references to analyses we have done on a selection of research papers from the many thousands of accounts in the scientific literature published through about the end of 1992, plus some additional 1993 papers as they have appeared. Most of the detailed analyses are contained in Polson and Heynick,[1] a Technical Report on the proposed Office of Naval Research (ONR) and Air Force High-Frequency Active Auroral Research Program Ionospheric Research Instrument (HAARP IRI).

The papers discussed are grouped under a set of RFR-bioeffects topics. The papers were selected to provide representative coverage of each topic. Endeavors were made to describe differences in findings of the papers within each topic, and where possible, to assess the quality of the research. Most of the papers were presumed to have undergone peer review before publication.

Some of the analyses and critiques were derived from a general review of the RFR-bioeffects literature by Heynick[2] prepared by SRI International for the Human Systems Division (AFSC), U.S. Air Force School of Aerospace Medicine, San Antonio, Texas (now the Armstrong Laboratory).

We studied other general reviews of the literature on RFR bioeffects, including the report by Elder and Cahill[3] for the U. S. Environmental Protection Agency (EPA). However, the conclusions of our analyses regarding possible effects of exposure of people to RFR were reached independently.

*Radiofrequency Standards*, Edited by B.J.
Klauenberg *et al.*, Plenum Press, New York, 1994

## OVERVIEW OF THE BIOLOGICAL EFFECTS OF RFR

Much of the data on the bioeffects of RFR was derived from experiments in which various mammals and nonmammals (e.g., birds, insects, bacteria, other microorganisms) were exposed to RFR. Also investigated were tissues such as excised organs and neurons artificially kept alive (*in vitro*), blood, single cells, cultures of cells, and subcellular components. However, such results are regarded as inferential because the life processes of such animals and preparations can differ importantly from those for humans. Evidence was also obtained from various epidemiologic and occupational studies. Such studies appear to provide more direct evidence relative to possible effects on human health, but the exposure frequencies, levels, and durations are rarely known with any degree of accuracy, rendering the findings questionable. The following sections summarize the various bioeffects topics treated in more detail in the documents cited above.

### Summaries of Human Studies and Related Animal Studies

**Epidemiologic/Occupational Studies.** Among the early epidemiologic studies done in Eastern European countries on possible detrimental effects associated with exposure to RFR were those of Pazderova,[4] Pazderova et al.,[5] Klimkova-Deutschova,[6] Kalyada et al.,[7] Sadchikova,[8] and Siekierzynski.[9] In general, mixed findings were reported, including symptomatology such as "asthenic syndrome" and "microwave sickness" that were not generally recognized or supported by subsequent epidemiologic studies in Western countries.

Among the examples of well conducted epidemiologic studies was a study by Robinette and Silverman,[10] in which the authors examined the decedence records of 19,965 Navy veterans of the Korean War classified as having had significant occupational exposure to RFR and compared them with the decedence records of 20,726 Naval men who had little occupational exposure to RFR. The data showed no statistically significant differences between exposed and control groups in the deaths from all disease, respectively 1.6% and 1.5%, both of which were significantly lower than for the age-specific general population. As another example, Lilienfeld et al.[11] had searched for possible effects of the irradiation of the U.S. Embassy in Moscow on its then resident personnel and their dependents; that study, performed by preeminent epidemiologists, who were careful to acknowledge its limitations, also yielded negative results.

Other epidemiologic studies were flawed for various reasons, such as by analysis of too-small population samples, use of mailed self-administered questionnaires to acquire the data, inappropriate statistical treatment of the data, or incorrect assembly of population data bases. This category includes studies by Hamburger et al.,[12] Lester and Moore,[13,14] Milham,[15,16] and Burr and Hoiberg.[17] The findings of such studies, whether positive or negative, is not strong evidence in either direction.

Epidemiologic studies were done that primarily sought ocular effects of RFR. In one of two studies, Cleary et al.[18] examined the records in Veterans Administration hospitals for patients with cataracts, and found only 19 of 2,644 veterans they classified by military occupation specialties (MOSs) as radar workers and 21 of 1,956 veterans classified as nonradar veterans. Thus, independent of how accurately they classified the veterans, the data yielded no evidence for RFR-cataractogenesis. Cleary and Pasternack[19] selected 736 workers employed at 16 microwave installations as occupationally exposed to RFR and 559 unexposed workers from the same locations as controls. By contrast with Cleary et al.,[18] the findings were unclear, in part because physiological aging of the lenses had occurred in both the exposed and the control groups, and the groups were not well matched in age distribution. In addition, the authors used an arbitrary, subjective scale for grading lens changes, and that scale did not represent reductions in visual acuity.

In the third of three ophthalmologic studies, Appleton et al.[20] examined the eyes of the personnel at various Army posts where various types of electronic equipment were under development, test, or use. The authors tabulated data on three possible indicators of eye damage. On the basis of histories of microwave exposure, the pooled totals were 605 experimental and 493 control personnel. For comparisons, both groups were divided into 5 subgroups for the 10-year age spans 20-29 through 60-69 years. The findings were negative regarding RFR involvement, but were open to question because the authors used a yes or no scale to score lens damage, and they did not provide any statistical treatment of the data.

Hollows and Douglas[21] examined the lenses of 53 radiolinemen who were occupationally exposed to RFR by erecting and/or maintaining radio, television, and repeater towers throughout Australia. The RFR frequencies ranged from 558 kHz to 527 MHz. Measurements of power density in and around work areas yielded values in the range 0.08 mW/cm$^2$ to 4,000 mW/cm$^2$. The examination results were compared with those for 39 age-matched controls from the same Australian states who had never been radiolinemen. Statistically significant differences in eye changes were found between exposed and control groups, but non-RFR factors could not be ruled out, notably the reported presence in both groups of nuclear sclerosis, a type of lens opacity possibly attributable to exposure to solar irradiation.

Distinct from epidemiologic studies, there have been cases of eye damage ascribed to chronic exposure at levels well below exposure guidelines, and cases of accidental exposure to relatively high RFR levels. As an example of the former, Zaret[22] described a case of a housewife who reported in 1961 that her near vision was becoming blurred. In 1972, her ophthalmologist found extensive opacities, which were more advanced in the right eye, after which cataract extraction was done on that eye. Use of a microwave oven acquired in 1966 was the suspected cause, but measurements in 1971 of that oven by the Los Angeles County Radiological Health Division indicated a maximum leakage of 2 mW/cm$^2$ during operation. In 1973, Zaret,[22] an ophthalmologist, examined the patient and found that the left lens showed an advanced stage of capsular cataract. He indicated that microwave-induced cataracts are quite distinctive in appearance relative to cataracts from other causes. However, his findings regarding RFR etiology of cataracts were disputed in several letters printed with the paper.

As an example of accidental RFR exposure, Hocking et al.[23] reported on the exposure of 9 radio linemen to the RFR from an inadvertently activated open waveguide. Two of the men had been exposed to 4.6 mW/cm$^2$ for up to 90 minutes and comprised the "high-exposure" group; the other 7 men had been exposed to less than 0.15 mW/cm$^2$ and comprised the "low-exposure" group. In subsequent ophthalmic examinations, various eye abnormalities were seen in both groups, but vision in none of the subjects was affected.

Collectively, the epidemiologic studies to date have not provided any reliable evidence that chronic exposure to RFR at levels within current U.S. exposure guidelines (such as ANSI/IEEE C95.1)[24] is hazardous.

**Congenital Anomalies.** Sigler et al.[25] had obtained results suggestive of an association between the occurrence of Down's syndrome in children with radar exposure of the fathers during military service. However, a study by Cohen et al.[26] with a larger data base yielded negative findings, superseding those of the earlier study. Similarly, the negative findings of a study by Burdeshaw and Schaffer[27] on the incidence of birth defects from proximity to military bases superseded those of two studies by Peacock et al.[28,29]

A cohort and case-control study by Källén et al.[30] on infants born to physiotherapists presumed to have been occupationally exposed to various agents such as chemicals, drugs, X-rays, RFR yielded fewer dead or malformed infants than in the general population. The data base for the cohort part of the study was large, thus yielding statistically credible negative findings. However, use of a questionnaire in the case-control part of the study renders questionable a finding of a weak association of malformed or perinatally dead infants with the use of shortwave equipment.

In conclusion, the studies on congenital anomalies or perinatal infant deaths have not yielded any scientifically valid evidence that such effects are caused by chronic exposure to RFR at levels below current U.S. exposure guidelines.

**Ocular Effects in Animals.** With the possible exception of the Kues et al.[31] study described below, all of the experiments with animals indicate that ocular damage by exposure to RFR is a gross thermal effect. Especially noteworthy are the findings of Guy and coworkers that exposure to RFR at levels that yield a temperature rise within the eye of about 5°C or more are necessary for thermal eye damage, and that no damage occurs from such RFR levels if the eye is cooled during exposure. Guy et al.[32] reported an average-power-density threshold for eye damage of roughly 150 mW/cm$^2$ for exposure durations of 100 minutes (or longer).

Stewart-DeHaan et al.[33] exposed excised lenses to pulsed 918-MHz RFR at specific absorption rates (SARs) in the range 10 to 1,300 W/kg. The results also comprise evidence for the thermal basis of RFR eye damage. Noteworthy is a similar study by Creighton et al.[34] with CW as well as pulsed 918-MHz RFR, because the pulsed RFR yielded almost five times greater depth of lens damage than the CW RFR under corresponding exposure conditions.

Kues et al.[31] reported increases in numbers of corneal lesions they observed by specular microscopy in the eyes of monkeys exposed to 2.45-GHz CW RFR at a SAR of 7.8 W/kg within the eye and not at lower SARs. The adequacy of the exposure technique and the use of the same monkeys in more than one aspect of the study have been questioned, as has the apparent reversibility of the corneal effect even though the primate corneal endothelium is not known to repair itself through cell division. Resolution of such points awaits further studies or replication in other laboratories.

Foster et al.[35] used 50% incidence of opacities as a threshold criterion in rabbits whose heads were exposed for 30 minutes to 2.45-GHz RFR at various input powers in a waveguide. The result was a whole-head SAR of 15.3 W/kg (for a 375-gram head).

Thus, taken collectively, the animal studies on eye damage from RFR did not yield scientific evidence that prolonged exposure to RFR at levels below the current U.S. exposure standard is likely to prove hazardous. The work of Kues et al. should be subjected to independent verification.

**Auditory Effect.** The RFR-auditory effect is the perception of RFR pulses by persons as apparent sound without any electronic aids. Frey and coworkers were first to study the effect in the U.S., but their hypothesis that the effect was caused by direct brain stimulation by the RFR pulses was disproved by later studies. Instead, much experimental evidence exists that RFR pulses of appropriate characteristics are transduced within the head into thermoelastic acoustic waves that propagate to the inner ear, where they are perceived as sound. Among such evidence is a study by White,[36] who demonstrated that RFR pulses can be used to generate thermoelastic acoustic waves in various media. Foster and Finch[37] confirmed White's findings in water, and proved that such waves are not generated in water at 4°C, at which its thermal expansion coefficient is zero. Olsen and Hammer[38] and Olsen and Lin[39] studied RFR-pulse transduction in spherical brain-equivalent models of the head and obtained results that support the thermoelastic theory for a homogeneous brain sphere with stress-free boundaries.

Taylor and Ashleman[40] demonstrated that the effect does not occur in cats whose cochleas are destroyed. Guy et al.[41] confirmed the latter results, as did Chou and Galambos.[42] Cain and Rissman,[43] using 3.0-GHz pulsed RFR, determined peak-power-density thresholds in human volunteers, and obtained a peak-power-density threshold of about 300 mW/cm$^2$ for pulse perception. Tyazhelov et al.,[44] studied the qualities of apparent sounds perceived by humans from exposure to 800-MHz pulsed RFR, and showed that pulse perception as sound could be modulated by the concurrent reception of acoustic tones.

In summary, the preponderance of experimental results indicates that perception of RFR pulses as sound results from induction of thermoelastic waves in the head, rather than by direct brain stimulation by the RFR. Also to be noted is that because individual pulses of specific characteristics can be perceived, it is not meaningful to calculate time-averaged power densities for two or more widely spaced pulses and thereby cite such values as evidence that the effect is nonthermal in nature.

**RFR Shock and Burn.** It is known that RFR can cause electric shock in the body or burns in tissue under certain circumstances, and specific exposure limits have been included in the ANSI/IEEE[24] guidelines, applicable in the frequency range 0.003 to 100 MHz. Those guidelines are based on the studies by Dalziel and Mansfield,[45] Dalziel and Lee,[46] Deno,[47] Bracken,[48] Rogers,[49] Gandhi and Chatterjee,[50] Guy and Chou,[51] and Chatterjee et al.[52]

Among the considerations in deriving such limits were measurements of the currents that yielded barely perceptible sensations (perception-threshold currents), and currents that caused discomfort (let-go currents) from touching or grasping metallic structures in the vicinity of antennas radiating in the frequency range above. Also considered were the currents induced in metallic objects of various sizes and configurations by nearby antennas radiating at such frequencies, and the potential hazards to humans should they touch or otherwise come in contact with such objects.

Basically, the shock and burn sections of the ANSI/IEEE[24] guidelines for "controlled (primarily occupational) environments" specify a maximum induced current of 200 mA through both feet or 100 mA through each foot, and a maximum contact current of 100 mA, both in the frequency range 0.1-100 MHz. In the range 0.003-0.1 MHz, the induced-current limits are given by 2000 f through both feet or 1000 f through each foot, and the contact-current limit is given by 1000 f, where f is the frequency in MHz. For "uncontrolled environments" [accessible by the general public], the limits in the frequency range 0.1-100 MHz are induced currents of 90 and 45 mA respectively in both feet or each foot and a contact of current 45 mA; in the frequency range 0.003-0.1 MHz, the induced current limits are given by 900 f and 450 f, respectively, and the contact current is given by 450 f.


### Studies of Nonhuman Species

**Mutagenesis, Cytogenetic Effects, and Carcinogenesis in Microorganisms and Fruit Flies.** Mutagenesis and carcinogenesis are considered to be related. In fact, many chemicals have been screened for carcinogenicity by testing whether they produce mutations in special mutant strains of bacteria. Similarly, mutagenic effects were sought in various plants and animals from exposure to RFR as an indication that RFR can cause or promote cancer. Various strains of yeast as well as the fruit fly are also commonly used for mutagenesis investigations.

As examples, when Blackman et al.[53] exposed cultures of *E. coli* (of a strain in which mutations can be detected readily) to either 1.7 GHz at 2 mW/cm$^2$ (3 W/kg) or 2.45-GHz RFR at 10 or 50 mW/cm$^2$ (15 or 70 W/kg) for 3 to 4 hours, they found no significant differences in genetic activity when culture temperature was held constant. Dutta et al.[54] got similar results with Salmonella cultures exposed to 2.45-GHz RFR at 20 mW/cm$^2$ (40 W/kg), as did Anderstam et al.[55] at 27.12 MHz and 2.45 GHz in *E. coli* or Salmonella.

Pay et al.[56] exposed male fruit flies for 45 minutes to 2.45-GHz RFR at 6 mW/cm$^2$ and found that subsequent matings with female fruit flies showed no significant differences between exposed and control groups in mean generation times or brood sizes. Hamnerius et al.[57] exposed fruit-fly embryos from a sex-linked, genetically unstable stock having light-yellow eyes instead of red eyes to 2.45-GHz RFR at an SAR of 100 W/kg (about 200 mW/cm$^2$) for 6 hours. Only 4 mutations in 7,512 exposed males (0.05%) and 2 mutations in 3,344 control males (0.06%) were seen, a nonsignificant difference. The authors also exposed fly embryos to X-rays as a positive control, and found that 1,000 rad yielded 29 mutations in 1,053 males (2.75%). They also noted that the chemical mutagen EMS yielded 444 mutations in 4,859 males (9.14%).

**Carcinogenesis in Mammals and Mammalian Tissues.** In a study by Prausnitz and Susskind,[58] 200 mice were exposed to 9.3-GHz pulsed RFR at an average power density of 100 mW/cm$^2$ [SAR about 45 W/kg] for 4.5 minutes per day, 5 days per week, for 59 weeks. The authors reported that some mice had developed leukosis, which they described as a "cancer of the white blood cells," and that leukosis incidence was higher in the exposed mice than the control mice. The effect was undoubtedly real, but its interpretation by the authors was probably faulty. In dictionaries of medicine and pathology, leukosis (also spelled leucosis) is defined as an abnormal rise in the number of circulating white blood cells, and is not regarded as a form of cancer. Various factors can give rise to leukosis, including stress, disturbances of the endocrine system, and infection. The observed liver abscesses may have been due to pneumonia in the mouse colony.

Roberts and Michaelson,[59] in reanalyzing the data of Prausnitz and Susskind,[58] with appropriate statistical treatment, found that the results do not support a link between exposure to RFR and cancer development. They also remarked that the greater longevity of the RFR-exposed mice could be taken equally plausibly as indicating that the RFR was beneficial.

Skidmore and Baum[60] exposed five pregnant rats during 17 days of gestation in a simulator of "EMP" (electromagnetic pulses resembling the RFR from a nuclear blast), with five unexposed pregnant rats as controls. The peak electric field was 447 kV/m. Following exposure, no gross abnormalities were found in the fetuses. Twenty female rats were exposed to the EMP for 38 weeks and were observed for possible development of mammary tumors, together with 20 controls. At age one year, no mammary tumors were found.

They also exposed 50 male mice of a strain known to be susceptible to spontaneous leukemia development between 6 and 12 months of age. After exposure for 33 weeks, 42 (84%) of the EMP-exposed mice and 24 (48%) of the unexposed control mice survived. Not clear is why a much higher percentage of the exposed mice survived than the control mice, a possible indication that uncontrolled non-RFR factors may have been present. Histologic examinations showed that 9 of the 42 exposed survivors (21%) and 11 of the 24 control survivors (46%) had developed leukemia. However, the sample sizes were too small to ascribe any statistical validity to that difference in percentages.

Szmigielski et al.[61] investigated whether RFR exposure: decreases the natural resistance of one mouse strain to lung cancer cells intravenously injected before exposure; increases the incidence of breast tumors in female mice of a strain known to have high spontaneous incidence of such tumors; and increases the incidence of skin cancer in mice of the first strain that were locally depilated and painted with the chemical carcinogen 3,4-benzopyrene (BP). The exposures were for 2 hours a day, 6 days a week, for up to 6 months to 2.45-GHz RFR at 5 mW/cm$^2$ (SAR 2-3 W/kg) or 15 mW/cm$^2$ (6-8 W/kg). Other mice were similarly sham exposed, and still others were raised under stress-inducing confinement.

In the lung cancer study, RFR-exposure at 15 mW/cm$^2$ for 3 months yielded a significantly larger mean number of neoplastic nodules (colonies originating from single cells) than exposure at 5 mW/cm$^2$. The mean number of nodules in those raised under confinement for 3 months was comparable to that for those exposed at 5 mW/cm$^2$, a possible indication that confinement stress alone may be carcinogenic and that the larger mean number of nodules in those exposed at 15 mW/cm$^2$ may have been due to the heat stress from the higher RFR level.

In the breast-cancer investigation, the cumulative numbers of mice with discernible tumors and their survival times were tabulated. By regression analysis, the results were summarized in terms of the mean cancer development time in 50% of the mice and the mean survival time of 50% of the mice. The mean cancer-development and survival times of those exposed at 5 mW/cm$^2$ were comparable to the respective means for confinement-stressed mice, and both were shorter than the corresponding values for sham exposure and longer than the corresponding times for 15 mW/cm$^2$.

In the skin-cancer experiments, cancer development from use of BP was determined by histopathologic examination and scored on a subjective 7-grade scale from 0 to 6. At score 4, small papillomas were found microscopically to contain cancer cells, so mice with scores of 4-6 were regarded as having skin cancer. Skin cancer occurred within 7-10 months in more than 85% of those treated with BP. As in the breast-tumor study, the numbers of mice affected by exposure at 5 mW/cm$^2$ or confinement stress were comparable. The mean cancer development times in 50% of those sham-exposed, confinement-stressed, exposed at 5 mW/cm$^2$, or exposed at 15 mW/cm$^2$ for 3 months before BP treatment were respectively 272, 201, 171, and 171 days. Not understandable is the lack of difference between the values for 5 and 15 mW/cm$^2$.

McRee et al.[62] found no significant effects of exposure of mice to 2.45-GHz RFR 8 hours per day for 28 days at 20 mW/cm$^2$ (about 27 W/kg) on the induction of sister chromatid exchanges, a sensitive technique for assaying genetic damage from mutagens and carcinogens, or on the rate of proliferation of bone-marrow cells.

Meltz et al.[63] investigated whether pulsed 2.45-GHz RFR alone can induce mutagenesis, chromosal aberrations, and sister chromatid exchanges in mammalian cells, and whether the RFR can alter the genotoxic damage induced by the chemical mutagen proflavin when the RFR is administered simultaneously with the mutagen. The authors found that exposure of cell cultures derived from a mouse leukemic cell line to the pulsed RFR at about 40 W/kg, either alone or in combination with the mutagen yielded negative findings: The RFR-mutagen combination produced no statistically significant increase in induced mutant frequency relative to the results for treatment with the mutagen alone. Moreover, RFR exposure alone yielded no evidence of mutagenic action.

In a comprehensive University of Washington study of chronic exposure,[64-72] 100 male rats were exposed within individual cylindrical waveguides to 2.45-GHz RFR at an average power density of about 0.5 mW/cm$^2$ under controlled-environmental and specific-pathogen-free conditions, a level selected to simulate, by scaling considerations, chronic exposure of humans to 450-MHz RFR at an SAR of about 0.4 W/kg (the basis of the 1982 ANSI guidelines[73]). The controls were 100 sham-exposed male rats. Both groups were concurrently treated in the same facility for 25 months (virtually their entire lifetimes), except for those withdrawn for interim tests and those that expired before the end of the exposure regimen. After 13 months, 10 each of the RFR-exposed and sham-exposed rats were euthanized (the interim kill), as were 10 of the 12 RFR-exposed and 10 of the 11 sham-exposed rats that survived to the end of the exposure regimen.

No primary malignancies were found at the interim kill. However, the most controversial finding was that among those older than one year, primary malignant lesions (of various kinds) were found in a total of 18 of the RFR-exposed rats but only 5 of the sham-exposed rats. Among the arguments given to discount the importance of this finding were that the numbers of rats that exhibited each specific type of malignancy were similar to those reported in the literature for untreated rats of the same strain, and that the differences in the numbers for each malignancy were statistically nonsignificant. Thus, apparent statistical significance could be attained only by combining those numbers, an oncologically dubious procedure.

Santini et al.[74] investigated whether B16 melanoma would develop in black mice and whether their survival times would be affected by exposure to RFR. They exposed one group of 15 mice to 2.45-GHz CW RFR at 1 mW/cm$^2$ (SAR 1.2 W/kg) for 6 daily sessions per week, each 2.5 hours a day, until death (up to 690 hours total). They similarly exposed another group to 2.45-GHz pulsed RFR at the same average power density. A third group was sham-exposed. No statistically significant differences were found among the three groups in either tumor development or survival.

Balcer-Kubiczek and Harrison[75] investigated whether exposure to RFR of mouse-embryo-fibroblast-cell cultures induces malignant transformation in such cells. They exposed cultures for 24 hours to 2.45-GHz RFR alone at 0.1, 1, or 4.4 W/kg or to the RFR at 4.4 W/kg before or after exposure to X-rays (as a tumor initiator) at 0.5, 1, or 1.5 Gy. Control cultures were sham-exposed. They then incubated the cultures with or without a known tumor promoter (TPA), and assayed them for incidence of neoplastic transformations by counting the transformed foci in each culture dish.

The results for RFR alone or TPA alone (at the dose used) showed no evidence of tumor promotion. However, the mean counts of transformed foci rose with SAR for the RFR-exposed cultures incubated with TPA. Those results were interpreted by the authors as indicating that RFR acts synergistically in a dose-dependent manner with TPA to promote (rather than initiate) neoplastic transformation.

Several points in the paper are open to question. First, a plot of mean neoplastic transformation incidence versus SAR indicated an apparently linear incidence rise with SAR (0.1, 1.0, 4.4 W/kg). That result can be misleading because, unlike what was done for the plots of incidence versus X-ray dose, in which linear scales were used for both variables, the authors used a linear scale for incidence and an exponential scale for SAR. If the three SAR points had been plotted on a linear scale also, the graph would have displayed a much sharper rise with SAR between 0.1 and 1.0 W/kg than between 1.0 and 4.4 W/kg. Specifically in such a plot, the slope of the line connecting the points for 0.1 and 1.0 W/kg is more than five times larger than for the line connecting the points for 1.0 to 4.4 W/kg. Thus, it is difficult to interpret the RFR dose-response relationship.

Second, use of a well characterized carcinogen that causes chromosomal damage as the tumor initiator such as benzopyrene, rather than X-rays, might have yielded more definitive results. Third, the numbers of foci relative to the numbers of dishes treated were small and the numbers of dishes used for each treatment differed considerably. This point raises the question as to whether the authors had augmented the number of dishes for each treatment until they obtained adequate percentages of foci for statistical significance. Fourth, not mentioned in the paper was whether those who counted the foci had prior knowledge of the treatment of each dish. Foreknowledge of the treatment can yield false positives and should be controlled for by use of "blind" procedures.

Nevertheless, in view of the importance of the reported findings, the study should be replicated, perhaps with a more standard chemical initiator than X-rays, and with a "blind" foci counting procedure to avoid possible subconscious bias.

The EPA issued a draft report[76] in which a potential link between cancer incidence and exposure to electromagnetic fields was indicated. Although the emphasis was on powerline fields, exposure to microwave fields was also included. However, EPA's Radiation Advisory Committee of the Science Advisory Board (SAB) had set up a subcommittee on

Nonionizing Electric and Magnetic Fields to review the draft report, and that subcommittee has issued a report, SAB,[77] in which the subcommittee suggested numerous changes in emphasis, coverage, and wording. It concluded that the draft report[76] will have to be rewritten to accommodate all of the suggestions and comments.

In summary, although questions have been raised regarding interpretation of the findings of the University of Washington[64-72] and the Balcer-Kubiczek and Harrison[75] studies, taken collectively the studies conducted to date provide no scientifically credible evidence that exposure to low RFR within ANSI/IEEE guidelines levels causes mutations or cytogenetic effects, or that such RFR induces or promotes any form of cancer in mammals or nonmammals. In addition, little credence can be given to the EPA[76] draft report, pending issuance of a revision thereof.

**Teratogenesis.** Rugh et al.[78,79] exposed pregnant mice to 2.45-GHz RFR at 123 mW/cm$^2$ (SAR 110 W/kg) for 2 to 5 minutes on what they describe as the gestation day of highest sensitivity to ionizing radiation. They stated that they could not find any teratogenesis threshold. However, a reanalysis of their data indicated the existence of a threshold: At mean total doses less than about 3 cal/g or power densities less than about 1 mW/cm$^2$, 100% of the fetuses examined were normal and significant numbers of abnormal fetuses were obtained at RFR levels above that threshold, but the dependence on dose was obscure.

Chernovetz et al.[80] found that absorption of about 5 cal/g of 2.45-GHz RFR is not teratogenic to mice, a threshold considerably higher than the 3-cal/g value above. They also found that the mean total lethality dose of the dams was about 5.7 cal/g, indicating that teratogenesis would occur in pregnant mice only at levels that are close to lethality for the dams.

Stavinoha et al.[81] exposed 4-day-old mice for 20 minutes to 10.5-MHz, 19.27-MHz, or 26.6-MHz RFR pulses at an electric field strength of 5.8 kV/m and weighed them daily up to age 21 days, with sham-exposed mice as controls. There were essentially no differences in post-exposure weights at corresponding ages. Berman et al.[82,83] did find consistently smaller mean body weights of live mouse fetuses from dams exposed to 2.45-GHz RFR at 28.0 mW/cm$^2$ (22.2 W/kg), but other researchers either could not confirm such findings or found that growth retardation was thermally induced. However, it is difficult to reconcile those growth retardation results of Berman et al.[82,83] in the mouse with the findings by Berman et al.[84] and Smialowicz et al.[85] of no retardation in rats at RFR levels capable of heating pregnant females to temperatures over 40°C. Clearly, studies of nonhuman primates would be much more definitive.

Lary et al.[86] observed teratogenic effects in rats exposed to 27.12-MHz fields at 55 A/m and 300 V/m (SAR about 11 W/kg), but of severity that increased with colonic temperature. The largest changes were seen for prolonged exposure to maintain colonic temperature at 42.0°C. The authors ascribed those effects to the hyperthermia induced by the RFR. A subsequent study by Lary et al.[87] with the same RFR indicated the existence of a colonic temperature threshold of 41.5°C for birth defects and prenatal death.

Tofani et al.[88] reported teratogenic effects in rats exposed to 27.12-MHz RFR at field strengths of 20 V/m and 0.05 A/m (equivalent power density 0.1 mW/cm$^2$; author-estimated SAR about 0.00011 W/kg). They had characterized the effects as nonthermal and due to long-term exposure.

Lu and Michaelson[89] took issue with the exposure methodology used. They questioned the absence of technical details, such as a description of the means for providing food and water and for removing wastes during continuous exposures of pregnant rats. They also questioned the reality of the positive findings at an average SAR far lower than the threshold for RFR bioeffects (3-4 W/kg) that was the basis for the ANSI[73] exposure guidelines. They also noted that the use, by Tofani et al.,[88] of Durney et al.[90] for their SAR estimation was inappropriate because that reference applied to spheroidal models in the far field, whereas the exposures in this study were in the near field. Last, as noted by Lu and Michaelson,[89] not discussed in the paper was whether RFR-absorbent materials were used in the exposure chamber to avoid multipath exposure, and the likelihood that the proximity of the rats to one another in the exposure boxes (spacings less than 1 wavelength) resulted in large dosimetric variations and uncertainties.

In their response, Tofani et al.[91] clarified several of the points raised by Lu and Michaelson,[89] but remarked that they chose to do the exposures in a room without any shielding or RFR-absorbing materials rather than in a transverse electromagnetic (TEM) cell within a shielded anechoic chamber available at their laboratory. As justification, they stated: "Our aim in this work is the evaluation of the biological effects due to a low-level, long-term exposure to a 27.12-MHz electromagnetic field in conditions as similar as possible to those people who usually are exposed (i.e., to near-field, multi-path radiation with distances between individuals shorter than a wavelength)." That response appears to beg the question, because the dose rates from such exposures could have varied considerably from rat to rat and with time for each rat.

Tofani et al.[91] also remarked: "Effects due to overcrowding ought to result in the sham-exposed group too, since that group was managed in the same way." That remark appears to miss the point that overcrowding probably introduced large spatial and temporal variations in RFR-exposure levels rather than directly causing the reported effects, so overcrowding *per se* in the sham-exposed rats would not be expected yield those effects. In summary of the foregoing, little credence can be given to the conclusion of Tofani et al.[91] that the effects they observed were nonthermal.

In a behavioral study of squirrel monkeys by Kaplan et al.,[92] an excess of unexpected infant deaths was found. However, that finding was not confirmed in a subsequent study involving enough monkeys for an adequate statistical treatment of the results.

Taken collectively, the studies above indicated that teratogenic effects can occur in both nonmammalian and mammalian subjects from RFR exposure, but only at levels that produce significant internal temperature rises. The results for mammals show that increases in maternal body temperature that exceed specific thresholds (41.5°C in rats) are necessary for causing teratogenic effects.

**Nervous System.** *Blood-Brain-Barrier Effects.* Early studies by Frey et al.[93] and Oscar and Hawkins[94] of RFR effects on the blood-brain barrier (BBB) were subsequently shown likely to have been caused by artifact in the methods used in their experiments. Also, Merritt et al.[95] and Ward and Ali[96] were unable to reproduce those findings, and Preston et al.[97] showed that certain specific RFR-induced changes in the brain could be interpreted wrongly as BBB alterations. Using a newer method by Rapoport et al.,[98] Preston and Préfontaine[99] and Gruenau et al.[100] found no evidence for RFR-induced alterations of the BBB. Four comprehensive studies by Williams et al.[101,102,103,104] with conscious unrestrained rats, in which several different tracers and methods were used for detecting BBB penetration, also yielded negative findings. Neilly and Lin[105] demonstrated that disruption of the rat BBB at high RFR levels is due to elevation of brain temperature. In addition, they found that high ethanol doses inhibit BBB disruption by moderating the increases in brain temperature produced by the RFR. In summary, the preponderance of negative experimental findings indicates that exposure to RFR at low levels has no effect on the permeability of the human BBB.

*Histopathology and Histochemistry of the Central Nervous System.* Albert et al.[106] had reported lower mean counts of Purkinje cells in 40-day-old rat pups exposed for 5 days *in utero* to 2.45-GHz RFR at 10 mW/cm$^2$ relative to counts for sham-exposed pups. The results are difficult to interpret in view of the large SAR variations (0.8 to 6 W/kg, 2 W/kg estimated mean) due to movement of the dams during exposure. Also, the findings of an experiment in which rat pups were exposed to the RFR and euthanized 40 days later did not show such differences. In addition, little credence can be given to results for pups euthanized right after exposure because Purkinje cells are immature in neonates. Albert et al.[107] did a similar study on squirrel monkeys previously exposed perinatally; no significant differences in Purkinje cell counts were seen between RFR-exposed and sham-exposed groups.

Merritt et al.[108] exposed 10 pregnant unrestrained rats to pulsed 2.45-GHz RFR for 24 hours per day during gestation days 2-18. The RFR level was held constant at 0.4 W/kg for each dam as its mass increased during the exposure period. Concurrently, 10 rats were sham-exposed. All 20 rats were weighed on day 2, and were reweighed every fourth day during treatment. On day 18, they were euthanized and the fetuses were removed and examined. After each fetus was weighed, its brain was excised, weighed, and assayed for RNA, DNA, and protein. The last three endpoints were expressed in terms of both mg/brain and μg/mg of brain tissue (thus comprising a total of 8 endpoints). The difference between groups for each endpoint was nonsignificant ($p > 0.05$). Also, no RFR-exposed litter was microencephalous.

Sanders et al.[109] used continuous fluorescence measurements *in vivo* to investigate whether RFR exposure of rat brain tissue stresses cells that inhibit respiratory chain function, thereby decreasing the concentrations of adenosine triphosphate (ATP) and creatine phosphate (CP). Such measurements were made through an aperture in the skull during exposure to 591-MHz RFR at 5 or 13.8 mW/cm$^2$ for durations of 0.5 to 5 minutes. The results indicated that such reductions in concentration by RFR occurred at levels characterized as not producing measurable brain hyperthermia (but with local hyperthermia not ruled out). Although some points are open to question, the positive findings appear to be valid, and are worthy of further study. Also, the experiments were performed on anesthetized rats, with consequent lowering of their brain temperatures; whether similar results would occur without anesthesia has not been determined.

Sanders et al.[110] performed similar experiments, but at 200 MHz and 2,450 MHz as well as 591 MHz. Noteworthy is that the effects were higher at 591 MHz than at 200 MHz but were not observed at 2.45 GHz, suggesting that the effect occurs only within a specific frequency range below 2.45 GHz.

Lai et al.[111] studied the effects of single, 45-minute exposures of rats to 2.45-GHz CW or pulsed RFR on choline uptake in several regions of the brain, the uptake being a measure of cholinergic activity. The exposures were done in two types of chambers (cylindrical waveguides and a miniature anechoic exposure chamber), and both were adjusted to obtain a whole-body SAR of 0.6 W/kg. Control rats were sham-exposed in the same chambers. Both positive and negative results were observed, but not clear were apparent inconsistencies in results for the two kinds of exposure chamber and between pulsed and CW RFR.

In summary, histopathologic and histochemical changes in the central nervous system were seen at relatively low SARs by Sanders et al.[109,110] and Lai et al.,[111] but their significance with regard to possible human health is not clear, and questions about those studies remain open. Overall, considering other experimental results in which effects were ascribed to brain temperature increases, it seems unlikely that subthermal levels would cause such changes in the human central nervous system.

*Changes in Electroencephalograms (EEGS) and Evoked Responses (ERS).* Various investigations have been done to ascertain the effects of RFR on the EEG or on the responses evoked by visual or auditory stimuli (ERs). As proved by Johnson and Guy,[112] the use of indwelling metallic electrodes, wires, or screws may be questioned as a procedure likely to induce artifactual effects in the animals under study as well as in the recordings themselves.

Thus, findings of early studies may be discounted because of such use. Also, EEG measurements done after and/or before RFR-exposure may present problems stemming from the time consumed in attaching the electrodes and variability in their placement. With such usage, moreover, any transient effects that may occur during exposure would not be detected.

Takashima et al.[113] sought changes in EEGs of rabbits exposed once (acute exposure) to frequencies in the range 1-30 MHz at fields of 0.5 to 1 kV/m, done between two spaced aluminum plates. They endeavored to minimize recording artifacts by using a preamplifier having a pass band of 3-100 Hz and a 60-Hz notch filter. In addition, they sampled, digitized, and converted time-domain EEG recordings taken before and after RFR-exposure into complex frequency spectra by fast-fourier-transform (FFT) and thence into power spectra. Typical pre-exposure power spectra from anesthetized rabbits showed frequency components between 5 and 15 Hz that varied during each sequence, indicating the absence of a dominant component, EEGs denoted "normal."

In one set of experiments, anesthetized rabbits in which stainless-steel electrodes had been implanted within the brain were exposed once to fields modulated at 60 Hz. Specific changes were seen in the EEGs recorded during exposure. In another set of experiments however, in which the EEG electrodes were removed before exposure and reinserted after exposure, the power spectra observed resembled normal EEGs as defined above. The authors attributed the previous EEG changes to the local field created by those metal electrodes.

Those authors also sought possible effects of chronic exposure: They exposed unanesthetized rabbits 2 hours/day for 6 weeks to 1.2-MHz fields modulated at 15 Hz, and recorded EEGs every 2 weeks with silver electrodes placed directly on the skull before and after exposure. Abnormal EEG patterns began to appear after 2 to 3 weeks of exposure. The authors constructed histograms from power-spectrum sequences derived from 4-week exposures and normal EEGs. The histogram for the exposed animals showed major peaks at 2 and 10 Hz, but the major peaks for "normal" (pre-exposure) EEGs were at 4.5, 8, and 11.5 Hz.

In this study, the rabbits were used as their own control group, in that the data taken during and after exposure were compared with the "normal" (pre-exposure) data from the same group. However, the lack of a similarly treated sham-exposed group renders it difficult to assess whether the reported EEG changes were the result of exposure *per se,* or of adaptation to the repetitive aspects of the experimental procedures, such as handling and recording. Also, techniques other than the FFT may be more appropriate for analysis of power spectra of EEGs recorded at short time intervals (3 minutes). Qualitatively, nevertheless, the chronic-exposure data appear to show an enhancement of low-frequency power-spectral components and reduction of high-frequency activity. On the other hand, comparison of data from the anesthetized animals used in the acute experiments with data from the unanesthetized animals used in the chronic experiments lacks analysis of the effects of anesthesia as a possible confounding factor.

In several studies, such as by Tyazhelov et al.[114] and Chou and Guy,[115] endeavors were made to minimize artifacts by designing electrodes and leads from materials having high resistivities comparable to those for tissue. When such electrodes were implanted before exposure and were present during exposure, no significant differences in EEGs or ERs between control and RFR-exposed animals were seen, as exemplified by the study of Chou et al.[116] In summary, with appropriate measures taken to avoid artifacts, there is no credible scientific evidence that low-level RFR affects the EEG.

*Calcium Efflux.* In studies by Bawin et al.,[117] calcium efflux was reported to occur in brain hemispheres of newly-hatched chicks that were excised and exposed to 147-MHz RFR amplitude-modulated at frequencies below about 100 Hz. One of the chick hemispheres was exposed between a pair of plane-parallel plates, with the other hemispheres serving as the controls. The effect was absent with unmodulated RFR at the same frequencies; it was highest for modulation with 16 Hz. Sheppard et al.[118] obtained similar results with amplitude-modulated 450-MHz RFR. Other results by Bawin and coworkers for chick brains poisoned with cyanide indicated that calcium efflux is not an effect that involves calcium transport across cell membranes.

Bawin and Adey[119,120] also performed experiments indicating that the effect was ascribable to the modulation itself rather than to the RFR carrier frequency. In those experiments, they exposed chick-brain preparations to sinusoidal fields at discrete frequencies of 1, 6, 16, or 32 Hz instead of to RFR fields amplitude-modulated at those frequencies. However, the effects at the sub-ELF and ELF frequencies were opposite in direction to those for the modulated RFR.

Blackman et al.[121] performed experiments toward reproducing the 147-MHz chick-brain results of Bawin et al.,[117] but used a TEM line for exposure instead of parallel plates. Two series of exposures were done. In one, power density was held constant at 0.75 mW/cm$^2$ and the modulation frequencies were 0, 3, 9, 16, and 30 Hz. In the other, the modulation frequency was held constant at 16 Hz and the power densities were 0, 0.5, 0.75, 1.0, 1.5, and 2.0 mW/cm$^2$. The results for the constant power density series were similar to those of Bawin et al.,[117] with maximum effect at 16 Hz. Regarding the 16-Hz modulation series, the authors stated:

"Our results indicate that the modulation-frequency window in which calcium-ion flux is enhanced only occurs within a restricted range of power densities...These results indicate a maximal power-density effect at 0.75 mW/cm$^2$ and no enhancement at levels plus or minus 0.25 mW/cm$^2$ of this value."

Shelton and Merritt[122] investigated whether RFR pulses at repetition rates comparable to the modulation frequencies used in the chick-brain studies would elicit alterations in calcium efflux from the rat brain. They excised brain hemispheres from rats and exposed the hemispheres for 20 minutes to 20-ms pulses of 1-GHz RFR modulated at 16 pps (pulse per second), at average power densities of 0.5, 1.0, 2.0, or 15 mW/cm$^2$ selected to search for the reported power-density window. In other experiments, the exposures were to 10-ms pulses, 32 pps, at 1.0 or 2.0 mW/cm$^2$. The brain hemispheres were then assayed for calcium efflux. Their findings were negative, but as they pointed out, no direct comparisons can be made between their results and those of investigators on chick brains because besides the differences in species and exposure durations, there were important differences between pulse modulation at 16 pps and sinusoidal modulation at 16 Hz.

In a subsequent study, Merritt et al.[123] performed experiments *in vivo* as well as *in vitro* on possible pulsed-RFR-induced alterations of calcium efflux from the rat brain. For both types of experiments, brain tissue was loaded with $^{45}Ca^{++}$ by injection directly into the right lateral ventricle of rats anesthetized with ether. In the *in vitro* experiments, samples of excised brain tissue were exposed for 20 minutes to 20-ms pulses, 16 pps, of 1-GHz RFR at 1 or 10 mW/cm$^2$ (0.29 or 2.9 W/kg) or 2.45-GHz RFR at 1 mW/cm$^2$ (0.3 W/kg). Radioactivities of the incubation media and the tissue samples were assayed in terms of mean disintegrations per minute per gram of tissue (DPM/g). The difference in mean DPM/g values between RFR-exposed and sham-exposed samples was not significant for any of the exposure conditions.

In the whole-animal exposures, 2 hours after the rats were injected with $^{45}Ca^{++}$, groups were exposed for 20 minutes to 2.06-GHz RFR with their long axes parallel to the E-field, one group each to CW at 0.5, 1.0, 5.0, or 10.0 mW/cm$^2$ and one group each to 10-ms pulses at 8, 16, or 32 pps at an average power density of 0.5, 1.0, 5.0, or 10.0 mW/cm$^2$. The normalized SAR was 0.24 W/kg per mW/cm$^2$. The rats were euthanized after exposure, and their brains were removed and processed appropriately for $^{45}Ca^{++}$ assays.

Statistical tests on the 17 treatment combinations (4x4 RFR-exposures, 1 sham-exposure) showed that the difference between the results for the sham group and the combined RFR groups, and the differences between the results for the sham group and each of the RFR groups were not significant.

Adey et al.[124] exposed paralyzed awake cats for 60 minutes to 450-MHz RFR, amplitude-modulated at 16 Hz, after incubating the brain cortex of each cat with $^{45}Ca^{++}$ through a hole in the skull. The average power density was 3.0 mW/cm$^2$ (SAR about 0.29 W/kg in the interhemispheric fissure). The authors used the experimental data for curve fitting, but did not present any actual data. They stated that there was no difference between the means of the exposed and control values, but that the standard deviations were greater for all of the exposed data points than for the controls. It is difficult to interpret these second-order statistical findings as indicating the existence of the effect.

Blackman et al.,[125] noted that the currents in the walls of the TEM line yielded an alternating magnetic component in addition to the TEM field, in contrast to the solely electric field produced between the parallel plates used by Bawin and coworkers, which led them to hypothesize that the magnetic component could influence changes in calcium efflux and to suggest that the earth's DC magnetic field might also be involved in the effect. Therefore, they did exposures in a transmission line that permitted use of an alternating electric field either alone or together with an alternating magnetic field. They also placed the transmission line within a pair of large Helmholtz coils oriented to produce a DC magnetic field parallel to the earth's field, that could be varied to alter the local magnitude and polarity of that field.

Mixed results were obtained, from which it was difficult to relate the occurrence of calcium efflux or its absence to changes in local geomagnetic field. More recently, Blackman et al.[126] reported that calcium efflux can be reduced, enhanced, or nullified by appropriately varying the temperature of the chick-brain samples before and during exposure. The authors interpreted these results as indicating the existence of a temperature window analogous to the RFR intensity window, but which may also be interpreted as an unrecognized artifact in this and previous studies.

In summary, although researchers in several laboratories have reported obtaining the calcium-efflux effect, researchers in other laboratories have been unable to confirm its existence. Several of the recent studies that report positive findings suggest that magnetic fields, primarily at powerline frequencies as well as the earth's DC field, could contribute significantly to the effect, with ion cyclotron resonance as the primary mechanism. However, quantitative analyses by others indicate that it is most unlikely that ion cyclotron resonance could occur under the proposed conditions of the cell membrane. In any case, there is no experimental evidence that the effect, if it does exist, would occur in or be harmful to humans or intact animals.

**Immunology and Hematology.** *Studies in Vitro.* Smialowicz[127] compared suspensions of mouse-spleen cells exposed to 2.45-GHz RFR at 10 mW/cm$^2$ (19 W/kg) for up to 4 hours with suspensions held at 37°C for the same durations as controls. After treatment, suspensions were cultured with or without each of four different mitogens. No significant differences in percentages of cells undergoing mitosis or in percentages of viable cells were found between RFR-exposed and control specimens treated for corresponding durations. Hamrick and Fox[128] also obtained negative results for rat lymphocytes cultured with or without the mitogen phytohemagglutinen (PHA) and exposed to 2.45-GHz RFR at up to 20 mW/cm$^2$ (2.8 W/kg) for up to 44 hours.

Roberts et al.[129] found that the viability of cultures of human mononuclear leukocytes was unaffected by exposure to 2.45-GHz RFR for 2 hours at SARs in the range 0.5 to 4 W/kg. There were also no significant effects on DNA, RNA, and total protein synthesis, or in assays for spontaneous production of interferon, influenza-virus-induced production of alpha-interferon, and mitogen-induced production of gamma-interferon. Subsequently, Roberts et al.[130] infected human mononuclear leukocyte cultures with influenza virus and then exposed them to 2.45-GHz RFR, either CW or pulsed at 60 or 16 Hz, all at 4 W/kg. No significant differences were found in leukocyte viability between RFR-exposed and sham-exposed virus-infected or uninfected cultures, or in DNA synthesis from mitogen stimulation.

Lyle et al.[131] sought effects for 60-Hz-amplitude-modulated 450-MHz RFR at 1.5 mW/cm$^2$ (SAR not stated) on the effectiveness of a specific class of rodent T-lymphocytes (effector cells) against a specific class of lymphoma (target) cells. Mixtures of T-lymphocytes and lymphoma cells exposed to the RFR showed reductions of 17-24% in mean effectiveness of T-lymphocytes against lymphoma cells relative to unexposed mixtures. Similar percentages (15-25%) were obtained in mixtures not exposed to the RFR but in which T-lymphocytes were exposed to RFR before mixing them with lymphoma cells, thus indicating that the effect of the RFR was on the T-lymphocytes. However, the effect on the T-lymphocytes decreased with elapsed time after exposure.

Effects were also sought for 450-MHz RFR modulated at 3, 16, 40, 80, and 100 Hz. The reduction of effectiveness was negligible with unmodulated RFR; nonsignificant with 3, 16, and 40 Hz; maximal with 60 Hz; and significant (but smaller) with 80 and 100 Hz, an indication that the effect was due to the amplitude modulation itself.

Sultan et al.[132] studied the effects of combined RFR-exposure and hyperthermia on the effectiveness of normal mouse B-lymphocytes against a specific antigen (anti-mouse immunoglobulin from the goat). They exposed suspensions of normal mouse B-lymphocytes at 37, 41, and 42.5°C to 2.45-GHz CW RFR for 30 minutes at levels in the range 5-100 mW/cm$^2$ (2.25-45 W/kg). Control suspensions at each temperature were sham-exposed. The effectiveness of B-lymphocytes against the antigen was more than 90% for cells heat-treated at 37°C, but less than 60% of those treated at 41°C, and less than 5% of those treated at 42.5°C. The authors concluded that the mechanisms involved are thermally based, with no apparent effects of the RFR *per se* if RFR-exposed and control samples are held at the same temperature.

Sultan et al.[133] reported similar negative findings with cell suspensions exposed for 30 minutes to 147-MHz RFR amplitude-modulated at 9, 16, or 60 Hz at average power densities from 0.1 to 48 mW/cm$^2$ (0.004-2.0 W/kg). Thus, their results showed no amplitude-modulation effect. They also found that for temperatures not exceeding 42°C, the effectiveness of mouse B-lymphocytes returned to normal 2 hours after heat treatment.

Cleary et al.[134] obtained negative findings on the viability and phagocytotic ability of rabbit neutrophils exposed to 100-MHz CW RFR for 30 or 60 minutes at electric field strengths ranging from 250 to 410 V/m (120 to 341 W/kg), or for 60 minutes to 100-MHz RFR amplitude-modulated at 20 Hz (331 W/kg). However, the credibility of these findings is diminished by the relatively large variabilities among the control groups in each case, an indication of the possible presence of uncontrolled non-RFR factors.

Kiel et al.[135] sought effects of RFR on oxidative metabolism of human peripheral mononuclear leukocytes. They noted that chemiluminescence (CL) occurs in the production of oxygen metabolites and that CL can be enhanced and used as a sensitive detector of such effects. They collected blood samples from human volunteers, and separated, washed, and resuspended the leukocytes. Samples were paired, and one of each pair was exposed for 30 minutes to 2.45-GHz CW RFR at 104 W/kg with the sample held at 37°C and the other was held at 37°C in an incubator without RFR exposure. Following treatment, half of each sample was used for measuring CL activity, and the other half for determining cell counts and viability. The differences between the RFR-exposed and sham-exposed samples were not significant.

In studies of possible effects of RFR interactions with samples of red blood cells (RBCs) taken from animals or humans, alterations of cell membrane function, particularly any effects on the movement of sodium ions ($Na^+$) and potassium ions ($K^+$) across the membrane were sought. In early studies done in Eastern Europe, increases in cell breakdown and efflux of $K^+$ from rabbit red blood cells (RBCs) were reported for exposure to 1-GHz or 3-GHz RFR at levels as low as 1 mW/cm$^2$.

Peterson et al.[136] found that heating suspensions of rabbit RBCs with 2.45-GHz RFR at 10-140 mW/cm$^2$ (46-644 W/kg) yielded higher hemoglobin (Hb) or $K^+$ losses than conventionally heating suspensions. However, in all experiments in which RFR-heated and conventionally-heated RBCs were warmed at the same rate to the same final temperature, both Hb and $K^+$ were lost in equal amounts, indicating the thermal basis for the effect. However, the authors also used 2.45-GHz RFR to heat samples of human RBCs to 37°C and maintain them there, and found no significant differences in either hemolysis or $K^+$ release relative to conventionally heated samples.

Brown and Marshall[137] sought nonthermal effects of RFR on growth and differentiation of the murine erythroleukemic (MEL) cell line. They exposed MEL cells cultured with HMBA (an inducer of MEL cells to differentiate and form hemoglobin) to 1.18-GHz RFR at SARs up to 69.2 W/kg, with incubation temperature held at 37.4°C. There were no significant differences in any of the endpoints between cultures exposed at each RFR level and corresponding control cultures.

*Studies in Vivo.* Huang et al.[138] exposed hamsters to 2.45-GHz RFR at levels in the range 5-45 mW/cm$^2$ (2.3-20.7 W/kg), and cultured leukocyte suspensions with or without the mitogen PHA. For cultures not stimulated with PHA, the percentage of transformed cells increased with RFR level to a maximum at 30 mW/cm$^2$ (13.8 W/kg), clearly a thermal effect. For unclear reasons, however, the percentage decreased at still higher RFR levels. There were no significant changes in differential leukocyte counts, thus supporting the contention that RFR does not cause lymphocytosis. For PHA-stimulated cultures exposed at 30 and 45 mW/cm$^2$ (13.8 and 20.7 W/kg), the percentage of cells in mitosis diminished significantly relative to the total number of lymphocytes.

Huang and Mold[139] exposed mice to 2.45-GHz RFR at 5-15 mW/cm$^2$ (3.7-11 W/kg), after which they cultured spleen cells with tritiated thymidine and with or without a T-lymphocyte mitogen or a B-lymphocyte mitogen. Then the cells were assayed for thymidine uptake, a measure of DNA synthesis during cell proliferation. The results for the RFR-exposed mice showed cyclic time variations of thymidine uptake for the mitogen-stimulated and nonstimulated cultures, but also for the sham-exposed mice, apparently due to factors other than RFR, thus rendering the findings of this study questionable.

Lin et al.[140] exposed mice to 148-MHz RFR at 0.5 mW/cm$^2$ (0.013 W/kg) for 10 weeks beginning on postpartum day 4, 5, 6, or 7, and weighed them periodically up to age 600 days. The mean weights of the mice did not differ significantly from those of sham-exposed mice at corresponding ages. In blood samples drawn periodically up to age 600 days, no significant differences were seen between RFR-exposed and sham-exposed mice in any of the blood parameters.

Wiktor-Jedrzejczak et al.[141] exposed mice to 2.45-GHz RFR at 14 W/kg in a single 30-minute session or in three such sessions, one per day, three days apart. Total T-lymphocyte populations were unaffected by either single-session or triple-session exposures. However, single sessions significantly increased the population of one subclass of splenic B-lymphocytes ($CR^+$) and not of another subclass ($Ig^+$), but triple sessions increased both subclasses. Also, splenic cells from mice given single or triple exposures and cultured with various T-cell or B-cell mitogens showed significant increases in blastic transformation of B-cells but nonsignificant effects on T-cells. Mice were inoculated with either of two types of antigens and were exposed to the RFR. Antibody production to either antigen was reduced by RFR-exposure, but only the difference for one of the antigens was statistically significant. Taken together, the results of this study show that thermogenic RFR levels (e.g., 14 W/kg) can have weak stimulatory effects on splenic B-lymphocytes but none on T-lymphocytes.

Sulek et al.[142] found that the threshold for increases in $CR^+$ B-cells was about 5 W/kg for 30-minute exposures to 2.45-GHz RFR, yielding an energy-absorption threshold of 10 J/g. They noted that multiple exposures at levels below that threshold were cumulative if done within one hour of one another, but not if spaced 24 hours apart. However, Schlagel et al.[143] presented results showing that RFR-induced increases in $CR^+$ B-cells depend on genetic factors. For example, mice of strain CBA/J that have a specific histocompatibility haplotype showed marked increases in $CR^+$ cells due to RFR-exposure, but mice having other haplotypes did not.

Wiktor-Jedrzejczak et al.[144] noted the findings of Schlagel et al.[143] and suggested that the effect might be mediated by a humoral factor. As a test, they performed experiments involving implantation of spleen cells derived from RFR-exposed CBA/J mice into sham-exposed CBA/J mice and vice versa, and assayed both groups for circulating $CR^+$ spleen lymphocytes. For both donor and recipient mice exposed to RFR, the counts of circulating $CR^+$ spleen lymphocytes were higher than for their respective controls, thus supporting their hypothesis.

Smialowicz et al.[145] exposed CBA/J mice 10-12 weeks old to 2.45-GHz CW RFR at 15-40 mW/cm$^2$ (11-29 W/kg) for 30 minutes. Six days later, assays of the percentages of $CR^+$ splenic cells and the numbers of nucleated splenic cells showed no significant differences in those two endpoints between sham-exposed mice and mice exposed at any of the RFR levels. Thus, these authors could not confirm the RFR-induced $CR^+$ increases found by Wiktor-Jedrzejczak and coworkers. Assuming that older mice may be more responsive, Smialowicz et al.[145] also exposed mice 14, 16, and 24 weeks old at 30 or 40 mW/cm$^2$. Only the 16-week-old mice exposed at 40 mW/cm$^2$ (29 W/kg) showed significantly higher percentages of $CR^+$ cells and smaller numbers of nucleated spleen cells.

Liburdy[146] exposed mice for 15 minutes to 26-MHz RFR at 80 mW/cm$^2$ (5.6 W/kg), which produced core-temperature rises of 2-3°C. Other mice were heated in an oven at 63°C for the same duration to obtain the same rises in core temperature. A decrease in the mean lymphocyte count (lymphopenia) and a rise in the mean neutrophil count (neutrophilia) were seen in the RFR-exposed mice, which persisted for 12 hours after exposure. Additional RFR-exposures at 3-hour intervals sustained the effects and prolonged the recovery period. By contrast, the oven-heated mice showed only slight effects. Those effects were absent for mice exposed to 26-MHz RFR at 50 mW/cm$^2$ or to 5-MHz RFR at 800 mW/cm$^2$, both of which yielded a whole-body SAR of about 0.4 W/kg (the basis for the 1982 ANSI standard[73]).

Smialowicz et al.[147] exposed 16 rats almost continuously for 69-70 consecutive days to 970-MHz RFR at 2.5 W/kg. Blood samples after exposure showed no significant differences from sham-exposed rats in erythrocyte count, total or differential leukocyte counts, mean cell volume of erythrocytes, hemoglobin concentration, or hematocrit. Splenic cells cultured with various mitogens exhibited no significant differences in responses from sham-exposed rats. However, the results of blood-serum analysis and the absence of changes in erythrocyte assays of the RFR-exposed group indicated that the rats may have been dehydrated. The authors noted that 2.5 W/kg is about half the basal metabolic rate of the rat, and that at 970 MHz, there probably were regions within the rat where local SARs were much higher than 2.5 W/kg, and that such higher SARs could have affected the endocrinologic system of the rat.

Smialowicz et al.[148] exposed pregnant mice to 2.45-GHz RFR at 28 mW/cm$^2$ (16.5 W/kg) for 100 minutes daily. At 3 and 6 weeks of age, the pups were assessed for development of primary immune response to an antigen (SRBC), proliferation of lymphocytes *in vitro* by stimulation with mitogens, and *in vitro* activity of natural killer (NK) cells against lymphoma cells. No consistent significant differences were found between RFR-exposed and sham-exposed mice in any of the endpoints.

Smialowicz et al.[149] exposed mice to CW or pulsed 425-MHz RFR at up to 8.6 W/kg. No differences were seen in mitogen-stimulated responses of lymphocytes or in primary antibody response to sensitization with SRBC or another antigen (PVP) between RFR-exposed and sham-exposed mice, or between mice exposed to the CW or pulse-modulated RFR.

Smialowicz et al.[150] exposed mice for 1.5 hours per day to 2.45-GHz RFR at several levels. For positive controls, other mice were injected with either hydrocortisone or saline. Splenic cells were then assayed *in vitro* for NK-cell activity by their cytotoxicity against mouse-lymphoma cells. NK-cell activity was significantly suppressed for 30 mW/cm$^2$ (21 W/kg), but activity returned to normal within 24 hours after the last RFR-exposure. However, this transient effect was not seen at 15 or 5 mW/cm$^2$ (10.5 or 3.5 W/kg). NK-cell activity was also assayed *in vivo*. Suppression of activity was seen in mice exposed at 30 mW/cm$^2$, but with return to normal several days after the last exposure. Hydrocortisone injection caused activity suppression both *in vitro* and *in vivo*.

Ortner et al.[151] exposed rats to 2.45-GHz RFR for 8 hours at 2 or 10 mW/cm$^2$ (0.44 or 2.2 W/kg). Within 5 to 15 minutes after exposure, peritoneal mast cells were extracted, and histamine releases from the cells induced by a chemical stimulant were determined. No significant differences from controls were observed in percentage of cell viability, percentage of cells, amount of histamine per cell, and cell diameter. For other rats similarly exposed, the counts of total red and white cells were not affected by exposure at either RFR level, nor were the blood-hemoglobin levels or percentages of lymphocytes and neutrophils relative to those for sham-exposed rats. The other types of cells were also unchanged by the RFR, and serum biochemistry parameters were not affected by either RFR level.

Wong et al.[152] sought possible effects of prolonged exposure of rats to low RFR levels in the HF band (3-30 MHz). In one of two experiments, 20 groups of 5 rats each were exposed to 20-MHz RFR at 1,920 mW/cm$^2$ (about 0.3 W/kg) for 6 hours per day, 5 days per week, and another 20 groups were sham-exposed as controls. After 8 days of treatment, 6 groups each of the exposed and control rats were euthanized and examined for histopathology. This was also done for 7 groups each after 22 days and for the remaining 7 groups each after 39 days. The results showed a significantly higher mean count of red blood cells and a significantly lower mean hemoglobin content for the rats terminated after 39 days of exposure than for the control group. However, statistical analysis showed that those differences were not RFR-related.

In the second experiment, 12 rats were exposed to the RFR and 12 other rats were sham-exposed, but each rat was housed separately and all rats were euthanized after 6 weeks. On termination, blood samples were collected and the routine counts and blood-chemistry assays were done, the spleens were excised and weighed, and suspensions of splenic cells were prepared. Also various other tissues were examined for histopathology. Unlike the previous results, no significant differences were seen in mean red blood cell count, hemoglobin, or any blood-chemistry parameters. All of the 24 rats were found to be histologically normal.

*Studies in Vivo On The Effects Of Chronic Exposure On Health, Longevity, And Resistance To Disease.* As discussed previously, Prausnitz and Susskind[58] exposed 100 mice to 9.3-GHz pulsed RFR at average power density of 100 mW/cm$^2$ (roughly 45 W/kg) for 4.5 minutes daily, which yielded a mean rise in body temperature of 3.3°C, 5 days per week for 59 weeks. Controls were 100 sham-exposed mice. Some deaths occurred in both groups, which were attributed to a pneumonia infection introduced accidentally into the colony. However, the death rate was found to be lower in the RFR-exposed mice than in the sham-exposed mice, a finding that could be ascribed to the protective effect of the daily rise in temperature ("fever") induced by the RFR. On necropsy, liver abscesses were found in some mice, but because of tissue breakdown, the relative incidence in RFR-exposed and control mice could not be determined. The authors mistakenly described the occurrence of leukosis (increase of circulating lymphocytes) as "cancer of the white blood cells."

Szmigielski et al.[153] injected mice with staphylococcal bacteria at a dose selected to yield a 60% survival rate on day 3 after injection. Before injection, the mice were sham-exposed or exposed to CW or pulsed 2.45-GHz RFR at 5 or 15 mW/cm$^2$ (2-3 or 6-9 W/kg) 2 hours per day for 6 or 12 weeks. For the mice exposed at 5 mW/cm$^2$ for 6 weeks, the survival rate was 80%; for those exposed at 5 mW/cm$^2$ for 12 weeks, it was 45%. The differences among those two RFR groups and the sham-exposed group were not statistically significant. The survival rates at 15 mW/cm$^2$ for 6 or 12 weeks respectively were 25% and 5%.

Liddle et al.[154] sought effects of RFR-exposure at various ambient temperatures on the survival of mice given an LD$_{50}$ dose of another strain of staphylococcus. They injected mice with that strain and exposed them to 2.45-GHz RFR at 10 mW/cm$^2$ (6.8 W/kg) for 5 days (4 hours per day) at 8 different ambient temperatures in the range 19-40°C. For temperatures up to 31°C, the percentages of RFR-exposed mice that survived the challenge were significantly higher than for sham-exposed mice. The results also indicated that most of the deaths of exposed animals were due to hyperthermia, and indicated that RFR-exposure may be beneficial to infected animals at low and moderate ambient temperatures, in consonance with the findings of several other studies.

In the previously discussed study at the University of Washington, 100 rats were exposed to 2.45-GHz RFR at 0.5 mW/cm$^2$ and sham-exposed 100 other rats for 25 month, and 10 each of the two groups were euthanized after 13 months (the interim kill), as were 10 of the 12 RFR-exposed rats and 10 of the 11 sham-exposed rats that survived the 25-month regimen (the final kill). In the immunologic aspects of the study, assays of suspensions of splenic cells at the interim kill showed significantly higher counts of T- and B-lymphocytes for the RFR-exposed rats than the sham-exposed rats, indicating that the RFR had stimulated the lymphoid system. However, there were no significant differences in T-cell and B-cell populations between the RFR and sham groups of the final kill, a possible indication of the onset of immunosenescence.

The values of CR$^+$ for the RFR groups of both kills were lower than for the sham groups, but the differences were not significant, indicating no differences between RFR and sham groups in lymphocyte maturation. For both kills, there were no significant differences in percentages of plaque-forming cells in response to immunization with sheep red blood cells. Stimulation of splenic-cell suspensions with various mitogens yielded mixed results at the interim kill. No mitogen results were obtained for the terminal kill because the lymphocyte cultures failed to grow and respond to any of the mitogens.

Blood samples drawn periodically from all of the rats were assayed for various hematologic parameters and serum chemistry. By multivariate analyses, there were no overall significant differences between RFR-exposed and sham-exposed rats in the hematologic parameters. Differences in thyroxine (T$_4$) levels between the RFR-exposed and sham-exposed rats were nonsignificant, indicating that the RFR had no effect on the hypothalamic-pituitary-thyroid feedback mechanism. As expected, however, the T$_4$ levels of both groups decreased significantly with age.

Toler et al.[155] implanted cannulas in the aortas of 200 rats. After the rats recovered, 100 of them were concurrently exposed to 435-MHz pulsed RFR. Exposures were at 1 mW/cm$^2$ average power density for about 22 hours daily, 7 days a week, for 6 months. The whole-body SARs varied with time, ranging from 0.04 to 0.4 W/kg, with a mean of about 0.3 W/kg. The other 100 rats were concurrently sham-exposed. Blood samples drawn cyclically without rat restraint or anesthesia were assayed for hormones ACTH, corticosterone, and prolactin in the plasma; for plasma catecholamines; and for hematologic parameters, including hematocrit and various blood cell counts. Heart rates and arterial blood pressure were also monitored. There were no significant RFR-induced differences between groups in any of the endpoints.

Overall, many of the early studies seeking possible effects of RFR on suspensions of various classes of leukocytes exposed *in vitro* suffered from the lack of adequate control of cell temperature during exposure. In later studies in which effective control over culture temperature was exercised, nonsignificant differences were obtained with exposed cultures held at the same temperature as control cultures for the same durations. In studies where elevation of culture temperature by RFR or conventional means did affect leukocytes adversely, the effects were clearly of thermal origin.

In early studies of RFR exposure of erythrocytes *in vitro*, hemolysis and potassium-ion (K$^+$) efflux were found for rabbit erythrocytes that were exposed at average power densities as low as 1 mW/cm$^2$. In subsequent investigations, however, the hemoglobin and K$^+$ losses from rabbit erythrocytes resulting from heating with RFR to 37°C did not differ significantly from losses from conventional heating; the threshold for effect was found to exceed 46 W/kg.

Studies seeking immunological effects of exposing animals to RFR *in vivo* yielded mixed results. Some investigators found that RFR-exposure of mammals increased the proliferation of leukocytes or the production of antibodies (relative to controls), but with few exceptions, the measured or estimated SARs were well in excess of 1 W/kg. More subtle effects on mammalian immune systems were sought in more recent studies, making use of significant advances in assay methods, and with attention to possible effects of non-RFR stress. Some of those investigations were directed toward the effects of RFR on the activity of natural killer (NK) cells, the results of which again showed that SARs much higher than 1 W/kg were necessary for effect. On the other hand, some studies indicated that animals exposed for short periods to relatively high RFR levels can withstand bacterial infection better than sham-exposed animals.

More directly relevant to possible effects of RFR on the human immune system would be studies in which animals are chronically exposed to RFR (preferably over virtually their entire lifetimes), to determine whether such exposure adversely affects their health, longevity, and resistance to natural disease or to experimental challenge with various microorganisms or toxins therefrom. The previously discussed University of Washington study, though with some findings open to question, is an example. Also noteworthy is the study of rats by Toler et al.[155] for the number of animals involved and the long exposure duration. As remarked by the authors, the absence of RFR-induced effects complement those of Guy and coworkers at the University of Washington. However, because of funding limitations, relatively few such studies have been carried out or repeated by other laboratories.

**Physiology and Biochemistry.** *Metabolism and Thermoregulation.* Bollinger[156] exposed rhesus monkeys to 10.5-MHz or 19.3-MHz RFR at successively higher power densities up to 600 mW/cm$^2$ (about 0.2 or 0.6 W/kg, respectively), or to 26.6-MHz RFR at up to 300 mW/cm$^2$ (0.6 W/kg). Colonic temperatures and electrocardiograms (EKGs) taken during exposure indicated no obvious indications of thermal stress, heart-rate increases, or other influences on the electrical events of the heart cycle due to the RFR. Also, rhesus monkeys were exposed to 10.5- or 26.6-MHz RFR for 1 hour at 200 or 105 mW/cm$^2$ (0.06

or 0.2 W/kg), or to 19.3-MHz RFR for 14 days (4 hours per day) at 115 mW/cm$^2$ (0.1 W/kg). Hematologic and blood-chemistry analyses done before and after exposure showed no significant differences between exposed and control monkeys for most of the cellular components of blood. Unrelated to RFR were significant differences in mean counts of monocytes and eosinophils. No abnormalities ascribable to exposure were seen in gross pathological and histopathological examinations.

Frazer et al.[157] exposed rhesus monkeys to 26-MHz RFR at up to 1,000 mW/cm$^2$ (2.0 W/kg) for 6 hours, during which skin and rectal temperatures were measured. The monkeys remained in thermal equilibrium even at the highest RFR level; their thermoregulatory mechanisms were able to efficiently dissipate the additional heat from the RFR. Krupp[158] exposed rhesus monkeys for 3 hours to 15 or 20 MHz RFR at levels up to 1,270 mW/cm$^2$ (1.3 W/kg). Again, the thermoregulatory mechanisms of the monkeys readily accommodated the additional heat. Krupp[159] followed up on 18 rhesus monkeys that had been exposed 1-2 years previously to 15-, 20- or 26-MHz RFR for up to 6 hours at least twice at levels in the range 500-1270 mW/cm$^2$. No RFR-related variations from normal values of hematologic and biochemical blood indices or of physical conditions were found.

Ho and Edwards[160] exposed mice for 30 minutes to 2.45-GHz RFR in a waveguide system that permitted measurement of oxygen-consumption rates and SAR during exposure. Such measurements were done at 5-minute intervals during exposure. Oxygen-consumption rates were also measured at 5-minute intervals for 30 minutes before and after exposure. The oxygen-consumption rates were converted into specific metabolic rates (SMRs) and expressed in the same units as the SARs (W/kg). At the highest RFR level, both the mean SAR and mean SMR had steadily decreased during exposure, thereby decreasing the total thermal burden of the mice. Apparently, they sought to diminish their thermal burdens by altering their body configurations during exposure to minimize their RFR-absorption rates, and they reduced their oxygen consumption. After exposure completion, oxygen consumption rates returned to normal.

To study voluntary thermoregulation, Stern et al.[161] trained fur-clipped rats in a cold chamber to press a lever that turned on an infrared lamp. The rats were then exposed to 2.45-GHz RFR for 15-minute periods at increasing levels, ranging from 5 to 20 mW/cm$^2$ (1-4 W/kg). As the RFR level was raised, the rats responded to maintain a nearly constant thermal state by decreasing the rate at which they turned on the lamp.

Adair and Adams[162] trained three squirrel monkeys to regulate their environmental temperature ($T_a$) behaviorally by adjusting the flows of air at various temperatures into an exposure chamber. The monkeys were then exposed to 2.45-GHz RFR for 10-minute periods at levels in the range 1-22 mW/cm$^2$ (0.15-3.3 W/kg). The monkeys were also exposed to infrared radiation (IR) of equivalent power densities while being sham-exposed to RFR. For the RFR at about 7 mW/cm$^2$ (1.05 W/kg) and higher, all were stimulated to select a lower $T_a$, indicating the existence of a threshold of 1.1 W/kg whole-body SAR or 20% of the resting metabolic rate of the squirrel monkey. Comparable reductions in selected $T_a$ did not occur for exposure to the IR.

Bruce-Wolfe and Adair[163] studied the ability of squirrel monkeys to vary the level of 2.45-GHz RFR as a thermal energy source. They trained four monkeys to successively select air streams having temperatures of 10°C and 50°C (30° ± 20°C, the thermoneutral temperature in the exposure chamber). The resulting mean $T_a$ was about 35°C. Then the 50-°C air source was replaced with 2.45-GHz RFR at 20 mW/cm$^2$ (3 W/kg) and 30-°C air. Thus, only the latter two sources were activated whenever the monkeys demanded heat, and only the 10-°C air source was activated whenever they demanded cooling. The results indicated that the monkeys were readily able to use the thermal energy from the RFR for thermoregulation instead of the 50-°C air source, and were thereby able to maintain normal rectal temperature.

Adair et al.[164] did similar experiments to determine the effects of long-term RFR exposure on thermoregulation. The exposures were for 15 weeks, 40 hours/week, to 2.45-GHz CW RFR at 1 or 5 mW/cm$^2$ (0.16 or 0.8 W/kg) at $T_a$s of 25, 30, or 35°C. Fourteen monkeys were trained to select a preferred $T_a$, and were treated concurrently in fours, one pair each for RFR-exposure and sham exposure. The results for 25°C or 30°C showed no changes in preferred $T_a$ during exposure at 1 mW/cm$^2$. However, at 35°C and 1 mW/cm$^2$ (0.16 W/kg), or at all three $T_a$s and 5 mW/cm$^2$ (0.8 W/kg), the monkeys selected cooler $T_a$s (1 to 3°C lower). Their colonic temperatures were not affected, but their skin temperatures varied with $T_a$ and RFR-exposure in an unreliable way.

Lotz and Saxton[165] studied the vasomotor and metabolic responses of five rhesus monkeys exposed to 225-MHz CW RFR with body axis parallel to the electric component. In the first of two protocols, each monkey was given 10-minute RFR exposures at successively higher levels, with enough time after each exposure for the monkey to return to its pre-exposure equilibrium, until a marked vasomotor response was evidenced by a rapid change in tail-skin temperature. RFR levels in the range 1.2-12.5 mW/cm$^2$ (0.3-3.6 W/kg) were used. At 20°C, metabolic heat production was not altered at 1.2 mW/cm$^2$ (0.3 W/kg) but declined with increasing RFR level. At 26°C, the rate of metabolic heat production before exposure was well below that at 20°C, and was not altered by the 10-minute RFR-exposures. The lowest RFR level that reliably altered metabolic heat production during such 10-minute exposures was in the range 5-7.5 mW/cm$^2$ (1.4-2.1 W/kg).

In the second protocol, the monkeys were equilibrated at 20°C and then given single 120-minute exposures at levels in the range 0-10 mW/cm$^2$ (0-2.9 W/kg). The monkeys were also similarly treated at 26°C, but at RFR levels up to 7.5 mW/cm$^2$ (2.1 W/kg). The mean metabolic heat production dropped sharply during RFR exposure at 20°C, but remained essentially unchanged during RFR exposure at 26°C. Also evident was progressive recruitment of metabolic and vasomotor responses at 20°C. At both ambient temperatures, the mean colonic temperature during the last 30 minutes of RFR-exposure was higher than for the last 30 minutes of sham-exposure, even at 2.5 mW/cm$^2$ (0.7 W/kg), which was below the threshold for thermoregulatory action. This result indicated that thermoregulatory responses could not fully compensate for the heat generated by the RFR even in the cooler environment (20°C).

In summary, the thermal basis for effects of RFR on the autonomic thermoregulatory systems of mammals and on their behavioral thermoregulatory responses to RFR is evident. Especially noteworthy are results for primates because of their far greater similarities to humans than other animals.

**Endocrinology.** Cairnie et al.[166] exposed unanesthetized male mice for 16 hours to 2.45-GHz RFR at 50 mW/cm$^2$ (60 W/kg) and determined their rectal and testis temperatures at exposure end. The mean rectal temperature was significantly higher than for sham-exposed mice, but the mean testis temperature did not differ significantly, showing that the thermoregulatory system of the testes was able to compensate fully for the increased thermal burden from RFR at close to lethal level. Also, conscious mice exposed for various durations to 2.45-GHz RFR in the range 21-37 mW/cm$^2$ exhibited no testicular cell damage or abnormal sperm counts. The corresponding ranges of whole-body and testicular SARs were 25.3-44.5 W/kg and 8.4-14.8 W/kg.

Lebovitz and Johnson[167] exposed unanesthetized male rats to 1.3-GHz RFR for 9 days (6 hours per day) at a whole-body SAR of 6.3 W/kg, resulting in a mean core-temperature rise of 1.5°C. On exposure completion, groups of rats were weighed and euthanized at intervals corresponding to 1/2, 1, 2, and 4 cycles of spermatogenesis. The RFR-exposed rats yielded 87.6% normal sperm after a half-cycle of spermatogenesis versus 95.8% for sham-exposed rats. The difference was significant, but one rat of the RFR-exposed group contributed most of their abnormal sperm, which rendered the finding suspect. There was no significant difference in the mean weight of seminal vesicles, indicating that exposure at 6.3 W/kg was not deleterious to testosterone production, a finding supported by histologic evaluations by light microscopy.

Lebovitz and Johnson[168] exposed unrestrained male rats for 8 hours to 1.3-GHz CW RFR at 9 W/kg, selected to yield a core-temperature rise of 4.5°C and stated to be lethal for chronic exposure. Subgroups at 1/2, 1, 2, and 4 spermatogenesis cycles following exposure were analyzed for testis mass and daily sperm production as in the previous study. Trunk blood was assayed for follicle-stimulating hormone (FSH) and leutinizing hormone (LH). There were no significant differences between RFR-exposed and sham-exposed rats in any of the endpoints, except for a decline in sperm count 2 cycles of spermatogenesis after RFR-exposure. However, the authors remarked that a single positive result among the negative results for all of the many other endpoints studied is highly questionable. They also noted that a differential sensitivity of germ cells at this stage of maturation had been reported for conventional heating of the testes.

Lotz and Michaelson[169] first "gentled" rats for 2 weeks before RFR exposure by weighing and handling them at least four times a week, and behaviorally equilibrating each rat by taking its colonic temperature and putting it into an exposure cage for 3-5 hours for several days before use. They had observed a rapid rise in colonic temperature and corticosterone (CS) levels in the blood of the rat in the first 30 minutes of occupancy in the exposure chamber, followed by return to baseline values by the end of 180 minutes, thus proving the need for such equilibration before exposure.

The authors then exposed unanesthetized gentled rats to 2.45-GHz RFR at up to 60 mW/cm$^2$ (9.6 W/kg) for up to 120 minutes and measured their colonic temperatures and CS levels after exposure. The mean colonic temperature showed a small but significant rise after 30 minutes at 13 mW/cm$^2$ (2.1 W/kg), with 30-minute exposures at higher levels yielding mean temperature rises approximately proportional to the RFR level. The mean CS level increased nonsignificantly for durations up to 120 minutes at 13 mW/cm$^2$ (2.1 W/kg), up to 60 minutes at 20 mW/cm$^2$ (3.2 W/kg), and 30 minutes at 30 mW/cm$^2$ (4.8 W/kg). The results yielded threshold values for adrenal-axis stimulation of 30-50 mW/cm$^2$ (4.8-8.0 W/kg) for 60-minute exposures and 15-20 (2.4-3.2 W/kg) mW/cm$^2$ for 120-minute exposures. The latter range is somewhat less than half the resting metabolic rate of the rat.

Lu et al.[170] investigated the effects of 2.45-GHz RFR on serum thyroxine (T$_4$) concentration in male Long-Evans rats from two suppliers, Blue-Spruce (BS) and Charles River (CR). The normalized SAR was 0.19 W/kg per mW/cm$^2$. The rats were acclimated and gentled before exposure. Seven protocols were used that involved exposures at various RFR levels up to 70 mW/cm$^2$ for up to 8 hours. At the end of treatment, the rats were euthanized and assayed for serum T$_4$ concentration.

The results for 1-hour exposures at up to 70 mW/cm$^2$ showed no dependence of T$_4$ concentration on RFR level. Although some significant changes were seen for 2-hour exposures, those results were supplier-dependent. Thus, when the T$_4$ levels for the RFR-exposed rats from each supplier were compared with the sham-exposed rats from the same supplier, no significant RFR-induced changes in T$_4$ concentration were seen.

For the 4-hour exposures, only the results for BS rats exposed at up to 20 mW/cm$^2$ were shown. Their T$_4$ levels, compared to those for sham-exposure, were significantly higher at 1 mW/cm$^2$, not significantly changed at 5 or 10 mW/cm$^2$, and were significantly lower at 20 mW/cm$^2$. The results for the other protocols showed no significant RFR-induced alterations in T$_4$ level except for CR rats given 3 consecutive daily 4-hour exposures at 40 mW/cm$^2$, for which the T$_4$ level was significantly lower than for shams.

Lu et al.[171] endeavored to determine the influence of confounding factors in studies of effects of RFR on the adrenal cortex, with serum CS concentration used as the index of adrenocortical function. Gentled rats were subjected to 10 protocols involving exposures to 2.45-GHz RFR for 2 or 4 hours at up to 55 mW/cm$^2$ (11 W/kg). Colonic-temperature and CS-concentration rises were found to be dependent on RFR level with distinct thresholds, but the effect diminished with repetition. For example, the threshold for change of CS concentration was 40 mW/cm$^2$ (8 W/kg) at the first exposure, but no changes were observed in rats exposed 10 times at levels up to 40 mW/cm$^2$ (8 W/kg).

Injection of ethanol lowered baseline colonic temperatures and raised CS concentrations, effects not observed for saline-injected controls. Ethanol injection after RFR-exposure at 10 or 20 mW/cm$^2$ yielded higher concentrations of CS than in rats injected with saline after exposure. Removal of hair from the rats did not affect the baseline colonic temperatures and concentrations of CS significantly, but it decreased the RFR-induced hyperthermia and CS stimulation. The authors concluded that no adrenal response to RFR is evident without a colonic temperature rise of at least 0.7°C (20 mW/cm$^2$ or 4 W/kg).

Lotz and Podgorski[172] collected blood samples hourly for 24 hours from 6 rhesus monkeys and assayed them for cortisol, $T_4$, and growth hormone (GH). At the same clock time during those 24 hours, they exposed each monkey for 8 hours to 1.29-GHz RFR at 20, 28, or 38 mW/cm$^2$ (2.1, 3.0, or 4.1 W/kg). Three sessions at each RFR level were alternated with sessions of sham-exposure at intervals of 10-14 days for recovery. The data collected for the corresponding clock periods of the three sessions at each RFR level were averaged, to yield a 24-hour temporal series of mean values for each condition.

The mean rectal temperature rose within 2 hours after the start of RFR exposure to plateaus that were dependent on RFR level, but returned to control level within 2 hours after exposure end. For sessions at 20 and 28 mW/cm$^2$, the mean cortisol levels did not differ significantly from those for sham-exposure sessions, but rose significantly during sessions at 38 mW/cm$^2$. The levels then diminished to control values, indicating that the effect was transient. The authors suggested the existence of a threshold between 28 and 38 mW/cm$^2$ (3.0 and 4.1 W/kg) for rises in cortisol levels and associated such rises with rectal-temperature elevations of about 1.7°C, thus supporting the hypothesis that adrenocortical effects of RFR are thermally induced. For all RFR levels, no significant differences in mean GH or $T_4$ levels were seen.

In summary, although some effects of exposure to RFR on the endocrine system seem to be predictable from physiological considerations, other, more subtle effects may be worthy of additional study, such as those related to the interactions among the pituitary, adrenal, thyroid, and hypothalamus glands and/or their secretions. Part of the problem appears to be related to the uncertainties about stress mechanisms and various accommodations to such mechanisms. Animals placed in novel situations are more prone to exhibit stress responses than those adapted to experimental situations.

Because the effects of RFR on the endocrine systems of animals are largely ascribable to increased thermal burdens, to stresses engendered by the experimental situation, or to both, there is no clear evidence that such effects would occur in humans exposed to RFR at levels that do not produce significant increases in body temperature.

**Cardiovascular Effects.** Frey and Seifert[173] exposed excised beating frog hearts to 10-μs pulses of 1.425-GHz RFR at 60 mW/cm$^2$ peak. The pulses were triggered at the peak of the electrocardiogram (EKG) P wave and at 100 and 200 ms after that peak. The results for zero and 100-ms delays were inconclusive, but a significant rise in heart rate (tachycardia) was seen for the 200-ms delay.

Clapman and Cain[174] were unable to obtain those results. They also exposed groups of frog hearts to 2-μs or 10-μs pulses of 3-GHz RFR at 5,500 mW/cm$^2$ peak; for one 2-μs group, the pulses were triggered at the initial rise of the EKG's QRS complex and another 2-μs group was exposed to unsynchronized pulses at 500 pps (5.5 mW/cm$^2$ average power density). Again, no significant differences in heart rate were seen between any of the RFR-exposed groups and a control group, in contrast with those of Frey and Seifert.[173] and Liu et al.[175] also sought similar effects with excised hearts, but their findings were negative. Those authors also opened the thorax of frogs to expose the heart *in situ* to 100-μs pulses of 1.42-GHz or 10-GHz RFR, and again obtained negative findings.

Galvin et al.[176] isolated cardiac muscle cells from the quail heart and exposed them in suspension to 2.45-GHz RFR at 37°C for 90 minutes at SARs up to 100 W/kg. After exposure, samples of the suspensions were examined for integrity of the cells by their exclusion of trypan blue, a vital stain. Cell integrity was unaffected by exposure at 1 W/kg, but suspensions exposed at 10, 50, and 100 W/kg yielded successively larger increases in percentages of cells permeable to trypan blue relative to control suspensions. Suspensions were assayed for release of the enzymes creatine phosphokinase (CPK) and lactic acid dehydrogenase (LDH). CPK release was unaffected at any SAR. Release of LDH rose with SAR, but the increases were nonsignificant relative to controls except at 100 W/kg. By electron microscopy, the structures of heart cells exposed at 1, 10, and 50 W/kg, as well as control cells, were normal. Cells exposed at 100 W/kg showed increased intracellular changes, but intercellular junctions remained intact.

Yee et al.,[177] concerned about possible electrode artifacts, tested several types of electrodes for recording beat rates from isolated frog hearts during RFR-exposure. Faster than the usual decreases in heart rate after excision were seen only with glass electrodes containing either potassium chloride or a metal wire, results showing that bradycardia can be induced by field intensification caused by such electrodes.

Yee et al.[178] mounted isolated frog hearts within a waveguide filled with physiologic (Ringer's) solution. Each heart's beat rate was monitored for 60 minutes at 5-minute intervals. For those exposed to RFR, exposure was begun 10 minutes into the monitoring period and was terminated 30 minutes later. At the end of the period, the mean heart rate of the group of control hearts decreased linearly to 67% of the initial rate.

Other hearts were exposed to trains of 10-μs 2.45-GHz pulses at SARs up to 200 W/kg, either triggered or not triggered by the EKG. Most of the RFR-exposed groups exhibited decreases in heart rates similar to those of the controls. Noteworthy were the results for two groups exposed to RFR, both continuously at 200 W/kg, but only one cooled by circulating bathing solution. The uncooled group yielded heart-temperature increases of 2.5, 5.5 and 8°C at 5, 15, and 30 minutes of exposure, with concomitant decreases in mean heart rate relative to controls. By contrast, the cooled RFR-exposed group showed no significant differences from controls. Moreover, a group heated with circulating bathing solution to obtain a temperature-versus-time rise similar to the 200-W/kg group showed a linear decrease in heart rate to final values comparable to those of that group.

Yee et al.[178] also noted that Schwartz et al.[179] had reported a 19% increase in calcium efflux from isolated frog hearts that were exposed for 30 minutes to 16-Hz-modulated, 1-GHz RFR at 0.15-3 W/kg, an effect not seen for CW RFR or RFR amplitude-modulated at natural heart-beat rates. They tried to reproduce that finding by exposed one group each to CW and pulsed 2.45-GHz RFR, both amplitude-modulated at 16 Hz, at 3 W/kg. Their results did not significantly differ from those for controls. They remarked that an increase in concentration of free $Ca^{++}$ in the cytoplasm of heart cells triggers rapid changes in heartbeat, so the absence of such changes in these groups does not support the findings of Schwartz et al.[179]

Yee et al.[180] did a study similar to their earlier one but with rat hearts. As in the frog-heart study, the beat rate of each heart was monitored for 60 minutes at 5-minute intervals, but exposure was begun 20 minutes into the monitoring period and was terminated at 50 minutes. The results led the authors to conclude that exposure to pulsed 2.45-GHz RFR at 2 or 10 W/kg had no specific influence on the myocardium or its neural components.

Presman and Levitina[181,182] had done two heart studies on the live rabbit, one with 2.4-GHz CW RFR at 7-12 mW/cm$^2$ and the other with 3-GHz pulsed RFR at 3-5 mW/cm$^2$ average power density, 4.3-7.1 W/cm$^2$ pulse power density. In both studies, rabbits were exposed for 20-minute periods each in various orientations. During each exposure and for 10 minutes before and afterward, the EKGs of the rabbits were recorded with plate electrodes.

Exposure of the entire top surface of the rabbit produced neither tachycardia nor bradycardia during RFR exposure. However, tachycardia was seen during the first 5 minutes postexposure, changing to bradycardia during the remaining 5 minutes postexposure. By contrast, exposing the top of the head only or the rear only produced significant tachycardia during exposure, with the head exposure yielding the greater effect. Exposing the underside of the rabbit yielded bradycardia during exposure, which was followed by returns toward normal heart rates during postexposure. The findings were difficult to assess because no data were given, only the relative differences among mean values. Also, artifact may have been present from use of metal electrodes.

Kaplan et al.[183] and Birenbaum et al.[184] tried to reproduce the results of those studies. Their findings were negative except for RFR levels that were clearly hyperthermic.

Phillips et al.[185] exposed rats to 2.45-GHz RFR for 30 minutes at 0, 4.5, 6.5, or 11.5 W/kg. Nonsignificant bradycardia was seen at 4.5 W/kg; mild but significant bradycardia developed within 20 minutes at 6.5 W/kg, followed by recovery in 2 hours; pronounced bradycardia occurred abruptly at 11.1 W/kg, after which heart rates rose to values well above those of controls, and they persisted at the higher levels to the end of the test period. The authors surmised that the heart block was caused by release of toxic materials, elevated serum potassium, or myocardial ischemia, all from excessive heat.

Galvin and McRee[186] studied the effect of RFR-exposure *in vivo* on the functioning of cat hearts with and without surgically produced myocardial ischemia (MI). One group of MI hearts was exposed to 2.45-GHz CW RFR at 30 W/kg with an applicator for 5 hours, and another MI group was sham-exposed. For comparison, two groups of non-MI hearts were similarly treated. Before and during the treatment, mean arterial blood pressure, cardiac output, heart rate, and EKG were measured, and blood samples were assayed for plasma protein concentration and creatine phosphokinase (CPK) activity. After the treatment, the hearts were excised and assayed for tissue CPK activity.

The results for both the MI and non-MI cats indicated no significant differences in mean arterial blood pressure, cardiac output, or heart rate between RFR-exposed and sham-exposed groups, and no synergism of ischemia and RFR-exposure for those cardiovascular indices. Also, the RFR and sham groups showed no significant differences in plasma or tissue CPK activity. Thus, localized exposure of the undamaged or ischemic heart to the RFR *in vivo* had no effect on the myocardium or its neural components, results at variance with those for excised hearts exposed to RFR.

Galvin and McRee[187] exposed conscious rats from below for 6 hours to 2.45-GHz CW RFR at 10 mW/cm$^2$ (3.7 W/kg) and assayed various cardiovascular, biochemical, and hematologic indices. No significant differences between RFR-exposed and sham-exposed rats were seen in any of the blood parameters. The initial mean heart rates of the two groups did not differ significantly, but the mean heart rate of the RFR group decreased to about 90% during the first hour and remained there (with smaller variations) during the rest of the period. Based on the results of a followup experiment, the authors surmised that the bradycardia was due to reduction of metabolic rate to compensate for the heat from the RFR.

In summary of the physiologic and biochemical effects of RFR, there are scientifically credible experimental data to show that the thermoregulatory systems of nonhuman primates can readily compensate for high RFR levels, an especially important finding because of the greater similarities in anatomy and physiology between human and nonhuman primates than between humans and the various species of laboratory mammals.

Most of the studies of possible effects of RFR on endocrine systems were conducted on rodents. Studies that reported positive findings also yielded indications that the effects were largely due to increases in the thermal burdens of the animals. In many studies, observed alterations in endocrine function may have been significantly influenced by stresses in the animals. For this reason, the results of those studies that involved stress reduction by acclimating animals to handling and the experimental situation are notable. Nevertheless, some of the more subtle effects are worthy of further study.

Regarding cardiovascular effects, the positive findings reported in early studies (bradycardia, tachycardia, or both) were suspect because of the use of attached or indwelling electrodes that probably introduced artifact. Various kinds of electrodes were investigated, and several special types were developed that were not perturbed by RFR or did not perturb the local RFR fields. Studies involving use of such electrodes showed that heart rates were altered only at RFR levels that produced rises in temperature or otherwise added thermal burdens to the animal. Also investigated were possible effects of RFR pulses at repetition rates synchronous with periodic characteristics of the EKG. Specifically, the authors of an early study reported induction of tachycardia in excised hearts by RFR pulses in synchrony with the EKG, but others could not confirm this finding in excised hearts or in live animals.

Several researchers showed that for CW RFR, levels well in excess of 1 mW/cm$^2$ or 1 W/kg were necessary for significant alterations of heart rate. Small decreases in heart rate were seen in equilibrated conscious rats exposed for 6 hours at 3.7 W/kg, a finding ascribed to a compensating reduction in metabolic rate. The results of another study indicated that the functioning of hearts damaged from other causes (e.g., rendered ischemic) is not affected by exposure to CW RFR at 10 mW/cm$^2$ or lower.

**RFR and Behavior.** *Behavioral Studies in Rodents.* Justesen and King[188] trained food-deprived rats to lick a nozzle 40 successive times to obtain a drop of dextrose-water solution. They then added the presentation of an audio tone at random intervals to signal availability of the reward. After such training, the rats were exposed to 2.45-GHz RFR at up to 1.5 mW/cm$^2$ (4.6 W/kg) in 1-hour sessions consisting of alternating 5-minute intervals of RFR and no RFR. The mean number of responses by the rats diminished with increasing RFR level, but the decreases at higher levels were related to the cessation of responding rather than lower licking rates, most likely associated with warming.

In a three-part study, Hunt et al.[189] exposed rats to 2.45-GHz RFR for 30 minutes. In the first part, each rat was exposed to the RFR at 6.3 W/kg or sham-exposed, after which its exploratory movements within a test chamber were recorded. The mean activities after either treatment decreased with time, but the values were generally lower during most of the period after RFR-exposure than after sham-exposure and became comparable for the treatments toward session end. The RFR-exposed rats were often seen sleeping during the middle parts of sessions.

In the second part, the authors trained rats to repeatedly swim a 6-meter channel for 24 hours, and scored each rat's performance versus time as its median swim speed for each block of 20 traverses. In one experiment, rats were sham-exposed or exposed at 6.3 W/kg for 30 minutes and tested immediately after treatment to determine any prompt effects.

Their mean performance was similar to that of their sham group for about 200 traverses, but was below mean control speed for about the next 100 traverses, after which both groups again performed comparably. In two other similar experiments, the RFR level was 11 W/kg, and the rats were tested immediately or after 1 day of delay. Measurements of colonic temperatures immediately after treatment showed that the rats had been rendered severely hyperthermic. The performance of the group tested right after exposure was clearly impaired by the hyperthermia, but the group tested 1 day later showed recovery and yielded results similar to those of the 6.3-W/kg group.

In the third part, the authors trained water-deprived rats to press a lever in a complex vigilance-discrimination task to obtain small quantities of saccharin-flavored water. On the first day, all groups were sham-exposed. During the next four days, each group was exposed for one 30-minute period each at 6.5 and 11 W/kg and two periods of sham-exposure. Performance was tested for 30 minutes after each treatment. The mean error rate 5 minutes after exposure at 6.5 W/kg was significantly higher than after sham-exposure, but dropped to the latter range at 15 and 25 minutes. After exposure at 11 W/kg, however, the mean error rates were all much higher than after sham-exposure. The mean number of responses diminished with increasing RFR level, but the decreases at higher levels were related to response cessation rather than to lower licking rates, an effect most likely associated with warming.

Monahan and Ho[190] exposed mice for 15 minutes within a waveguide to 2.45-GHz CW RFR at forward powers up to 4.8 W in an ambient temperature held at 24°C by air flow through the waveguide. During exposure, the mean SAR and the mean percentage of forward energy absorbed were measured at 5-minute intervals. In another experiment, exposures were limited to 10 minutes and the absorptions were recorded at 12-second intervals. The mice could not be watched within the waveguide, but the results of both experiments showed that they had oriented themselves to reduce the percentages of RFR energy absorbed and the SARs when the forward power was 1.7 W (initial SAR 28 W/kg) or higher.

Lin et al.[191] sham-exposed or exposed food-deprived rats to 918-MHz RFR at 10, 20, or 40 mW/cm$^2$ (2.1, 4.2, or 8.4 W/kg) during 30-minute sessions. The rat holder was a truncated cone of rods designed to allow the rat to poke its head through the narrower end and move it freely. A small upward head movement interrupted a horizontal light beam, thereby registering a count. The rat was required to do 30 such movements rapidly and regularly for a food pellet. A downward head movement gave access to the pellet delivered.

One of three rats was exposed for 30 minutes each at 2.1, 4.2, and 8.4 W/kg on consecutive days, another rat was exposed on alternate days at the same levels, and a third was given 30-minute sessions of sham-exposure. No significant effects on performance were seen at 2.1 or 4.2 W/kg. At 8.4 W/kg, the performance of the two RFR-exposed rats did not change during the first 5 minutes. However, both rats displayed heat stress and diminished performance during the remaining 25 minutes. Yet another rat was exposed at successively higher levels up to 32 mW/cm$^2$ (6.7 W/kg), at which it exhibited similar signs of heat stress. Its performance rates indicated a threshold between 30 and 40 mW/cm$^2$ (6.3 and 8.4 W/kg).

Schrot et. al.[192] trained 3 rats to respond to auditory stimuli with four presses on three levers in a specific sequence that was changed for each session. Just before each session, the rats were sham-exposed or exposed to pulsed 2.8-GHz RFR at a specific average power density in the range 0.25-10 mW/cm$^2$ (0.04-1.7 W/kg) for 30 minutes. Sessions were conducted daily, 5 days a week. Exposure at 10 mW/cm$^2$ (1.7 W/kg) of all three rats yielded higher error-responding rates, lower sequence-completion rates, and alterations in the normal acquisition pattern. Similar effects were seen at 5 mW/cm$^2$ (0.7 W/kg) but to a lesser extent. Below 5 mW/cm$^2$, most data points were within the control range, but a few were outside that range. The significance of the latter points is uncertain.

In a study by Gage and Guyer,[193] they trained rats to perform on a reinforcement schedule in which the opportunity to obtain a food pellet was presented on the average of once each minute in a preplanned sequence of intervals without cueing. After training, groups were exposed to 2.45-GHz RFR for 15.5 hours at 8 or 14 mW/cm$^2$ (1.6 or 2.8 W/kg) in 22°, 26°, or 30°C ambient temperature. The response rates at each temperature diminished directly with increasing RFR level, but the effects of ambient temperature itself were not consistent.

Lebovitz[194] exposed groups of unrestrained rats to 1.3-GHz pulsed RFR in individual waveguides. In each waveguide were a vertical displacement bar (behavioral operandum), a means for illuminating the operandum as a cue, and a means for delivering food pellets. Before exposure, groups of food-deprived rats were trained to press the bar for food pellets until they learned to press the lever 5 successive times to obtain a pellet (a fixed-ratio-5, or FR-5 schedule). Those that performed at the highest and most stable rates were trained further to respond only when the operandum was illuminated (a multiple fixed-ratio, extinction schedule of reinforcement). During training, the FR schedule was gradually raised to FR-25. The responses when the operandum was illuminated (SD) and when it was not illuminated (Sd, which yielded no pellets) were counted separately. Exposures were for 3 hours daily, 5 days a week. Behavioral sessions were initiated 15 minutes after the start of exposure and were halted 15 minutes before exposure end. Each behavioral session was divided into 6 equal sequential blocks to evaluate intrasession changes.

The results for 8 weeks of exposure at 1.5 W/kg showed no significant SD differences between the RFR and sham groups. Slight declines in rates during sessions were seen in both groups. The response rates for Sd were more variable than for SD. The intrasession declines in Sd rates, which were also evident for baseline and recovery weeks, was sharper than for SD, but there were no significant differences between the RFR and sham groups.

Groups exposed for 9 weeks at 3.6 W/kg also showed no significant differences in SD response rates. Marginally significant changes in weekly rates were seen, and the rate of intrasession decline was significant. The Sd response rates of both groups showed sharp intrasession declines again. The results for 6 weeks of exposure at 6.7 W/kg showed no significant differences between groups in overall SD response rates.

From the negative results for SD and Sd at 1.5 W/kg and the doubtfully significant decline in Sd rate at 3.6 W/kg, the author suggested that the latter RFR level could be the approximate threshold for modifying the rat behavioral paradigm studied. The author also indicated that 6.7 W/kg is the approximate resting metabolic rate for a rat, so that RFR level represents a doubling of the heat dissipation requirements of the rat. They therefore concluded that thermal factors were likely involved in the positive results.

Lebovitz[195] exposed similarly trained groups of rats to CW or pulsed 1.3-GHz RFR and tabulated their SD and Sd response rates, called S+ and S- in this paper. The results for 3.6 W/kg and 5.9 W/kg were consonant with those of the previous study. Also, the S+ and S- rates for pulsed RFR at 6.7 W/kg were similar to those with CW RFR at 5.9 W/kg. The author concluded that the differences in rates between the pulsed-RFR and sham groups could not be ascribed to the pulsed character of the RFR *per se*. Core temperatures were measured in other rats. Exposures to CW or pulsed RFR at 3.5 W/kg, the approximate threshold above, yielded no significant differences in rectal-temperature changes compared with rats similarly sham-exposed. However, exposures at 6.3 W/kg yielded increases of 0.5° to 1°C, with no significant duration-dependent differences. Thus, the thermal basis for the behavioral changes above is evident.

D'Andrea et al.[196] exposed a group of 14 chamber-adapted rats to 2.45-GHz CW RFR at 0.5 mW/cm$^2$ (0.14 W/kg) 7 hours daily for 90 days. Body masses and intake of food and water were measured daily. Each rat was tested monthly for its threshold reactivity to footshock by observing movements of its paws in response to electric shocks of varied intensity within a gridded-floor chamber. Differences in body mass, food and water intake, or threshold footshock reactivities relative to those for the 14 sham-exposed rats were not significant. Right after such treatment, 7 rats of each group were assessed for open-field behavior, shuttlebox performance, and lever pressing for food pellets on an interresponse time schedule. The rest of the rats were examined for gross pathology; no significant differences ascribable to the RFR were seen. Major changes were observed in the open-field tests, but none of the differences were related to RFR-exposure. Open-field tests done 60 days after treatment yielded similar results.

Shuttlebox performance was tested for responses to an electric shock given right after presentation of a tone and white light as a warning. The rat could prevent presentation of the shock by crossing to the other side of the shuttlebox during 10 seconds of the warning (an avoidance response), or could cross while receiving the shock (an escape response). The time lags (latencies) for avoidance and escape responses were recorded. Overall, there were no significant differences between RFR-exposed and sham-exposed rats. A shuttlebox test done 60 days after treatment showed no significant differences in mean latencies or their variances.

Two days after the shuttlebox testing, the rats were deprived of food and trained daily to press a lever twice for a food pellet, with a specific time interval between the presses. The training was rendered progressively more difficult until the rats were required to do the second press only between 12 and 18 seconds after the first press to obtain a food pellet. During the training, the RFR group earned fewer pellets than the sham group, but the differences at corresponding times and overall were not significant.

The results of this study and of two similar studies in the same laboratory[197,198] were not fully consistent and showed little if any statistically significant differences between RFR-exposed and sham-exposed rats, but suggested that the threshold for behavioral responses to 2.45-GHz RFR in rats may be in the range 0.5-2.5 mW/cm$^2$ (0.14-0.70 W/kg).

Mitchell et al.[199] exposed chamber-adapted rats to 2.45-GHz CW RFR at 10 mW/cm$^2$ (2.7 W/kg) for 7 hours. Right after treatment, each rat was tested for spontaneous locomotor activity, acoustic startle response, and retention of a shock-motivated passive avoidance task. Lower spontaneous activity was seen in RFR-exposed rats than in sham-exposed rats. In the startle-response test, each rat was subjected to 20 intense acoustic pulses at variable intervals in the range 20-60 seconds, and the response of the rat during each acoustic pulse was determined. The startle responses of the RFR-exposed rats were significantly lower than for the sham-exposed rats. Each rat was then placed in the lighted smaller part of a gated, two-chamber shuttle box, the larger part of which was dark and equipped to deliver an electric shock. The gate was opened after 1 minute, and if the rat moved into the larger chamber within 2 minutes, it was given a shock, but if it remained in the smaller chamber for more than 2 minutes, it was removed and not tested further. One week later, retention of the shock experience was tested in the box by allowing each rat 5 minutes (instead of 2 minutes) within the smaller chamber to react. The differences in passive avoidance activity between the RFR and sham groups were not significant.

Akyel et al.[200] trained groups of 4 rats each on three different behavioral schedules to obtain food pellets. After training, each rat was exposed once a week for 10 minutes to 10-µs pulses of 1.25-GHz RFR at 1-MW peak forward power, with its long axis parallel to the electric component of the RFR. During sessions, the average forward power was held constant at 4, 12, 36, or 108 W, obtained by using a pulse repetition frequency of 240, 720, 2160, or 6480 pps. Each rat was administered all four RFR levels in a quasi-random weekly order. The whole-body specific absorptions (SAs) and whole-body specific absorption rates

(SARs) respectively ranged from 0.5 to 14.0 kJ/kg and 0.84 to 23.0 W/kg. Each rat was tested starting right after exposure end.

At the three lower RFR levels, no significant differences in any of the three behavior schedules were seen. At the highest level (14.0 kJ/kg, 23.0 W/kg), however, the rats trained on two of the schedules failed to reach baseline performance, and those on the third schedule exhibited variable effects. Exposures at that level caused an average colonic-temperature rise of 2.5°C, and the rats did not respond at all for about 13 minutes after exposure completion. The authors concluded that those behavioral changes were thermally induced.

*Behavioral Studies in Nonhuman Primates.* Galloway[201] trained rhesus monkeys to press one of three levers when that lever was lit in order to obtain a food pellet (discriminative behavior). After the training, the head of each monkey was exposed with an applicator to 2.45-GHz RFR at estimated mean head SARs of 7, 13, 20, 27, and 33 W/kg, and the effects on their performance were examined. The RFR was administered for 2 minutes just before each behavioral session but was stopped earlier if the monkey began to convulse. Convulsions occurred for all exposures at 33 W/kg, and often at 20 W/kg. Each monkey was exposed at least twice at each level during a 9-month period. Also, three of them were exposed at 13 W/kg for 5 daily 1-hour schedules of 2 minutes on and 1 minute off, totaling 40 minutes of exposure per day. No effects on that discriminative task were evident for either exposure regimen.

In a repeated-acquisition test, the three levers were lit sequentially four times in a specific order, and each monkey had to press the levers in the correct sequence to obtain a pellet. Daily sessions of 60 trials each were conducted before exposure, with the correct sequence changed each day. In the sessions just preceding RFR-exposure, a slight learning trend (decrease in error rate) was seen, but the changes were too small to ascribe significance. This was also true for the results at all RFR levels except 33 W/kg, for which the error rate at session start was highest. Thus, except possibly for the latter result, the RFR had no effect on this behavioral paradigm.

Cunitz et al.[202] inserted the head of a 3-kg or a 5-kg rhesus monkey through a hole in the bottom of a 383-MHz resonant cavity, with the monkey facing a diamond array consisting of the ends of four light pipes mounted through the cavity's side wall. Each monkey was trained to move a lever to the left, right, up, or down when any of the pipes was lit, to indicate the position of the lit pipe end in the array. For criterion performance, the monkey had to do 100 correct lever presses to obtain a food pellet. During testing, the pipes were lit in random order. A correct response caused the lighting of another pipe plus a tone for 0.75 second. An incorrect response yielded a 3-second timeout during which all of the lights were off and lever movements had no consequences.

In each session, the monkey's head was exposed to 383-MHz RFR for 2 hours at a specific input to the cavity in the range 0-15.0 W. The head SARs were estimated to range up to 33 W/kg and 20 W/kg respectively for the 3-kg and 5-kg monkey. The larger one was also exposed at 17.5 W (23 W/kg). Each monkey was restrained in a chair during the first hour, and the testing was done for the second hour. Sessions were conducted on 5 consecutive days at each successive level, with sham-exposure sessions before the RFR was raised to the next level. The lowest head SARs for diminished performance by the two monkeys were about the same: 22 and 23 W/kg.

Scholl and Allen[203] trained 3 rhesus monkeys in a visual-tracking task that required each monkey to move a lever so as to hold a continuously moving spot within a prescribed clear area on the screen of a display monitor. The spot was moved electronically in a specific pattern, and the responses generated continuous difference signals (errors). The central 15% of the screen was clear and comprised the on-target area. The monkeys received a brief electric shock for each 1 second accumulated outside the on-target area.

After training, the monkeys were exposed to horizontally polarized, 1.2-GHz CW RFR at 10 and 20 mW/cm$^2$ (measured at the center of the head in the absence of the monkey) for 2 hours each at one level and at the other level 2 days later. This polarization and frequency were chosen to provide half-wave resonant absorption in the monkey head. The corresponding head SARs were 0.8 and 1.6 W/kg. The data on mean tracking errors clearly showed that their performance was not diminished by the RFR-exposures.

De Lorge[204] trained five rhesus monkeys to perform the following task while seated: Each monkey was required to press a lever in front of its right arm, thus producing either a low-frequency tone to signal that no food pellet will be coming, or a higher-frequency tone for which the monkey had to press a lever in front of its left arm to receive a pellet. During 1-hour training sessions, pellets were made available at variable intervals averaging about 30 seconds. During 2-hour training sessions, pellets were made available at about 60-second intervals. The monkeys were exposed frontally to 2.45-GHz RFR at levels in the range 4-72 mW/cm$^2$ at head height. The estimated head SARs were 0.4-7.2 W/kg (0.1 W/kg per mW/cm$^2$).

After the monkeys achieved stable performance, they were tested on the variable 30-second delivery schedule in 1-hour sessions during which each was exposed to the RFR at 4 or 16 mW/cm$^2$ (0.4 or 1.6 W/kg head SAR) for 30 minutes. Their performances were not affected by either RFR level. Only three of them were tested on the variable 60-second delivery schedule in 2-hour sessions, during which they were exposed for 1 hour at levels in the range 16-72 mW/cm$^2$ (1.6-7.2 W/kg head SAR). One of them was also exposed at 16 mW/cm$^2$ during entire 2-hour test sessions. The performances of all three monkeys showed no significant departures from control rates for up to 52 mW/cm$^2$ (5.2 W/kg) and for two of them at 62 mW/cm$^2$ (6.2 W/kg); at the latter level, the performance of the third monkey was about 80% of its mean control rate. At 72 mW/cm$^2$ (7.2 W/kg), all three performed at about 50% of their mean control values. Those results suggested that the monkeys had reacted to body heating by the RFR at the higher levels, which diminished their performance.

De Lorge[205] trained four squirrel monkeys to press either the right or the left lever on top of a chair to obtain a food pellet. At first, a red light and a blue light were turned on alternately with each successive press of the levers. Next, the contingencies were changed such that presses of the right lever continued to alternate the red and blue lights (without reward) but a press of the left lever was rewarded only when the blue light was on. Each of the next stages of training required a higher number of right-lever presses to turn on the blue light. The last stage was a schedule in which each right-lever press yielded either a half-second of red light or 10 seconds of blue light, with a left-lever press during the latter yielding a pellet.

Each monkey was then exposed from above to 2.45-GHz RFR at levels in the range 10-75 mW/cm$^2$. SARs were estimated to have been 0.5 to 3.75 W/kg. No consistent behavioral changes occurred below 50 mW/cm$^2$ (2.5 W/kg); above that level, the effects increased with RFR level. During the 1-hour sessions, only the rate of right-lever responses exhibited a slight trend toward lower rates with increasing RFR level. The results for the 2-hour sessions were similar, but more pronounced. The right-lever-response rate versus RFR level varied widely among the monkeys, but at 60 mW/cm$^2$ (3.0 W/kg), all showed decrements to about 60%.

The author concluded that the observed behavioral changes were temporary and clearly related to hyperthermia. Consistent changes were seen when rises in rectal temperature exceeded 1°C, which corresponded to a threshold between 40 and 50 mW/cm$^2$ (2.0-2.5 W/kg). The author noted that similar results had been obtained with rhesus monkeys tested for the same behavioral task during exposure to 2.45-GHz RFR, but with a threshold 10 to

20 mW/cm$^2$ higher, and suggested that RFR-induced behavioral changes in different species may be scaled on the basis of body mass.

The findings of this study, reinforced by the similar results with rhesus monkeys, are important because the measurements of performance of a complex behavioral task during exposure to RFR were carried out with two species much closer to human physiology and intelligence than more commonly used nonprimate laboratory animals, and because reasonably accurate RFR thresholds for each primate species were determined.

De Lorge[206] similarly trained rhesus monkeys to perform a task in which each monkey was to press a lever in front of its right hand (called an observing response), which produced a brief low tone to signal that no food pellet will be delivered, or a longer high tone to signal the availability of a pellet. If the monkey pressed a lever in front of its left hand while the high tone was on (a detection response), the tone would cease and a pellet would be delivered. A response on the left lever at other times caused a 5-second interval in which presses of the right lever yielded only the low tone.

After stable performance was reached, each monkey was frontally exposed, during 1-hour sessions, to vertically polarized 225-MHz CW RFR (near the whole-body resonant frequency), or to pulsed RFR at 1.3 GHz or 5.8 GHz (both above whole-body resonance). The exposures to 225 MHz were at 5-11 mW/cm$^2$ (2.0-4.4 W/kg); those to 1.3-GHz RFR were at 20-95 mW/cm$^2$ average (2.6-12.4 W/kg); and those to 5.8-GHz RFR were at 11-150 mW/cm$^2$ (0.34-4.7 W/kg). The results yielded average-power-density thresholds for behavioral effects that increased with frequency: 7.5 mW/cm$^2$ at 225 MHz, 63 mW/cm$^2$ at 1.3 GHz, and 140 mW/cm$^2$ at 5.8 GHz. However, the corresponding whole-body SARs varied up and down with frequency: respectively 3.0, 8.2, and 4.3 W/kg, presumably because of differences in penetration depth. The detection-response rate on the food lever was not consistently affected by RFR-exposure at any of the frequencies: No effect was observed for 225 MHz or 5.8 GHz; a decreased response rate was occasionally seen for 1.3 GHz, but only at 83 mW/cm$^2$ (10.8 W/kg) or higher.

Exposure to 5.8-GHz RFR at 150 mW/cm$^2$ (4.7 W/kg) also produced minor burns on the faces of three of the monkeys, with the worst burns occurring between the eyes and along the orbitonasal area. The erythema disappeared within a few days except in one monkey who continually irritated the burned skin by removing scabious material. No burns occurred at 140 mW/cm$^2$ (4.4 W/kg), the behavioral threshold for this frequency, or at the highest levels of the other frequencies. The small penetration depth for 5.8 GHz (about 0.8 cm) probably was an important factor.

D'Andrea et al.[207] trained five rhesus monkeys to operate three levers (left, right, center) in various sequences to obtain food pellets. The sessions were 60 minutes long. The task during each session comprised three successive 10-minute schedules of lever presses, followed by a repeat of the same three schedules.

In the first 10-minute schedule, the monkey was required to withhold responses for 8 seconds after the start of an audio tone, and then to respond only within the next 4 seconds; the correct response during those 4 seconds was two presses of the left lever within 2 seconds of each other. The authors called this an interresponse-time (IRT) schedule. The second 10-minute period involved a time-discrimination (TD) schedule in which a press of the center lever in the presence of blue light yielded white light of either short or long duration in random fashion; at the end of either duration, the white light was replaced with red and green light. For the monkey to obtain a pellet when the red and green light was present, it had to press the right lever if the preceding white light was of short duration or the left lever if the preceding white light was of long duration.

During the third 10-minute period, a fixed-interval (FI) schedule was used: The monkey was presented with a continuous high tone, and its first press of the right lever after 55 seconds yielded a pellet.

During the 60-minute behavioral test sessions, each monkey was sham-exposed or exposed from above to 3-µs pulses of 1.3-GHz RFR at a root-mean-square pulse power density of 131.8 W/cm$^2$. The peak SAR was 15.0 W/kg in the head and 8.3 W/kg whole-body. The pulse repetition rate was 2, 4, 8, 16, or 32 pps, with corresponding average power densities of 0.92, 1.85, 3.70, 7.40, or 14.80 mW/cm$^2$.

The results showed no significant differences between sham-exposures and RFR-exposures in any of the behavioral responses. The authors noted that the the energy absorbed in the head by each pulse (280 mJ/kg) was well above the threshold for the RFR-auditory effect, and they remarked that if such auditory stimulation did occur, it produced no obvious effect on the trained behavior.

In summary, many of the studies on avoidance behavior indicate that RFR could be a noxious or unpleasant stimulus. There is much evidence, however, that changes in behavioral patterns induced by RFR are responses by their thermoregulatory systems, either to minimize absorption of heat in normal or warm ambient environments (including high levels of humidity) or to obtain warmth in relatively cold environments. Thus, other than possible auditory perception of RFR pulses, animals do not appear to directly sense RFR.

The results of studies on disruption of performance or learned behavior by RFR were variable; however, most of the findings showed that the behavioral changes were ascribable to the added thermal burden imposed by the RFR, and were significant at whole-body SARs well in excess of 1 W/kg.

It is worth emphasizing that the behavioral findings of the primate studies are more relevant than those with the other animal species with regard to possible effects of RFR on human behavior, because the tasks the primates had to learn were far more complex, and their physiologies and intelligence are much closer to those of humans. It is also noteworthy that reasonably accurate thresholds for RFR-induced behavioral changes were determined for each primate species studied, and that those thresholds served as a basis for the ANSI/IEEE[24] human-exposure standard.

*RFR and Drugs.* Various studies have been conducted on possible interactive effects of exposure to RFR and medications or other drugs taken or administered. Those discussed below are representative.

Thomas et al.[208] trained rats on a fixed-interval, 1-minute (FI-1) schedule to press a bar for a pellet. After stable baseline patterns were achieved, an effect-versus-dose function for chlordiazepoxide, a psychoactive drug (tradename *Librium*), given 30 minutes before a session, was established. This function showed that the responding rate rose with increased drug dose up to 10 mg/kg of body weight, attaining 2 to 3 times the baseline rate at that dose. At still higher doses, the responding rate decreased; it attained zero at 40 mg/kg.

After training, the rats were exposed to 2.45-GHz RFR at 1-W/cm$^2$ peak, 1-mW/cm$^2$ average (0.2 W/kg) during the 30 minutes before each bar-pressing session. RFR-exposure yielded the same shape of effect-versus-dose function, but the magnitudes were generally higher by a factor of about 2. By contrast, RFR-exposure without drug injection produced no difference in responding rate. The results of this study are unequivocal, but the mechanisms are obscure. For example, although the average power density and whole-body SAR were low, local SARs in brain regions that are target areas for central actions of chlordiazepoxide may have been high enough for a thermally potentiating effect. It is also possible that the pulse parameters produced the RFR-hearing effect during the 30-minute exposure preceding the bar-pressing session. If so, it is not clear whether or how this effect would have influenced the testing sessions themselves, during which RFR was absent.

Thomas and Maitland[209] trained six rats to depress a lever on a schedule in which a second response at least 18 seconds after a first response was rewarded with a pellet, but a second response in less than that interval reset the timing period. After such training, effects of exposure to the pulsed RFR (0.2 W/kg) used in the previous study were sought on the dose-versus-response function of the psychoactive drug d-amphetamine.

Three of the rats were dosed with the drug once per week and exposed for 30 minutes (single-exposure condition). Their behavior was studied for 1 hour right after exposure for any direct drug-RFR interaction. To detect possible cumulative action of RFR, the other three rats were dosed with d-amphetamine once a week and exposed for 4 days a week, 30 minutes daily (multiple-exposure condition), except on drug-injection days. On the latter days, their behavior was observed for the 30 minutes after injection. The sessions were conducted for 13 weeks, and included sham-exposures and saline injections.

For the single-exposure condition, the mean response rates after saline injection and sham-exposure and after saline injection and RFR-exposure were comparable to baseline performances. When those rats were given d-amphetamine and sham-exposed, their mean response rates rose with drug dose to a maximum at 2.0 mg/kg, with consequent reductions in the frequency of correct responses that yielded reinforcement. At higher drug doses, the mean response rates dropped sharply, to zero for 4.5 mg/kg. By contrast, the mean response rate of drug-injected rats and exposed to the RFR rose to values significantly higher than for the corresponding drug doses and sham-exposure, with maximum response rate at 0.5 mg/kg. Above 0.5 mg/kg, the mean response rate declined sharply, to zero for 1.5 mg/kg. Those results show that exposure to the RFR after injection of a given dose of d-amphetamine yielded behavior similar to that obtained with a larger dose without the RFR exposure.

The dose-response functions of the rats with and without multiple RFR-exposures were qualitatively similar to those with and without single RFR-exposures. With the multiple sham-exposures, maximum responses were obtained for 2.0 mg/kg, with a sharp decline to zero for 4.5 mg/kg. The maximum responses for the multiple RFR-exposures were obtained with 0.5 mg/kg, and the responses declined sharply to zero for 2.0 mg/kg.

Thomas et al.[210] did similar research with the drugs diazepam and chlorpromazine. Diazepam (*Valium*) has been widely used as a tranquilizer and muscle relaxant. Chlorpromazine is a sedative and an antiemetic. Four rats each of two different strains were trained on a fixed-interval, 1-minute (FI-1) schedule of reinforcement. After training, the dose-effect functions for diazepam were determined in one strain and for chlorpromazine in the other strain. For each drug, one dose was injected into rats 30 minutes before each session, and their response rates were compared with their respective baseline performances. Chlorpromazine diminished the performance of all four rats for doses above about 1 mg/kg. For those given diazepam, the drug caused slight increases in response rate at up to 2.5 mg/kg and declines at higher doses.

After determining the dose-effect functions for the drugs, the authors exposed each rat for 30 minutes to 2.8-GHz pulsed RFR at 0.2 W/kg right after administering each drug, and tested the rats at exposure end. The RFR did not alter the effects of either chlorpromazine or diazepam, in contrast with the results with chlordiazepoxide and d-amphetamine. Such differences in findings are difficult to reconcile.

Ashani et al.[211] sought possible changes of the hypothermic effects of anticholinesterase drugs in rats and the influence of antidotes for such drugs. Mixed results were obtained for 10-minute exposures to 2.8-GHz RFR at 10 mW/cm$^2$ (about 2 W/kg). Since few people ingest such drugs or their antidotes, there appears to be no direct significance of the findings with regard to possible effects of RFR on human health.

Pappas et al.[212] conducted experiments to determine whether RFR alters seizures induced by the drug pentylenetetrazol (PTZ), and to study the effects of RFR on the efficacy of chlordiazepoxide (CDZ) for counteracting such seizures. In one experiment, rats were exposed for 30 minutes to 2.7-GHz pulsed RFR (2-μs pulses at 500 pps) at average power densities up to 20 mW/cm$^2$ (2.25 W/kg). After exposure, the rats were given PTZ at doses up to 80 mg/kg, and their seizure activity was studied. In another experiment, each rat was injected with anti-seizure CDZ at doses up to 15.0 mg/kg before exposure. After exposure, 70 mg/kg of PTZ was injected, and the degree of inhibition of PTZ-induced seizures by the CDZ was studied.

In both experiments, the rats were watched for signs of seizure activity for 8 minutes after injection of PTZ. The latency interval to the onset of the first sign was recorded, and the seizure intensity was rated from 0 (no seizure, normal exploratory activity) to 4 (wild running and convulsions with 99% mortality).

In the first experiment, the latency times to seizure following PTZ injection decreased with increasing PTZ dose for all RFR levels. The rats exposed at 15 mW/cm$^2$ (2.25 W/kg) exhibited significantly shorter latencies than the sham-exposed rats. The mean score of seizure intensity increased with PTZ dose, with no apparent effect of RFR level. Although the mean score was significantly higher for 15 mW/cm$^2$ (2.25 W/kg) than 5 mW/cm$^2$ (0.75 W/kg), the values for the RFR groups did not differ significantly from those of the sham group. The authors suggested that the decreases in seizure latency and the increases in seizure intensity at 15 mW/cm$^2$ could have resulted from local brain hyperthermia rather than alteration of the PTZ effect on brain neuronal activity.

The results of the second experiment were unclear. Rats given 7.5 mg/kg of CDZ had longer latencies after exposure at 15 mW/cm$^2$ (2.25 W/kg) than at lower levels, and the latencies of rats given 15 mg/kg were shorter after exposure at 5 mW/cm$^2$ (0.75 W/kg) than for sham-exposure or exposure at 10 or 15 mW/cm$^2$ (1.5, or 2.25 W/kg). The authors suggested that those positive findings may be ascribable to a random Type II statistical error. Therefore, they did a third experiment, the results of which did not support the earlier findings that 15 mW/cm$^2$ (2.25 W/kg) or even 20 mW/cm$^2$ (3 W/kg) enhanced PTZ seizures and increased the antiseizure protection of CDZ at 7.5 mg/kg. Thus, they regarded the few apparently positive results as spurious, Type II errors, but did not rule out possible experimenter bias.

Lai et al.[213] exposed unrestrained rats to circularly-polarized pulsed 2.45-GHz RFR in cylindrical waveguides for 45 minutes at a spatially-averaged power density of 1 mW/cm$^2$ (0.6 W/kg whole-body SAR). The authors noted that at this SAR, the range of average power densities for linearly polarized RFR would be 3-6 mW/cm$^2$. Control rats were sham exposed. The effects of each of several drugs given right after exposure were investigated.

In the first experiment, the effect of the RFR on the stereotypy induced by subcutaneous apomorphine injection was studied: Rats were scored on which of five different forms of behavior they exhibited during 5-minute observation periods right after apomorphine injection and after four 15-minute intervals. The results yielded a higher mean score for RFR-exposed rats than for sham-exposed rats, indicating that RFR-exposure produced more intense stereotypy, with biting and clawing, than did sham-exposure. Also studied was the effect of the RFR on the hypothermia induced by apomorphine: Colonic temperatures were measured right after RFR- or sham-exposure, after which the rats were injected with apomorphine and their colonic temperatures were recorded four more times at 15-minute intervals. At 15 minutes after drug injection, the mean temperature of the RFR group had decreased by 0.95°C, but the decrease was only 0.63°C for the sham group, a significant difference showing that the hypothermic effect of apomorphine was enhanced by the RFR.

Next, the effect of the RFR on stereotypy induced by d-amphetamine was studied. Each rat was injected with the drug right after exposure and was watched for the presence of any of three normal behaviors and three abnormal behaviors. Starting 4 minutes after injection, the rats were observed for 1 minute every 5 minutes for an hour. The difference in average score between the RFR and sham groups for the six behaviors was nonsignificant. The effect of the RFR on the hyperthermia induced by d-amphetamine was also recorded: Colonic temperatures were measured right after exposure as before, but at 15-minute intervals for 90 minutes after drug injection. The results were shown as the mean colonic temperature change for each group versus time interval after injection. At zero time, the mean temperature was 38.3°C for the RFR group and 38.2°C for the sham group. The mean temperature of the sham group reached its maximum at 45 minutes, at which time it had increased by 1.4°C, and then it diminished to about 39.3°C at the end of the 90-minute period. By contrast, the temperature of the RFR group reached its maximum at 60 minutes, an increase of 1.2°C, and diminished to about 39.2°C at 90 minutes. These differences in time-dependent increases were significant.

In the final experiment, rats were injected with morphine at doses of 1 to 20 mg/kg right after exposure. The number of rats that exhibited catalepsy (general muscular rigidity and a certain posture for more than 1 minute) 30 minutes after injection and the number of rats that died within 2 hours after injection were recorded.

The differences in the percentages of RFR-exposed and sham-exposed rats that exhibited catalepsy were reported to be significant. However, in an erratum,[214] the authors indicated that the statistical method used was inappropriate. Their use of another statistical method showed that the responses of the sham group increased with drug dose, but the responses of the RFR group were not dose-dependent. Also, the original paper reported that no deaths had occurred in the RFR or sham group at doses of 1 or 5 mg/kg, but at the larger doses, the percentages of deaths were significantly higher in the RFR group. In the erratum, however, the authors noted that they found no significant differences in deaths due to the RFR.

It is interesting that the mean colonic temperatures of the apomorphine-injected and amphetamine-injected groups immediately after RFR-exposure were both 0.1°C higher than for their respective sham-exposed groups (38.3 versus 38.2°C in both cases). Not clear was the influence of thermoregulation on those results. Would similar results be obtained if pre-injection colonic temperatures were raised by the same amount by an agent other than RFR?

Lacking in this investigation were data on animals administered saline instead of the drugs. Such control data might have more clearly delineated subtle non-RFR factors that may have influenced the results.

In a similar study, Lai et al.[215] examined the effects of the same RFR on the actions of pentobarbital in the rat. In the first of two series, unanesthetized, unrestrained rats were exposed to the RFR. Each rat's colonic temperature was taken right after exposure, after which the rat was injected with sodium pentobarbital at a dose sufficient to induce surgical anesthesia. After the rat lost its righting reflex, its colonic temperature was recorded at 15-minute intervals for 150 minutes, and the time interval after injection to regain the righting reflex was noted. A group of rats was sham-exposed and similarly treated.

In the second series, baseline colonic temperatures were measured, and the rats were injected with pentobarbital. After 15 minutes, by which time all of the rats had lost their righting reflex, some rats were exposed to the RFR for 45 minutes anteriorly (head toward source) and other rats posteriorly (rear toward source). Their colonic temperatures were recorded for 90 minutes after exposure as were their time intervals for regaining the righting reflex.

The (conscious) rats in the first series did not show any preferred orientation during RFR-exposure and there was no significant difference in mean colonic temperature between the RFR and sham groups immediately after exposure (37.8°C for both). The mean colonic-temperature changes for each group versus time after pentobarbital injection indicated that both groups reached maximal hypothermia (about -3°C) at 75 minutes. Mean temperature depressions of the RFR group did not differ significantly from those of the sham group at corresponding times up to 105 minutes. During the interval from 105 to 150 minutes, the mean depressions for the RFR group were significantly larger than for the sham group, indicating that the recovery of the RFR group from the hypothermia was slower. Also, the mean time to righting-reflex recovery for the RFR group was significantly longer than for the sham group.

In the second series, the baseline mean colonic temperatures of the pentobarbital-injected rats before RFR- or sham-exposure were 37.9°C. Right after anterior-exposure or posterior-exposure, the mean temperatures of the RFR groups were respectively 34.6 and 34.7°C, and was 34.1°C for the sham group. The difference between the two RFR groups was not significant, but both values were significantly higher than for the sham group. All three groups attained maximal hypothermia 30 minutes after exposure (45 minutes after injection). The changes at that time for the posterior-RFR, anterior-RFR, and sham groups were respectively about -3.8, -4.2, and -4.5°C, but only the difference between the posterior-RFR and sham groups was significant.

From 30 to 90 minutes, all three groups showed recovery toward baseline temperatures. However, the posterior-RFR group recovered from the hypothermia earlier and recovered their righting reflex more quickly than the anterior-RFR and sham groups. The authors surmised that those findings were due to the differences in local energy deposition in the two orientations, which could yield differences in drug metabolism or kinetics.

Lai et al.[216] also did experiments to determined the effects of the same RFR on ethanol-induced hypothermia and consumption of ethanol. For the ethanol-hypothermia experiment, 15 rats were RFR-exposed and 14 were sham-exposed for 45 minutes. Right after exposure, each rat was removed from its waveguide, its colonic temperature was measured, and it was injected with a solution of ethanol (3 μg/kg of body weight in 25% of water by volume). After injection, the colonic temperatures of the rats were measured at 15-minute intervals for 120 minutes.

The mean colonic temperature of the RFR-exposed rats at exposure end did not differ significantly from that of the sham-exposed rats. Ataxia developed within 5 minutes of ethanol injection, but righting reflex remained intact. The mean colonic-temperature changes versus time after ethanol injection for the two groups showed that hypothermia had occurred in the RFR group at a significantly slower rate than the sham group, but the temperature depressions of both groups became about the same 90 minutes after injection.

In the ethanol-consumption part, rats were given 90-minute sessions in the waveguides daily for 9 days. On the first three days, the rats were placed within the waveguides for the first 45 minutes with the RFR source on "standby." At this time, a bottle of 10% sucrose solution was inserted in each waveguide and the amount consumed during the remaining 45 minutes was measured. On the fourth day, the procedure was the same but half the rats (24) were exposed to the RFR for the full 90 minutes, and the other half were sham-exposed. The procedure was the same on the next three days, but a bottle containing 15% ethanol + 10% sucrose was used instead (the latter to render the ethanol more palatable). On the eighth day, half the rats were exposed to the RFR for 90 minutes, the remaining were similarly sham-exposed, and the amounts of sucrose-ethanol solution consumed were measured. On the ninth day, the roles of the two groups were reversed: the first group was sham-exposed, and the second was RFR-exposed. The results indicated that the RFR had no apparent effect on sucrose consumption, but that it increased the consumption of the sucrose + ethanol solution.

It should be noted that there were significant differences in the mean baseline temperatures among the groups given different dosages, and that there was an apparent discrepancy in the mean baseline temperatures between the sham-saline groups. Such differences among control groups, as well as the significant colonic-temperature rises during both RFR- and sham exposure, indicate the presence of large uncontrolled factors and render questionable most of the findings of this study.

Lai et al.[217] exposed rats to the same pulsed RFR (2-μs pulses of 2.45-GHz RFR at 500 pps) at a whole-body SAR of 0.6 W/kg either once for 45 minutes or for 10 daily 45-minute sessions to determine the effects of the RFR on the concentration and affinity of benzodiazepine receptors in the cerebral cortex, hippocampus, and cerebellum. The results of the single exposures showed a significant increase of receptor concentration in the cerebral cortex but not in the hippocampus or cerebellum. There were no significant changes in the binding affinity of the receptors in any of the three regions. The multiple exposures yielded no change in receptor concentration right after the last exposure, a possible indication of adaptation to repeated exposures.

In another experiment, rats adapted first to the experimental situation and then exposed once to the RFR showed significantly higher benzodiazepine-receptor concentrations in the cerebral cortex than did similarly treated sham-exposed rats. Those results led the authors to suggest that because such receptors are responsive to anxiety and stress, low-level RFR can be a source of stress.

In overall summary, the studies on possible synergism between RFR and psychoactive drugs such as diazepam, chlorpromazine, chlordiazepoxide, and dextroamphetamine, yielded unclear or inconsistent results. In some studies, the changes in drug dose-response relationship were subtle and not necessarily induced by the RFR. In most of the studies that yielded RFR-induced changes in drug response, whole-body SARs of 0.6 W/kg or average power densities of 1 mW/cm$^2$ or higher coupled with relatively high drug dosages were necessary. In still other studies, the results were negative (no effects). At relatively low RFR levels, the role of thermoregulation in the results is unclear. Also the occurrence of relatively high local SARs in the brain cannot be ruled out, a point applicable to the studies above by Lai et al., in which the whole-body SARs were only about 0.6 W/kg.

In general, it seems unlikely that the effects of psychoactive drugs prescribed by physicians or the effects of recreationally consumed alcohol would be altered by exposure to environmental levels of RFR.

**Cellular and Subcellular Effects.** Webb and Dodds[218] sought effects of RFR at specific frequencies above 30 GHz (in the millimeter-wave region) on growth of *E. coli* bacteria. Their results appeared to show that bacterial growth was inhibited by 136-GHz RFR, but from examination of their data, the presence of non-RFR factors was likely. Webb and Booth[219] also reported the absorption of RFR by *E. coli* cells and by preparations of protein, RNA, and DNA derived from *E. coli* at specific (resonant) frequencies within the range 65-75 GHz. The latter findings were difficult to evaluate because the information on methodology, instrumentation, and statistical treatment was inadequate.

Several investigations sought to confirm predictions by Fröhlich[220] of resonances above 30 GHz. For example, Webb and Stoneham[221] reported the detection of resonances in the range 70-5000 GHz in active cells of *E. coli* and B. megaterium, using laser Raman spectroscopy. The authors found no resonances in resting cells, cell homogenates, or nutrient solutions, and they therefore associated the observed active-cell resonances with metabolic processes. Cooper and Amer[222] disputed those findings. They indicated that cell suspensions yield spurious Raman lines in the frequency range of interest under certain conditions, notably by Mie scattering from cell clumps, and that they thereby were able to reproduce many of the spectra.

Gandhi et al.[223] used a stable computer-controlled system to measure RFR absorption in various biological specimens at discrete frequencies in the range 26.5-90.0 GHz in small steps. They studied solutions of DNA from salmon sperm and RNA from whole yeast and yeast-like fungi, and suspensions of *E. coli* cells and baby-hamster-kidney cells transformed with mouse sarcoma virus. They found no resonances at any of the frequencies sampled, and they strongly suggested that none of the biological materials studied absorb significant RFR energy in that frequency range.

Swicord and Davis[224] used a novel method for measuring absorption by optically transparent liquids and for studying interactions between cellular constituents and RFR at frequencies below (as well as above) the millimeter-wave range. They measured RFR absorption in the range 8-12 GHz by aqueous solutions of DNA extracted from *E. coli*. A plot of attenuation coefficient for DNA versus frequency showed no resonances, but the attenuation increased linearly with frequency and its values were much higher than for physiologic (Ringer's) solution or deionized water at the same frequencies.

Edwards et al.[225] noted that biochemical analysis of the DNA solution used in the Swicord and Davis[224] study indicated the presence of significant amounts of RNA and protein impurities, and that the DNA had been sheared extensively by improper handling. In addition, the enhanced absorption observed for such samples in the range 8-12 GHz was absent for carefully prepared DNA samples of high molecular weight that were free of protein and RNA.

Gabriel et al.[226] described the efforts in two laboratories (London and Uppsala), that respectively used a reflection technique and a transmission technique to detect resonances in the range 1-10 GHz for aqueous solutions of circular DNA molecules of the form studied by Edwards et al.[225] Gabriel et al.[226] noted that a most important feature of both of those techniques is the use of a reference sample to normalize the reflection or transmission coefficients to eliminate systematic experimental artifacts, such as slight impedance mismatches. Their plots of relative permittivity and loss factor of a plasmid DNA solution versus frequency yielded values close to those for pure water. Moreover, similar plots of attenuation coefficient and incremental attenuation coefficient relative to water did not show any of the resonances reported by Edwards et al.[225]

Foster et al.[227] also tried to reproduce the findings of Edwards et al.[225] They suggested that the apparent resonances might be due to a reflection artifact from the coaxial connector to the probe used by Edwards et al.[225] for the measurements and to their lack of consideration of possible radiation from the probe. Accordingly, Foster et al.[227] carried out such measurements with and without the use of a time-domain-gating procedure for removing connector artifacts. Without using time-domain gating, they obtained reflection-coefficient oscillations crudely resembling the resonances reported by Edwards et al.[225] in DNA solutions of threefold higher concentration than the latter authors had used. Moreover, the oscillations were eliminated by time-domain gating. In addition, they saw no resonances when radiation from the probe was eliminated.

Sagripanti et al.[228] reported that plasmid DNA, when exposed to low-levels of RFR in the frequency range 2.00 to 8.75 GHz, exhibited both single-strand and double-strand breaks, but only if small quantities of copper ions (cuprous but not cupric) were present. The samples consisted of 10 µg of plasmid DNA in 28 µl of buffer within a 1.5-ml micro test tube. For exposure, a coaxial probe with both inner and outer conductors of copper was immersed in each sample. Attenuation and standing-wave ratio were measured, to determine the maximum and minimum SARs ($SAR_{max}$ and $SAR_{min}$). Those data indicated that $SAR_{max}$ was about five times larger than $SAR_{min}$, with $SAR_{true}$ somewhere between the two. The experimental results were referenced to the values of $SAR_{max}$.

First, samples were sham-exposed or exposed for 20 minutes to 2.55-GHz RFR at an $SAR_{max}$ of 10 W/kg. Those results showed that the mean number of double-strand breaks in the RFR-exposed samples was significantly higher than in the sham-exposed samples. The authors characterized such exposures as nonthermal, because of the large surface-to-volume ratio of the samples and their consequent ability to dissipate heat readily. They also noted that exposures at levels of about 1 kW/kg were needed to detect any significant temperature rises in the samples.

Next, in experiments toward seeking frequency specificity of the effect, samples were exposed to 8.75 GHz-RFR, which was previously found by Edwards et al.[225] to be one of the frequencies of maximum resonant absorption by DNA. They also exposed samples to 2.00-GHz, 3.45-GHz, and 7.64-GHz RFR, which were frequencies of minimum absorption reported by Edwards et al.[225] Their results showed no variations in double-strand breaks attributable to resonant absorption by DNA.

For statistical analysis, the authors pooled data on 12 experiments at the five frequencies above. The results showed a significantly higher mean percentage of double-strand breaks for the RFR-exposed samples than the sham-exposed samples. However, the mean percentage of double-strand breaks for the sham-exposed samples was also significantly higher than for control samples, for which the copper probe was close to the sample but not in contact with it. When the probe was covered with a thin plastic coating, the difference between sham-exposed and control samples vanished, but also no strand breaks were detected in RFR-exposed samples.

In other experiments, samples were incubated in either cupric or cuprous chloride, or in the storage buffer (controls), and not RFR- or sham-exposed. The results indicated that only incubation in cuprous chloride mimicked the strand breaking seen with RFR-exposure. Based on linear increases of damage with exposure duration, the authors concluded that the presence of cuprous chloride caused the strand breaking and that the RFR increased the effect.

In summary, many of the early studies on microorganisms and subcellular preparations yielded results that were taken as evidence of nonthermal effects of RFR. The existence of resonances at frequencies above about 30 GHz was postulated on theoretical grounds, and several studies were done that appeared to confirm that hypothesis. However, subsequent studies with the use of more sophisticated engineering and biological techniques and in which artifacts were reduced markedly, yielded results that did not confirm earlier findings of resonances or other evidence of nonthermal effects at such frequencies.

Specifically, the apparent absorption resonances in the range 2-9 GHz reported for aqueous solutions of DNA molecules derived from *E. coli* were regarded as indicative of direct action of RFR with such molecules. Later endeavors to reproduce such findings, however, yielded negative results. In addition, analytical and experimental results were obtained indicating that such resonances were most likely artifactual, associated with the probes and measurement methodology used.

In general, research on possible RFR effects on microorganisms or of RFR-exposure *in vitro* of cell preparations derived from macroorganisms is important toward eliciting possible mechanisms of direct interaction of RFR with such biological entities or their constituents at levels that can be characterized as nonthermal. However, whatever the findings of such research, their relevance to possible effects of exposure of intact animals to RFR, and ultimately the significance of such findings with regard to possible hazards of RFR to humans would have to be established.

## UNRESOLVED ISSUES

The potential biological effects of RFR at frequencies up to 300 GHz have been assessed from representative peer-reviewed studies published in the scientific literature.

The preponderance of evidence indicates that chronic exposure to the RFR levels generally prevailing in the environment is not hazardous to human health. Nevertheless, there are several basic uncertainties, summarized below, regarding biological effects of RFR.

(1) Many of the epidemiologic studies on possible bioeffects of RFR were extensive and well done, but contained defects or uncertainties in varying degrees, such as imprecise assignment of individuals to exposure and control groups; difficulties in obtaining accurate medical records, death certificates, or responses to health questionnaires for individuals included in both the exposure and control groups; and most important, the large uncertainties about the frequencies, levels, and exposure durations for those selected for inclusion in exposure groups and the amount of exposure received by those selected for inclusion in control groups.

(2) Applying results on laboratory animals to humans, though essential, is an expedient that contains fundamental problems and uncertainties due to the basic differences between humans and other species. Investigations with nonhuman primates may narrow some of the interspecies gaps considerably, but at costs that are often prohibitive. Thus, it seems unlikely that major reductions in such uncertainties will occur in the near future.

(3) The results of many studies indicate the existence of threshold RFR levels for various bioeffects, thus providing confidence that exposure to levels that are appreciably below the thresholds are most unlikely to be deleterious. However, most experimental data that indicate the existence of thresholds were obtained by the use of single or repetitive exposures of relatively short durations. Although it is hard to conceive of mechanisms whereby RFR-exposures at well below threshold values over a long time are cumulative, very few investigations have been done that involve essentially continuous exposure of animals to low-level RFR (below threshold levels or those that can cause significant heating) during most of their lifetimes. The high costs of such chronic studies and the low probability that any positive effects will be found are major reasons why such studies are not given high priority by funding agencies.

(4) Regarding basic mechanisms of interaction between RFR and various biological entities, many important discoveries have been made, notably by exposure of cells and subcellular structures and constituents *in vitro* to relatively low RFR levels. The effects on such entities can be characterized as subthermal, but the gap between such effects and possibly hazardous effects on intact humans or animals from exposure to such RFR levels is enormous. Factors such as large body masses, penetration depth and internal field distributions, and changes in body orientation during exposures to RFR *in vivo* can vastly moderate such interactions or remove them entirely. Moreover, life processes *per se* are extremely complex. For these reasons, this gap is not likely to be reduced to any great extent.

It is necessary to distinguish between a bioeffect and a hazard. For example, a person's metabolism can be increased harmlessly by mild exercise. Analogously, an effect produced at RFR intensities that yield heat that can be easily accommodated within the thermoregulatory capabilities of an individual may not necessarily be deleterious. Moreover, any effects produced thereby are generally reversible. However, the thermoregulatory capabilities of any given species may be exceeded at high RFR intensities, so compensation for such effects may be inadequate. Thus, exposure at such intensities can cause thermal distress or even irreversible thermal damage.

It is scientifically impossible to guarantee that low levels of RFR that do not cause deleterious effects for relatively short exposures will not cause the appearance of deleterious effects many years in the future. As indicated previously, however, the weight of the present scientific evidence indicates the existence of threshold levels for various RFR bioeffects and that low-level RFR-exposures are not cumulative.

# REFERENCES

1. P. Polson and L.N. Heynick, "Technical Report: Analysis of the Potential for Microwave Bioeffects to Result From Operation of the Proposed ONR and Air Force High-Frequency Active Auroral Research Program (HAARP) Ionospheric Research Instrument (IRI): General Analysis. (Technical Report prepared for U.S. Air Force Geophysics Directorate Laboratory, Hanscom Air Force Base (1992).

2. L.N. Heynick, Critique of the literature on bioeffects of radiofrequency radiation: a comprehensive review pertinent to Air Force operations. USAF School of Aerospace Medicine, Brooks AFB, TX, Report USAFSAM-TR-87-3 (1987).

3. J.A. Elder and D.F. Cahill, Biological effects of radiofrequency radiation. Final Report EPA-600/8-83-026F, Environmental Protection Agency, NC 27711 (1984).

4. J. Pazderová, Workers' state of health under long-term exposure to electromagnetic radiation in the VHF band (30-300MHz), Pracovni Lokarctui (in Czech), Vol. 23:8 265-271, English translation:JPRS No. UDC 616-001.228.1-057-07 (1971).

5. J. Pazderová, J. Picková, and V. Bryndová,Blood proteins in personnel of television and radio transmitting stations, *in:* "Biologic Effects and Health Hazards of Microwave Radiation," P. Czerski et al., eds., Polish Medical Publishers, Warsaw, pp. 281-288, (1974).

6. Klimkova-Deutschova, Neurologic findings in persons exposed to microwaves, *in:* Biologic Effects and Health Hazards of Microwave Radiation, P. Czerski et al., eds., Polish Medical Publishers, Warsaw, pp. 268-272, (1974).

7. T.V. Kalyada, P.P. Fukalovak, and N.N. Goncharova, Biologic effects of radiation in the 30-300MHz range, *in:* Biologic Effects and Health Hazards of Microwave Radiation, P. Czerski et al., eds., Polish Medical Publishers, Warsaw, pp. 52-57 (1974).

8. M.N. Sadchikova, Clinical manifestations of reactions to microwave irradiation in various occupational groups, *in:* Biologic Effects and Health Hazards of Microwave Radiation, P. Czerski et al., eds., Polish Medical Publishers, Warsaw, pp.261-267, (1974).

9. M. Siekierzynski, A study of the health status of microwave workers, *in:* Biologic Effects and Health Hazards of Microwave Radiation, P. Czerski et al., eds., Polish Medical Publishers, Warsaw, pp.273-280, (1974).

10. C.D. Robinette and C. Silverman, Causes of death following occupational exposure to microwave radiation (radar) 1950-1974, *in:* "Symposium on Biological Effects and Measurement of Radiofrequency/Microwaves," D.G. Hazzard, ed., Dept. of Health, Education, and Welfare, Washington, D.C., HEW Publication No. (FDA) 77-8026 (1977).

11. A.M. Lilienfeld, J. Tonascia, S. Tonascia, C.H. Libauer, G.M. Cauthen, J.A. Markowitz, and S. Weida, Foreign service health status study: evaluation of status of foreign service and other employees from selected Eastern European posts. Final Report, July 31, 1978, Contract No. 6025-619073, Dept. of Epidemiology, School of Hygiene and Public Health, The Johns Hopkins University, Baltimore, MD (1978).

12. S. Hamburger, J.N. Logue, and P.M. Silverman, Occupational exposure to non-ionizing radiation and an association with heart disease: an exploratory study. *J. Chron. Dis.,* Vol. 36, No. 11, pp. 791-802 (1983).

13. J.R. Lester and D.F. Moore Cancer mortality and Air Force bases. *J. Bioelectricity,* Vol. 1, No. 1, pp. 77-82 (1982).

14. J.R. Lester and D.F. Moore, Cancer incidence and electromagnetic radiation, *J. Bioelectricity,* Vol. 1, No. 1, pp. 59-76 (1982).

15. S. Milham, Jr., Occupational mortality in Washington state: 1950-1979. DHHS (NIOSH) Publication No. 83-116, Contract No. 210-80-0088, U.S. Department of Health and Human Services, National Institute for Occupational Safety and Health, Cincinnati, Ohio, (October 1983).

16. S. Milham, Jr., Increased mortality in amateur radio operators due to lymphatic and hemopoietic malignancies, *Am. J. Epidem.,* Vol. 127, No. 1, pp. 50-54 (1988).

17. R.G. Burr and A. Hoiberg, Health profile of U.S. Navy pilots of electronically modified aircraft, *Aviat., Space, and Environ. Med.* (February 1988).

18. S.F. Cleary, B.S. Pasternack, and G.W. Beebe, Cataract incidence in radar workers, *Arch. Environ. Health,* Vol. 11, pp. 179-182 (1965).

19. S.F. Cleary and B.S. Pasternack, Lenticular changes in microwave workers, *Arch. Environ. Health,* Vol. 12, pp. 23-29 (1966).

20. B. Appleton, S.E. Hirsh, and P.V.K. Brown Microwave lens effects: II. Results of five-year survey, *Acta Ophthal.,* Vol. 93, pp. 257-258 (1975).

21. F.C. Hollows and J.B. Douglas, Microwave cataract in radiolinemen and controls, *Lancet,* No. 8399, Vol. 2, pp. 406-407 (18 August 1984).

22. M.M. Zaret, Cataracts following use of microwave oven, *N.Y. State J. Med.,* Vol. 74, No. 11, pp. 2032-2048 (1974).

23. B. Hocking, K. Joyner, and R. Fleming, Health aspects of radiofrequency radiation accidents; Part I: Assessment of health after a radiofrequency radiation accident, *J. Microwave Power EE,* Vol. 23, No. 2, pp. 67-74 (1988).

24. ANSI/IEEE C95.1-1991: IEEE Standard for safety levels with respect to human exposure to radio frequency electromagnetic fields, 3 kHz TO 300 GHz. The Institute of Electrical and Electronics Engineers, New York, NY 10017 (1992).

25. A.T. Sigler, A.M. Lilienfeld, B.H. Cohen, and J.E. Westlake, Radiation exposure in parents of children with mongolism (Down's syndrome), *Bull. Johns Hopkins Hosp.,* Vol. 117, pp. 374-395 (1965).

26. B.H. Cohen, A.M. Lilienfeld, S. Kramer, and L.C. Hyman, Parental factors in Down's syndrome-results of the second Baltimore case-control study, *In* "Population Genetics-Studies in Humans," E.G. Hook and I.H. Porter, eds., Academic Press, New York, pp. 301-352 (1977).

27. J.A. Burdeshaw and S. Schaffer , Factors associated with the incidence of congenital anomalies: a localized investigation. Final Report, Report No. XXIII, 24 May 1973-31 March 1976, Contract No. 68-02-0791, EPA 600/1-77-016 (March 1977).

28. P.B. Peacock, J.W. Simpson, C.A. Alford, Jr., and F. Saunders, Congenital anomalies in Alabama, *J. Med. Assoc. Ala.,* Vol. 41, No. 1, pp. 42-50 (1971).

29. P.B. Peacock, S.R. Williams, and E. Nash, Relationship between the incidence of congenital anomalies and the use of radar in military bases. Final Report, Report No. III, Project No. 3118, Contract No. 68-02-0791 submitted by Southern Research Institute to EPA (Nov. 1973).

30. B. Källén, G. Malmquist, and U. Moritz, Delivery outcome among physiotherapists in Sweden: is non-ionizing radiation a fetal hazard? *Arch. Environ. Health,* Vol. 37, No. 2, pp. 81-85 (1982).

31. H.A. Kues, L.W. Hirst, G.A. Lutty, S.A. D'Anna, and G.R. Dunkelberger, Effects of 2.45-GHz microwaves on primate corneal endothelium, *Bioelectromagnetics,* Vol. 6, No. 2, pp. 177-188 (1985).

32. A.W. Guy, J.C. Lin, P.O. Kramar, and A.F. Emery, Effect of 2450-MHz radiation on the rabbit eye, *IEEE Trans. Microwave Theory Tech.,* Vol. 23, No. 6, pp. 492-498 (1975).

33. P.J. Stewart-DeHaan, M.O. Creighton, L.E. Larsen, J.H. Jacobi, M. Sanwal, J.C. Baskerville, and J.R. Trevithick, *In vitro* studies of microwave-induced cataract: reciprocity between exposure duration and dose rate for pulsed microwaves, *Exp. Eye Res.,* Vol. 40, pp. 1-13 (1985).

34. M.O. Creighton, L.E. Larsen, P.J. Stewart-DeHaan, J.H. Jacobi, M. Sanwal, J.C. Baskerville, H.E. Bassen, D.O. Brown, and J.R. Trevithick, *In vitro* studies of microwave-induced cataract. II Comparison of damage observed for continuous wave and pulsed microwaves, *Exp. Eye Res.,* Vol. 45, pp. 357-373 (1987).

35. M.R. Foster, E.S. Ferri, and G.J. Hagan, Dosimetric study of microwave cataractogenesis, *Bioelectromagnetics,* Vol. 7, No. 2, pp. 129-140 (1986).

36. R.M. White, Generation of elastic waves by transient surface heating, *J. Appl. Phys.,* Vol. 34, No. 12, pp. 3559-3567 (1963).

37. K.R. Foster and E.D. Finch, Microwave hearing: evidence for thermoacoustic auditory stimulation by pulsed microwaves, *Science,* Vol. 185, pp. 256-258 (19 July 1974).

38. R.G. Olsen and W. C. Hammer, Evidence for microwave-induced acoustical resonances in biological material, *J. Microwave Power,* Vol. 16, Nos. 3 & 4, pp. 263-269 (1981).

39. R.G. Olsen and J. C. Lin, Microwave pulse-induced acoustic resonances in spherical head models, *IEEE Trans. Microwave Theory Tech.,* Vol. 29, No. 10, pp. 1114-1117 (1981).

40. E.M. Taylor and B.T. Ashleman, Analysis of central nervous system involvement in the microwave auditory effect, *Brain Res.,* Vol. 74, pp. 201-208 (1974).

41. A.W. Guy, C.-K. Chou, J.C. Lin, and D. Christensen, Microwave-induced acoustic effects in mammalian auditory systems and physical materials, *In* "Ann. N.Y. Acad. Sci.," P. W. Tyler, ed., Vol. 247, pp. 194-218 (1975).

42. C.-K. Chou and R. Galambos, Middle-ear structures contribute little to auditory perception of microwaves, *J. Microwave Power,* Vol. 14, No. 4, pp. 321-326 (1979).

43. C.A. Cain and W.J. Rissman, Mammalian auditory responses to 3.0 GHz microwave pulses, *IEEE Trans. Biomed. Eng.,* Vol. 25, No. 3, pp. 288-293 (1978).

44. V.V. Tyazhelov, R.E. Tigranian, E.O. Khizhniak, and I.G. Akoev, Some peculiarities of auditory sensations evoked by pulsed microwave fields, *Radio Sci.,* Vol. 14, No. 6S, pp. 259-263 (1979).

45. C.F. Dalziel and T.H. Mansfield, Effect of frequency on perception currents, *Trans. AIEE,* Vol. 69, Pt. II, pp. 1162-1168 (1950).

46. C.F. Dalziel and W.R. Lee, Lethal electric currents, *IEEE Spectrum,* Vol. 6, pp. 44-50 (1969).

47. D.W. Deno, Calculating electrostatic effects of overhead transmission lines, *IEEE Trans. Power App. Syst.,* Vol. 93, pp. 1458-1471 (1974).

48. T.D. Bracken, Field measurements and calculations of electrostatic effects of overhead transmission lines, *IEEE Trans. Power App. Syst.,* Vol. 95, pp. 494-504 (1976).

49. S.J. Rogers, Radiofrequency burn hazards in the MF/HF band, *in* "USAFSAM Aeromedical Review 3-81, Proceedings of a Workshop on the Protection of Personnel Against RFEM," J.C. Mitchell, ed., pp. 76-89 (1981).

50. O.P. Gandhi and I. Chatterjee, Radiofrequency hazards in the VLF to MF band, *Proc. IEEE,* Vol. 70, No. 12, pp. 1462-1464 (1982).

51. A.W. Guy and C.-K. Chou, Very low frequency hazard study. USAF School of Aerospace Medicine, Brooks AFB, Texas; Final Report on Contract F33615-83-C-0625, May 1985, submitted by U. of Washington, Seattle WA (1985).

52. I. Chatterjee, D. Wu, and O.P. Gandhi, Human body impedance and threshold currents for perception and pain for contact hazard analysis in the VLF-MF band, *IEEE Trans. Biomed. Eng.,* Vol. 33, No. 5, pp. 486-494 (1986).

53. C.F. Blackman, M.C. Surles, and S.G. Benane, The effect of microwave exposure on bacteria: mutation induction, *in* "Biological Effects of Electromagnetic Waves," C.C. Johnson and M. Shore eds., U.S. Dept. of Health, Education, and Welfare, Washington, D.C., HEW Publication (FDA) 77-8010, pp. 406-413 (1976).

54. S.K. Dutta, W.H. Nelson, C.F. Blackman, and D.J. Brusick, Lack of microbial genetic response to 2.45-GHz CW and 8.5- TO 9.6-GHz pulsed microwaves, *J. Microwave Power,* Vol. 14, No. 3, pp. 275-280 (1979).

55. B. Anderstam, Y. Hamnerius, S. Hussain, and L. Ehrenberg, Studies of possible genetic effects in bacteria of high frequency electromagnetic fields, *Hereditas,* Vol. 98, pp. 11-32 (1983).

56. T.L. Pay, E.C. Beyer, and C.F. Reichelderfer, Microwave effects on reproductive capacity and genetic transmission in drosophila melanogaster, *J. Microwave Power,* Vol. 7, No. 2, pp. 75-82 (1972).

57. Y. Hamnerius, H. Olofsson, A. Rasmuson, and B. Rasmuson, A negative test for mutagenic action of microwave radiation in Drosophila melanogaster, *Mutation Res.,* Vol. 68, No. 2, pp. 217-223 (1979).

58. S. Prausnitz and C. Susskind, Effects of chronic microwave irradiation on mice, *IRE Trans. Bio-Med. Electron.,* pp. 104-108 (1962).

59. N.J. Roberts, Jr. and S. M. Michaelson, Microwaves and neoplasia in mice: analysis of a reported risk, *Health Phys.,* Vol. 44, No. 4, pp. 430-433 (1983).

60. W.D. Skidmore and S.J. Baum, Biological effects in rodents exposed to 10 million pulses of electromagnetic radiation, *Health Phys.,* Vol. 26, No. 5, pp. 391-398 (1974).

61. S. Szmigielski, A. Szudzinski, A. Pietraszek, M. Bielec, M. Janiak, and J.K. Wremble, Accelerated development of spontaneous and benzopyrene-induced skin cancer in mice exposed to 2450-MHz microwave radiation, *Bioelectromagnetics,* Vol. 3, No. 2, pp. 179-191 (1982).

62. D.I. McRee, G. MacNichols, and G.K. Livingston, Incidence of sister chromatid exchange in bone marrow cells of the mouse following microwave exposure, *Radiat. Res.,* Vol. 85, pp. 340-348 (1981).

63. M.L. Meltz, P. Eagan, and D.N. Erwin, Proflavin and microwave radiation: absence of a mutagenic interaction, *Bioelectromagnetics,* Vol. 11, No. 2, pp. 149-157 (1990).

64. C.-K. Chou, A.W. Guy, and R.B. Johnson, Effects of long-term low-level radiofrequency radiation exposure on rats: Volume 3. SAR in rats exposed in 2450-MHz circularly polarized waveguide. USAF School of Aerospace Medicine, Brooks AFB, TX, Report USAFSAM-TR-83-19 (1983).

65. A.W. Guy, C.-K. Chou, R.B. Johnson, and L.L. Kunz, Effects of long-term low-level radiofrequency radiation exposure on rats: Volume 1. Design, facilities, and procedures. USAF School of Aerospace Medicine, Brooks AFB, TX, Report USAFSAM-TR-83-17 (1983).

66. A.W. Guy, C.-K. Chou, and B. Neuhaus, Effects of long-term low-level radiofrequency radiation exposure on rats: Volume 2. average SAR and SAR distribution in man exposed to 450-MHz RFR. USAF School of Aerospace Medicine, Brooks AFB, TX, Report USAFSAM-TR-83-18 (1983).

67. A.W. Guy, C.-K. Chou, L.L. Kunz, J. Crowley, and J. Krupp, Effects of long-term low-level radiofrequency radiation exposure on rats: volume 9. Summary. USAF School of Aerospace Medicine, Brooks AFB, TX, Report USAFSAM-TR-85-64 (1985).

68. R.B. Johnson, D. Spackman, J. Crowley, D. Thompson, C.-K. Chou, L.L. Kunz, and A.W. Guy, Effects of long-term low-level radiofrequency radiation exposure on rats: Volume 4. Open-field behavior and corticosterone. USAF School of Aerospace Medicine, Brooks AFB, TX, Report USAFSAM-TR-83-42 (1983).

69. R.B. Johnson, L.L. Kunz, D. Thompson, J. Crowley, C.-K. Chou, and A.W. Guy, Effects of long-term low-level radiofrequency radiation exposure on rats: Volume 7. Metabolism, growth, and

development. USAF School of Aerospace Medicine, Brooks AFB, TX, Report USAFSAM-TR-84-31 (1984).

70. L.L. Kunz, K.E. Hellstrom, I. Hellstrom, H.J. Garriques, R.B. Johnson, J. Crowley, D. Thompson, C.-K. Chou, and A.W. Guy, Effects of long-term low-level radiofrequency radiation exposure on rats: Volume 5. Evaluation of the immune system's response. USAF School of Aerospace Medicine, Brooks AFB, TX, Report USAFSAM-TR-83-50 (1983).

71. L.L. Kunz, R.B. Johnson, D. Thompson, J. Crowley, C.-K. Chou, and A.W. Guy, Effects of long-term low-level radiofrequency radiation exposure on rats: Volume 6. hematological, serum chemistry, thyroxine, and protein electrophoresis evaluations. USAF School of Aerospace Medicine, Brooks AFB, TX, Report USAFSAM-TR-84-2 (1984).

72. L.L. Kunz, R.B. Johnson, D. Thompson, J. Crowley, C.-K. Chou, and A.W. Guy, Effects of long-term low-level radiofrequency radiation exposure on rats: Volume 8. evaluation of longevity, cause of death, and histopathological findings. USAF School of Aerospace Medicine, Brooks AFB, TX, Report USAFSAM-TR-85-11 (1985).

73. ANSI C95.1-1982: Safety levels with respect to human exposure to radio frequency electromagnetic fields, 300 kHz TO 100 GHz. Published by the Institute of Electrical and Electronics Engineers, NY (1982).

74. R. Santini, M. Hosni, P. Deschaux, and H. Pacheco, B16 Melanoma development in black mice exposed to low-level microwave radiation, *Bioelectromagnetics,* Vol. 9, No. 1, pp. 105-107 (1988).

75. E.K. Balcer-Kubiczek and G.H. Harrison, Neoplastic transformation of C3H/10T-1/2 cells following exposure to 120-HZ modulated 2.45-GHz microwaves and phorbal ester tumor promoter, *Radiat. Res.,* Vol 126, No. 1, pp. 65-72 (1991).

76. EPA, Evaluation of the potential carcinogenicity of electromagnetic fields. U.S. Environmental Protection Agency, Office of Health and Environment Assessment, Washington, DC 20460, Workshop Review Draft Report EPA/600/6-90/005B (1990).

77. SAB Potential carcinogenicity of electric and magnetic fields. U.S. Environmental Protection Agency Report EPA-SAB-RAC-92-013 (January 1992).

78. R. Rugh, E.I. Ginns, H.S. Ho, and W.M. Leach, Are microwaves teratogenic? *In* "Biological Effects and Health Hazards of Microwave Radiation," P. Czerski et al,. eds., Polish Medical Publishers, Warsaw, pp. 98-107 (1974).

79. R. Rugh, E.I. Ginns, H.S. Ho, and W.M. Leach, Responses of the mouse to microwave radiation during estrous cycle and pregnancy, *Radiat. Res.,* Vol. 62, pp. 225-241 (1975).

80. M.E. Chernovetz, D.R. Justesen, N.W. King, and J.E. Wagner, Teratology, survival, and reversal learning after fetal irradiation of mice by 2450-MHz microwave energy, *J. Microwave Power,* Vol. 10, No. 4, pp. 391-409 (1975).

81. W.B. Stavinoha, A. Modak, M.A. Medina, and A.E. Gass, Growth and development of neonatal mice exposed to high-frequency electromagnetic waves. USAF School of Aerospace Medicine, Brooks AFB, Texas; Final Report SAM-TR-75-51 on Contract F41609-74-C-0018, submitted by University of Texas Health Science Center, San Antonio, Texas (1975).

82. E. Berman, J.B. Kinn, and H.B. Carter, Observations of mouse fetuses after irradiation with 2.45 GHz microwaves, *Health Phys.,* Vol. 35, pp. 791-801 (1978).

83. E. Berman, H.B. Carter, and D. House, Reduced weight in mice offspring after *in utero* exposure to 2450-MHz (CW) microwaves, *Bioelectromagnetics,* Vol. 3, No. 2, pp. 285-291 (1982).

84. E. Berman, H.B. Carter, and D. House, Observations of rat fetuses after irradiation with 2450-MHz (CW) microwaves, *J. Microwave Power,* Vol. 16, No. 1, pp. 9-13 (1981).

85. R.J. Smialowicz, Hematologic and immunologic effects of nonionizing electromagnetic radiation, *in* "Symposium on Health Aspects of Nonionizing Radiation," W.D. Sharpe, ed., Bull. N.Y. Acad. Med., Vol. 55, No. 11, pp. 1094-1118 (1979) [Review].

86. J.M. Lary, D.L. Conover, P.H. Johnson, and J.R. Burg, Teratogenicity of 27.12-mhz radiation in rats is related to duration of hyperthermic exposure, *Bioelectromagnetics,* Vol. 4, No. 3, pp. 249-255 (1983).

87. J.M. Lary, D.L. Conover, P.H. Johnson, and R.W. Hornung, Dose-response relationship between body temperature and birth defects in radiofrequency-irradiated rats, *Bioelectromagnetics,* Vol. 7, No. 2, pp. 141-149 (1986).

88. S. Tofani, G. Agnesod, P. Ossola, S. Ferrini, and R. Bussi, Effects of continuous low-level exposure to radiofrequency radiation on intrauterine development in rats, *Health Phys.,* Vol. 51, No. 4, pp. 489-499 (1986).

89. S.-T. Lu and S.M. Michaelson, Comments on "Effects of continuous low-level exposure to radiofrequency radiation on intrauterine development in rats," *Health Phys.,* Vol. 53, No. 5, p. 545 (1987).

90. C.H. Durney, C.C. Johnson, P.W. Barber, H.W. Massoudi, M.F. Iskander, J.L. Lords, D.K. Ryser, S.J. Allen, and J.C. Mitchell, Radiofrequency radiation dosimetry handbook [2nd edition]. USAF School of Aerospace Medicine, Brooks AFB, TX, Report SAM-TR-78-22 (1978).

91. S. Tofani, G. Agnesod, P. Ossola, S. Ferrini, and R. Bussi, Reply to Lu and Michaelson regarding effects of continuous low-level exposure to radiofrequency radiation, *Health Phys.,* Vol. 53, No. 5, pp. 546-547 (1987).

92. J. Kaplan, P. Polson, C. Rebert, K. Lunan, and M. Gage, Biological and behavioral effects of prenatal and postnatal exposure to 2450-MHz electromagnetic radiation in the squirrel monkey, *Radio Sci.,* Vol. 17, No. 5S, pp. 135-144 (1982).

93. A.H. Frey, S.R. Feld, and B. Frey, Neural function and behavior: defining the relationship, *in* "Ann. N.Y. Acad. Sci.," P.W. Tyler, ed., Vol. 247, pp. 433-439 (1975).

94. K.J. Oscar and T.D. Hawkins, Microwave alteration of the blood-brain barrier system of rats, *Brain Res.,* Vol. 126, pp. 281-293 (1977).

95. J.H. Merritt, A.F. Chamness, and S.J. Allen, Studies on blood-brain barrier permeability after microwave-radiation, *Rad. and Environm. Biophys.,* Vol. 15, pp. 367-377 (1978).

96. T.R. Ward and J.S. Ali, Blood-brain barrier permeation in the rat during exposure to low-power 1.7-GHz microwave radiation, *Bioelectromagnetics,* Vol. 6, No. 2, pp. 131-143 (1985).

97. E. Preston, E.J. Vavasour, and H.M. Assenheim, Permeability of the blood-brain barrier to mannitol in the rat following 2450 MHz microwave irradiation, *Brain Res.,* Vol. 174, pp. 109-117 (1979).

98. S.I. Rapoport, K. Ohno, W.R. Fredericks, and K.D. Pettigrew, A quantitative method for measuring altered cerebrovascular permeability. Radio Sci., Vol. 14, No. 6S, pp. 345-348 (1979).

99. E. Preston and G. Préfontaine, Cerebrovascular permeability to sucrose in the rat exposed to 2,450-MHz microwaves, *J. Appl. Physiol.: Respiratory, Environmental, and Exercise Physiol.,* Vol. 49, No. 2, pp. 218-223 (1980).

100. S.P. Gruenau, K.J. Oscar, M.T. Folker, and S.I. Rapoport, Absence of microwave effect on blood-brain barrier permeability to C14-Sucrose in the conscious rat, *Exper. Neurobiol.,* Vol. 75, pp. 299-307 (1982).

101. W.M. Williams, W. Hoss, M. Formaniak, and S.M. Michaelson, Effect of 2450 MHz microwave energy on the blood-brain barrier to hydrophilic molecules. A. Effect on the permeability to sodium fluorescein, *Brain Res. Rev.,* Vol. 7, pp. 165-170 (1984).

102. W.M. Williams, M. del Cerro, and S.M. Michaelson, Effect of 2450 MHz microwave energy on the blood-brain barrier to hydrophilic molecules. B. Effect on the permeability to HRP, *Brain Res. Rev.,* Vol. 7, pp. 171-181 (1984).

103. W.M. Williams, J. Platner, and S.M. Michaelson, Effect of 2450 MHz microwave energy on the blood-brain barrier to hydrophilic molecules. C. Effect on the permeability to [14C] sucrose, *Brain Res. Rev.,* Vol. 7, pp. 183-190 (1984).

104. W.M. Williams, S.-T. Lu, M. del Cerro, and S.M. Michaelson, Effect of 2450 MHz microwave energy on the blood-brain barrier to hydrophilic molecules. D. Brain temperature and blood-brain barrier permeability to hydrophilic tracers, *Brain Res. Rev.,* Vol. 7, pp. 191-212 (1984).

105. J.P. Neilly and J.C. Lin, Interaction of ethanol and microwaves on the blood-brain barrier of rats, *Bioelectromagnetics,* Vol. 7, No. 4, pp. 405-414 (1986).

106. E.N. Albert, M.F. Sherif, N.J. Papadopoulos, F.J. Slaby, and J. Monahan, Effect of nonionizing radiation on the purkinje cells of the rat cerebellum, *Bioelectromagnetics,* Vol. 2, No. 3, pp. 247-257 (1981).

107. E.N. Albert, M.F. Sherif, and N.J. Papadopoulos, Effect of nonionizing radiation on the Purkinje cells of the uvula in squirrel monkey cerebellum, *Bioelectromagnetics,* Vol. 2, No. 3, pp. 241-246 (1981).

108. J.H. Merritt, K.A. Hardy, and A.F. Chamness, *In utero* exposure to microwave radiation and rat brain development, *Bioelectromagnetics,* Vol. 5, No. 3, pp. 315-322 (1984).

109. A.P. Sanders, D.J. Schaefer, and W.T. Joines, Microwave effects on energy metabolism of rat brain, *Bioelectromagnetics,* Vol. 1, No. 2, pp. 171-181 (1980).

110. A.P. Sanders and W.T. Joines, The effects of hyperthermia and hyperthermia plus microwaves on rat brain energy metabolism, *Bioelectromagnetics,* Vol. 5, No. 1, pp. 63-70 (1984).

111. H. Lai, A. Horita, and A.W. Guy, Acute low-level microwave exposure and central cholinergic activity: studies on irradiation parameters, *Bioelectromagnetics,* Vol. 9 No. 4, pp. 355-362 (1988).

112. C.C. Johnson and A.W. Guy, Nonionizing electromagnetic wave effects in biological materials and systems, *Proc. IEEE,* Vol. 60, No. 6, pp. 692-718 (1972).

113. S. Takashima, B. Onara, and H.P. Schwan, Effects of modulated RF energy on the EEG of mammalian brains, *Rad. and Environm. Biophys.,* Vol. 16, pp. 15-27 (1979).

114. V.V. Tyazhelov, R.E. Tigranian, and E.P. Khizhniak, New artifact-free electrodes for recording of biological potentials in strong electromagnetic fields, *Radio Sci.,* Vol. 12, No. 6S, pp. 121-123 (1977).

115. C.-K. Chou and A.W. Guy, Carbon electrodes for chronic EEG recordings in microwave research, *J. Microwave Power,* Vol. 14, No. 4, pp. 399-404 (1979).

116. C.-K. Chou, A.W. Guy, J.B. McDougall, and L.-F. Han, Effects of continuous and pulsed chronic microwave exposure on rabbits, *Radio Sci.,* Vol. 17, No. 5S, pp. 185-193 (1982).

117. S.M. Bawin, L.K. Kaczmarek, and W.R. Adey, Effects of modulated VHF fields on the central nervous system, *in* "Ann. N.Y. Acad. Sci.," P. W. Tyler ed., Vol. 247, pp. 74-81 (1975).

118. A.R. Sheppard, S.M. Bawin, and W.R. Adey Models of long-range order in cerebral macromolecules: effects of sub-ELF and of modulated VHF and UHF fields, *Radio Sci.,* Vol. 14, No. 6S, pp. 141-145 (1979).

119. S.M. Bawin and W.R. Adey, Interactions between nervous tissues and weak environmental electric fields, *in* "Biological Effects of Electromagnetic Waves," C.C. Johnson and M. Shore, eds., U.S. Dept. of Health, Education, and Welfare, Washington, D.C., HEW Publication (FDA) 77-8010, pp. 323-330 (1976).

120. S.M. Bawin and W.R. Adey, Sensitivity of calcium binding in cerebral tissue to weak environmental electric fields oscillating at low frequencies, *Proc. Nat. Acad. Sci.,* Vol. 73, No. 6, pp. 1999-2003 (1976).

121. C.F. Blackman, J.A. Elder, C.M. Weil, S.G. Benane, D.C. Eichinger, and D.E. House, Induction of calcium-ion efflux from brain tissue by radio-frequency radiation: effects of modulation frequency and field strength, *Radio Sci.,* Vol. 14, No. 6S, pp. 93-98 (1979).

122. W.W. Shelton, Jr. and J. H. Merritt, *In vitro* study of microwave effects on calcium efflux in rat brain tissue, *Bioelectromagnetics,* Vol. 2, No. 2, pp. 161-167 (1981).

123. J.H. Merritt, W.W. Shelton, and A.F. Chamness, Attempts to alter $^{45}CA^{2+}$ binding to brain tissue with pulse-modulated microwave energy, *Bioelectromagnetics,* Vol. 3, No. 4, pp. 475-478 (1982).

124. W.R. Adey, S.M. Bawin, and A.F. Lawrence, Effects of weak amplitude-modulated microwave fields on calcium efflux from awake cat cerebral cortex, *Bioelectromagnetics,* Vol. 3, No. 3, pp. 295-307 (1982).

125. C.F. Blackman, S.G. Benane, J.R. Rabinowitz, D.E. House, and W.T. Joines, A role for the magnetic field in the radiation-induced efflux of calcium ions from brain tissue *in vitro, Bioelectromagnetics,* Vol. 6, No. 4, pp. 327-337 (1985).

126. C.F. Blackman, S.G. Benane, and D.E. House, The influence of temperature during electric- and magnetic-field-induced alteration of calcium-ion release from *in vitro* brain tissue, *Bioelectromagnetics,* Vol. 12, No. 3, pp. 173-182 (1991).

127. R.J. Smialowicz, The effect of microwaves (2450 MHz) on lymphocyte blast transformation *in vitro, in* "Biological Effects of Electromagnetic Waves," C.C. Johnson and M. Shore, eds., Vol. I, U.S. Department of Health, Education, and Welfare, HEW Publication (FDA) 77-8010, pp. 472-483 (1976).

128. P.E. Hamrick and S.S. Fox, Rat lymphocytes in cell culture exposed to 2450 MHz (CW) microwave radiation, *J. Microwave Power,* Vol. 12, No. 2, pp. 125-132 (1977).

129. N.J. Roberts, Jr., S.-T. Lu, and S.M. Michaelson, Human leukocyte functions and the U. S. Safety Standard for exposure to radio-frequency radiation, *Science,* Vol. 220, 15 April 1983, pp. 318-320 (1983).

130. N.J. Roberts, Jr., S.M. Michaelson, and S.-T. Lu, Mitogen responsiveness after exposure of influenza virus-infected human mononuclear leukocytes to continuous or pulse-modulated radiofrequency radiation, *Radiat. Res.,* Vol. 110, No. 3, pp. 353-361 (1987).

131. D.B. Lyle, P. Schechter, W.R. Adey, and R.L. Lundak, Suppression of T-lymphocyte cytotoxicity following exposure to sinusoidally amplitude-modulated fields, *Bioelectromagnetics,* Vol. 4, No. 3, pp. 281-292 (1983).

132. M.F. Sultan, C.A. Cain, and W.A.F. Tompkins, Effects of microwaves and hyperthermia on capping of antigen-antibody complexes on the surface of normal mouse B lymphocytes, *Bioelectromagnetics,* Vol. 4, No. 2, pp. 115-122 (1983).

133. M.F. Sultan, C.A. Cain, and W.A.F. Tompkins, Immunological effects of amplitude-modulated radio frequency radiation: B lymphocyte capping, *Bioelectromagnetics,* Vol. 4, No. 2, pp. 157-165 (1983).

134. S.F. Cleary, L.-M. Liu, and F. Garber, Viability and phagocytosis of neutrophils exposed *in vitro* to 100-MHz radiofrequency radiation, *Bioelectromagnetics,* Vol. 6, No. 1, pp. 53-60 (1985).

135. J.L. Kiel, L.S. Wong, and D.N. Erwin, Metabolic effects of microwave radiation and convection heating on human mononuclear leukocytes, *Physiological Chemistry and Physics and Medical NMR,* Vol. 18, pp. 181-187 (1986).

136. D.J. Peterson, L.M. Partlow, and O.P. Gandhi, An investigation of the thermal and athermal effects of microwave irradiation on erythrocytes, *IEEE Trans. Biomed. Eng.,* Vol. 26, No. 7, pp. 428-436 (1979).

137. R.F. Brown and S.V. Marshall, Differentiation of murine erythroleukemic cells during exposure to microwave radiation, *Radiat. Res.,* Vol. 108, No. 1, pp. 12-22 (1986).

138. A.T. Huang, M.E. Engle, J.A. Elder, J.B. Kinn, and T.R. Ward, The effect of microwave radiation (2450 MHz) on the morphology and chromosomes of lymphocytes, *Radio Sci.,* Vol. 12, No. 6S, pp. 173-177 (1977).

139. A.T. Huang and N.G. Mold, Immunologic and hematopoietic alterations by 2,450-MHz electromagnetic radiation, *Bioelectromagnetics,* Vol. 1, No. 1, pp. 77-87 (1980).

140. J.C. Lin, J.C. Nelson, and M.E. Ekstrom, Effects of repeated exposure to 148-MHz radio waves on growth and hematology of mice, *Radio Sci.,* Vol. 14, No. 6S, pp. 173-179 (1979).

141. W. Wiktor-Jedrzejczak, A. Ahmed, P. Czerski, W.M. Leach, and K.W. Sell, Immune response of mice to 2450-MHz microwave radiation: overview of immunology and empirical studies of lymphoid splenic cells, *Radio Sci.,* Vol. 12, No. 6S, pp. 209-219 (1977).

142. K. Sulek, C.J. Schlagel, W. Wiktor-Jedrzejczak, H.S. Ho, W.M. Leach, A. Ahmed, and J. N. Woody, Biologic effects of microwave exposure: I. threshold conditions for the induction of the increase in complement receptor positive (CR+) mouse spleen cells following exposure to 2450-MHz microwaves, *Radiat. Res.,* Vol. 83, pp. 127-137 (1980).

143. C.J. Schlagel, K. Sulek, H.S. Ho, W.M. Leach, A. Ahmed, and J.N. Woody, Biologic effects of microwave exposure. II. Studies on the mechanisms controlling susceptibility to microwave-induced increases in complement receptor-positive spleen cells, *Bioelectromagnetics,* Vol. 1, No. 4, pp. 405-414 (1980).

144. W. Wiktor-Jedrzejczak, C.J. Schlagel, A. Ahmed, W.M. Leach, and J.N. Woody, Possible humoral mechanism of 2450-MHz microwave-induced increase in complement receptor positive cells, *Bioelectromagnetics,* Vol. 2, No. 1, pp. 81-84 (1981).

145. R.J. Smialowicz, P.L. Brugnolotti, and M.M. Riddle, Complement receptor positive spleen cells in microwave (2450-MHz)-irradiated mice, *J Microwave Power,* Vol. 16, No. 1, pp. 73-77 (1981).

146. R.P. Liburdy, Radiofrequency radiation alters the immune system: modulation of T- and B-lymphyocyte levels and cell-mediated immunocompetence by hyperthermic radiation, *Radiat. Res.,* Vol. 77, pp. 34-46(1979).

147. R.J. Smialowicz, C.M. Weil, P. Marsh, M.M. Riddle, R.R. Rogers, and B.F. Rehnberg Biological effects of long-term exposure of rats to 970-MHz radiofrequency radiation, *Bioelectromagnetics,* Vol. 2, No. 3, pp. 279-284 (1981).

148. R.J. Smialowicz, M.M. Riddle, R.R. Rogers, and G.A. Stott, Assessment of immune function development in mice irradiated *in utero* with 2450-MHz microwaves, *J. Microwave Power,* Vol. 17, No. 2, pp. 121-126 (1982).

149. R.J. Smialowicz, M.M. Riddle, C.M. Weil, P.L. Brugnolotti, and J.B. Kinn, Assessment of the immune responsiveness of mice irradiated with continuous wave or pulse-modulated 425-MHz radiofrequency radiation, *Bioelectromagnetics,* Vol. 3, No. 4, pp. 467-470 (1982).

150. R.J. Smialowicz, R.R. Rogers, R.J. Garner, M.M. Riddle, R.W. Luebke, and D.G. Rowe, Microwaves (2,450 MHz) suppress murine natural killer cell activity, *Bioelectromagnetics,* Vol. 4, No. 4, pp. 371-381 (1983).

151. M.J. Ortner, M.J. Galvin, and D.I. McRee, Studies on acute *in vivo* exposure of rats to 2450-MHz microwave radiation--I. Mast cells and basophils, *Radiat. Res.,* Vol. 86, pp. 580-588 (1981).

152. L.S. Wong, J.H. Merritt, and J.L. Kiel, Effects of 20-MHz radiofrequency radiation on rat hematology, splenic function, and serum chemistry, *Radiat. Res.,* Vol. 103, No. 2, pp. 186-195 (1985).

153. S. Szmigielski, W. Roszkowski, M. Kobus, and J. Jeljaszewicz, Modification of experimental acute staphylococcal infections by long-term exposure to non-thermal microwave fields or whole body hyperthermia, *Proc. URSI Int. Symposium on Electromagnetic Waves and Biology,* Paris, France, June-July 1980, pp. 127-132 (1980).

154. C.G. Liddle, J.P. Putnam, and O.H. Lewter, Effects of microwave exposure and temperature on survival of mice infected with streptococcus pneumoniae, *Bioelectromagnetics,* Vol. 8, No. 3, pp. 295-302 (1987).

155. J. Toler, V. Popovic, S. Bonasera, P. Popovic, C. Honeycutt, and D. Sgoutas, Long-term study of 435 MHz radio-frequency radiation on blood-borne end points in cannulated rats--part II: methods, results and summary, *J. Microwave Power EE,* Vol. 23, No. 2, pp. 105-136 (1988).

156. J. N. Bollinger, Detection and evaluation of radiofrequency electromagnetic radiation-induced biological damage in Macaca mulatta. Final report submitted by Southwest Research Institute, San Antonio, Texas, to the USAF School of Aerospace Medicine, Brooks AFB, Texas (1971).

157. J.W. Frazer, J.H. Merritt, S.J. Allen, R.H. Hartzell, J.A. Ratliff, A.F. Chamness, R.E. Detwiler, and T. McLellan, Thermal responses to high-frequency electromagnetic radiation fields. USAF School of Aerospace Medicine, Brooks AFB, Texas, Report SAM-TR-76-20 (1976).

158. J.H. Krupp, Thermal response in Macaca mulatta exposed to 15- and 20-MHz radiofrequency radiation. USAF School of Aerospace Medicine, Brooks AFB, Texas, Report SAM-TR-77-16, (Sept. 1977).

159. J.H. Krupp, Long-term followup of Macaca mulatta exposed to high levels of 15-, 20-, and 26-MHz radiofrequency radiation. USAF School of Aerospace Medicine, Brooks AFB, Texas, Report SAM-TR-78-3, (Jan. 1978).

160. H.S. Ho and W.P. Edwards, The effect of environmental temperature and average dose rate of microwave radiation on the oxygen-consumption rate of mice, *Radiat. Environ. Biophys.,* Vol. 16, pp. 325-338 (1979).

161. S. Stern, L. Margolin, B. Weiss, S.-T. Lu, and S.M. Michaelson Microwaves: effect on thermoregulatory behavior in rats, *Science,* Vol. 206, pp. 1198-1201, (7 Dec. 1979).

162. E.R. Adair and B.W. Adams, Microwaves modify thermoregulatory behavior in squirrel monkey, *Bioelectromagnetics,* Vol. 1, No. 1, pp. 1-20 (1980).

163. V. Bruce-Wolfe and E.R. Adair, Operant control of convective cooling and microwave irradiation by the squirrel monkey, *Bioelectromagnetics,* Vol. 6, No. 4, pp. 365-380 (1985).

164. E.R. Adair, D.E. Spiers, R.O. Rawson, B.W. Adams, D.K. Shelton, P.J. Pivirotto, and G. M. Akel, Thermoregulatory consequences of long-term microwave exposure at controlled ambient temperatures, *Bioelectromagnetics,* Vol. 6, No. 4, pp. 339-363 (1985).

165. W.G. Lotz and J.L. Saxton, Metabolic and vasomotor responses of rhesus monkeys exposed to 225-MHz radiofrequency energy, *Bioelectromagnetics,* Vol. 8, No. 1, pp. 73-89 (1987).

166. A.B. Cairnie, D.A. Hill, and H.M. Assenheim, Dosimetry for a study of effects of 2.45-GHz microwaves on mouse testis, *Bioelectromagnetics,* Vol. 1, No. 3, pp. 325-336 (1980).

167. R.M. Lebovitz and L. Johnson, Testicular function of rats following exposure to microwave radiation, *Bioelectromagnetics,* Vol. 4, No. 2, pp. 107-114 (1983).

168. R.M. Lebovitz and L. Johnson, Acute, whole-body microwave exposure and testicular function of rats, *Bioelectromagnetics,* Vol. 8, No. 1, pp. 37-43 (1987).

169. W.G. Lotz and S.M. Michaelson, Temperature and corticosterone relationships in microwave-exposed rats, *J. Appl. Physiol.: Respiratory, Environmental, and Exercise Physiol.,* Vol. 44, No. 3, pp. 438-445 (1978).

170. S.-T. Lu, N. Lebda, S.M. Michaelson, and S. Pettit, Serum-thyroxine levels in microwave-exposed rats, *Radiat. Res.,* Vol. 101, pp. 413-423 (1985).

171. S.-T. Lu, S. Pettit, S.-J. Lu, and S.M. Michaelson, Effects of microwaves on the adrenal cortex, *Radiat. Res.,* Vol. 107, No. 2, pp. 234-249 (1986).

172. W.G. Lotz and R.P. Podgorski, Temperature and adrenocortical responses in rhesus monkeys exposed to microwaves, *J. Appl. Physiol.: Respiratory, Environmental, and Exercise Physiol.,* Vol. 53, No. 6, pp. 1565-1571 (1982).

173. A.H. Frey and E. Seifert , Pulse modulated UHF energy illumination of the heart associated with change in heart rate, *Life Sci.,* Vol. 7, No. 10, Part II, pp. 505-512 (1968).

174. R.M. Clapman and C.A. Cain, Absence of heart-rate effects in isolated frog heart irradiated with pulse modulated microwave energy, *J. Microwave Power,* Vol. 10, No. 4, pp. 411-419 (1975).

175. L.M. Liu, F.J. Rosenbaum, and W.F. Pickard, The insensitivity of frog heart rate to pulse modulated microwave energy, *J. Microwave Power,* Vol. 11, No. 3, pp. 225-232 (1976).

176. M.J. Galvin, C.A. Hall, and D.I. McRee, Microwave radiation effects on cardiac muscle cells *in vitro,* *Radiat. Res.,* Vol. 86, pp. 358-367 (1981).

177. K.-C. Yee, C.-K. Chou, and A W. Guy, Effect of microwave radiation on the beating rate of isolated frog hearts, *Bioelectromagnetics,* Vol. 5, No. 2, pp. 263-270 (1984).

178. K.-C. Yee, C.-K. Chou, and A.W. Guy, Effects of pulsed microwave radiation on the contractile rate of isolated frog hearts, *J. Microwave Power & EM Energy,* Vol. 21, No. 3, pp. 159-165 (1986).

179. J.L. Schwartz, J. Delorme, and G.A.R. Mealing, [Abstract]. Effects of low-frequency amplitude modulated radiofrequency waves on the calcium efflux of the heart , *Biophys. J.,* Vol. 41, p. 295a (1983).

180. K.-C. Yee, C.-K. Chou, and A.W. Guy, Influence of microwaves on the beating rate of isolated rat hearts, *Bioelectromagnetics,* Vol. 9, No. 2, pp. 175-181 (1988).

181. A.S. Presman and N.A. Levitina, Nonthermal action of microwaves on cardiac rhythm--communication I. A study of the action of continuous microwaves, *Bull. Exp. Biol. Med.,* Vol. 53, No. 1, pp. 36-39, (Engl. Transl. of pp. 41-44 of 1962a Russ. publ.) (1963).

182. A.S. Presman and N.A. Levitina, Nonthermal action of microwaves on the rhythm of cardiac contractions in animals--Report II. Investigation of the action of impulse microwaves, *Bull. Exp. Biol. Med.,* Vol. 53, No. 2, pp. 154-157 (Engl. Transl. of pp. 39-43 of 1962 Russ. publ.) (1963).

183. I.T. Kaplan, W. Metlay, M.M. Zaret, L. Birenbaum, and S.W. Rosenthal, Absence of heart-rate effects in rabbits during low-level microwave irradiation, *IEEE Trans. Microwave Theory Tech.,* Vol. 19, No. 2, pp. 168-173 (1971).

184. L. Birenbaum, I.T. Kaplan, W. Metlay, S.W. Rosenthal, and M.M. Zaret , Microwave and infrared effects on heart rate, respiration rate and subcutaneous temperature of the rabbit, *J. Microwave Power,* Vol. 10, No. 1, pp. 3-18 (1975).

185. R.D. Phillips, E.L. Hunt, R.D. Castro, and N.W. King, Thermoregulatory, metabolic, and cardiovascular response of rats to microwaves, *J. Appl. Physiol.,* Vol. 38, No. 4, pp. 630-635 (1975).

186. M.J. Galvin and D.I. McRee, Influence of acute microwave radiation on cardiac function in normal and myocardial ischemic cats, *J. Appl. Physiol: Respiratory, Environmental, and Exercise Physiol.,* Vol. 50, No. 5, pp. 931-935 (1981).

187. M.J. Galvin and D.I. McRee, Cardiovascular, hematologic, and biochemical effects of acute ventral exposure of conscious rats to 2450-MHz (CW) microwave radiation, *Bioelectromagnetics,* Vol. 7, No. 2, pp. 223-233 (1986).

188. D.R. Justesen and N.W. King, Behavioral effects of low level microwave irradiation in the closed space situation, *in* "Biological Effects and Health Implications of Microwave Radiation," S.F. Cleary, ed., U.S. Dept. of Health, Education, and Welfare, Washington, D.C., HEW Publication BRH/DBE 70-2, pp. 154-179 (1970).

189. E.L. Hunt, N.W. King, and R.D. Phillips, Behavioral effects of pulsed microwave radiation. 104, *in* "Ann. N.Y. Acad. Sci.," P. W. Tyler, ed., Vol. 247, pp. 440-453 (1975).

190. J.C. Monahan and H.S. Ho, Microwave induced avoidance behavior in the mouse, *in* "Biological Effects of Electromagnetic Waves," C.C. Johnson and M. Shore, eds., Vol. I, U.S. Dept. of Health, Education, and Welfare, Washington, D.C., HEW Publication (FDA) 77-8010, pp. 274-283 (1976).

191. J.C. Lin, A.W. Guy, and L.R. Caldwell, Thermographic and behavioral studies of rats in the near field of 918-MHz radiations, *IEEE Trans. Microwave Theory Tech.,* Vol. 25, No. 10, pp. 833-836 (1977).

192. J. Schrot, J.R. Thomas, and R.A. Banvard, Modification of the repeated acquisition of response sequences in rats by low-level microwave exposure, *Bioelectromagnetics,* Vol. 1, No. 1, pp. 89-99 (1980).

193. M.I. Gage and W.M. Guyer, Interaction of ambient temperature and microwave power density on schedule-controlled behavior in the rat, *Radio Sci.,* Vol. 17, No. 5S, pp. 179-184 (1982).

194. R.M. Lebovitz, Prolonged microwave irradiation of rats: effects on concurrent operant behavior, *Bioelectromagnetics,* Vol. 2, No. 2, pp. 169-185 (1981).

195. R.M. Lebovitz, Pulse modulated and continuous wave microwave radiation yield equivalent changes in operant behavior of rodents, *Physiology and Behavior,* Vol. 30, No. 6, pp. 891-898 (1983).

196. J.A. D'Andrea, J.R. DeWitt, O.P. Gandhi, S. Stensaas, J.L. Lords, and H.C. Nielson, Behavioral and physiological effects of chronic 2,450-MHz microwave irradiation of the rat at 0.5 mW/cm$^2$, *Bioelectromagnetics,* Vol. 7, No. 1, pp. 45-56 (1986).

197. J.A. D'Andrea, J.R. DeWitt, R.Y. Emmerson, C. Bailey, S. Stensaas, and O.P. Gandhi, Intermittent exposure of rats to 2450 MHz microwaves at 2.5 mW/cm$^2$: Behavioral and physiological effects, *Bioelectromagnetics,* Vol. 7, No. 3, pp. 315-328 (1986).

198. J. R. DeWitt, J.A. D'Andrea, R.Y. Emmerson, and O.P. Gandhi, Behavioral effects of chronic exposure to 0.5 mW/cm$^2$ of 2,450-MHz microwaves, *Bioelectromagnetics,* Vol. 8, No. 2, pp. 149-157 (1987).

199. C.L. Mitchell, D.I. McRee, N.J. Peterson, and H.A. Tilson, Some behavioral effects of short-term exposure of rats to 2.45-GHz microwave radiation, *Bioelectromagnetics,* Vol. 9, No. 3, pp. 259-268 (1988).

200. Y. Akyel, E.L. Hunt, C. Gambrill, and C. Vargas, Jr., Immediate post-exposure effects of high-peak-power microwave pulses on operant behavior of Wistar rats, *Bioelectromagnetics,* Vol. 12, No. 3, pp. 183-195 (1991).

201. W.D. Galloway, Microwave dose-response relationships on two behavioral tasks, *in* "Ann. N.Y. Acad. Sci.," P.W. Tyler, ed., Vol. 247, pp. 410-416 (1975).

202. R.J. Cunitz, W.D. Galloway, and C.M. Berman, Behavioral suppression by 383-MHz radiation, *IEEE Trans. Microwave Theory Tech.,* Vol. 23, No. 3, pp. 313-316 (1975).

203. D.M. Scholl, and S.J. Allen, Skilled visual-motor performance by monkeys in a 1.2-GHz microwave field, *Radio Sci.,* Vol. 14, No. 6S, pp. 247-252 (1979).

204. J.O. de Lorge, The effects of microwave radiation on behavior and temperature in rhesus monkeys, *in* "Biological Effects of Electromagnetic Waves," C.C. Johnson and M. Shore, eds., U.S. Dept. of Health, Education, and Welfare, Washington, D.C., HEW Publication (FDA) 77-8010, pp. 158-174 (1976).

205. J.O. de Lorge, Operant behavior and rectal temperature of squirrel monkeys during 2.45-GHz microwave irradiation, *Radio Sci.,* Vol. 14, No. 6S, pp. 217-225 (1979).

206. J.O. de Lorge, Operant behavior and colonic temperature of Macaca mulatta exposed to radio frequency fields at and above resonant frequencies, *Bioelectromagnetics,* Vol. 5, No. 2, pp. 233-246 (1984).

207. J.A. D'Andrea, B.L. Cobb, and J.O. de Lorge, Lack of behavioral effects in the rhesus monkey: high peak microwave pulses at 1.3 GHz, *Bioelectromagnetics,* Vol. 10, No. 1, pp. 65-76 (1989).

208. J.R. Thomas, L.S. Burch, and S.S. Yeandle, Microwave radiation and chlordiazepoxide: synergistic effects on fixed-interval behavior, *Science,* Vol. 203, pp. 1357-1358 (1979).

209. J.R. Thomas and G. Maitland, Microwave radiation and dextroamphetamine: evidence of combined effects on behavior of rats, *Radio Sci.,* Vol. 14, No. 6S, pp. 253-258 (1979).

210. J.R. Thomas, J. Schrot, and R A. Banvard, Behavioral effects of chlorpromazine and diazepam combined with low-level microwaves, *Neurobehav. Toxicol.,* Vol. 2, pp. 131-135 (1980).

211. Y. Ashani, F.H. Henry, and G.N. Catravas, Combined effects of anticholinesterase drugs and low-level microwave radiation, *Radiat. Res.,* Vol 84, pp. 496-503 (1980).

212. B.A. Pappas, H. Anisman, R. Ings, and D.A. Hill, Acute exposure to pulsed microwaves affects neither pentylenetetrazol seizures in the rat nor chlordiazepoxide protection against such seizures, *Radiat. Res.,* Vol. 96, No. 3, pp. 486-496 (1983).

213. H. Lai, A. Horita, C.-K. Chou, and A.W. Guy, Psychoactive-drug response is affected by acute low-level microwave irradiation, *Bioelectromagnetics,* Vol. 4, No. 3, pp. 205-214 (1983).

214. H. Lai, A. Horita, C.-K. Chou, and A. W. Guy, Erratum to Lai et al. (1983), *Bioelectromagnetics,* Vol. 6, No. 2, p. 207 (1985).

215. H. Lai, A. Horita, C.-K. Chou, and A.W. Guy, Effects of acute low-level microwaves on pentobarbital-induced hypothermia depend on exposure orientation, *Bioelectromagnetics,* Vol. 5, No. 2, pp. 203-211 (1984).

216. H. Lai, A. Horita, C.-K. Chou, and A. W. Guy, Ethanol-induced hypothermia and ethanol consumption in the rat are affected by low-level microwave irradiation, *Bioelectromagnetics,* Vol. 5, No. 2, pp. 213-220 (1984).

217. H. Lai, M.A. Carino, A. Horita, and A.W. Guy, Single vs. Repeated microwave exposure: effects on benzodiazepine receptors in the brain of the rat, *Bioelectromagnetics,* Vol. 13, No. 1, pp. 57-66 (1992).

218. S.J. Webb and D.D. Dodds, Inhibition of bacterial cell growth by 136 GC microwaves, *Nature,* Vol. 218, pp. 374-375 (27 April 1968).

219. S.J. Webb and A.D. Booth, Absorption of microwaves by microorganisms, *Nature,* Vol. 222, pp. 1199-1200 (21 June 1969).

220. H. Fröhlich, Evidence for bose condensation-like excitation of coherent modes in biological systems, *Phys. Lett.,* Vol. 51A, No. 1, pp. 21-22 (1975).

221. S.J. Webb and M.E. Stoneham, Resonances between 100 and 1000 GHz in active bacterial cells as seen by laser Raman spectroscopy, *Phys. Lett.,* Vol. 60A, No. 3, pp. 267-268 (1977).

222. M.S. Cooper and N.M. Amer, The absence of coherent vibrations in the Raman spectra of living cells, *Phys. Lett.,* Vol. 98A, No. 3, pp. 138-142 (1983).

223. O.P. Gandhi, M.J. Hagmann, D.W. Hill, L.M. Partlow, and L. Bush, Millimeter wave absorption spectra of biological samples, *Bioelectromagnetics,* Vol. 1, No. 3, pp. 285-298 (1980).

224. M.L. Swicord and C.C. Davis, An optical method for investigating the microwave absorption characteristics of DNA and other biomolecules in solution, *Bioelectromagnetics,* Vol. 4, No. 1, pp. 21-42 (1983).

225. G.S.Edwards, C.C. Davis, J.D. Saffer, and M.L. Swicord, Microwave-field-driven acoustic modes in DNA, *Biophys. J.,* Vol. 47, pp. 799-807(1985).

226. C. Gabriel, E.H. Grant, R. Tata, P.R. Brown, B. Gestblom, and E. Noreland, Microwave absorption in aqueous solutions of DNA, *Nature,* Vol. 328, pp. 145-146, (9 July 1987).

227. K.R. Foster, B.R. Epstein, and M.A. Gealt , "Resonances" in the dielectric absorption of DNA? *Biophys. J.,* Vol. 52, pp. 421-425 (1987).

228. J.-L. Sagripanti, M.L. Swicord, and C.C. Davis Microwave effects on plasmid DNA, *Radiat. Res.,* Vol. 110, No. 2, pp. 219-231 (1987).

**Session H:  Public Health Policy--Risk Communication**

**Chair: E. Grant**

How New RFR Standards Will Impact the Broadcast and
Telecommunications Industries
*Richard A. Tell*


How the Popular Press and Media Influence Scientific
Interpretations and Public Opinion
*Clifford J. Sherry*


Impact of Public Concerns about Low-Level
Electromagnetic Fields on Interpretation of Electromagnetic
Fields/Radiofrequency Database
*John M. Osepchuk*


Communicating Risk of Electromagnetic Fields/
Radiofrequency Radiation
*B. Jon Klauenberg*

# HOW NEW RFR STANDARDS WILL IMPACT
# THE BROADCAST AND TELECOMMUNICATIONS INDUSTRIES

Richard A. Tell

Richard Tell Associates, Inc.
8309 Garnet Canyon Lane
Las Vegas, Nevada 89129-4897
USA

## INTRODUCTION

Despite the controversial nature of whether radiofrequency radiation (RFR) poses the potential for adverse health effects in exposed individuals, there has been no lack of development of RFR exposure standards in recent years. During the last ten years, no less than an average of one new guideline or standard has been issued each year. Regulatory activities in the United States,[1,2] Canada,[3] the United Kingdom,[4] Australia,[5] Germany,[6] China,[7] and Czechoslovakia[8] exemplify this prolific generation of new and, very commonly, more stringent and complex standards.

An obvious trend of these modern RFR exposure standards is the move toward more restrictive exposure limits with lower permitted RFR field strengths and power densities. Along with lower permitted RFR field levels has emerged a concern on the part of the industry of potential impact that these new standards could have on broadcasting and telecommunications in particular. In some instances, for example, these concerns have precipitated extensive studies of the possible economic consequences that alternative regulatory exposure limits might have on broadcasting activities.[9] Similar reactions have also occurred within government over the possibility that proposed new standards might impact adversely on military and other government operations.[10] The purpose of this paper is to address the specifics of how these new RFR standards may impact broadcasting and telecommunications.

## SOME ELEMENTS OF NEW STANDARDS THAT POSE
## POTENTIAL COMPLIANCE PROBLEMS FOR INDUSTRY

Several aspects of recently developed RFR standards suggest an increased probability of adverse impact on the broadcasting and telecommunications industries. While broadcasting typically makes use of much higher transmitted power levels than those used by communications services, the assumption that telecommunications operations are less likely to be adversely affected may not be accurate. This has recently been brought to light in the United States where the use of portable cellular telephones has been questioned in connection with the development of brain tumors.[11,12,13] Some selected aspects of recent RFR standards will be addressed from the perspective of how they may contribute to impact on the industry. The recently published recommendations of the Institute of Electrical and Electronics Engineers (IEEE),[1] which constitutes a major revision of the older American National Standards Institute (ANSI) C95.1-1982 RF protection guide,[14] will be used as an example for this discussion.

### Field Strength/Power Density Limits

The most apparent difference among more recent RFR standards has been the general reduction in maximum recommended exposure levels for humans compared to those values permitted on older standards. Certainly, this move toward more stringent limits on RF fields is reflected in the new IEEE standard[1] that contains two sets of field limits, depending on the type of exposure environment. The new limited field strengths, or power densities, are, at the same time, both more and less restrictive than the older ANSI standard[14]. For example, in the most critically controlled frequency range corresponding to the body resonance frequencies and also the very high frequency (VHF) band for FM radio and television broadcasting, the maximum permissible exposure (MPE) in controlled environments is set at 1.0 milliwatts per square centimeter ($mW/cm^2$) as averaged over any 6-minute period; in uncontrolled environments, the MPE is 0.2 $mW/cm^2$ as averaged over any 30-minute period. The uncontrolled environment limit in this frequency range is a factor of five times lower than the older ANSI standard that the new IEEE standard replaces. Lower permitted field strengths or power densities mean that potentially larger areas near broadcast towers may exhibit RF fields that exceed the new limits. The 0.2 $mW/cm^2$ power density limit has become near universal in many of the standards published during the past 10 years and has been adopted widely at the state, county and city levels, within the United States.[15,16,17,18] Standards even more stringent than this figure, such as those of Czechoslovakia in which the MPE is set at the equivalent of 0.0011 $mW/cm^2$ for continuous exposure of the general public in the VHF range, have substantially more potential for impact on broadcasting and telecommunications.

### Two-Tiered Standards

Newer standards for RFR often take a two-tiered approach to the recommended exposure limits, distinguishing between occupational and general public populations or between controlled and uncontrolled environments. This process ultimately reflects a difference in the margin of safety implied by the standard based on either the population exposed or the exposure area and, hence, a difference in the perceived risk of the exposure that might be received by individuals in different circumstances. For example, the two-tier structure of the recent IEEE standard[1] really implies three relevant regions in which RFR exposures may occur: (a) those areas with fields less than the limits for an uncontrolled environment; (b) those areas with fields greater than the limits for an uncontrolled environment (i.e., a controlled environment); and (c) restricted areas with fields greater than the controlled environment limits in which no exposure is permitted.

The requirement to distinguish among these three areas introduces substantially more complexity from a standards implementation perspective than does a single-tier standard, resulting, generally, in greater costs for compliance. Also, the greater difficulty of understanding and accepting an apparent difference in relative risk of exposure by RFR workers, compared to other non-RFR workers and the general public, commonly brings with it a greater probability of concerns and grievances among employees. These worries can and do bring about added effort on the part of employers in dealing with these issues. In some cases, companies have elected to develop internal company policies that actually limit employee exposures to the lesser of the two exposure values to eliminate or reduce such concerns or controversy. So, in some circumstances, the two-tiered standards, while providing for virtually no difference in RFR exposures that individuals in controlled environments may receive when compared to older standards, have resulted in the application of more stringent rules for employees.

## Induced-Current Limits

Several new RFR standards contain limits on RF currents that are induced by exposure to RF fields. Both induced body currents, as measured via the feet, and contact currents, as determined when touching objects that result in a current flow, are relatively new features with which operators of transmitting equipment must become familiar. The IEEE standard[1] specifies a maximum value of contact current of 100 milliamperes (mA) and 45 mA for controlled and uncontrolled environments respectively in the frequency range of 0.1 to 100 MHz. In Canada, these current limits are 40 mA and 15 mA for frequencies from 0.1 MHz to 30 MHz.[3]

The introduction of these new limits on induced currents has significantly increased the complexity of determining compliance with the standards since the instrumentation required for proper measurements is not yet widely available, standardized measurement protocols for assessing induced currents have not been developed, and it is simply not practical to make induced current measurements in certain occupational situations. A case in point is the problem of conducting appropriate contact current measurements on broadcast antenna towers where workers may be located from time-to-time. Additional technical problems arise when RFR sources, such as several broadcast stations, may be using the same antenna tower; in this case, an accurate assessment of contact currents may require some form of narrowband instrumentation, making the task that more involved and difficult. When such a measurement requirement exists in conjunction with such a complex and awkward measurement situation, the likelihood of a proper hazard survey to assess compliance is lessened.

Of particular concern in compliance determinations is the frequency range over which induced current limits apply; in the case of the IEEE[1] standard, the current limits extend to 100 MHz. In the United States, this represents the middle of the FM radio broadcast band (88-108 MHz). For situations with several FM stations operating from the same tower, the issue may arise as to why currents must be measured for a station operating on 99.9 MHz but not for a station on 100.1 MHz (a 0.2 MHz separation exists for FM broadcasters in the United States). Whether such a distinction can be justified scientifically is one question, but the practical matter of ascertaining compliance of stations at a multi-transmitter broadcast site can become extremely formidable and subject to debate. This condition can render broadband current instrumentation ineffective for the task. This type of complication will drive up the costs of conducting the required measurements for determining compliance with standards.

The additional requirement to assess induced body currents in general area RFR surveys is also made all the more difficult with more restrictive current limits. For example, the current induced in a standing individual 1.8 meters (6 feet) tall in good electrical contact with ground is given approximately by:

$$I_{sc} \text{ (mA)} = 0.35 \ E(V/m)f(MHz)$$

where E is the electric field strength parallel with the standing individual and f is the frequency of the RF field. This expression, derived from,[19] indicates that an electric field

393

strength of only 1.4 V/m is sufficient to induce a body current of 15 mA at a frequency of 30MHz, the current limit in the new Canadian RFR standard[3] for individuals in the general population. Electric fields of this low magnitude are widely prevalent near shortwave broadcasting stations[20,21,22] and, hence, there is a strong indication that substantial additional work will be required during field investigations to characterize the extent of areas not in compliance with appropriate guidelines.

## Spatial Averaging of RF Fields

A substantially new approach to specifying RFR exposure limits used in the IEEE standard[1] is that of spatial averaging of the squares of field strength or power density over an area equivalent to the vertical cross-section of the human body being exposed. While this methodology represents a more accurate way of assessing the energy absorption rates for the body as a whole, restrictive language within the standard prohibits such averaging if the eyes or testes may be exposed. Such language, in a practical sense, renders the opportunity to spatially average RF fields problematical at best. The practical upshot of the confusion that could result from environmental studies of RFR in which spatial averaging is used is that the risk of applying spatial averaging methods invites the possibility of technical challenges in court. As a consequence, while the standard was originally intended to apply to spatially averaged fields, reality is that in most cases only the spatial peak values of fields will be used to conservatively estimate actual exposures. Hence, what could have served as an ameliorating factor in the application of the new more stringent RFR limits does not really apply in most cases, meaning that the lower field strength/power density limits will logically imply greater impact.

## Low Power Device Exclusions

Similar to the earlier ANSI standard,[14] the new IEEE standard[1] provides for excluding certain low-power devices such as "walkie-talkies" and other portable RFR sources. With a concern over a more restrictive limit on specific absorption rates (SARs) in uncontrolled environments (in which the spatial peak SAR in the head may not exceed 1.6 watts per kilogram [W/kg] of tissue), the IEEE standard excludes low-power devices on a frequency selective basis. For frequencies between 100 kHz and 450 MHz, the stated MPE may be exceeded if the radiated power is no greater than 1.4 W. At frequencies between 450 MHz and 1.5 GHz, the MPE may be exceeded if the radiated power is no greater than 1.4(450/f)W where f is the frequency in MHz. However, this exclusion does not apply to devices with the "radiating structure maintained within 2.5 cm of the body." Again, similar to the ambiguity with respect to spatial averaging of fields, the issue of whether this clause can legally be interpreted to exclude low-powered communications devices, such as cellular telephones and "walkie-talkies," is open to debate since, for such sources, the entire unit, including the case, becomes part of the radiating structure. Hence, since the case is held by the user, the 2.5 cm spacing criteria of the clause is not upheld and qualification of such devices would appear to require a laboratory study of the SARs associated with their use. Such qualification testing could add significantly to the cost of the units. Nonetheless, many manufacturers may elect to perform these costly evaluations rather than defend against law suits brought for non-compliance with the exposure standard. Indeed, there is the possibility that Federal telecommunications regulators might, at some point, require the equivalent of type certification for handheld radio transmitters.[1]

---

[1]Personal communication with Dr. Robert F. Cleveland, Federal Communications Commission, Washington, DC, February, 1993.

### Definitions of Exposure Environments

A final problematical area of new RFR standards that can lead to additional impact on the broadcasting and telecommunications industries is the definition of exposure environments. In the IEEE standard,[1] environments, not population groups, are distinguished including those that are controlled and uncontrolled. According to the IEEE standard,

> "Controlled environments are locations where there is exposure that may be incurred by persons who are aware of the potential for exposure as a concomitant of employment, by other cognizant persons, or as the incidental result of transient passage through areas where analysis shows the exposure levels may be above those shown in Table 2 but do not exceed those in Table 1, and where the induced currents may exceed the values in Table 2, Part B, but do not exceed the values in Table 1, Part B."

**Table 1.** Summary of numbers and percentages of FM broadcast stations projected to produce alternative RFR power densities on the ground. Data taken from Reference 35.

| Power Density | Single stations on ground[1] | | Multiple sites on ground[2] | | Single stations on buildings | | Multiple sites on buildings | | All stations | |
|---|---|---|---|---|---|---|---|---|---|---|
| ($\mu$W/cm$^2$) | No. | % | No. | % | No. | % | No. | % | No. | % |
| 10,000 | 3 | 0.1 | 0 | 0.0 | 14 | 3.5 | 6 | 37.5 | 23 | 0.6 |
| 5,000 | 15 | 0.5 | 6 | 4.0 | 29 | 7.2 | 9 | 56.3 | 59 | 1.6 |
| 2,000 | 59 | 1.9 | 19 | 12.8 | 51 | 12.7 | 11 | 68.8 | 140 | 3.8 |
| 1,000 | 116 | 3.7 | 27 | 18.1 | 76 | 18.9 | 13 | 81.3 | 232 | 6.3 |
| 900 | 124 | 4.0 | 28 | 18.8 | 83 | 20.6 | 14 | 87.5 | 249 | 6.8 |
| 800 | 142 | 4.6 | 32 | 21.5 | 88 | 21.9 | 14 | 87.5 | 276 | 7.5 |
| 700 | 158 | 5.1 | 32 | 21.5 | 99 | 24.6 | 15 | 93.8 | 304 | 8.3 |
| 600 | 188 | 6.1 | 35 | 23.5 | 107 | 26.6 | 15 | 93.8 | 345 | 9.4 |
| 500 | 225 | 7.3 | 41 | 27.5 | 116 | 28.9 | 15 | 93.8 | 397 | 10.8 |
| 400 | 280 | 9.0 | 44 | 29.5 | 134 | 34.3 | 15 | 93.8 | 473 | 12.9 |
| 300 | 400 | 12.9 | 48 | 32.2 | 154 | 38.3 | 15 | 93.8 | 617 | 16.8 |
| 200 | 560 | 18.1 | 57 | 38.3 | 173 | 43.0 | 15 | 93.8 | 805 | 22.0 |
| 100 | 877 | 28.3 | 82 | 55.0 | 195 | 48.5 | 15 | 93.8 | 1169 | 31.9 |
| 75 | 982 | 31.7 | 88 | 59.1 | 211 | 52.5 | 15 | 93.8 | 1296 | 35.4 |
| 50 | 1206 | 39.0 | 105 | 70.5 | 227 | 56.5 | 15 | 93.8 | 1553 | 42.4 |
| 20 | 1916 | 61.9 | 117 | 78.5 | 275 | 68.4 | 15 | 93.8 | 2323 | 63.4 |
| 10 | 2469 | 79.8 | 132 | 88.6 | 325 | 80.8 | 15 | 93.8 | 2941 | 80.3 |
| 1 | 2905 | 93.9 | 147 | 98.7 | 389 | 96.8 | 15 | 93.8 | 3456 | 94.4 |
| Total | 3095 | | 149 | | 402 | | 16 | | 3662 | |

[1] Single stations on ground refers to sites having only one FM station operating from a ground mounted antenna tower;

[2] Multiple sites on ground refers to sites having more than one FM stations operating from a ground mounted antenna tower;

[3] Single stations on buildings refers to sites having only one FM station operating from a building mounted antenna tower;

[4] Multiple sites on buildings refers to sites having more than one FM station operating from a building mounted antenna tower.

Two key elements of this definition, other than for workers, are "cognizance," or awareness that one is being potentially exposed to RFR and "transient passage." This terminology leaves open to question the precise legal interpretation of whether certain areas that may be subject to intense RFR levels would be considered as controlled or uncontrolled environments. Such determinations can play a vital role in standards compliance studies; in the interest of avoiding conflicting interpretations, many broadcasters may elect to expand controlled access areas rather than rely on other, less expensive means of making individuals passing through fields that exceed uncontrolled environment limits cognizant of their elevated exposures. The end result of more stringent power density limits is that larger areas will eventually become controlled access areas, resulting in greater costs to the industry.

**Table 2.** Summary of analysis to evaluate different mitigation strategies for FM radio broadcast stations through a change in antenna type. Adapted from Reference 35.

| Power density ($\mu$W/cm$^2$) | Numbers of stations exceeding RFR power density levels | | |
| --- | --- | --- | --- |
| | Without modification | With change of antenna | With 1/2 wavelength spacing |
| 20,000 | 1 | 0 | 0 |
| 10,000 | 3 | 0 | 0 |
| 5,000 | 15 | 0 | 0 |
| 2,000 | 59 | 9 | 0 |
| 1,000 | 116 | 22 | 0 |
| 900 | 124 | 28 | 0 |
| 800 | 142 | 32 | 0 |
| 700 | 158 | 38 | 0 |
| 600 | 188 | 43 | 0 |
| 500 | 225 | 57 | 0 |
| 400 | 280 | 72 | 0 |
| 300 | 400 | 97 | 1 |
| 200 | 560 | 150 | 3 |
| 100 | 878 | 267 | 16 |
| 75 | 983 | 330 | 25 |
| 50 | 1,206 | 477 | 41 |
| 20 | 1,917 | 1,007 | 112 |
| 10 | 2,472 | 1,535 | 201 |
| 1 | 2,908 | 2,340 | 1,259 |

## RFR SOURCES AND THE RELEVANCY OF IMPACT CONCERNS

Numerous sources of RFR are used within the broadcasting and telecommunications industries. All of these sources present the potential of producing elevated RFR levels in their immediate vicinities, some with more potential for being impacted by new RFR standards than others. In the broadcast area, AM radio stations in the medium frequency (MF) band and FM radio stations in the VHF band and TV stations in the VHF and UHF bands represent those sources having the greatest likelihood of being impacted from new, more stringent RFR exposure standards. Beyond the area nearest broadcasting antenna towers, environmental RFR levels are very low. A national study of RFR fields within United States metropolitan areas estimated that 99 percent of the population is exposed to broadcasting related fields with power densities less than 1 microwatt per square centimeter (0.001 mW/cm$^2$), i.e., only 1 percent of the population is routinely exposed to greater power densities.[23] This one percent of the population exposed to greater than 1 microwatt per square centimeter is generally associated with residences very near high power broadcast stations. Figure 1, adapted from Reference 23 illustrates the accumulative percentage of the U.S. population estimated to be exposed to RFR levels equal to or less than a range of power densities.

In another study, measured RFR levels of up to 0.1 mW/cm$^2$ were found in tall buildings situated adjacent to major broadcast facilities.[24] Yet other investigations that have focused more specifically on the region immediately near the base of broadcast towers, accessible to the public, have reported significantly higher RFR levels, in some cases in the 1-10 mW/cm$^2$ range.[25, 26, 27]

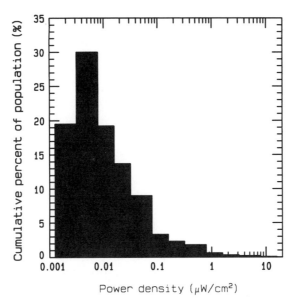

**Figure 1.** Cumulative percentage of population projected to be exposed to RFR at power densities equal to or less than various values (on x-axis) from FM radio and VHF and UHF TV stations. Adapted from Reference 23.

A particular compliance problem for operators of AM broadcasting stations is tower repainting. Since the entire tower structure is energized during broadcasting, tower climbers have direct contact with the radiating antenna and are generally exposed to very high field strengths. At issue is whether such tower work, which can take significant periods of time, can be carried out while the station remains on-the-air and at the same time be in compliance with applicable RFR exposure standards. In this case, the field strength values generally always exceed the exposure limits so a more detailed approach to exposure evaluation must be used. Measurements of contact currents have been investigated as one alternative method for compliance assessments.[28] Through knowledge of the currents flowing in the wrists, for example, the local SAR values may be estimated. Properly documenting compliance with occupational exposures on energized antenna towers poses a significant obstacle in the broadcasting industry.

Shortwave radio broadcasting facilities, while small in total number within any given nation relative to domestic broadcast stations, are potent sources of RFR in their immediate vicinity. Of particular concern to station operators is the new requirement to evaluate induced currents that may exist in individuals near the station. The issue of contact currents becomes even more troublesome since it is possible that significant currents may flow into the body when touching various conductive structures that are exposed to even low field levels.

Satellite communications earth stations, used for both broadcast and telecommunications purposes, operate with extremely high effective radiated powers but, due to exceptionally directive antennas, do not create significant RFR levels in their vicinity, except within the main beam of radiation. Point-to-point microwave communications systems, principally in the form of low-power repeater stations located on tall towers, also do not represent sources that can produce intense RFR at ground level.[29, 30]

Cellular telephone base stations (cell sites) use relatively low effective radiated powers (100 W per channel) and produce very weak ground-level RFR power densities. Nonetheless, considerable public opposition to the installation of new cell sites has become a major concern for cellular telephone companies. The fact that the RF fields produced by the base stations at locations of the public are less than any of the world's RFR exposure standards has not reduced the concern of many members of the public.

Hand-held, portable radio transmitters, such as cellular telephones and "walkie-talkies," while low powered, have come under scrutiny because of public concerns over whether these devices may be linked with the development of brain tumors. Detailed investigation of the local SARs that result from portable cellular telephones are considerably more time consuming than simple field strength measurements.[31] Similar concerns developed last year in the U.S. over the use of portable police traffic radar guns.[32] Traffic radar guns, although they operate at very low power levels (tens of milliwatts radiated power) and produce low RFR levels,[33,34] are not covered by the low-power exclusion clause of many RFR standards since they operate at high microwave frequencies, well above the nominal 1.5 GHz upper limit for excluding such devices. Continued concerns driven by public reaction to anecdotal reports of police officers' use of radar and cancer could well result in such devices having to be certified as to peak SARs in users. Additional testing means additional complexity and cost to the industry.

## EFFORTS TO ASSESS POTENTIAL IMPACT ON THE INDUSTRY

### The Civilian Broadcast Service

Impact of any standard is usually envisioned to mean economic impact. The question of how much impact new RFR standards have, or may have in the future, on the broadcasting and telecommunications industries is extremely difficult to answer in a definitive way. Three components of this question are: impact from an environmental perspective; impact in terms of occupational exposure; and, impact that derives from a concern about exposure of users of products that emit RFR. Seeking quantitative answers to each of these components requires specific methodologies tailored to each question. In an environmental context, the only effort to derive estimates of the costs associated with RFR regulations at a national level was that of the U.S. Environmental Protection Agency during a period when the agency was considering the issuance of recommendations on public exposure to RFR.[35] That effort represented a substantial investment of resources and included a very detailed engineering analysis of RF fields associated with radio and television stations.[36]

Using a database of broadcasting stations in the U.S., a computer model was developed to calculate the expected RFR levels as a function of distance from a substantial number of stations. For FM radio stations, the field calculation model was based on extensive measurements of the radiation pattern of elements that comprise typical FM broadcast antennas. Detailed information on the height of each antenna above ground, effective radiated power of the station, and the number of elements in the antenna were used in the analysis. The number of stations projected to exceed a series of alternative power densities was determined for different classes of FM radio stations. The results of this analysis are presented in Table 1.

Table 1 illustrates the not so surprising result that the lower the RFR level, the greater the number of **potentially** impacted stations. It is important to emphasize in studies of this type that results similar to those found in Table 1 can only indicate that there is a **potential** for some form of impact, not that **actual** impact would exist.

An ancillary analysis, conducted by the EPA in 1984, consisted of administering a detailed questionnaire, to 1,118 FM broadcast stations, seeking information about their transmitting facilities that might reveal a more accurate insight to the likelihood of actual impact that new RFR regulations might impose.[36] One question sought to determine the number of stations that had transmitter sites already fenced to preclude access by the public; in particular, the stations were asked whether their antenna towers had fences around them at a sufficient distance to prevent entrance to the area within that the projected power density would exceed a value of 1 mW/cm$^2$. Based on a response of 52 percent of all the stations queried, it was determined for the 116 single FM stations projected to produce ground level power densities greater than 1 mW/cm$^2$, that:

58 of those stations exceeding 1 mW/cm$^2$ (50%) answered the questionnaire;

51 (or 88%) had inadequate fence boundaries;

15 (or 26%) had inadequate property boundaries.

In each case, the specific power density projections for that station providing a response to the questionnaire were used in a comparison with supplied answers. The implications of these data are that most FM broadcast stations, at the time of the survey, did not own or have control over a sufficiently large property parcel to preclude access to the region of concern and, of those that did have fences, most of the fences were not at sufficiently great distances from the towers to prohibit access to the elevated power density areas. There are no comparable data from other countries so it is not possible to develop an insight to whether similar results might be obtained elsewhere. Clearly, however, the fact that the above area access exclusion investigation was performed for a relatively high power density value, 1 mW/cm$^2$, compared to more recent recommendations for exposure in uncontrolled environments that are up to five times more stringent (0.2 mW/cm$^2$), suggests that considerably greater potential exists for impact on broadcasters with the implementation of new, more stringent RFR standards.

A significant finding in the earlier EPA work was that the current design approach most popular in America for FM broadcast antennas tended to exacerbate the problem of RFR standards compliance. The full-wavelength spacing typically used for multiple element FM broadcast antennas commonly produces a substantial grating lobe downward toward the base of the tower, resulting in significantly elevated power densities. It was found, however, through the use of altered element spacing in the antennas, the regions about an FM antenna tower exhibiting high-power densities could be substantially reduced in size or eliminated altogether.[37] This study also revealed considerable differences among several of the commercially available broadcast antennas, some producing significantly more RFR in their grating lobes than others. Hence, one mitigating factor, for some stations, could be to simply replace the existing antenna with another one having superior downward radiation characteristics. These insights were used[36] to examine the potential for correcting high RFR levels near FM broadcast stations to achieve compliance with a range of hypothetical, alternative RFR levels. These results are summarized in Table 2 that shows the number of stations projected to exceed certain power densities without any modification of their antenna, with simply changing to another commercially available antenna type and changing to a more expensive half-wavelength spaced element antenna that greatly reduces RFR on the ground near the tower.

These data illustrate the dramatic impact that simply choosing an alternative antenna model can have on potential impact of RFR standards on broadcasters. Even for projected power densities of 200 $\mu W/cm^2$, a different antenna type reduces the number of potentially noncompliant stations by 73%. Through the use of specially designed, non-standard element spacing in the antenna, the number of non-compliant stations at the 200 $\mu W/cm^2$ level would be reduced by 99%.

By applying a detailed set of different compliance measures, the EPA then analyzed the potential economic impact that establishing environmental RFR rules for public exposure might have on the broadcast industry.[9] The compliance measures used in the analysis included such provisions as posting of areas for certain remote transmitter sites, replacement of antennas with more efficient models and altered element spacing models, leasing antenna space on existing taller towers, construction of new taller towers either on the present site or on another site, and prohibiting public access. In some instances, it was deemed that certain of these measures were not applicable. In no case was an assumption made that a broadcaster would have to reduce their broadcast power to achieve compliance.

Different forms of economic analysis were pursued but perhaps the most interesting results were in terms of the estimated cost to the broadcast industry as a whole and the estimated cost to the average broadcast station. The analysis was carried out for a series of 18 alternative RFR field level combinations for AM and FM radio stations and VHF and UHF TV stations. RFR power densities for FM and TV stations varied from as low as 1 $\mu W/cm^2$ to as high as 10,000 $\mu W/cm^2$; for AM radio stations, electric field strengths used in the analysis varied from 10 volts per meter (V/m), in the most stringent case, to 1,000 V/m at the upper end of the range. Three different cost scenarios were assumed including a low, medium and high cost option based on extensive engineering input on various costs associated with different compliance measures. For example, it was assumed that virtually all stations would have to pay for an environmental survey to assess RFR levels at the transmitter site.

Table 3 summarizes the results of part of the EPA economic impact study of the potential impact of new RFR standards on the broadcast industry. These data are graphically presented in Figure 2. The figures are in terms of 1985 U.S. dollars. The indicated potential costs to the broadcast industry as a whole are seen to range between $6.9 million and $414.6 million depending on the compliance measure cost scenario applied in the analysis and, very importantly, the RFR exposure limit. While the costs are roughly constant across a wide range of RFR levels, these costs, not surprisingly, rise dramatically at the most stringent end of the RFR range.

When viewed in terms of the estimated potential <u>average</u> cost on a per station basis for broadcasters to comply with the above described RFR levels, the cost figures range from as little as $1,300 to as much as $285,300 depending on the type of station, the compliance cost scenario assumed and the RFR level. Table 4 summarizes these data from the EPA study.

So, while the total industry cost estimated to bring nonconforming stations into compliance with a 200 $\mu W/cm^2$ RFR standard is as much as $20.3 million, for example for FM broadcasters, the cost to the average nonconforming station could be as much $21,400. One observation that is evident from inspection of the above data is that, for most of the RFR levels investigated (but not all), it is the FM service that is responsible for almost half of the total industry cost. Generally, the AM service carries the least cost burden, based on the analysis.

**Table 3.** Summary of estimates of potential cash flow cost of the cost of compliance for the U.S. broadcast industry (FM radio, AM radio, VHF and UHF TV) for alternative RFR levels computed in terms of 1985 U.S. dollars (millions) for low to high cost scenarios. Adapted from Reference 9.

| VHF power density ($\mu$W/cm$^2$) | FM | AM | TV | Total |
|---|---|---|---|---|
| 10,000 | 3.0-5.1 | .2-5.4 | 0.7-1.2 | 6.9-11.6 |
| 5,000 | 3.1-5.3 | 3.2-5.6 | 0.7-1.2 | 7.0-12.1 |
| 2,000 | 3.4-6.4 | 3.6-6.7 | 0.7-1.2 | 7.7-14.3 |
| 1,000 | 4.0-8.2 | 33.6-6.8 | 0.8-1.3 | 8.4-16.3 |
| 900 | 4.1-8.5 | 3.6-6.9 | 0.8-1.3 | 8.6-16.7 |
| 800 | 4.3-9.0 | 3.6-6.9 | 0.8-1.3 | 8.6-16.7 |
| 700 | 4.4-9.4 | 3.6-6.9 | 0.8-1.3 | 8.9-17.7 |
| 600 | 4.8-10.4 | 3.7-7.0 | 0.8-1.4 | 9.3-18.9 |
| 500 | 5.2-11.6 | 3.7-7.1 | 1.0-1.7 | 9.9-20.4 |
| 400 | 5.6-13.0 | 4.2-8.9 | 1.1-1.8 | 11.0-23.7 |
| 300 | 6.6-16.0 | 4.2-9.0 | 1.3-2.1 | 12.2-27.1 |
| 00 | 8.0-20.3 | 4.5-9.8 | 1.7-2.7 | 14.2-32.8 |
| 100 | 10.9-29.2 | 4.7-10.5 | 3.4-5.4 | 19.0-45.1 |
| 75 | 12.0-33.0 | 4.8-11.0 | 4.2-6.8 | 21.0-50.8 |
| 50 | 14.0-39.4 | 5.0-11.6 | 6.2-10.5 | 25.2-61.5 |
| 20 | 20.8-61.2 | 5.6-13.8 | 13.9-40.5 | 53.7-135.6 |
| 10 | 26.1-79.2 | 6.3-15.9 | 21.3-40.5 | 53.7-135.6 |
| 1 | 59.9-149.1 | 10.9-31.5 | 89.8-233.9 | 160.5-414.6 |

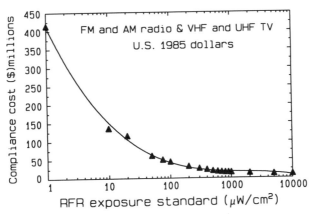

**Figure 2.** Estimated potential cost of compliance for the U.S. broadcast industry (FM radio, AM radio, VHF and UHF TV) for alternative RFR levels computed in terms of 1985 U.S. dollars (millions) for a high-cost scenario. Derived from Reference 9.

**Table 4.** Summary of estimates of potential cash flow cost of the cost of compliance for the average U.S. broadcast station (FM radio, AM radio, VHF and UHF TV) for alternative RFR levels computed in terms of 1985 U.S. dollars (thousands) for low to high cost scenarios. Adapted from Reference 9.

| VHF power density ($\mu$W/cm$^2$) | FM | AM | TV |
|---|---|---|---|
| 10,000 | 4.8-13.5 | 1.3-2.3 | 0.7-1.1 |
| 5,000 | 5.0-14.0 | 1.3-2.3 | 0.7-1.1 |
| 2,000 | 5.8-16.5 | 1.3-2.4 | 0.7-1.1 |
| 1,000 | 6.4-18.3 | 1.3-2.4 | 23.1-34.0 |
| 900 | 6.5-18.8 | 1.3-2.4 | 23.1-24.0 |
| 800 | 6.6-18.9 | 1.3-2.4 | 23.1-34.0 |
| 700 | 6.7-19.3 | 1.3-2.4 | 23.1-34.0 |
| 600 | 6.9-20.0 | 1.3-2.4 | 23.1-34.0 |
| 500 | 7.1-20.4 | 1.3-2.4 | 30.5-44.9 |
| 400 | 7.1-20.5 | 1.3-2.4 | 30.7-45.2 |
| 300 | 7.2-20.9 | 1.3-2.4 | 30.9-45.5 |
| 200 | 7.4-21.4 | 1.4-2.6 | 33.8-50.6 |
| 100 | 7.5-22.0 | 1.4-2.6 | 45.6-70.3 |
| 75 | 7.6-22.8 | 1.4-2.6 | 48.1-77.1 |
| 50 | 7.7-22.9 | 1.5-2.8 | 54.5-91.8 |
| 20 | 8.0-24.3 | 1.8-3.3 | 72.6-124.4 |
| 10 | 8.1-25.1 | 2.0-3.9 | 84.5-160.8 |
| 1 | 16.2-40.6 | 4.0-7.9 | 109.4-285.3 |

**Non-broadcast RFR Sources**

A much more difficult task, in terms of estimating potential economic impact, is determining the impact that new RFR standards might have on non-broadcast sources. Efforts to look into this matter are often fraught with considerable uncertainty, first, in accurately accounting for all potentially important sources, and, secondly, establishing reasonable compliance cost scenarios. This is especially so for government-operated sources, such as in the military, since impact on systems used in the national defense of a country may be viewed in a different light than less critical systems. Also, at least within the U.S., defense-related RFR sources, such as radars and other high-powered systems, are commonly evaluated for potential hazards in terms of the absolute worst-case possibility of the source causing a high RFR level. Because of this tendency to establish so-called "hazard zones" around such equipment, based on a "stopped beam" method for assessing power density at a point in the vicinity, it can often be the case that actual RFR exposures near the sources do not approach the theoretically predicted values or those measured under abnormal "stopped beam" conditions. These common practices can be very conservative and have a tendency to promote a heightened sensitivity on the part of system operators to the possibility that new RFR standards may impact their systems.

One study to investigate possible impact on government-operated RFR sources, again by the U.S. EPA,[10] provides useful information on this subject. The study, performed by the Electromagnetic Compatibility Analysis Center (ECAC) for EPA, used a government master file and Federal Communications Commission frequency assignments as the data base for the study. Emitters from these files meeting certain minimum power requirements were assembled into a reduced data base and subjected to mathematical models to determine the electromagnetic fields produced by the sources at various distances. Non-broadcast sources included the following categories: (1) satellite earth station main beam;

(2) earth station off-axis; (3) civilian fixed-beam radar; (4) civilian scanning beam radar; (5) military fixed-beam radar; (6) military scanning-beam radar; (7) other (10 kHz to >960 MHz). Detailed analysis results from this study indicated the number of frequency assignments found by ECAC that were potentially capable of producing various power densities at different distances from the sources.

As an example of this study, all unclassified radar frequency assignment records were examined to, first, identify those emitters located at geographically unique sites (i.e., to eliminate record duplication in the ECAC data base). Power densities for each emitter,

**Table 5.** Summary of analysis of number of unclassified radars potentially capable of producing greater than 0.5 mW/cm$^2$ at 100 m.

| Category | # of frequency assignments | # of unique sites |
|---|---|---|
| Civil fixed beam | 22 | 2 |
| Civil scanning | 0 | 0 |
| Military fixed beam | 12 | 7 |
| Military scanning | 5 | 5 |

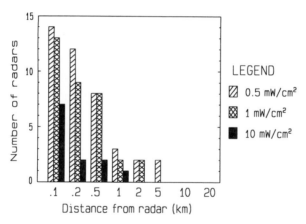

**Figure 3.** Estimated number of military fixed beam radars potentially capable of producing 0.5, 1 and 10 mW/cm$^2$ at various distances from radars in the U.S. and possessions.

prepared by the ECAC, were assessed to determine the number of such emitters that had the potential to produce electromagnetic fields exceeding a value of 0.5 mW/cm$^2$, one of several power density values being considered at that time as a possible environmental limit by the EPA, at a distance of 100 m from the source. The results of this analysis, for radars, is summarized in Table 5 below as adapted from a draft background information document prepared by the EPA.[38] The data for military fixed-beam radars indicate the highest potential power densities in their vicinity and these data are plotted in Figure 3.

While the data base provided to EPA by ECAC consisted of only unclassified frequency assignments, a separate analysis was conducted to estimate the effect of including both unclassified and classified emitters in the power-density analysis. This exercise resulted in the finding that an additional two sites with military fixed-beam radars and four sites for military scanning radars would be expected to be added to the numbers listed in the above table for the U.S.

The data on radars was also organized to indicate the maximum power densities that were projected to be possible at various distances from the radar and Tables 6, 7, 8 and 9 below summarize this way of viewing the data. Another analysis indicated the number of radar sites potentially capable of exceeding 0.5 mW/cm$^2$ at different distances from the source. These data are summarized in Table 10 below.

**Table 6.** Number of Civil Fixed Beam Radars Projected to be Capable of Producing Power Densities of 0.5, 1 and 10 mW/cm$^2$ at Various Distances.

| Distance (km) | 0.5 mW/cm$^2$ | 1 mW/cm$^2$ | 10 mW/cm$^2$ |
|---|---|---|---|
| 0.1 | 21 | 21 | 0 |
| 0.2 | 21 | 21 | 0 |
| 0.5 | 0 | 0 | 0 |

**Table 7.** Number of civil scanning beam radars projected to be capable of producing power densities of 0.5, 1 and 10 mW/cm$^2$ at various distances.

| Distance (km) | 0.5 mW/cm$^2$ | 1 mW/cm$^2$ | 10 mW/cm$^2$ |
|---|---|---|---|
| 0.1 | 0 | 0 | 0 |

**Table 8.** Number of military fixed beam radars projected to be capable of producing power densities of 0.5, 1 and 10 mW/cm$^2$ at various distances.

| Distance (km) | 0.5 mW/cm$^2$ | 1 mW/cm$^2$ | 10 mW/cm$^2$ |
|---|---|---|---|
| 0.1 | 14 | 13 | 7 |
| 0.2 | 12 | 9 | 2 |
| 0.5 | 8 | 8 | 2 |
| 1 | 3 | 2 | 1 |
| 2 | 2 | 2 | 0 |
| 5 | 2 | 0 | 0 |
| 10 | 0 | 0 | 0 |

**Table 9.** Number of military scanning beam radars projected to be capable of producing power densities of 0.5, 1 and 10 mW/cm$^2$ at various distances.

| Distance (km) | 0.5 mW/cm$^2$ | 1 mW/cm$^2$ | 10 mW/cm$^2$ |
|---|---|---|---|
| 0.1 | 10 | 7 | 0 |
| 0.2 | 7 | 0 | 0 |
| 0.5 | 0 | 0 | 0 |

**Table 10.** Summary of analysis of number of unclassified radars potentially capable of producing greater than 0.5 mW/cm$^2$ at various distances.

| Category | Distance from radar (km) | | | | | | |
|---|---|---|---|---|---|---|---|
| | 0.1 | 0.2 | 0.5 | 1 | 2 | 5 | 10 |
| Civil fixed beam | 2 | 2 | 0 | 0 | 0 | 0 | 0 |
| Civil scanning | 0 | 0 | 0 | 0 | 0 | 0 | 0 |
| Military fixed beam | 7 | 6 | 5 | 4 | 1 | 1 | 0 |
| Military scanning | 5 | 1 | 0 | 0 | 0 | 0 | 0 |

In conclusion, the above data illustrate that the total number of radar systems in the country, in 1982, projected to be capable of producing elevated microwave radiation levels in their vicinity is small, particularly when viewed in the context of the nation.

EPA also prepared an analysis of the number of sources that could potentially produce RFR in excess of one particular hypothetical exposure option defined by the following criteria:

If $F < 6$ MHz, then the RFR limit $= 2.65$ mW/cm$^2$

If $F \geq 6$ MHz and $F < 30$ MHz, then the RFR limit $= 95.5/F^2$ mW/cm$^2$

If $F \geq 30$ MHz and $F < 1$ GHz, then the RFR limit $= 0.106$ mW/cm$^2$

If $F \geq 1$ GHz, then the RFR limit $= 0.5$ mW/cm$^2$

For sources using highly directional antennas, such as earth stations and radars, RFR fields were calculated using a 30 dB sidelobe reduction factor; for shortwave communications sources, a 10 percent duty cycle was assumed. When the data base was analyzed in terms of the number of unique RFR sites capable of potentially exceeding the above exposure criteria, it was found that only 56 emitters would be expected to exceed the criteria at a distance of 100 m. The results of the entire analysis for all distances are summarized in Table 11.

**Table 11.** Summary of analysis of non-broadcast source RFR sites potentially capable of exceeding one particular set of RFR exposure criteria considered by the EPA Analysis assumes 30 dB sidelobe reduction for reflector antennas and 10% duty cycle for shortwave antennas.

| Source category | Distance from RFR source (meters) | | | | | | | |
|---|---|---|---|---|---|---|---|---|
| | 100 | 200 | 500 | 1000 | 2000 | 5000 | 10000 | 20000 |
| Civilian fixed beam radar | 2 | 2 | 0 | 0 | 0 | 0 | 0 | 0 |
| Civilian scanning beam radar | 0 | 0 | 0 | 0 | 0 | 0 | 0 | 0 |
| Military fixed beam radar | 4 | 4 | 2 | 2 | 1 | 1 | 0 | 0 |
| Military scanning beam radar | 0 | 0 | 0 | 0 | 0 | 0 | 0 | 0 |
| Other (10 kHz-100 kHz) | 0 | 0 | 0 | 0 | 0 | 0 | 0 | 0 |
| Other (100 kHz-30.6 MHz) | 5 | 2 | 1 | 0 | 0 | 0 | 0 | 0 |
| Other (30.6 MHz-960 MHz) | 32 | 8 | 1 | 1 | 0 | 0 | 0 | 0 |
| Other (>960 MHz) | 7 | 6 | 3 | 3 | 1 | 1 | 0 | 0 |
| Earth stations off-axis | 6 | 6 | 5 | 5 | 5 | 5 | 4 | 4 |
| Totals | 56 | 28 | 12 | 11 | 7 | 7 | 4 | 4 |

Given the qualifications listed above and the hypothetical RFR control levels stated earlier in equation form, this analysis showed a relatively small impact on non-broadcast sources. When the distance from the sources is increased, the potential for impact is substantially reduced. Exact interpretation of these results, however, is difficult. The amount of land controlled around each of the RFR sources considered was not available to the EPA and it is therefore unknown whether the production of RFR exceeding the particular levels used in the analysis at any specific distance would present a problem. Even if the land were controlled to the necessary distance, there would be no assurance that all persons entering the area would be workers and this would represent an additional complication for implementing any standard that distinguishes between occupational and general public exposure. A military base, for example, may be routinely visited by non-military personnel. Determining the cost of or practicality of modifications to non-compliant systems if land access restrictions were not possible would be a complex task. The best that can be concluded is that only a small number of RFR sources, representing an extremely small percentage of non-broadcast sites would likely be affected, even by relatively stringent guidelines on exposure.

## CONCLUSIONS

The advent of new RFR standards brings with it a requirement for lower field strengths, at least for parts of the RFR spectrum, than have been common in the past. The lower field requirements in and of themselves will pose a greater potential impact on operators of RFR sources, in particular broadcasters. Beyond more stringent limits on fields, added requirements to measure induced RF currents in exposed individuals will result in additional labor during on-site RFR hazard surveys. This added complexity of new RFR standards brings with it a greater potential for varied interpretation and, hence, more room for controversy over the exact meaning of a particular standards compliance study. Such interpretation can mean substantial legal fees for defending against suits over liability for safe operation of RF radiating systems. Operators of RFR sources are now having to put greater resources into dealing with the public over issues of RFR safety in the neighboring environment, in part, because of the more involved nature of newer RFR standards and public concerns driven by controversy over the possible adverse consequences of, for example, portable cellular telephones, police radar guns and VDTs. This heightened public sensitivity to RFR issues is leading to an increasing demand for environmental impact reports to be filed for projects requesting authorization to install new or modified RF systems.

New RFR standards do not necessarily imply greater impact; even though field strength limits can, in some instances, be more stringent than those in previous standards, relaxed averaging times can sometimes off-set the potential for impact. But in general, it must be concluded that newer RFR standards must pose at least some additional impact on source operators.

Organizations considering the development of new RFR standards should give serious thought to a proper balance between a scientifically defensible rationale and the ability of the standard to be practically implemented with a minimum of interpretation. Achieving such a balance will reduce the potential for impact on users of the spectrum.

## REFERENCES

1. IEEE Standard for Safety Levels with Respect to Human Exposure to Radio Frequency Electromagnetic Fields, 3 kHz to 300 GHz. IEEE C95.1-1991. Institute of Electrical and Electronics Engineers, Inc., 345 East 47th Street, New York, NY 10017 (1991).

2. ANSI Standard for Safety Levels with Respect to Human Exposure to Radio Frequency Electromagnetic Fields, 3 kHz to 300 GHz. ANSI/IEEE C95.1-1991. American National Standards Institute, Inc., 345 East 47th Street, New York, NY 10017 (1992).

3. Safety Code-6. Limits of Exposure to Radiofrequency Fields at Frequencies from 10 kHz-300 GHz. Minister of National Health and Welfare, Canada, Publication EHD-TR-160 (1991).

4. Guidance as to Restrictions on Exposures to Time Varying Electromagnetic Fields and the 1988 Recommendations of the International Non-Ionizing Radiation Committee. National Radiological Protection Board,UK, Report NRPB-GS11, May, p 22 (1989).

5. Maximum Exposure Levels - Radio-Frequency Radiation - 300 kHz to 300 GHz, Part 1. Australian Standard 2772.1-1990, published by the Standards Association of Australia, Standards House, 80 Arthur Street, North Sydney, N.S.W., Australia (1990).

6. Sicherheit bei elektromagnetischen Feldern Grenzwerte für Feldäen zum Schutz von Personen im Frequenzbereich von 0 bis 30 kHz Änderung 1. VDT 0848 Teil 4 A1, Deutsche Elektrotechnische Kommission im DIN und VDE (DKE), Germany, November (1990).

7. Regulation on Protection from EM Radiation. Issued by the State Environmental Protection Bureau, China, Document GB 8702-88 (1988).

8. Order of the Ministry of Health of the Czech Republic dated 3 October 1990 concerning the protection of health from the adverse effects of electromagnetic radiation, Czechoslovakia, No. 408/1990 Sb (1990).

9. C.H. Hall, An Estimate of the Potential Costs of Guidelines Limiting Public Exposure to Radiofrequency Radiation from Broadcast Sources. Technical report prepared for the U. S. Environmental Protection Agency by Lawrence Livermore National Laboratory under EPA-DOE Interagency Agreement A-89-F-2-803-0. Report No. EPA 520/1-85-025, UCRL Report 53562, July, 2 volumes (1985).

10. Support to the Environmental Protection Agency, Nonionizing Radiation Source Analysis. Prepared by the Electromagnetic Compatibility Analysis Center, Annapolis, Maryland 21401. ECAC Report ECAC-CR-82-093 prepared by F. H. Tushoph and V. P. Nanda, September (1982).

11. Cellular telephone radiation blamed for brain tumor, *Microwave News*, Vol. XII, No. 3, pp 1, 11, May/June (1992).

12. M. Ryan, Cellular scare could delay wireless-comm growth, FCC to revise EMF standards, *Electronic Engineering Times*, February 8 (1993).

13. J.J. Keller, McCaw to study cellular phones as safety questions affect sales, *The Wall Street Journal*, January 29 (1993).

14. ANSI. Safety levels with respect to human exposure to radiofrequency electromagnetic fields, 300 kHz to 100 GHz. American National Standard C95.1-1982, American National Standards Institute, September 1 (1982).

15. Regulations governing fixed facilities which generate electromagnetic fields in the frequency range of 300 kHz to 100 GHz and microwave ovens. 105 CMR 122.000, *Code of Massachusetts Regulation*, Vol. 2/3, No. 379. Effective October 1, 1983 (1983).

16. Nonionizing radiation frequency range from 10 kHz to 100 GHz. 440 CMR 5.00, Department of Labor and Industries, Commonwealth of Massachusetts. Recently redesignated 453 CMR 5.00 (1986).

17. An ordinance amending the Zoning Ordinance regarding radio and television transmission towers. *Ordinance No. 330*, Multnomah County, Oregon, adopted July 20, 1982, effective August 19, (1982).

18. Radio and television broadcast facilities ordinance. *Ordinance No. 160049*, the City of Portland, Oregon, adopted August 19, 1987, effective September 19, (1987).

19. O.P. Gandhi, I. Chatterjee, D. Wu and Y-G Gu , Likelihood of high rates of energy deposition in the humans legs at the ANSI recommended 3-30-MHz RF safety levels, *Proceedings of the IEEE*, Vol. 73, pp. 1145-1147(1985).

20. K. Jokela and L. Puranen, Theoretical and measured electric and magnetic field strengths around the dipole curtain antennas at the Pori short-wave station, Finnish Centre for Radiation and Nuclear Safety, P.O. Box 268, SF-00101 Helsinki, Finland (1988).

21. E.D. Mantiply and N.N. Hankin, Radiofrequency Radiation Survey in the McFarland, California Area. U.S. Environmental Protection Agency Report EPA/520/6-89/022, Office of Radiation Programs, Las Vegas, Nevada 89193, September (1989).

22. K. Tokushige, Y. Kamimura, Y. Yamanaka and Y. Shimizu, Measurements and estimations of electromagnetic field strength of radiation on the ground near a HF broadcast station. Transactions of the Japan Institute of Radio Engineers, Vol. 34, No. 173, December, pp. 211-220 (1988).

23. R.A. Tell and E.D. Mantiply, Population exposure to VHF and UHF broadcast radiation in the United States. *Proceedings of the IEEE*, Vol. 68, No. 1, January, pp. 6-12 (1980).

24. R.A. Tell and N.N. Hankin. Measurements of Radiofrequency Field Intensity in Buildings with Close Proximity to Broadcast Stations. U. S. Environmental Protection Agency, Office of Radiation Programs, Las Vegas, Nevada 89114. Report ORP/EAD-78-3, August. (1978).

25. An Investigation of Radiofrequency Radiation Levels on Lookout Mountain, Jefferson County, Colorado September 22-26, 1986. U. S. Environmental Protection Agency, Office of Radiation Programs, Las Vegas, Nevada 89114. Report prepared for the Office of Engineering and Technology, Federal Communications Commission, February (1987).

26. Radiofrequency Electromagnetic Fields and Induced Currents in the Spokane, Washington Area June 29-July 3, 1987. U. S. Environmental Protection Agency, Office of Radiation Programs, Las Vegas, Nevada 89193. Report EPA/520/6-88/008 prepared for the Office of Engineering and Technology, Federal Communications Commission, (June 1988).

27. An Investigation of Radiofrequency Radiation Levels on Healy Heights Portland, Oregon July 28-August 1, 1986. U. S. Environmental Protection Agency, Office of Radiation Programs, Las Vegas, Nevada 89114. Report prepared for the Office of Engineering and Technology, Federal Communications Commission, January (1987).

28. R.A. Tell, Induced Body Currents and Hot AM Tower Climbing: Assessing Human Exposure in Relation to the ANSI Radiofrequency Protection Guide. Report prepared by Richard Tell Associates, Inc., 6141 W. Racel Street, Las Vegas, Nevada 89131, for Federal Communications Commission, Office of Engineering and Technology, Washington, DC 20554, October 7, [National Technical Information Service accession number PB92-125186] (1992).

29. R.C. Petersen, Levels of electromagnetic energy in the immediate vicinity of representative microwave radio relay towers, *in:* Proceedings of the International Conference on Communications. Boston, Massachusetts, June 10, pp. 31.5.1-31.5.5 (1980).

30. R.C. Petersen, Microwave frequency radiation power densities from 4 GHz and 6 GHz radio relay, *in:* Proceedings of the International Conference on Communications. Boston, Massachusetts, June 10, pp. 31.1.1-31.1.3 (1980).

31. K.H. Joyner, V. Lubinas, M.P. Wood, J. Saribalas, and J.A. Adams. Radio frequency radiation (RFR) exposures from mobile phones, *in:* Proceedings of Eighth International Congress of the International Radiation Protection Association, Montreal, Canada, May 17-22, pp. 779-782 (1992).

32. Testicular cancer cluster among police radar users, *Microwave News*, Vol. XII, No. 2, pp.7-8, 10, March/April issue (1992).

33. R.C. Baird, R.L. Lewis, D.P. Kremer, and S.B. Kilgore. Field Strength Measurements of Speed Measuring Radar Units. National Bureau of Standards report NBSIR 81-2225 prepared for the National Highway Traffic Safety Administration, Washington, DC, May (1981).

34. P.D. Fisher, Microwave exposure levels encountered by police traffic radar operators. *IEEE Transactions on Electromagnetic Compatibility*, Vol. 35, No. 1, February, pp. 36-45 (1993).

35. EPA. Federal Radiation Protection Guidelines; Proposed Alternatives for Controlling Public Exposure to Radiofrequency Radiation; Notice of Proposed Recommendations. *Federal Register*, July 30, 1986, Vol. 51, No. 146, pp. 27318-27339 (1986).

36. P.C. Gailey and R.A. Tell , An Engineering Assessment of the Potential Impact of Federal Radiation Protection Guidance on the AM, FM and TV Broadcast Services. U.S. Environmental Protection Agency report EPA 520/6-85-011. Office of Radiation Programs, Las Vegas, Nevada 89193, April, [National Technical Information Service accession number PB85-245868] (1985).

37. R.W. Adler and S. Lamont , Numerical Modeling Study of Gain and Downward Radiation for Selected FM and VHF-TV Broadcast Antenna Systems. Report prepared for U.S. Environmental Protection Agency, Nonionizing Radiation Branch, Las Vegas, NV 89114, by AGL, Inc., P.O. Box 253, Pacific Grove, CA 93950. EPA technical report EPA-520/6-85-018, March 15 (1984).

38. EPA DRAFT Background Information Document for Radiation Protection Guidance to Limit Exposure of the General Public to Radiofrequency Radiation. Draft report prepared by the Office of Radiation Programs, U.S. Environmental Protection Agency, Washington, DC, December 6 (1983).

# HOW THE POPULAR PRESS AND MEDIA
# INFLUENCE SCIENTIFIC INTERPRETATIONS
# AND PUBLIC OPINION

Clifford J. Sherry

Systems Research Laboratories
P. O. Box 35505
San Antonio, TX  78235-5301

## TECHNICAL JOURNALS

Scientists and engineers tend to think of publication in terms of refereed journals. The editor of such a journal typically sends the manuscript to two or more referees. The referees, who are, theoretically, impartial experts in the field, evaluate the manuscript to determine if it is worthy of publication in the journal. The principal reasons for rejection are lack of originality and/or inappropriate methods and/or interpretations. If the referees recommend rejection, they must explain the problems with the manuscript and/or the experiment(s). The authors have an opportunity to respond to the criticisms and resubmit the manuscript.

## POPULAR MEDIA

It is sometimes difficult for scientists and engineers to understand that the components of the popular media, both print and non-print, are businesses, and the principal reason for their existence is to make a profit. In terms of 1988 dollars, the market value of the popular media was 480 billion dollars, that is, almost half a trillion dollars. Magazines, newspapers, and newsletters make a profit via circulation and advertisements, while book publishers make a profit by selling books. Television and radio stations make a profit selling commercial time.

In sharp contrast to refereed scientific journals, the final decision about what to publish in books, magazines, newspapers, and newsletters is in the hands of the editor, generally with input from fact checkers and attorneys. The role of these professionals is to help the publication avoid being sued. The typical reasons for litigation are libel, defamation, invasion of privacy, and plagiarism. It is important to note that these causes for action (libel, defamation, invasion) do not deal with truth, *per se*, but rather truth as it applies to a person and his or her reputation. Plagiarism is the use of someone else's words without giving them credit.

*Radiofrequency Standards,* Edited by B.J.
Klauenberg *et al.,* Plenum Press, New York, 1994

## PRINT MEDIA--BOOKS

Books are one form of print media. Traditionally, there are three classes of books: 1) trade books (paperbacks and hardbacks); 2) mass market paperbacks; and 3) reference, technical, and text books. Most publishers specialize in one or another of the these classes, although some large publishers may publish books in each category.

Trade books are sold in traditional book stores. Increasingly, trade books are also sold by other retailers (e.g., airport book stalls). These books are designed for lay audiences, and most of the nonfiction falls into the following categories: self-help, how-to, and biography. The decision to publish a manuscript is generally in the hands of a single individual or a small committee and is based on potential sales of the book and legal issues (see above). Fiction includes both mainstream and general (mysteries, romance, etc.).

Mass-market paperbacks are sold in traditional book stores, as well as in a variety of other retail outlets, such as discount stores, supermarkets, drugstores, etc. Relatively few original mass-market, nonfiction books are published. Those published fall into the categories described above. Most mass market nonfiction is a reprint of a trade hardback. The same criteria are used to determine whether to publish one of these books, original or reprint, as to publish a trade book (see above). Mass market fiction consists of trade reprints and some original fiction such as romances.

Reference books such as dictionaries, atlases, thesaurus are sold in traditional book stores. Specialized reference books, textbooks, technical books (including most scientific and engineering books), are typically not sold in traditional book stores. These books are designed for specialized audiences and thus, theoretically, there is more editorial scrutiny. But, the bottom line is still sales.

Desk-top word processors and high quality printers have led to the desk-top or self-publishing industry. Theoretically, anyone with the appropriate equipment can publish a book. The only other requirement is an outlet. Many self-published books are sold by direct mail. Self-published books have virtually no outside editorial scrutiny.

Some authors must pay to have their manuscripts published. This type of publisher is commonly called a vanity press. Obviously, there is no editorial scrutiny. Books published by vanity presses have few, if any, outlets.

Recent surveys suggest that the average adult does not obtain much information from books. Thirty percent of adults report that they have not purchased a book in the last year and 67% have not purchased more than one book. Nineteen percent have not read a book in the last year.

## PRINT MEDIA--PERIODICALS

The other forms of print media are periodicals. As with books, the decision of what to publish in a periodical is in the hands of a single individual or a small group of individuals and is driven by the desire for profits and to avoid litigation. The components of the print media--periodicals are listed in Table 1. Tabloids are not included in Table 1. Tabloids are generally sold at supermarket checkout counters. They cover controversial topics, such as UFOs, Bigfoot, sightings of Elvis or JFK, etc. The *National Enquirer*, the oldest and best known, has a weekly circulation of > 4 million, making it the 14th best selling periodical. *Star*, one of its competitors, has a weekly circulation of > 3.5 million, making it the 16th best-selling periodical. *Time, Newsweek,* and *U.S. News & World Report* are the 12th, 19th, and 23rd best selling periodicals, respectively. The *Wall Street Journal, USA Today, New York Times, Los Angeles Times, Washington Post,* and *Chicago Tribune* all have smaller circulations than the *National Enquirer*. Unfortunately, with the possible exception of tabloids, periodicals are also not a principal source of information for most adults in the

Unites States. Less than 50% of the adults read a newspaper every day and only 19% prefer reading to watching television. According to recent surveys, less than 29% of the words consumed by adults comes from print media.

**Table 1.** Components of the Print Media--Periodicals.

- Newspapers
    - 1,100 Afternoon dailies
    - 500 Morning dailies
- Magazines
    - 11,566 Consumer and Trade Magazines
    - 3,700 Technical and Professional Journals
- Newsletters
    - 13,500 Subscription Newsletters

## NON-PRINT MEDIA--RADIO AND TELEVISION

The components of the non-print media are shown in Table 2. Most adults (and children) obtain the bulk of their information from non-print sources. For example, a recent survey of 100,000 French school children reported that 7 out of 10 youngsters obtain most of their scientific information from watching television, rather than their school lessons.

**Table 2.** Non-Print Media.

- Radio Stations
    - 8,763 AM and FM Radio Stations
    - 214 News/Talk Format
    - 23 All-News Format
- Television Stations
    - 1,030 Television Stations
    - 8,500 Cable Television Systems

At least 98.2% of all households in the United States have a television set and 62.6% have 2 or more sets. On the average, Americans have their televisions sets on for 6 hours and 56 minutes each day. Less than 3% of all adults never watch television, while 25% spend 2 hours/day watching television and 12% spend 6 or more hours/day watching television. Watching television is the single most time-consuming leisure activity. Television provides 51.7% of all of the words consumed. Adults tend to listen to the radio during "drive time" (that is, the time spent driving to and from work), and radio provides 18.8% of all words consumed.

As with the print media, the decision of what to air on a radio or television show is in the hands of one (e.g., program director or station manager) or a small number of individuals. Again, as with the print media, the final decision of what to air is based on avoiding litigation and making a profit.

## POTENTIAL PROBLEMS WITH THE POPULAR MEDIA

There are a number of troubling changes in print media. The number of independent book publishers, especially the small, specialty publishers, is apparently decreasing. This is due to two phenomena--mergers and purchase of publishing companies by large conglomerates, such as the purchase of Simon & Schuster by Gulf & Western or Holt, Rinehart, & Winston by CBS. Second, although the number of bookstores has remained relatively stable, the number of independent, privately owned bookstores has been decreasing. They are being replaced by chains, such as Walden, B. Dalton, etc. Walden has more than 750 bookstores. Since competition for shelf space is fierce (often the shelf life of a book is several weeks or less), these chains have a good deal of power to determine what they will display and ultimately what is published. Taken together, these changes mean that the decision about which books to publish and which books to display are in the hands of fewer and fewer individuals.

Similar changes are occurring with periodicals. The number of daily newspapers has been decreasing steadily over the past decade. There is little or no competition in some markets, which means that the decision of what is "news" is in the hands of fewer and fewer people. Until recently the same individual or corporation could not own a newspaper and a radio and/or television station in the same market. This was changed by a recent Federal court decision.

Although not a new trend, many trade magazines have controlled circulation. This means that the publication is supplied to selected readers at little or no cost. The cost of publication and distribution is covered by advertisers. Therefore, advertisers potentially have a good deal of control over the editorial content of the magazine.

Desktop publishing has allowed virtually anyone with the appropriate equipment and outlets to publish a book or newsletter. These publications are published with virtually no editorial oversight, other than by the authors themselves. The newest trend in newsletters is to eliminate printing entirely. The newsletter publisher provides his editorial content to his customers via a computer bulletin board or downloads it directly to the customer's computer.

One problem with the non-print media is its impermanent nature. No matter how complex the content of the story, the viewer or listener can see or hear it once and then it is gone. It is not possible to review the content of a story once it is gone. Further, a typical story lasts for about 30 seconds and consists of 70-75 words. During this time period, the viewer's attention is divided between the auditory portion of the story and rapidly changing visual images. Few stories last for as long as three minutes (about 450 words).

Several recent developments in the non-print media also are potentially troubling. The number of radio stations stations devoted to "talk" has increased in the last decade. These stations feature listener "call in". That is, listeners can call in and express their views about the topic under discussion. The are a number of popular "call-in" shows that are carried nation-wide. On the surface, it seems that these call-in shows provide a relatively open forum for the discussion of important and potentially controversial issues. But, on most, if not all, of these shows, the calls are carefully screened. If a caller presents an unpopular viewpoint, the host can stop the discussion by going to a commercial, station break, or news update. Typically, the hosts of these shows tend to be either conservative or liberal-- often well to the left or right of center--and have little interest in the viewpoint of the other side. Thus, only one side of a controversy is presented. Often these nationwide shows provide a toll-free (800) telephone number. It is generally unclear who funds the costs associated with the 800 telephone number and why they provide the funds.

Talks shows and "magazine"-type shows have increased in popularity on television. The talk shows are typically presented in the late morning or early afternoon each weekday and each network has at least one, and sometimes several. Therefore, there is considerable competition for topics and guests. The format of these shows is fairly standardized. Several people with a specific "problem" and an "expert" talk about the "problem" with the host, audience, and in some cases, telephone call-ins. These shows generally provide a forum for controversial subjects. Unfortunately, often one side of a controversy is presented with little or no airing of opposing or alternate viewpoints.

The "magazine"-type shows appear to hold themselves out as presenting "hard" news. Each network has at least one "magazine" type show. These shows devote 15-20 minutes or more to a story and often present it as hard news. Unfortunately, they often go beyond the standard "who, what, where, when, and how" and blend news and opinion. Several recent cases suggest that their principal purpose is entertainment and presenting "hard" news is secondary. For example, *Dateline NBC* recently presented a story titled "Waiting To Explode," that dealt with the safety of a General Motors truck. According to the *Dateline* story, the GM truck tended to explode into flames when it was involved in a collision. According to reports in the press, overzealous staffers at *Dateline NBC*, eager to please a hierarchy intent on making the news "sizzle" (their words, no pun intended), planted tiny rocket engine ignitors on the trucks to cause the trucks to burst into flame on cue. GM objected to the story and, based on threats of litigation, NBC publicly apologized.

Entertainment shows, such as situation comedies and dramas, increasingly include "comments" about controversial subjects that do not have anything to do with the main story line of the program. For example, in the opening scene of a recent episode of *In The Heat Of The Night*, a popular detective drama based on characters created by John Hall, several of the supporting actors are examining a hand-held police speed detector. They comment back and forth about changing from radar- to laser-based detectors because radar detectors have caused cancer in a number of police officers. This scene had little or nothing to do with the main story line, but was an "editorial comment" inserted by someone in the hierarchy of the network or the program. It is likely that members of the audience will remember the statements about radar and cancer, but not necessarily remember the source of the statements, that is, an entertainment show rather than a news show.

Cable television companies are required to provide one or more community access channels. These channels provide a forum for virtually anyone. Because of concerns about First Amendment rights, there is little or no control over the editorial content of these shows.

## THE IMPACT OF POPULAR MEDIA ON SCIENCE

It is relatively difficult to determine how much impact the popular media have on science and science policy. One relatively objective method is to determine how often material that has appeared in the popular media (articles in newspapers, magazines, trade books, or discussed on television/radio) are cited in scientific articles that have appeared in refereed scientific/engineering journals.

The work of Mr. Paul Brodeur is a good example. He is a long-time staff writer for the *New Yorker* magazine and the author of articles and books that deal with a variety of relatively controversial and complex technical subjects. These subjects are as diverse as the biological and behavioral effects of microwaves;[1] health effects of asbestos;[2] and most recently the biological and behavioral effects of extremely low frequency (ELF).[3]

Besides being a well-published author, it is unclear what Mr. Brodeur's educational and professional background is, but he is well cited in both lay publications, where he is treated as an expert (e.g., Blumberg[4]) and in refereed scientific journals. According to a recent search of *Science Citation Index*, Mr. Brodeur's work dealing with asbestos received 4 citations in 1986, 14 in 1987, and approximately an equal number of citations in 1988, 1989, and 1990. It is important to note that the citing articles appeared in refereed scientific journals, such as *The American Journal of Public Health, American Journal of Industrial Medicine*; *Medicine*; *Japanese Journal of Physiology*, etc., while the cited articles appeared in *The New Yorker* and in trade books.

Mr. Brodeur's work dealing with microwaves and ELF has received fewer citations in the scientific literature, but they are striking. For example, in the discussion of a presentation by Wingren and Karlsson[5] at a symposium dealing with cancer in industrial countries, Davis makes the following comment, "In an earlier meeting just before this one in Bologna, Paul Brodeur reported that he thinks that the most likely factors are not electrical but magnetic, and that gauss meter readings are important..."

Or consider the comments of Dr. Pendleton Tompkins, the editor emeritus of the American Fertility Society Journal: "I now believe that electromagnetic fields are hazardous. If you also believe so after reading Brodeur, then I think it worthwhile to write an editorial in our journal to alert obstetricians to this new risk."[6] Or the comments of Carney[7] on the safety of magnetic resonance imaging. Even a journal as old and respected as *The Lancet* has not been immune (see Sibbison[8]). It is important to also read the responses to these comments (see Jauchem,[9,10,11]) respectively, as well as Ashley,[12] and Jauchem.[13]

## REFERENCES

1. P. Brodeur. "The Zapping of America: Microwaves, Their Deadly Risk, and the Cover-Up," Norton, New York (1977).
2. P. Brodeur, Annals of law: the asbestos industry on trial., *The New Yorker.*, June 8, 15, 22, and July 1, (1985).
3. P. Brodeur, Annals of radiation, *The New Yorker,* June 12, 19, 26 (1989); July 9, (1990); December 7, (1992).
4. P. Blumberg, Paul Brodeur's war on electromagnetic fields, *Washington Journalism Review,* 40-44, January/February (1991).
5. G. Wingren and M. Karlsson. "Mortality Trends for Leukemia In Selected Countries. Trends in Cancer Mortality in Industrial Countries," *in:* D. L. Davis and D. Hoel, eds., Annals Of The New York Academy of Sciences, New York, 609, 280-88 (1990).
6. P. Tompkins, Hazards of electromagnetic fields to human reproduction, *Fertility and Sterility,* 53, 185 (1990).
7. A.L. Carney, Magnetic resonance imaging (MRI): Is it safe? *Clinical Electroencephalography* 20, XI (1990).
8. J.B. Sibbison, USA: Danger from electromagnetic fields, *The Lancet ,* 336, 106 (1990).
9. J.R. Jauchem, Hazards of electromagnetic fields to human reproduction: What information is in the scientific literature, *Fertility and Sterility,* 54, 955 (1990a).
10. J.R. Jauchem, Magnetic resonance imaging: Is it safe? *Clinical Electroencephalography*, 21, IX-XI (1990b).
11. J.R. Jauchem, Electromagnetic fields: Is there a danger.? *The Lancet,* 336, 884 (1990c).
12. J.R. Ashley, Book Review (Currents Of Death), *IEEE Antennas and Propagation Society Magazine,* 32, 45-48 (1990).
13. J.R. Jauchem, Epidemiologic studies of electric and magnetic fields and cancer: A case study of distortions by the media, *J. Clin. Epidemiol* 45, 1137-42 (1992).

# IMPACT OF PUBLIC CONCERNS ABOUT LOW-LEVEL
# ELECTROMAGNETIC FIELDS (EMF) ON INTERPRETATION
# OF EMF/RADIOFREQUENCY (RFR) DATABASE

John M. Osepchuk

Raytheon Research Division
131 Spring Street
Lexington, MA 02173

## ABSTRACT

A review of the history of setting standards for safe exposure to electromagnetic energy shows a strong reliance on science and the associated finding of thresholds for effects and hazards. The reliance on science has been attacked by a number of journalists, historians and social scientists over the last 15 years resulting in calls for "prudent avoidance" and the abandonment of reliance on established science. Accompanying this trend has been much misinformation and miseducation by the media leading to *electrophobia*. There are strong indications, however, that society is beginning to realize that emergent technologies and economic well-being are dependent on rational science-based standards. Thus, in the development of new standards, the need for two tiers of exposure limits in response to public concerns is justified only in an <u>uncontrolled environment</u> and only in those areas of the spectrum where a significant uncertainty exists about the scientific database. This resolves down to environmental exposures only, the frequency range around the resonance of the human body, and for long-term exposures only. Whether this two-tier philosophy should be extended to the extremely low frequency (ELF) range is debatable.

## HISTORICAL REVIEW OF STANDARDS

Electromagnetic (EM) energy is a concomitant of the modern technologies of electricity and electronics that are at the base of today's advanced standard of living. The side effects of EM energy include radiofrequency interference (RFI) and potential hazards of bodily exposure. RFI occurs at low levels of EM energy whereas bodily hazards occur only at much higher levels of exposure. Thus RFI standards involve low-level limits. Similarly, <u>emission</u> limits[1] that protect consumers from exposure to EM energy that "leaks" from electronic products also involve low-level limits of exposure--i.e., a very high safety factor. Safe exposure limits for humans generally have been based on applying a reasonable safety factor, e.g., 10-50, between thresholds for injury and the exposure limits. As shown in Figure 1, the conventional and worldwide basis for exposure-standards setting is the existence of some respected or authoritative body that periodically revises the standards . (This is analogous to periodic decisions of a court or jury.) These standards are based on threshold effects described in existing science. Thus in contradistinction to standards for safe exposure to ionizing radiation and some chemicals, exposure below limits for exposure to EM energy is <u>safe</u> and should be of no concern to people.

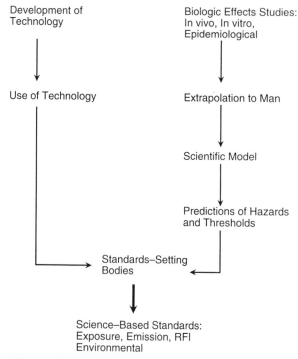

**Figure 1.** Normal Progression of EM Technology Assessment..

Standards have been developed for 25 years predominantly above 3 kHz up to the lower reaches of infrared (300 GHz). Some standards, however, were extended down to direct current (DC). An informative review of worldwide standards was recently presented by Petersen.[2] Perhaps the most detailed of such standards is the ANSI/IEEE C95.1-1991/1992 standard.[3] It also represents the broadest consensus of the scientific community. Important details include: frequency dependence; averaging time; relaxation for partial body exposure; and a two-tier set of limits in a band centered around the resonance frequency range for man, 30-300 MHz. The role of averaging time is depicted in Figure 2. We see that for exposure times greater than the averaging time, the limits are in terms of power, power density ($mW/cm^2$) or specific absorption rate (SAR). For periods of exposure less than the averaging time, the limits are in terms of energy, energy density ($mW \cdot hour/cm^2$) or SAR. In this standard the two tiers apply only for periods of exposure greater than the averaging time. The upper and lower tiers are labelled <u>controlled</u> and <u>uncontrolled</u> environment, respectively. Although elsewhere some interpret the two tiers as applying to <u>occupational</u> and <u>general public</u>, in the C95 standard this is not so. The distinction between controlled and uncontrolled environment in the C95 standard relates more to the voluntary (informed) and involuntary (uninformed) natures of exposures in the two environments, respectively.

Although standards have been developed for safe exposure in the ELF range, in the U.S. these have been stymied by the concept of "prudent avoidance" (PA) associated with the current fear of a link between cancer and exposure to ELF fields.[4] We have argued that PA is tantamount to the abandonment of science. Exponents of PA like to say that "the jury is out." We believe "juries" (scientists) have spoken on ELF safe limits. Proponents of PA in effect reject this existing science. We have argued[5] that rational public policy should be based on a full-spectrum approach and related to existing science--i.e., what is known and not what is unknown.

## ROLE OF "PUBLIC" IN STANDARDS SETTING

As Figure 1 suggests, traditional and perhaps the preferred mode of setting standards

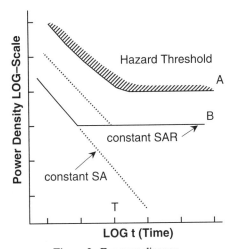

**Figure 2.** Exposure diagram.

involves an accepted authoritative body that interprets science to derive the standards. The purpose has been and is to protect humans from harm by setting exposure limits with a reasonable safety factor. We say reasonable in that in a rational society it is desirable not to impose unduly stringent exposure limits that could hamper technology, the economy, and the freedom to explore beneficial uses of EM energy.

Beginning in the 1970s, perhaps in parallel with analogous anti-science trends in other areas of technology, this science-based standards rationale has been challenged by a number of journalists and social scientists. Paul Brodeur[6] in the late 70s popularized the idea that the establishment is covering up the hazards of microwaves and that people should find the truth from him and the scientists he espoused. In 1984, under a government grant, the historian Steneck[7] criticized the C95 standards-setting process as follows:

> ... The scientific bias of all of ANSI's RF standards activities is clear . . ..
> C95 members could have spent their time discussing how low standards
> could be set without infringing on economics development instead of
> figuring out how standards could be set to avoid injury . . .. The lack of
> conservatism assumed in C95.1-1982 is aggravated by its strict scientific
> bias (a value).

This was later followed by the "risk communication" concept[8] of Sandman wherein "risk = hazard + outrage." Therefore, if "outrage" dominates over real "hazard," then it could be that science is irrelevant and some type of accommodation with those "exposed" is required using the art of "risk communication."

Finally, the 1989 OTA Report[4] and related documents precipitated a media campaign spreading fears of all types of ELF field sources at low levels of exposure--e.g., from distribution lines. In this process existing standards were ignored and the concept of "prudent avoidance" was born.

We have argued[9] against basing public policy on PA or "risk communication" and urged a restoration of policy based on existing science-based standards.

The proponents of PA say, "The jury is still out."

We say, "Juries have spoken and will continue to speak out."


## MEDIA MISINFORMATION AND PUBLIC EDUCATION

Proponents of irrational views on non-ionizing EM energy have tried, over the last 25 years, to link the latter with society's genuine concerns about particular, highly potent chemical agents or with ionizing radiation. The myths and misconceptions about microwaves are legendary, and we have touched[10] on some of the main ones. The origins of Electrophobia are the existence of controversial or even weak science[11] that fuels the appetites of the media for sensation and the lawyers for litigation. The existence of standards is anathema to those who exploit Electrophobia, and it is not uncommon for a journalist to dismiss standards as worthless based on his "expert" assessment.

The legitimate scientific community has fought back through activities of the Institute of Electrical and Electronics Engineers (IEEE) Committee on Man and Radiation (COMAR) and organizations like the Electromagnetic Energy Policy Alliance (EEPA). Available from these groups are fact sheets and detailed rebuttals of journalists such as Currents of Death: Rectified[12] by Eleanor Adair. In reviewing the history of this "propaganda" war it is interesting to note that the science heroes of some journalistic works of the 70s are no longer in fashion and new names have appeared. Likewise the buzzwords of the 70s, like "blood-brain barrier" and "microwave sickness", have largely disappeared in the writings of scientists and even journalists.

Though the actions of COMAR and EEPA help, the problem of public education remains. Not only must complicated science be assessed and explained, but even more fundamental problems remain. The language and terms used by the media and even scientists to explain EM questions to the public are often ambiguous, illogical, contradictory, and confusing--e.g., the use of terms like "energy," "EMF," "exposure," "power density," etc. We have complained[9] about this and suggested corrective steps. An example of a positive response to this appeal is the revised educational brochure by Yost.[13]

Closer to home is the fact that our standards, particularly the newer ones, are not well understood. The occasions on which misunderstanding of "averaging time" occur are multiple.

Lastly, there are the psychological overtones of fear associated with words like "radiation" or even "microwaves" that make it difficult to present a neutral rational picture of EM questions to many in the general public.


## REVERSAL OF TRENDS IN PUBLIC VIEWS ON TECHNOLOGY

Over the last 25 years Electrophobia developed in spirit with the prevailing trend of anti-technology, anti-establishment, and environmentalist (green) views. Now there is the beginning of the reversal of this trend--perhaps encouraged by world-wide economic crises and a popular realization that modern standards of living are slipping. For example, the scientist, Dixey Lee Ray, in 1990, voiced the need to reverse this trend in her book, Trashing the Planet.[14] Now prominent science journalists continue the theme for trend reversal. Ronald Bailey in Eco-Scam[15] and Michael Fumento in Science Under Siege[16] decry past and present scares about Alar, dioxin, Love Canal, and even EMFs, and call for a return to views based on hard science. Even an environmentalist, Martin Lewis, in Green Delusions[17]) has written a critique of the excesses of radical environmentalism. (See also Reference 18.)

Ron Petersen and I have written a paper[19] pointing out the critical role of rational standards in emerging technologies, ranging from near-range cellular radios and PCN (personal communication network) devices to more distant microwave power transmission and comfort heating technologies.

It is time to balance Steneck's move to protect the public from "possible" hazards of technology by making standards more realistic and science-based, so that the benefits of technology and accompanying economic benefits are not denied all of us as citizens and human beings. The jobs manufacturing police radar in small companies in Kansas need to be protected against irrational fears of microwaves and ill-based litigation--so reasoned a recent newspaper editorial.[20] Recently, benefits to a church that would accrue by renting space in their tower (steeple) for a cellular radio transmitter were denied (at least temporarily) by opposition from those in the surrounding communities even though church members accepted this exposure as safe.[21] This is a disturbing example of many cases where benefits for a segment of the public are denied by the blind opposition of a more distant public (when a land owner or town accepts a technological installation).

In the more distant future there will be more proposals for use of microwaves in beneficial power applications. The Soviet physicist Kapitsa is reported[22] to have said:

> It is worth noting that, before electrical engineering was pressed into service by power engineering, it was almost exclusively occupied with electrical communication problems (telegraphy, signaling, and so on). It is very probable that history will repeat itself: At present, electronics is used mainly in radio communications, but its future lies in solving major problems in power engineering.

In this future society those in the general public who elect to use microwaves for comfort heating[23] should not be denied this benefit.

Clearly, there is a recognition to protect the richness of future society for succeeding generations that could not arrive if our society remains stifled by fear of technology including EM energy.

## THE COMPLEX SOCIETAL BACKGROUND OF
## MODERN STANDARDS-SETTING

In Figure 1 we depicted the conventional process of standards setting as if the standards-setting body is an authority or at least accepted by society as an authority. In this view, then, standards are developed somewhat in private by experts, much as one expects a report by the College of Cardinals to be prepared. In fact, standards setting for safe exposure to EM energy now is done in the midst of a complex set of interrelationships among various interested parties in modern society. This is depicted in Figure 3.

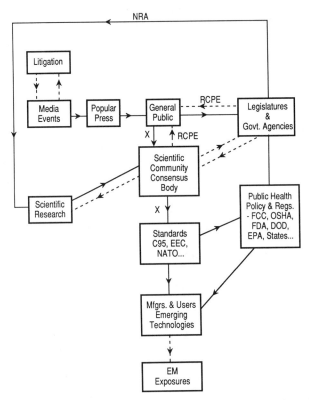

**Figure 3.** Diagram of interacting relationship of principal players in the process of standards developed for protection of humans from effects of EMF/RFR: US Model.

Most of this diagram is self-explanatory, but a few comments are in order. First of all, this is primarily a U.S. model in which a <u>broad</u> <u>consensus</u> body such as the SCC 28 committee and subcommittees of the IEEE represent the whole scientific community. We can hope that the activities within EEC and NATO may also be based on a broad consensus.

The principal interrelationships are denoted by solid arrows, and weak secondary relationships are marked by dashed arrows. The arrows marked by RCPE denote the processes of "risk communication" and "public education." The arrow marked NRA refers to the National Research Agenda in the U.S. The interrelationship between legislatures and the scientific consensus body is realistically assessed as weak. Although scientific research is a prime input to this body, at present little influence on scientific research is achieved by the consensus body for the purposes of enhancing future standards (i.e., increasing applied research targeted for settling of questions encountered in the standards-setting process.)

The reason for showing this chart is not only to illustrate the complex background in which we operate, but also to focus on what steps relate to the subject of this talk. The arrows marked X relate to how public concerns may affect our "interpretations of the EMF/RFR database" and how we may set standards--i.e., in particular, the two-tier structure. The upper arrow X relates to the influence of social scientists (or the C95 Working Group on Societal Aspects) on the consensus body, and the lower X relates to the actual way two tiers are incorporated into the standard.

In the U.S. two tiers were introduced for the first time in ANSI/IEEE C95.1-1991/1992, but with certain limitations designed not to inhibit beneficial uses of EM energy among the general public, especially in the future.


## REVIEW OF THE NEED FOR TWO TIERS

Prior to 1983 no US standards incorporated two tiers of safe exposure limits. Then the Commonwealth of Massachusetts public health department (DPH) issued a regulation[24] limiting exposures of the "general public" to long-term limits (MPE), one fifth (1/5) of the MPE's in ANSI C95.1-1982. Later the Massachusetts Division of Labor adopted ANSI C95.1-1982 as an occupational exposure standard. Thus a two-tier system was adopted for the first time. Note, however, that the averaging time for the lower tier was 30 minutes, so the two tiers are identical for short exposure times of less than 6 minutes. The reason for adopting the lower tier was a recognition of public concern about "long-term chronic" exposures in the environment, especially from radar stations in various towns in Massachusetts. This public concern was stimulated by journalists such as Brodeur and rumors about the microwave irradiation of U.S. embassy in Moscow. Even though the DPH regulation was called a regulation on exposure of the general public, it primarily was intended as an environmental limit.

In 1986 the U.S. National Council on Radiation Protection and Measurement (NCRP) issued a report[26] with two-tier exposure limits essentially identical to those adopted by Massachusetts. The NCRP rationale placed more emphasis on potential differences of susceptibility to injury among the general public vs. an occupational personnel population. As in the Massachusetts standard, however, a prime factor supporting the lower second tier was the fact that the general public may be exposed 168 hours a week to a transmitter in the environment, whereas occupational exposures at most are 40 hours a week. The difference is roughly a factor of five, which exists between the two tiers.

Thus in the past the main reason for two tiers was the fact of continuous environmental exposure to fixed-site transmitters such as radar stations or broadcast transmitters. Another factor underlying the Massachusetts regulation is the recognition that generally environmental exposures are involuntary. Even if it is safe, should a home owner, in principle, be allowed to irradiate his neighbor (including children) at occupational exposure levels. Note that occupational exposure levels in some cases (e.g., above X band) can be sensed and in other cases (e.g., in the VHF range) are large enough to cause severe "incidental interference" in home electronics and even burn out some electronic circuits.

In both the Massachusetts and NCRP regulations, it is recognized that even Eastern Europe and the West do not greatly dispute safe exposure levels over short periods of time. There is more argument about the desirable long-term limits. This debate, to the degree it has any validity, could be cited as justification for a lower tier MPE for the general public.

If the MPE is truly only used for an environmental limit, the averaging time concept may be somewhat academic--e.g., home electronics are more likely to be burned out if environmental levels are allowed to rise high above long-term MPEs for short periods of time. On the other hand, if the lower tier is there only to protect from bodily effects of exposure to RF radiation, then it does make sense to retain the time-averaging.

In the development of ANSI/IEEE C95.11-1991/1992, a special working group studied the two-tier question. They concluded, and the SCC 28 Subcommittee 4 agreed, that two tiers are needed only around the resonance range for man because that is where some uncertainty exists about long-term chronic exposure effects. Therefore, there should be no difference in the tiers as one approaches the infrared range or DC.

On the other hand, even in the resonance frequency range, it was felt wrong to suggest that voluntary exposure of the general public is hazardous and to be discouraged. For example, radio amateurs and even possibly users of cellular phones should not be discouraged from higher levels of exposure if it is associated with beneficial improvement in performance of their equipment and intended functions. In the future the voluntary exposures in the Pound heating system should not be discouraged if, in fact, they are well below hazardous levels and, in particular, safe limits for occupational exposure (i.e., upper tier). Then there is a secondary argument that occupational personnel may resent a higher MPE than that imposed on people not occupationally exposed.

For all these reasons the concepts of controlled and uncontrolled environment were introduced to describe the different exposure situations for the two tiers.

The definitions, as written in the new ANSI/IEEE,[3] of controlled and uncontrolled exposures are as follows:

> Controlled Environment. Controlled environments are locations where exposure may be incurred by persons who are aware of the potential for exposure as a concomitant of employment, by other cognizant persons, or as the incidental result of transient passage through areas where analysis shows the exposure levels may be above those shown in Table 2 but do not exceed those in Table 1, and where the induced currents may exceed the values in Table 2, Part B, but do not exceed the values in Table 1, Part B.
>
> Uncontrolled Environment. Uncontrolled environments are locations where there is the exposure of individuals who have no knowledge or control of their exposure. The exposures may occur in living quarters or workplaces where there are no expectations that the exposure levels may exceed those shown in Table 2 and where the induced currents do not exceed those in Table 2, Part B. Transitory exposures are treated in 4.1.1.

Some may cite the recent proposals for "prudent avoidance" at low frequencies as a reason for adopting a lower tier for the general public at any frequency. At low frequencies in the absence of any beneficial effects of the exposure or unavoidable exposure during a beneficial operation like cellular phones, it may make sense to adopt two tiers. But even here we argue that the concepts of controlled and uncontrolled environment make sense.

There are some that may take the extreme position at any frequency that until we are certain about the science and our degree of safety factor, we should set no standards or urge or even mandate reduction of all exposures to as low as reasonably achievable (ALARA). Clearly, this would open Pandora's box, stifle industry, cause an increase in public anxiety, and increase ill-based litigation.

There are those who would argue that uncertainties demand a second tier. We would agree only where these are significant and only if we distinguish voluntary and involuntary exposures through the use of concepts like those used in ANSI/IEEE C95.1-1991/1992.

Instead of just talking about vague uncertainties, let us look at how the C95 standard has changed in 25 years. In Figure 4 we show the capsule guide for

ANSI/IEEE C95.1-1991/1992. We see there are many complex details in terms of frequency dependence of MPE's, averaging times, induced current limits, SAR limits, etc. Not shown are the complex rules on low power exemption, partial-body exposure relaxation, peak power limits, and measurement conditions. The complication of two-tiers is apparent in Figure 3.

Now look at ANSI C95.1-1966 as depicted on the same capsule guide chart as used in Figure 4. This is presented in Figure 5. Very simple! One MPE and one averaging time. No frequency dependence or any of the other complications in the 1991/1992 standard.

In Table 1 we tabulate the additional features in successive standards adopted in 1974, 1982, and 1991 that changed C95 from that in Figure 5 to Figure 4.

**Table 1**. New features in succeeding C95 standards.

| 1966 | **MPE:** 10 mW/cm$^2$ |
|---|---|
| | **Averaging Time:** 6 minutes |
| | **Frequency Range:** 10 MHz to 100 GHz |
| | |
| 1974 | Revised wording of text. |
| | |
| 1982 | Introduce SA, SAR dosimetry rules and frequency dependence of MPE. |
| | Extend frequency range to 300 kHz. |
| | Introduce rules on measurement conditions. |
| | Introduce low-power and SAR exemption rule. |
| | |
| 1991 | Extended frequency range to 3 kHz to 300 GHz. |
| | Relax H-field limits at low frequency. |
| | Introduce induced current limits. |
| | Reduce averaging time at high frequencies. |
| | Introduce two tiers of MPE's in resonance range. |
| | Introduce temporal peak exposure limits. |
| | Introduce frequency dependence in low power exemption rules. |
| | Revise measurement conditions. |
| | Introduce rules for relaxation of limits under partial-body exposure conditions. |

Sometimes these changes led to more stringent MPEs like the drop in the resonance range in the '82 standard. Sometimes the MPE was relaxed as in the low frequency H limit (MPE) in the '91 standard. Sometimes the change was more subtle though not less important in improving safety. An example of this is the reduction of averaging time at high frequencies (e.g., 100 GHz) in the '91 standard. This eliminated the defect in the '82 standard of permitting exposure conditions close to burn thresholds for short periods of time. Now averaging time at 300 GHz matches those derived in infrared laser standards-- i.e., ~10 seconds. Already we see inadequacies in IEEE C95.1-1991, e.g., the inappropriate caveat on exposure of eyes and testes and excessive averaging time for H fields at low frequencies.

It is apparent that each standard is imperfect, but each succeeding revision becomes more exact in its correspondence to scientific validity. Does this mean we would have been better off with no standard in the past? The answer is No! Each standard had a beneficial effect of imposing concern and responsibility on manufacturers and users. The ample safety factor mitigated against the eruption of injuries in practice.

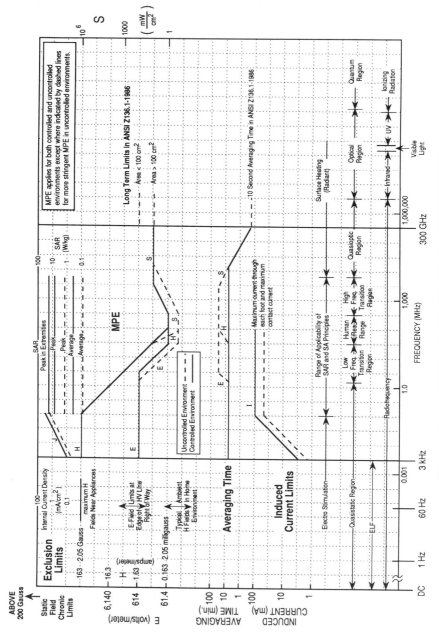

**Figure 4.** A Capsule Guide to the Final Draft Revision: IEEE C95.1-1991.

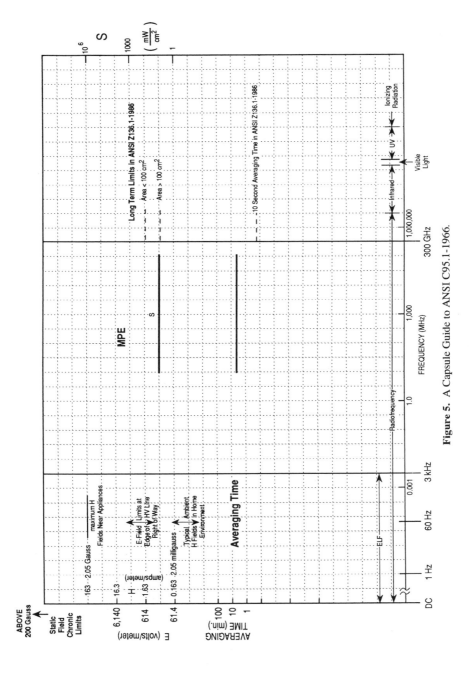

**Figure 5.** A Capsule Guide to ANSI C95.1-1966.

As in all aspects of life, we have the choice of not moving at all in fear because of the alarmists, or we can encourage intelligent growth and exploration of uses of EM energy with a rational safety standard.

We should adopt a lower tier only in cases of involuntary exposures at frequencies where there is significant scientific controversy as judged by a scientific consensus body such as IEEE SCC 28.

## CONCLUSIONS

We can see that public concerns have led to the adoption of two tiers of safe exposure limits for EM energy. We believe the concepts of <u>controlled</u> and <u>uncontrolled</u> environments in the C95 standard are a good model for further development. It is hoped that international harmonization with EEC and NATO will be encouraged through communication and cooperation between the various communities. Increased financial support of the standards community is also desirable in order to achieve a broad consensus in a timely manner.

## REFERENCES

1. Public Law 90-602, Radiation Control for Health and Safety Act of 1968, Approved by the U.S. Congress (October 18, 1968).
2. R.C. Petersen, Radio frequency/microwave protection guides, *Health Physics,* 61:59-67 (1991).
3. IEEE, IEEE Standard for Safety Levels with Respect to Human Exposure to Electromagnetic Fields, 3 kHz to 300 GHz, IEEE C95.1-1991, Institute of Electrical and Electronics Engineers, New York, 1992. Approved by ANSI (1992).
4. I. Nair, M.G. Morgan ,and H.K. Florig, Biological Effects of Power-Frequency Electric and Magnetic Fields, Report OTA-BP-E-53, U.S. Government Printing Office, Washington, D.C. (May 1989).
5. J.M. Osepchuk, The need for science-based standards in EMF policy, *in:* "Proc. of the First World Congress on Electromagnetics in Biology and Medicine," Orlando, FL ( June 1992).
6. P. Brodeur. "The Zapping of America," New York: William Morrow (1978).
7. N.H. Steneck. "The Microwave Debate," Cambridge, MA: MIT Press, 234-235 (1984).
8. P.M. Sandman, Risk communications: facing public outrage, *EPA Journal,* 21-22, (November 1987).
9. Q. Balzano, J.A. Bergeron, J.M. Osepchuk, and R.C. Petersen, On the need for a full-spectrum EMF policy: reliance on science-based standards for the safe use of electromagnetic energy, *Forum* (in press, 1993).
10. J.M. Osepchuk, Some misconceptions about electromagnetic fields and their effects and hazards, *in:* "Biological Effects and Medical Applications of Electromagnetic Energy," O.P. Gandhi, ed., Englewood Cliffs, New Jersey: Prentice Hall, ch. 20 (1990).
11. K.R. Foster and W.F. Pickard, The risks of risk research, *Nature,* 330:531-532 (1987).
12. E.R. Adair, Currents of death: rectified, Paper commissioned by IEEE-COMAR (1990).
13. M. Yost. "Non-Ionizing Radiation Questions and Answers," San Francisco: *San Francisco Press* (1993). (Revision of 1988 version.)
14. D.L. Ray. "Trashing the Planet," New York: Regnery Gateway (1990).
15. R. Bailey. "Eco-Scam.," New York: St. Martins (1993).
16. M.Fumento. "Science Under Siege," New York: Morrow (1993).
17. M.W. Lewis. "Green Delusions." Durham, North Carolina: Duke University Press (1992).
18. L.L.Schloesser, Green guidelines, *Reasons,* 33-34 (April 1993).
19. J.M. Osepchuk and R.C. Petersen, The role in emerging technologies played by standards for exposure to electromagnetic energy, *IEC* (1993, in press.)
20. Editorial, Jury should be commended, *Wilson County Citizen.* Fredonia, Kansas (January 25, 1993.
21. F. Burgos, Cellular radiation safety is a hot issue in Wilmette, *Chicago Sun-Times* (March 11, 1993).
22. V.A. Vanko, V.M. Lopukhin, and V.L. Sarvin, Satellite solar power systems, *Sov. Phys. Usp.,* 20(12): 989-1001 (December 1977).
23. R.V. Pound, Radiant heat for energy conservation, *Science,* 208:494-495, May 2, 1980.
24. Massachusetts Department of Public Health, "105 CMR 122.000, Fixed Facilities Which Generate Electromagnetic Fields in the Frequency Range of 300 kHZ to 100 GHz and Microwave Ovens," (1983).
25. NCRP, "Biological Effects and Exposure Criteria for Radio Frequency Electromagnetic Fields," National Council on Radiation Protection and Measurements, Bethesda, Maryland, NCRP Report No. 86 (1986).

# COMMUNICATING RISK OF ELECTROMAGNETIC FIELDS/ RADIOFREQUENCY RADIATION (EMF/RFR)

B. Jon Klauenberg[*]

Radiofrequency Radiation Division
Occupational and Environmental Health Directorate
U.S. Air Force Armstrong Laboratory
Brooks Air Force Base, Texas 78235-5324

## POLICY IMPACT ON PUBLIC

In general, the public has become very risk conscious, believing that it is exposed to more risks today than in the past and that it will encounter more in the future.[1] Media accounts and inflammatory headlines have angered and frightened the public. Since it is not newsworthy to report that the sky is *not* falling, viewpoints suggesting that the readership may be in danger are highlighted.

The public has been warned to be "prudent" and to avoid sources of EMF when the individual cost of doing so is acceptable. Policy proposals of "Prudent Avoidance,"[2] intended as guidance for individual decision making, have been incorporated into proposals for governmental regulation. "Aggressive Prudent Avoidance,"[3] concerning utilities, and "Super Prudent Avoidance,"[4] concerning cellular telephones, are but a few that have been forwarded. That government agencies have proposed such policy[2] has heightened public anxiety and has contributed to regulatory action at the local level. Electric power companies have been ordered to redesign and reroute transmission lines and manufacturers of household appliances are introducing devices that emit reduced EMF. These are but a few of many policy actions directed at controlling EMF/RFR emissions that have already impacted the public.

## SETTING POLICY IN THE FACE OF UNCERTAINTY

How can policy be established in view of large uncertainty in the science? Examination of the current policy climate suggests that conservative assumptions, differing policy

---

[*] These views and opinions are those of the author and do not necessarily state or reflect those of the U.S. Government.

standards, and public perceptions markedly influence risk assessment, risk management, regulatory decisions, and ultimately public health policy. Policy judgments continue to routinely influence risk assessment and scientific design and analysis decisions.[5] While scientists tend to demand solid evidence before acceptance, public health officials often adopt a standard of taking action on the basis of suggestive but still inconclusive findings, if no stronger evidence has been developed.[6] A few people adopt the view that, given any suggestion of hazard, we should assume an agent is risky until proven safe. As a result, public health policy is viewed as being promulgated based upon social and political based considerations and not solely upon science.

The Relative Risk Reduction Strategies Committee (RRRSC) of the U.S. Environmental Protection Agency Science Advisory Board (EPA-SAB) noted[7] that:

"Some view risk reduction in the face of incomplete or uncertain risk assessment as a kind of insurance premium, since the risks of postponing action can be greater than the risks entailed in taking inefficient or unnecessary action."

Furthermore, the RRRSC charged the EPA "... not to limit research to areas where risks to human health already are recognized... but to develop an ability to predict the potential future risks of emerging problems (e.g., low-level exposures to electromagnetic fields)."[7] However, in view of the increasing technical capability for measuring smaller and smaller quantities of environmental agents and the current EPA use of non-threshold risk assessment models, extreme care must be exercised to avoid costly efforts at reducing vanishingly smaller and smaller risks.


## POLICY ASSUMPTIONS DISTORT RISK ASSESSMENT

Policy judgments routinely influence risk assessment and scientific design and analysis decisions.[5] The uncertainties inherent in risk assessment are especially evident in the assessment of chronic health effects due to low-level exposures. Because of these uncertainties, policy considerations often enter into risk assessment through a selection of conservative assumptions. Conservative practices in risk assessment are viewed by some as necessary to provide adequate margins of safety. However, these practices may bias and distort the database on which risk management decisions are made. Moreover, risk managers frequently impose additional conservative safety margins when setting regulatory policy, further distorting the estimate of risk. Consequently, conservatism in risk assessment distorts the regulatory priorities of the Federal Government, directing societal resources to reduce what are often trivial risks while failing to address more substantial threats to life and health.[5]

One policy that distorts risk assessment is the reliance on conservative (worse-case) assumptions, yielding estimates that may overstate likely risks by several orders of magnitude.[5] The cumulative effect of a series of conservative policy choices, each designed to emphasize the high-risk end of the range of plausible risks may yield final risk estimates that are quite improbable.[8] Additional conservatism enters the process when data are weighted inequitably. Even though both the Office of Science and Technology Policy (OSTP) and the EPA risk assessment guidelines[9,10] recommend that relevant animal studies should be considered irrespective of outcome, positive studies routinely are given more weight than studies that fail to find effects. In fact, overcoming a classification as a B-2 carcinogen based on a single positive animal bioassay requires, at a minimum, two essentially identical studies showing no such relationship. Thus, once a positive result has been obtained in an animal bioassay, a substance often will be provisionally classified as a probable human carcinogen. The burden of proof then shifts to the no effect hypothesis.

However, because the null hypothesis can never be proven, a single positive study establishes a "... virtually irrebuttable presumption in favor of carcinogenesis."[5] Furthermore, negative studies often are not published or repeated. Publication of no effects studies (nothing falling from the sky) is traditionally difficult, and researchers tend not to continue research in areas where positive effects are hard to come by. Thus, the number of published negative studies is not as large as it could be. This results in an artificial positive skewing of the weight of evidence.

Conservative policies include the use of the hypothetical "Maximally Exposed Individual" (MEI) who is assumed to be exposed continuously for every minute of a 70 year lifetime. To model this hypothetical individual, Maximally Tolerated Doses (MTD) are typically administered to Maximally Sensitive Subjects (MSS) and extrapolated to low dose with linear, one-hit, logistic regression, absolute risk, relative risk, or some other mathematical model. The EPA uses a linear/non-threshold model, even though supra- or subthreshold cases have been documented. It is problematic that MTD procedures usually conducted in MSS are predisposed to produce apparent carcinogenic effects that may be indirectly due to direct toxic effects of high doses.[11]

## ADVOCACY DISTORTS ASSESSMENT

A growing number of scientists have taken to the public arena to advocate their beliefs before a true scientific consensus has emerged. Scientists who lobby strongly for one side in a debate risk losing their objectivity. Some scientists have recently advocated making bold and dramatic statements to appeal to the public through the media.[12] They justify moving beyond the available level of data analysis and the state of scientific consensus to that of public advocacy based upon their perception that their information is too important to wait for scientific consensus to develop.

Policy advocates frequently compare one risk with another. An advocate for the existence of a risk will emphasize similarities with other known hazards, while advocates arguing against the existence of a risk will attempt to emphasize the differences with known hazards and the similarities with low-level risks. There are several limitations of the risk comparison approach.[13] Risk estimates based on conservative assumptions may be inappropriately compared to estimates based on liberal assumptions. High-quality data may be indiscriminately compared to data of questionable scientific validity. Since risk assessments adopt assumptions ranging from conservative to liberal, it is difficult to prioritize risk for regulatory action.

Unfortunately, scientist-advocates garner the lion's share of media attention and consequently have greater impact on public opinion than their less vocal colleagues. Agency priorities become skewed toward the advocate's position, since they are influenced by public perceptions. To help avoid distortion of the risk assessment process by advocacy, a clearly communicated science policy process that establishes assessment guidelines must precede risk assessment.[14]

## RISK ASSESSMENT MODELS

The EPA's Comparative Risk Assessment (CRA) process[7] proposes that environmental priorities be based on expert scientific opinion regarding the greatest and/or most cost-effective opportunities for reducing risks. However, it has been shown that scientists' interpretations of facts are likely to be influenced by personal values and experiences.[15] This influence appears especially when the data are unclear and when social and political stakes are high.[16,17] In fact, the National Academy of Science's National Research Council has noted that it may not be possible to separate science and values or policy considerations, but that "a clear conceptual distinction between assessment of risks and the consideration of risk management alternatives" needs to be established.[18]

The EPA appears to be in the process of changing strategy to guide its risk reduction efforts. Recent statements indicate that the EPA does not intend to adhere to a "hard" version of the CRA process[19] and will integrate science with values and democracy. In "hard" CRA, experts calculate aggregate risk to populations and cost of reducing risk and then set priorities by allocating resources to interventions that have the lowest cost per unit of risk reduction. In contrast, in the "soft" version of CRA experts and the public would negotiate in an attempt to reach consensus on risk rankings that reflect the magnitude of possible hazard reductions and other social and political considerations that affect perception of risks. To fully develop the "soft" version, the public must be informed and scientific data must be decoded and expressed in language that the public can evaluate.

Others have proposed a "solution-based paradigm"[20] wherein solutions to environmental problems and not the problems themselves would be prioritized. This approach has appeal from a short-term perspective in that the greatest number of risks that can be reduced with current technologies will be acted on in the shortest span of time. The major drawback to this approach is that large risks for which effective or efficient solutions are not readily available will be placed on the "back burner" until a technology is developed or until sufficient drivers are brought to bear on the development of necessary technologies. Furthermore, without strong drivers this "technology forcing" will not succeed. The debate over "hard" versus "soft" CRA continues. A model has recently been proposed[21] that embraces the "soft," more democratic version and emphasizes public perceptions and concerns as critical to success. Yet it incorporates a "hard" central foundation of sound science.

## IMPACT OF PUBLIC PERCEPTION ON POLICY AND PRIORITIES

Scientists, engineers, and government regulators have had difficulty in understanding public opposition to technological development. Many technical experts believe that this opposition results from an overestimation by the public of the risks of technology, caused by lack of technical education, misinformation in the media, or even irrationality.[22] Controversy exists over whether science alone or additional factors such as social and political values should be considered in the Risk process. Scientific consensus on health risk is only one of the variables currently involved.[23] The EPA in the recent past has stated that ... "solid, scientifically based risk assessment is critically important to our ability to set risk-based priorities"[24] and that risk assessment must be based on scientific evidence and nothing else.[25] Similarly, the Regulatory Program of the United States Government states that risk assessments under federal oversight should be based on science only.[5] However, the RRRSC recently concluded that "There has been little correlation between the relative resources dedicated to different environmental problems and the relative risks posed by those problems."[7] In fact, the EPA report "Unfinished Business,"[26] concluded that the only factor that correlated with EPA programmatic priorities of budget and staff allocations was apparent public perception of risk. Thus, public perception, rather than scientific understanding, appears to drive Congressional legislation that in-turn provides EPA resources. The RRRSC deemed this appropriate stating that "because they experience risks first hand, the public should have a substantial voice in establishing risk-reduction priorities."[7] Unfortunately, the public's perception about which risks are significant threats are often quite different from those identified by scientists.[26]

## PUBLIC PERCEPTION, FEAR, AND CONCERN

Understanding public attitudes about risks is a necessary element of risk communication and may affect policy. Public perceptions impact the risk assessment, risk management, risk communication processes. Involved in the risk characterization is more than the single dimension of hazard assessment; it directly involves public values focusing on the fear and concern[27,28] perceived by the general public. Successful risk communication requires meeting the public's (and often the scientists' and regulators') value sets. Risk assessment must be communicated cogently to assure the negotiated criteria are based not only on sound science but also on recognition of the public's values. This approach has risk characterization and risk communications occurring simultaneously.

Over 20 variables that can affect the magnitude of public fear and concern have been identified.[27,28] Most of these variables are directly applicable to risk assessment of possible EMF/RFR field health effects. For example, while our use of electricity has historically been viewed as voluntary and beneficial, a section of the public is now very concerned over the possibility that they may be exposed to EMF/RFR involuntarily. Furthermore, the public is forced to accept possible EMF exposure since electrical power is pervasive in modern life. Thus, the universally accepted use of electric power, which previously fell into the categories of voluntary, familiar, and well known, is becoming reclassified in part of the public consciousness as unknown high technology that the public must accept involuntarily to obtain the indispensable benefits.

Public fear and worry have little to do with scientific or technical data. In fact, "educating" the public with mailings on EMF/RFR, as many public utilities have attempted, may increase fear and concerns. This is especially true when the risk is poorly understood, as is the case for EMF/RFR. Public concerns increase when there is uncertainty (conflicting scientific data or conflicting scientific opinion), possibly delayed effects (as in cancer), or irreversible hazards. Lack of trust in industry and government magnifies public concerns. The government tends to be believed only when it suggests that the public is at risk; denial of the existence of risk is viewed as a cover-up. Man-made EMF/RFR create more concern than naturally occurring EMF/RFR. As with radon, the public will accept EMF/RFR that normally occurs in nature, even though it may be of greater magnitude and more pervasive than the man-made fields.

Risk perception may also be affected by the visibility of the risk. Just as the 55 gallon drum has become a visible symbol of toxic waste,[27] the word RADIATION carries the same type of negative symbolism of dread to the non-ionizing EMF/RFR, increasing fear and concern. The media has repeatedly linked the health hazards of exposure to ionizing radiation with non-ionizing radiation. A blatant example appeared on the cover of the August 1990 issue of MacWorld that examined the question of EMF/RFR in video display terminals (VDT). Sitting on top of a VDT is a bottle labeled with the familiar IONIZING radiation warning symbol!

## MODELS OF RISK COMMUNICATION

Most models of communication presume a single communications event is occurring; however in practice, at least four separate scenarios are occurring simultaneously. First, risk assessment that has the objective of reasonably protecting health must be conveyed to the lay public. Risk communicators have been criticized for incorrectly assuming the public is naive about environmental and health issues.[29,30] The public can be expected to have members representing the full spectrum of risk awareness, subject knowledge, acceptance/avoidance, and bias. Second, the risk must be conveyed to the environmental activists and government/industry special interests at the ends of the spectrum. Third, the risk must be conveyed to the media. Lastly, the risk must be communicated to the scientific community and the regulators/lawyers/public policy officials who have objectives of protecting the public. Each of the scenarios involves audiences with varying levels of existing expertise, commitment, fear, and concern. The scenarios also involve varying objectives, from seeking knowledge to seeking a participatory role in the discussion process to blatant, prepared attempts to obfuscate[31] the process.

The National Research Council's suggestions that the process of communication, not just the message, should be important and that communication should be two-way[32] are central to effective risk communication. Meaningful risk communication should also involve significant interaction between risk communication and risk management.[33] Citizen criteria and citizen-initiated management suggestions need to be taken into account by decision makers. Organizational amplification of risk communication can enhance an organization's ability to reduce risks.[33] An open two-way communication information flow between external as well as internal stakeholders and those responsible for risk management is necessary to establish credibility and maintain trust. Thus, risk communication may be viewed as integral to risk reduction performance rather than merely a mouthpiece for it.

Risk communication must include evaluation of choices people have, their present cognitive set about the risk, their beliefs and values, and what information they have and need.[30] This "Mental Models" approach suggests that understanding the "intuitive thought processes" the public has will facilitate communication of information that is needed for making informed judgments. This cognitive approach[30] attends to individual costs including those incurred by thinking about hazards. The concept of "Prudent Avoidance"[2] springs from the stress on the individual and private management of hazard, when possible, and when costs to the individual are judged acceptable.

Fiorino has described two models of risk analysis and risk communication.[33] The "Old Technical Model" is elitist, whereas, the "New Democratic Model" is participatory. Characteristics of the Technical Model are science-advisory boards, the search for elite consensus and objective standards, and a reliance on formal quantitative analytic techniques. The objective of Technical risk communication is to present expert formulations objectively and persuasively to non-experts.

The Technical Model of risk analysis is a one-dimensional approach that concentrates on policy outcomes (fatalities avoided) and ignores process. The model assumes that the general public interest is achieved when governmental policy is in accord with the judgment of elites. In contrast, the Democratic Model is two-dimensional "in which interest is conceived not just in the form of gains in material well-being, power, status... but in personal satisfaction and growth attained from active engagement in the political process."[34]

## STAKEHOLDER INVOLVEMENT

Current perspectives on risk communication are tied to requirements for increased involvement of stakeholders. Facility neighbors often feel excluded from important environmental decisions that may affect their future, the health of their families, and the value of their property. The "Decide, Announce, and Defend" (DAD) mode of communicating with the public is no longer acceptable. The authoritarian DAD approach to informing the citizens failed to adequately involve the stakeholders. Communication was principally based on science and technical factors and was essentially unidirectional, giving little more than lip service to "including the public as legitimate partners." This technical risk communication is impeded by the public's perception of risks, the public's difficulty in understanding intricate probability models, the technical difficulty of the subject matter, and communications loaded with scientific jargon. The sheer volume of required data reporting and the bureau-technocratic format of the documents impedes understanding. The data blizzard burdening the public essentially "transforms the right to know to the right to be confused."[35] Current policy directions suggest that failing to provide understandable risk messages eventually may lead to liability from the courts interpreting legislative regulatory statutes.

The Democratic Model embodies characteristics of political oversight of administration, negotiated problem-solving, and "popular epidemiology." The "New" model incorporates the non-expert public perceptions of risk, which are rooted in cultural and group contexts. The model focuses on greater use of negotiation and mediation in consensus seeking. Some current mechanisms for reconciling technical and democratic values include citizen participation by way of public hearings, the ballot initiative or referendum, the citizen advisory panel, policy juries, and formal institutions such as the various "activist or rights groups."

Technical risk communication is evolving into a Democratic Model that incorporates the public into the process as an active and necessary component. The "New" - democratic risk communication emphasizes dialogue among parties and participation in the decision process.[34,35] Democratic risk communication strives to make information available and understandable in order that a meaningful dialogue involving both the public and the risk communicator can occur. Unfortunately, Democratic risk communication is not fully developed and is impeded by lack of, or difficulty in, establishing participatory institutions.[35,38]

Thus, risk assessment, risk management and risk communication can move in either of two directions. Continued practice of the "Old Elitist Technical" model will separate risk analysis from democratic controls by reinforcing the separation of scientific assessment from the political evaluation of risk, relying further on cost-benefit and other analytical models for making risk decisions, and failing to accept the merit and legitimacy of lay as compared to expert judgments about risk.[34] A possible outcome is that the public will respond with a further loss of trust and confidence and renewed efforts to remove control from technical and administrative elites through the political process.[34] Current thought suggests that a synthesis of the Democratic and Technical models will prove most effective. This will require accepting the legitimacy of lay judgments, and studying communications in a two-way model where communication to experts and the policy-makers is examined, as well as the usual communication process of technical expert informing the lay public.

Fiorino suggests that an "important perceptual shift would be to view risk analysis as a political process informed by expert judgment, rather than as an expert process in which lay public can occasionally be expected to intervene... moving risk analysis beyond the notion that democratic process begins where technical consensus ends."[34] The synthesis of democratic and technical values must occur throughout the policy process if it is to be politically acceptable and ethically defensible.


## GOAL CANNOT BE ZERO RISK

The public has developed a utopian expectation that all risks should and can be eliminated. However, ours is not a zero risk society and the economy will not allow attacking every suspected problem. Public health officials must place health risks in perspective and be able to more clearly articulate such risks and priorities.[39] Risk assessment attempts to provide the risk perspective necessary for identifying cost-effective priorities to maximize health and safety. However, risk assessment methodologies must be comparable before any meaningful perspective can be obtained.

The RRRSC recommended that the EPA should target its environmental protection efforts on the basis of opportunities for the greatest risk reduction. Recent review of the success and failures of environmental practices cautions that Society will "face heavy costs" ... "if finite resources are expended on low priority problems at the expense of high priority risks."[7] The Environmental Risk Reduction Act (S.110, originally Bill S.2132) calls for the SAB to generate a cost/benefit/effectiveness analysis of options because "funds can only be used effectively when they protect the largest number of people from the most egregious harm" and that "effective and efficient strategies to reduce risks must quantify significant costs and benefits to the greatest extent possible to attain the greatest risk reduction possible with resources available." The Bill calls for the EPA to identify a prioritized list of risks, evaluate the public awareness of each risk, determine alternative options for reducing risks with estimates of costs and time required, and identify uncertainty associated with the assessment process.

Cost of risk reduction must be a consideration in any prioritization. When the bearers of risk do not share in the costs of reduction, extravagance is likely.[40] Therefore, the individual burden of any proposed changes as well as the cost of additional massive research programs must be communicated to the public as well as the total overall costs. What costs will individuals bear to avoid EMF/RFR? If the costs per capita are substantial, the perceived relative risk will diminish.

Scientists, risk assessors and policy officials must recognize that policy and regulatory actions are based on many variables and that solid scientific consensus is not necessarily the most heavily weighted of these variables.[23] To meet the demands of changing policy, we must first meet head-on the dilemma addressed in the EPA's "Unfinished Business" and "Reducing Risks;"[26,7] that is, the gap between the public's expectation and policy.

## ACKNOWLEDGMENT

The author appreciates and gratefully adknowledges the review and helpful suggestions provided by Dr. Vincent Covello, Columbia University, that have greatly improved this paper. The author, however, takes full responsibility for the paper.

## REFERENCES

1. P. Slovic, Perception of risk. *Science*. 236:280-285 (1987).
2. I. Nair, M.G. Morgan, H.K. Florig, Biological effects of power frequency electric and magnetic fields. Background paper prepared for the Congress of the United States, Office of Technology Assessment. OTA-BP-E-53. (May 1989).
3. A.C. Brown, Commissioner-Public Utilities Commission of Ohio. Panel Discussion: How will government action on EMF affect business. *The Business Of EMF*. Washington International Energy Group. Washington D.C. (June 27, 1991).
4. Toward Utility Rate Normalization. Comments of Toward Utility Normalization: Before the Public Utilities Commission of the State of California. (April 1991).
5. Executive Office of The President, Office of Management and Budget, "Current regulatory issues in risk assessment and risk management,"*Regulatory Program of the United States Government* (April 1, 1990- March 31, 1991).
6. G.M. Morgan, Exposé treatment confounds understanding of a serious public health issue. *Scientific American*. April:118-123 (1990).
7. U.S. Environmental Protection Agency Science Advisory Board: Relative Risk Reduction Strategies Committee. Reducing risk: Setting priorities and strategies for environmental protection. SAB-EC-90-021, (September 1990).

8. J.V.Rodricks, S.L. Brown, R. Putzrath and D. Turnbull, An industry perspective: Invited presentation, Use of risk information in regulation of carcinogens. Presented at the December 16, 1987 Workshop on Determination of No Significant Risk Under Proposition 65, 19-41.

9. U.S. Office of Science and Technology Policy, "Chemical carcinogens: A review of the science and its associated principles. 50 FR 10378 (March 14, 1985).

10. U.S. Environmental Protection Agency. Guidelines for carcinogen risk assessment. 51 FR 34001 (September 24, 1986).

11. B.N. Ames and L.S. Gold, Too many rodent carcinogens: Mitogenesis increases mutagenesis. *Science.* 249:97--971 (1990).

12. R. Pool, Struggling to do science for society. News and Comment. *Science.* 248:672-673, 1990.

13. V.T. Covello, Communicating right-to-know information on chemical risks. *Environ. Sci. Technol.* 23:1444-1449 (1989).

14. B.D.,Goldstein, Risk assessment and the interface between science and law. *Columbia J. Environ Law.* 14:343-355 (1989).

15. G.L. Carlo, N.L. Lee, K.G. Sund and S.D. Pettygrove, The interplay of science, values, and experiences among scientists asked to evaluate the hazards of dioxin, radon, and environmental tobacco smoke. *Risk Analysis*, 12:37-43 (1992).

16. A. Whittemore, Facts and values in risk analysis for environmental toxicants, *Risk Analysis*, 3, 23-33 (1983).

17. D. Robins and R. Johnson, The role of cognitive and occupational differentiation in scientific controversies. *Social Studies of Science*, 6:349-368 (1976).

18. National Academy of Sciences. *Risk Assessment in the Federal Government:Managing the Process*, Washington, DC: National Academy Press (1983).

19. F.H. Habicht II., "National Environmental Priorities: The EPA Risk-Based Paradigm and Its Alternative." Conference held in Annapolis MD 16-17 Nov, 1992 (Organized by A. Finkel, Center for Risk Management, Resources for the Future. Reported in *RISK Newsletter*, 13,1 (1993).

20. N. Ashford, "National Environmental Priorities: The EPA Risk-Based Paradigm and Its Alternative." Conference held in Annapolis MD 16-17 Nov, 1992 (Organized by A. Finkel, Center for Risk Management, Resources for the Future. Reported in *RISK Newsletter*, 13,1 (1993).

21. B.J. Klauenberg and E.K. Vermulen, Role for risk communication in closing military waste sites, *Risk Analysis*, 14:351-356 (1994).

22. G.T. Gardner and L.C. Gould, Public perceptions of the risks and benefits of technology, *Risk Analysis* 9:225-242 (1989).

23. B.J. Klauenberg, Does public policy require scientific consensus? *Health Physics Newletter*, 19:25-29 Oct (1991).

24. F.H. Habicht II, EPA Assessment Program Featured Speaker, Society for Risk Analysis Annual Meeting (1991).

25. W.D. Ruckelshaus, Science, risk, and public policy. *Vital Speeches of the Day*, 49, 20:612-615, (August 1, 1983).

26. U.S. Environmental Protection Agency. Office of Policy Analysis. *Unfinished Business: A Comparative Assessment of Environmental Problems.* Vol 1. Overview Report. U.S. Government Printing Office: Washington D.C. (February, 1987).

27. P.M. Sandman, Risk communication: Facing public outrage. *EPA Journal.* pp. 21-22, Nov 1987.

28. V.E. Covello, P.M. Sandman and P. Solvic, eds., "Risk Communication, Risk Statistics, and Risk Comparisons: A Manual for Plant Managers," Chemical Manufactures Association, Washington DC (1988).

29. H. Otway, Experts, risk communciation, and democracy, *Risk Analysis* 7:125-129 (1987).

30. M.G. Morgan, B. Fischoff, A. Bostrom, L. Lave and C.J. Atman, Communicating risk to the public: first, learn what people know and believe, *Environ. Sci. Technol.* 26:2048-2056 (1992).

31. I.E. Kornfield, W. Subra and W. Collette, *How to Win in Public Hearings,* Center for Environmental Justice. Citizen's Clearinghouse for Hazardous Wastes, Inc., (1990).

32. National Research Council. *Improving Risk Communication.* National Academy Press, Washington D.C. (1989).

33. C. Chess, A. Saville, M. Tamuz and M. Greenberg, The organizational links between communication and risk management: The case of sybron chemicals inc. *Risk Analysis*, 12:431-438 (1992).

34. D.J. Fiorino, Technical and democratic values in risk analysis, *Risk Analysis* 9:293-299 (1989).

35. S.G. Hadden, Institutional barriers to risk communication, *Risk Analysis* 9:301-308 (1989).

36. B. Hance, C. Chess, and P. Sandman, *Improving Dialog with Communities: A Short Guide For Government Risk Communication.* Department of Environmental Protection Trenton New Jersey (1988).

435

37. Seven Rules for Risk Communication, *in:* "Risk Communication, Risk Statistics, and Risk Comparisons: A Manual for Plant Managers" V.E. Covello, P.M. Sandman, and P. Solvic, eds., Chemical Manufactures Association, Washington DC (1988).

38. R.E. Kasperson, Six propositions on public participation and their relevance for risk communication. *Risk Analysis* 6:275-281 (1986).

39. L. Gordon, Risk communication and environmental health priorities. *J. Environ Health*, 52:134 (1989).

40. R.J. Zeckhauser and W.K. Viscusi, Risk within reason. *Science*, 248:559-564 (1990).

# Appendix

# APPENDIX

# RECOMMENDATIONS OF ARW WORKING GROUPS

## WORKING GROUP 1: COORDINATING WITH INTERNATIONAL BODIES AND EUROPEAN COMMUNITY

**MEMBERS:** *Chairs*--M. Repacholi and J. Osepchuk, P. Bernardi, C. Gabriel, G. Mariutti, J. Merritt, B. Veyret

**RECOMMENDATION #1:** Involve the following international organizations in the review of the draft standard.

**Organization Contact:** ICNIRP; ILO; COST (DG12); CENELEC (Committee IIIb); IEEE COMAR; IRPA; URSI; WHO; CEC (DG5); IEC, TC78; IEEE SCC28; BEMS; EBEA; ICOH; ETSI, CEN and other DGs in CEC through CENELEC

**RECOMMENDATION #2:** In order to provide input from international organizations in the drafting process, we recommend the participation of representatives of these international organizations in the AGARD/ASI lecture series.

**RECOMMENDATION #3:** In order not to limit review of the draft standard we recommend that these international organizations be encouraged to achieve as wide a review as possible of the draft standard within their member countries.

**RECOMMENDATION #4:** In order to speed up the process we recommend that these international organizations be involved in early dissemination of the draft standard.

**RECOMMENDATION #5:** In order to facilitate the process, we recommend that WG1 (membership listed above) be assigned the task of coordinating communication with the above listed organizations, receiving and processing comments, and delivering a report to the standard drafting committee.

**RECOMMENDATION #6:** It is recommended that NATO collaborate with ICNIRP on the development of a peace-time RF standard. Development of this standard will take into consideration the current IRPA, IEEE and other appropriate standards.

## WORKING GROUPS II/III: ASI/AGARD PLANNING COMMITTEE

Working Groups II and III combined for discussion of Advanced Studies Institute and AGARD siting and funding.

**MEMBERS:** Chairs: B.J. Klauenberg, M. Grandolfo, (WG-II); K. Hofmann, R. Petersen, (WG-III) P. Pissis, A. Chiabrera, R. Woolnough, C. Sherry, J. Jauchem, P. Polson, M. Murphy, G.A. Mickley, J.A. Leonowich, M. Bornhausen, J. Kiel, P. Semm, C. Poole, M. Stuchly, R. Tell, E. Elson

**RECOMMENDATION # 1:** ARW Working Group II/III recognizes the need for additional emphasis on implementation and control problems. However, these issues are not central to the development of a revised STANAG for RFR. Working Group II recommendations that a proposal be submitted to NATO for a one week ARW on "Measurement and Practical Problems Associated with the Implementation of a BioRadhaz Control Standard." R. Woolnough, R. Tell, and R. Petersen volunteered to prepare the proposal.

**RECOMMENDATION #2:** Working Group II/III recommends that a 10 day Advanced Studies Institute be held for purpose of dissemination and education in support of the NATO STANAG.

**RECOMMENDATION #3:** Working Group II/III recommends that the following organizations be solicited for funding of the ASI and AGARD series and costs associated with these efforts: NATO, USAF, USAF-EOARD, US ARMY, US NAVY, USDOE, IEEE, USFDA, USOSHA, NSF, USEPA, NIEHS, ARPA, WHO, ILO, and Industry.

## WORKING GROUP IV:
## STANDARDS DRAFTING

**MEMBERS:** Chairs: O.P. Gandhi, E.R. Adair, M. Frei, L. Court, J. D'Andrea, J. Lin, R.F. Gardner, G.D'Inzeo, J. de Lorge, M. Bini, J. Bernhardt, A.W. Guy, P. Vecchia, M. Meltz, J. Elder, S. Tofani, R. Adair

**RECOMMENDATION #1:** ARW Working Group IV recognizes that most RF exposure guidelines have a basic similarity. The Working Group recommends the ANSI/IEEE C95.1-1992 as a basis for a NATO STANAG. The Group further believes that modifications consistent with the spirit of the standard may be made for the purpose of the STANAG.

**RECOMMENDATION #2:** ARW Working Group IV further recommends the adoption of the ANSI/IEEE C95.3-1991 Measurement Standard as a companion to the ANSI/IEEE C95-1-1992 RF Exposure Guideline as a basis for a NATO STANAG.

**RECOMMENDATION #3:** ARW Working Group IV recommends a second NATO Advanced Research Workshop be held for the purpose of fine-tuning the Draft NATO STANAG. If possible, this group shall consist of experts (operating personnel, physiologists, etc.) from all NATO nations.

**RECOMMENDATION #4:** ARW Working Group IV recommends that an Advanced Study Institute (ASI) and a series of AGARD lectures be considered for the purpose of information dissemination and education in support of the NATO STANAG.

In addition to these formal recommendations, several suggestions were made as topics for further discussion at the next NATO ARW. These include, but are not necessarily limited to:

    a. Measurement distance from a source.
    b. Inclusion of airspace in 1 gm of tissue (in the shape of a cube) for local SARs.
    c. Interpretation of guidelines for mixed far-field/near-field sources.
    d. Averaging time for induced currents and partial body exposure.
    e. Definition of grasping contact.
    f. Hazard Assessment.

# Attendance List

# SPEAKERS

Eleanor R. Adair, Ph.D.
John B. Pierce Laboratory, Inc.
290 Congress Avenue
New Haven, CT 06519
UNITED STATES

Robert Adair, Ph.D.
Physics Department
P.O. Box 6666
Yale University
New Haven, CT 06511-8167
UNITED STATES

Professor Jürgen H. Bernhardt
Bundesamt für Strahlenshutz
Institut für Strahlenhygiene
Ingolstädter Landstrasse 1
DW-8042 Neuherberg
GERMANY

Michael Bornhausen, M.D.*
Institute of Toxicology
GSF-Frochungszentrum
  für Umwelt und Gesundheit
D-8042 München-Neuherberg
GERMANY

John A. D'Andrea, Ph.D.
NAMRL
Naval Air Station, Code 24 A
Pensacola, FL 32508-1046
UNITED STATES

John de Lorge, Ph.D.
Head, Naval Aerospace
Medical Research Lab
Code 02
Pensacola, FL 32508-1046
UNITED STATES

Joe A. Elder, Ph.D.*
Health Effects Res Lab
US EPA (MD-51)
Research Triangle Park, NC 27711
UNITED STATES

Colonel Edward C. Elson, M.D.
WRAIR
Department of Microwave Research
(SGRD-UWI-D)
Washington, DC 20307-5100
UNITED STATES

David N. Erwin, Ph.D.**
Radiofrequency Radiation Division
Armstrong Laboratory (AL/OER)
Bldg 1184, 8308 Hawks Road
Brooks Air Force Base, TX 78235-5324

Melvin R. Frei, Ph.D.
Radiofrequency Radiation Division
Armstrong Laboratory
Bldg 1184, 8308 Hawks Road
Brooks Air Force Base, TX 78235-5324
UNITED STATES

Camelia Gabriel, Ph.D.
Department of Physics
King's College
Strand, London WC2R 2LS
UNITED KINGDOM

Om P. Gandhi, Ph.D.
Dept of Electrical Engineering
University of Utah
3080 Merrill Engineering Bldg.
Salt Lake City, UT 84112
UNITED STATES

Robert C. Gardner
Directorate of Defence Health and Safety
Ministry of Defence
Golf Aquila Road
Bromley BRI 2JB
UNITED KINGDOM

Martino Grandolfo, Ph.D.**
Director, Physics Laboratory
National Institute of Health
Viale Regina Elena 299
00161 Rome
ITALY

Arthur W. Guy, Ph.D.
Emeritus Professor
18122 60th Pl NE
Seattle, WA 98155
UNITED STATES

Klaus W. Hofmann, Ph.D.
Research Institute of
  High Frequency Physics
Research Establishment for Applied Science
Neuenahrer Str. 20 D-53343
5307 Wachtberg-Werthoven
GERMANY

James Jauchem, Ph.D.
Radiofrequency Radiation Division
Armstrong Laboratory (AL/OERB)
Bldg 1184, 8308 Hawks Road
Brooks Air Force Base, TX 78235-5324
UNITED STATES

Leeka Kheifets, Ph.D.
EPRI
3412 Hillview Avenue
P.O. Box 10412
Palo Alto, CA 94304
UNITED STATES

Johnathan L. Keil, Ph.D.
Radiofrequency Radiation Division
Armstrong Laboratory (AL/OERT)
Bldg 175E, 2503 D Drive
Brooks Air Force Base, TX 78235-5102
UNITED STATES

B. Jon Klauenberg, Ph.D.**
Radiofrequency Radiation Division
Armstrong Laboratory (AL/OERB)
Bldg 1184, 8308 Hawks Road
Brooks Air Force Base, TX 78235-5324
UNITED STATES

John A. Leonwich
Battelle Pacific Northwest Laboratory
Mail Stop K3-70
P.O. Box 999
Richland, WA 99352
UNITED STATES

James C. Lin, Ph.D.
College of Engineering (M/C 154)
Room 1030 Science & Engineering Offices
University of Illinois at Chicago
851 South Morgan St.
Chicago, IL 60607
UNITED STATES

Gianni F. Mariutti, Ph.D.**
National Institute of Health
Physics Laboratory
Viale Regina Elena, 299
00161 Rome
ITALY

Martin Meltz, Ph.D.
Department of Radiology
University of Texas Health Science Cente
7703 Floyd Curl Drive
San Antonio, TX 78284-7800
UNITED STATES

James H. Merritt
Radiofrequency Radiation Division
Armstrong Laboratory (AL/OERB)
Bldg 1182, 8308 Hawks Road
Brooks Air Force Base, TX 78235-5324
UNITED STATES

G. Andrew Mickley, Ph.D.
Department of Psychology
Baldwin-Wallace College
75 Eastland Road
Berea, OH 44017-2008
UNITED STATES

Michael Murphy, Ph.D.
Radiofrequency Radiation Division
Armstrong Laboratory (AL/OER)
Bldg 1184, 8308 Hawks Road
Brooks Air Force Base, TX 78235-5324
UNITED STATES

John M. Osepchuk, Ph.D.
Raytheon Research Division
Department 9143, Mail Stop L116
131 Spring Street
Lexington, MA 02173
UNITED STATES

Ronald C. Petersen
AT&T Bell Laboratories
600 Mountain Ave., Room 1F101C
Murray Hill, NJ 07974
UNITED STATES

Peter Polson, Ph.D.
18985 Tuggle Avenue
Cupertino, CA 95014-3658
UNITED STATES

Charles Poole, Sc.D.
114 Pleasant Street
Cambridge, MA 02139
UNITED STATES

Michael H. Repacholi, Ph.D
Chairman, ICNIRP
Australian Radiation Laboratory
Lower Plenty Road
Yallambie, Victoria
AUSTRALIA, 3085

Teresa M. Schnorr, Ph.D.
Industrywide Studies Branch
Division of Surveillance, Hazard
  and Field Studies
NIOSH
4676 Columbia Parkway, R-15
Cincinnati, OH 45226
UNITED STATES

Professor Peter Semm
Zool Institut
AG Magnetoneurobiologie
Universität Frankfort
Siesmayerstr. 70
6000 Frankfort 11
GERMANY

Cliff Sherry, Ph.D.
Systems Research Laboratories, Inc.
P.O. Box 35505
Brooks Air Force Base, TX 78235-5301
UNITED STATES

David Sliney, Ph.D. (read by R. Petersen)
U.S. Environmental Hygiene Agency
Laser Branch, Laser Microwave Division
HSHB-MR-LL
Aberdeen Proving Ground, MD 21005
UNITED STATES

Maria A. Stuchly, Ph.D.
Dept. of Electrical & Computer Engineerin
University of Victoria
P.O. Box 3055
Victoria, British Columbia
CANADA V8W 3P6

Richard A. Tell
Richard Tell Associates, Inc.
8309 Garnet Canyon Lane
Las Vegas, NV 89129-4897
UNITED STATES

Rick Woolnough
RADHAZ/Laser Safety Specialist
Systems Development
Department:S/EWDG
Location W6A
British Aerospace Defence
Military Aircraft Division
Warton Aerodrome
Preston PR 4 1 AX
Lancashire, England
UNITED KINGDOM

# PARTICIPANTS

Professor Paolo Bernardi
Department of Electronics
University of Rome
18, Via Eudossiana
00184 Rome
ITALY

Marco G. Bini, Ph.D.
IROE
National Research Council
64 Via Panciatichi
50127 Florence
ITALY

Professor Alessandro Chiabrera
DIBE-Universita di Genoa
Via Opera Pia 11A
16145 Genova
ITALY

Louis Court, Ph.D., M.D.
Médecin Général Inspecteur
Directeur CRSSA
Ministere de la Défense
24 avenue des Marquis du Gresivaudan
B.P. 87-38702
La Tronche Cedex
FRANCE

Professor Edward H. Grant
Head, Department of Physics
King's College London, Strand
London WC 2 R 2 LS
UNITED KINGDOM

Professor Guglielmo D'Inzeo
Department of Electronics
University of Rome "La Sapienza"
Via Eudossiama 18
00184 Roma
ITALY

Brigadier General Gianfranco Pecci**
Divisione Aerea Studi
Ricerche e Sperimentazioni
00040 Aeroporto Practica di Mare
(Pomezia)
ITALY

Polycarpos Pissis, Professor
National Technical University of Athens
Department of Physics
Zografou Campus, GR157 80
Athens
GREECE

Santi Tofani, Ph.D.
Laboratorio di Sanità Pubblica
Sezione Fisica
Via Lago S. Michele, 11
10015, IVREA (TO)
ITALY

Paolo Vecchia, Ph.D.**
Laboratorio di Fisica
Istituto Superiore di Sanità
Via le Regina Elena, 299
00161 Rome
ITALY

Bernard Veyret, Ph.D.
Laboratoire PIOM, ENSCPB
Groupe de BioElectromagnétisme
Université Bordeaux I
33405 Talence Cedex
FRANCE

---

\* Presented paper at meeting but did not submit for publication

\*\* Member of organizing committee

KEY TO OFFICIAL PHOTOGRAPH NATO ADVANCED RESEARCH
WORKSHOP: DEVELOPING A NEW STANDARDIZATION AGREEMENT (STANAG)
FOR RADIOFREQUENCY RADIATION
Pratica di Mare Italian Air Force Base, Rome, Italy
16-24 May, 1993

1 John de Lorge, Ph.D.
2 Polycarpos Pissis, Professor
3 Camelia Gabriel, Ph.D.
4 Peter Polson, Ph.D.
5 Bernard Veyret, Ph.D.
6 Martin L. Meltz, Ph.D.
7 Patrick J. Karshis, Sgt
8 Leeka Kheifets, Ph.D.
9 Michael H. Repacholi, Ph.D.
10 Rick Woolnough
11 Professor Jürgen H. Bernhardt
12 Cliff Sherry, Ph.D.
13 Om P. Gandhi, Ph.D.
14 Johnathan L. Kiel, Ph.D.

15 Joe A. Elder, Ph.D.
16 Ronald C. Petersen
17 Michael Murphy, Ph.D.
18 James Jauchem, Ph.D
19 James H. Merritt
20 John A. Leonowich
21 Richard A. Tell
22 LitCol Andrew Mickley, Ph.D.
23 Santi Tofani, Ph.D.
24 John M. Osepchuk, Ph.D.
25 Professor Paolo Bernardi
26 Arthur W. Guy, Ph.D.
27 Professor Peter Semm
28 Professor Guglielmo D'Inzeo
29 Maria A. Stuchly, Ph.D.

30 Marco G. Bini, Ph.D.
31 B. Jon Klauenberg, Ph.D.
32 Louis Court, Ph.D, M.D.
33 Professor Alessandro Chiabrera
34 David N. Erwin, Ph.D.
35 Martino Grandolfo; Ph.D.
36 Brig Gen Gianfranco Pecci
37 Alma Paoluzi (Sec. to Dr Grandolfo)
38 Michael Bornhausen, M.D.
39 Marco Sabatini (Technical Assistant)
40 Robert Adair, Ph.D.
41 Teresa M. Schnorr, Ph.D.
42 Eleanor R. Adair, Ph.D.

43 Gianni F. Mariutti, Ph.D.
44 Charles Poole, Sc.D.
45 Klaus W. Hofmann, Ph.D.
46 James C. Lin, Ph.D.
47 Colonel Edward C. Elson
48 John A. D'Andrea, Ph.D.
49 Colonel Giovanni Cardini
50 Robert C. Gardner
51 Melvin R. Frei. Ph.D.
52 Enrico Mariutti (Mascot, and son of Gianni)
53 Paolo Vecchia, Ph.D.
54 Mimmo Monteleone (Technical Assistant)
55 Tenent Pio Curti

# INDEX

Adrenocorticotropin, 224
Air Force Occupational Safety and Health (AFOSH)
  RF Standard 161-9, 147
ALARA principle, 422
American Conference of Governmental Industrial
  Hygienists (ACGIH), 25, 83-84, 119-
  120,147-149
American National Standard for Safe Use of Lasers,
  (ANSI Z136.1-1993), 84, 119-120
American National Standards Institute, 4,
  *see also* ANSI
Ames test , 332
Amplitude modulation, 96, 197-203, 330, 333, 351
ANSI, 4, 25, 31-33, 51-52, 84, 90, 93, 115-120, 127,
  147, 254, 257-259, 328, 333, 339-345, 353,
  370, 392-394, 416-417, 421-423
ANSI C95.1-1966, 423
ANSI C95.1-1973, 4
ANSI C95.1-1982, 90, 93-94, 115, 147, 392, 418, 421
ANSI C95.1-1991/1992, 33, 51-52, 89-93, 115, 118-
  120, 147-149, 258, 328, 416, 421-423
ANSI C95.2, 127
ANSI C95.3, 98-99
ANSI C95.3-1973, 90
ANSI C95.5-1981, 90
ANSI C95.3-1991, 89-95, 117
ANSI Z136.1-1993, 84, 119-120
Antenna, 329, 393-406
Asthenic syndrome, 338
Auditory effect
  cochlea, 274, 340
  thermoelastic acoustic waves, 340

Basic restrictions, 16, 19-20, 23-27, 38, 104-106, 133
Behavioral effects, 51-57, 60, 71-72, 76, 145, 226,
  273-274, 305, 363-370, 375, 413, 422
  acoustic startle response, 366
  anxiety, 226, 375, 422
  behavior schedule, 366
  complex vigilance discrimination task, 364
  exploratory movements, 363
  memory, 52-53, 305
  observing response, 370
  repeated-acquisition, 367
  shock-motivated passive avoidance task, 366
  shuttlebox, 366
  time perception, 52-53
  visual-tracking task, 367

Best-fit two-relaxation-constant Debye equations, 55
Biological clocks and rhythms
  circadian rhythms, 52-53, 281, 318, 333
  circannual rhythms, 279
  Zeitgeber, 287
  Zugunruhe 281, 285
Broadcasting, 391-398

Calcium
  calcium efflux, 198-205, 224, 332, 348,
    349-350, 361
  calcium uptake, 203-204
Cancer
  brain cancer, 171-172, 184, 332
  breast cancer, 172, 184, 316
  breast tumors, 317, 342
  childhood cancer, 158, 171
  hormonally mediated cancer, 172
  leucosis, 312, 342
  leukemia 45-46, 156-157, 160, 171-172,
    184, 312, 332, 342
  leukocytopoiesis, 3
  leukosis, 342, 355
  melanoma, 344
  skin cancer, 139, 311, 316-317, 342, 343
Carcinogenesis
  initiation, 37, 251-255, 332
  progression, 332
  promotion, 328, 332, 344
Catecholamine, 224
Cell
  B cell, 321
  cell membrane, 204, 332, 349-352
  cellular metabolism, 204
  T-lymphocyte, 60, 103, 203, 237-239, 282-283,
    321-322, 351-354, 361, 375
Cellular telephone, 392-394, 398, 406, 427
Central nervous system
  benzodiazepine receptors, 226, 375
  benzodiazepines, 226
  beta-endorphin, 224
  blood-brain-barrier, 346
  brain slices, 201
  brain synaptosomes, 202
  cat brain, 321
  chick brain, 198, 201, 204, 349
  choline uptake, 226, 347
  cholinergic, 285, 347